EXERCISE WORKBOOK

for

Advanced AutoCAD®

2007

by
Cheryl R. Shrock
Professor
Drafting Technology
Orange Coast College, Costa Mesa, Ca.

Autodesk
Authorized Author

INDUSTRIAL PRESS
New York

ISBN 0-8311-3303-1

Printed in the United States of America

Industrial Press Inc.
989 Avenue of the Americas
New York, NY 10018

10 9 8 7 6 5 4 3 2

This book is dedicated to the students that want to use AutoCAD productively, especially 3D, but do not have the time or need to learn every command.

Exercise Workbooks written by Cheryl R. Shrock:

Advanced AutoCAD **2000**	ISBN 0-8311-3193-4
Beginning AutoCAD **2000, 2000i & LT**	ISBN 0-8311-3194-2
Advanced AutoCAD **2000, 2000i & LT**	ISBN 0-8311-3195-0
Beginning AutoCAD **2002**	ISBN 0-8311-3196-9
Advanced AutoCAD **2002**	ISBN 0-8311-3197-7
Beginning AutoCAD **2004**	ISBN 0-8311-3198-5
Advanced AutoCAD **2004**	ISBN 0-8311-3199-3
Beginning AutoCAD **2005**	ISBN 0-8311-3200-0
Advanced AutoCAD **2005**	ISBN 0-8311-3201-9
Beginning AutoCAD **2006**	ISBN 0-8311-3213-2
Advanced AutoCAD **2006**	ISBN 0-8311-3214-0
Beginning AutoCAD **2007**	ISBN 0-8311-3302-3
Advanced AutoCAD **2007**	ISBN 0-8311-3303-1

*For information about these workbooks,
visit www.industrialpress.com*

*For information about Cheryl Shrock's online courses,
visit www.shrockpublishing.com*

Table of Contents

Lesson 12

Lesson 13

Lesson 14

Lesson 15

Lesson 16

Lesson 17

Lesson 18

INTRODUCTION

About this workbook
This workbook is designed to follow ***Exercise Workbook for Beginning AutoCAD 2007***. It is excellent for classroom instruction or self-study. There are 21 lessons and 3 *on-the-job* type projects in Architectural, Electro-mechanical and Mechanical.
13 Lessons continue your education in basic 2D commands.
8 Lessons introduce you to many basic 3D commands.

Each lesson starts with step-by-step instructions followed by exercises designed for practicing the commands you learned within that lesson. The *on-the-job* projects are designed to give you more practice in your desired field of drafting.

Exercises that include printing are designed for a Hewlett Packard 4MV LaserJet printer (11 x 17 sheet size) and a Hewlett Packard 500 or 600 Design Jet (18 x 24 & 24 x 36 sheet size). You should configure your system to recognize these plotter specifications in order to *preview* the exercises. *Note: your computer does not have to be attached to a printer to configure it. Only the printer specifications and limitations will be loaded.* Refer to **Appendix A "Add a Printer".**

2007 vs. 2007 LT
The LT version of AutoCAD has approximately 80 percent of the capabilities of the full version. It was originally created to fit on the small hard drives on Laptops. Hence, the name LT. In order to reduce the size of the program AutoCAD removed some of the high-end capabilities, such as Solid Modeling. Through out the workbook I point out which commands are not available to LT users. Consider this an opportunity to see the commands that you are missing and you can determine if you feel it necessary to upgrade.

IMPORTANT:
Three files are required for this workbook:
"2007-3D demo.dwg" or "2007LT demo.dwg" and "2007-3D Setup".
These files may be downloaded from:
www.industrialpress.com or www.shrockpublishing.com websites.

About the Author
Cheryl R. Shrock is a Professor and Chairperson of Computer Aided Design at Orange Coast College in Costa Mesa, California. She is also an Autodesk® registered author / publisher. Cheryl began teaching CAD in 1990. Previous to teaching, she owned and operated a commercial product and machine design business where designs were created and documented using CAD. This workbook is a combination of her teaching skills and her industry experience.

"Sharing my industry and CAD knowledge has been the most rewarding experience of my career. Students come to learn CAD in order to find employment or to upgrade their skills. Seeing them actually achieve their goals, and knowing I helped, is a real pleasure. If you read the lessons and do the exercises, I promise, you will not fail."

Cheryl R. Shrock

CREATE A TEMPLATE

Note: If you have already created the template "1Workbook Helper.dwt", required in the "2007 Beginning Workbook", skip to Intro-5.

The first item on the learning agenda is how to create a template file from the **"Workbook Helper.dwg".** Go to the website: **http://www.industrialpress.com** or **www.shrockpublishing.com** to download the files for workbook 2007 and save them to your "Desktop".

1. Start AutoCAD:

2. Select **File / Open**

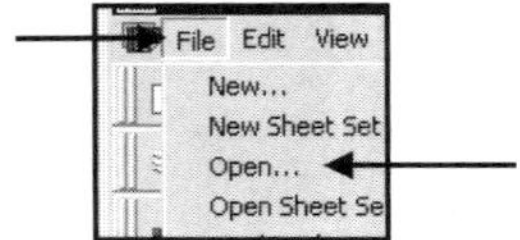

3. Select the **Directory** in which you downloaded the files. (Click on the ▾)

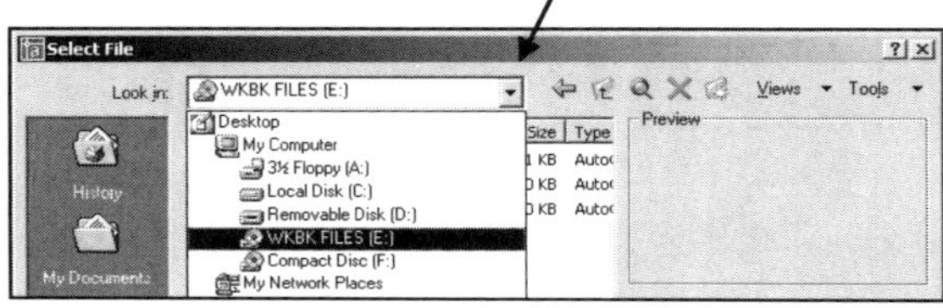

4. Select the file "**Workbook Helper.dwg**" and then "**Open**" button.

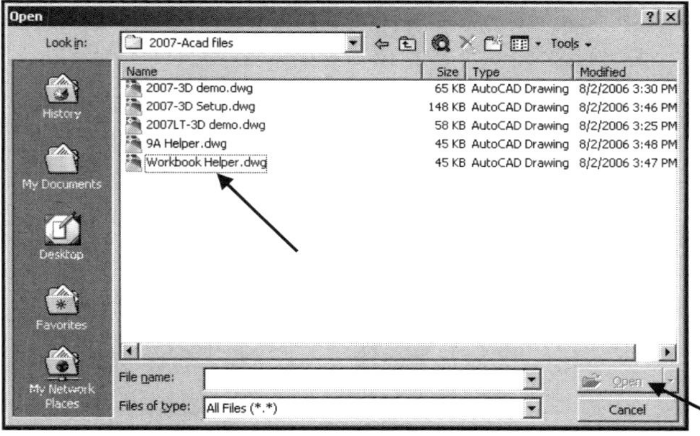

Notice the 3 letter extension for a "drawing" file is ".dwg".

5. Select **"File / Save As..."**

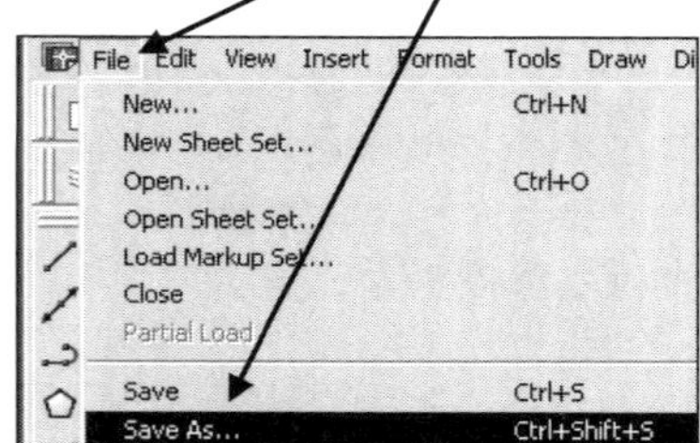

6. Select the "**Files of type**:" down arrow ▼ to display different saving formats. Select "**AutoCAD Drawing Template (*.dwt)**".

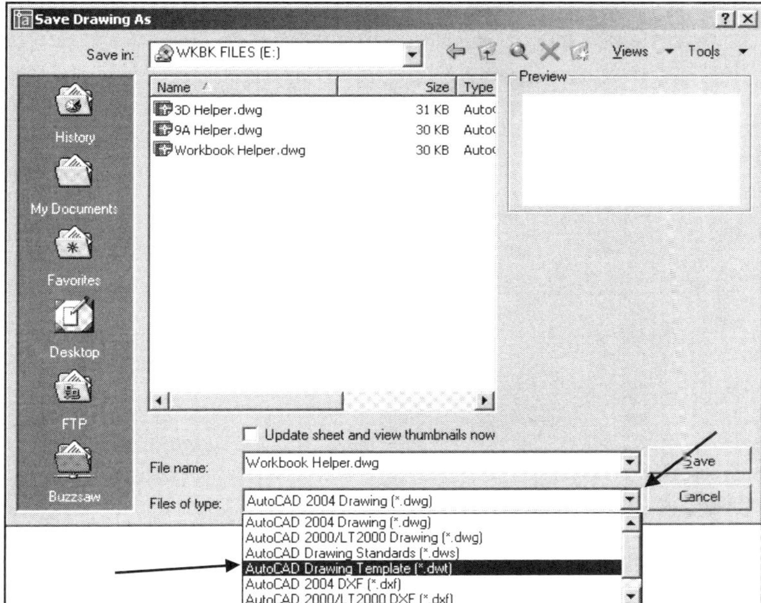

Notice the 3 letter extension for Template is ".dwt".

A list of all the AutoCAD templates will appear. (Note: Your list may be different)

7. Type the new name "**1Workbook Helper**" in the "**File name**:" box and then select the "**Save**" button.

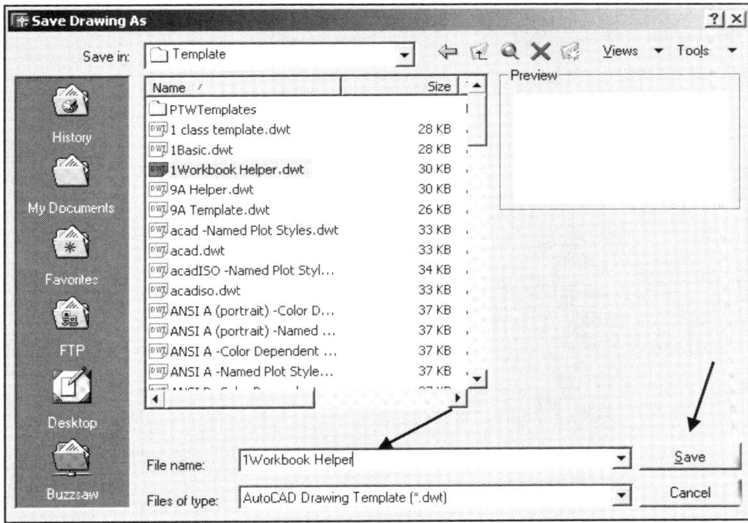

Note: The "1" before the name will place the file at the top of the list.
AutoCAD displays numerical first and then alphabetical.

Notice it was not necessary to type the extension .dwt because "Files of type" was previously selected.

8. Type a description and the select the "**OK**" button.

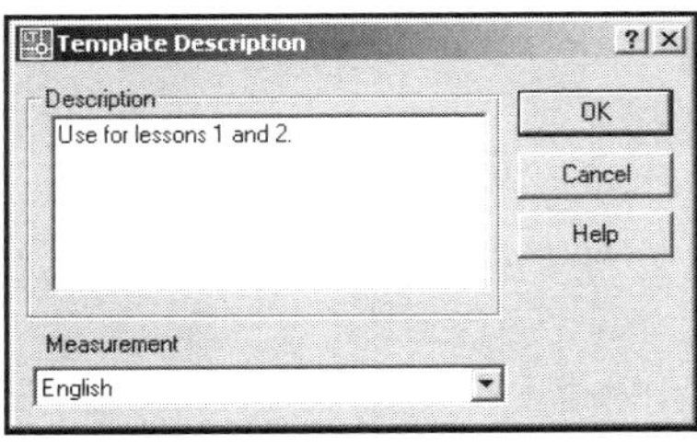

Now you have a template to use for lessons 1 and 2.

OPENING A TEMPLATE

The template that you created on the previous page will be used for lessons 1 and 2 only. It will appear as a blank screen but there are many variables that have been preset. This will allow you to start drawing immediately. (*If you have not completed the exercises in "Exercise Workbook for Beginning AutoCAD 2007", I strongly suggest that you do. It will make learning AutoCAD less confusing and more fun.*)

Let's start by opening the "1Workbook Helper.dwt" template.

1. Select **FILE / NEW**.

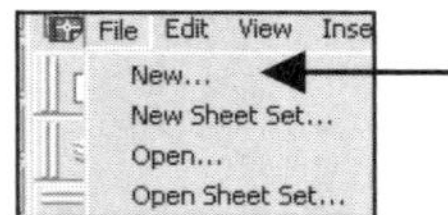

2. Select the **Use a Template** box (third from the left).

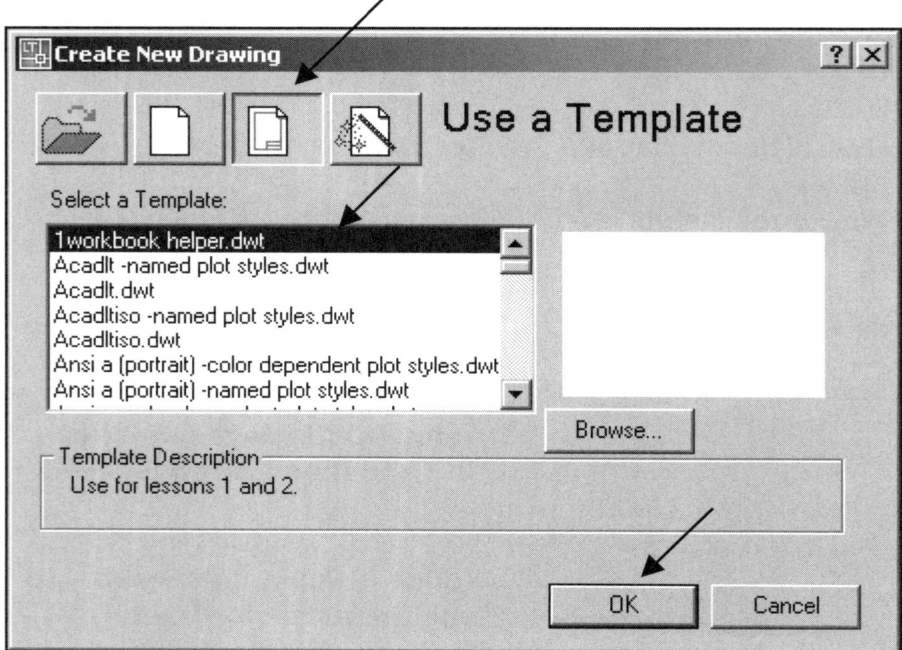

3. Select **1workbook helper.dwt** from the list of templates.

   (NOTE: *If you do not have this template, refer to page Intro-2.)*

4. Select the **OK** button.

Configuring your system

Note: If you have already configured your system for the 2007 "Beginning Workbook" you may skip to Lesson 1.

While you are using this workbook, it is necessary for you to make some simple changes, to your configuration, so our configurations are the same. This will ensure that the commands and exercises work as expected. The following instructions will guide you through those changes.

A. First start AutoCAD®

B. At the bottom of the screen there is a white rectangular area called the "*Command Line*". Type: ***Options*** then press the **<enter>** key. (not case sensitive)

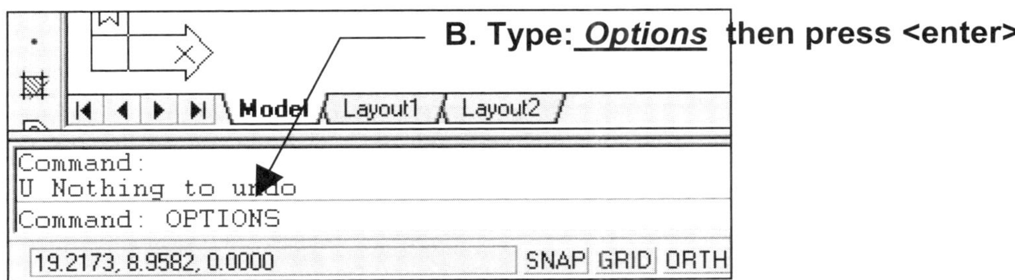

C. Select the ***Display*** tab and change the settings on your screen to match the dialog box below. Pay special attention to the settings with an ellipse around it.

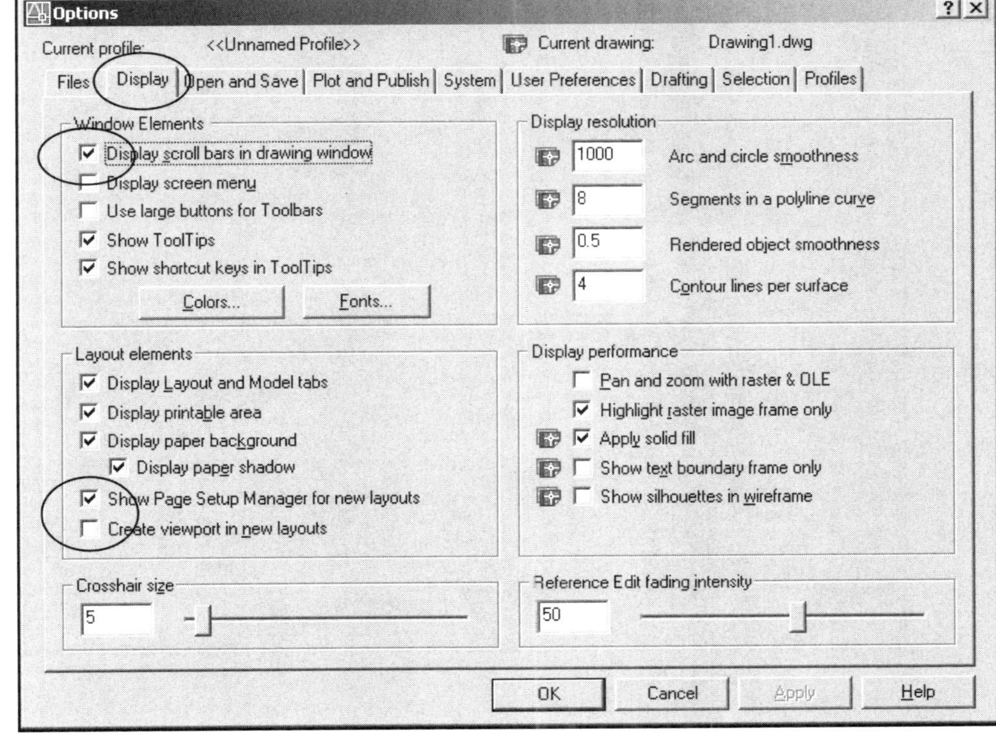

2007 LT

Options

Current drawing: Drawing1.dwg

Files | Display | Open and Save | Plot and Publish | System | User Preferences | Drafting | Selection

Window Elements
- ☑ Display scroll bars in drawing window
- ☐ Use large buttons for Toolbars
- ☑ Show ToolTips
- ☑ Show shortcut keys in ToolTips

Colors... Fonts...

Layout elements
- ☑ Display Layout and Model tabs
- ☑ Display printable area
- ☑ Display paper background
  - ☑ Display paper shadow
- ☑ Show Page Setup Manager for new layouts
- ☐ Create viewport in new layouts

Display resolution
- 1000 Arc and circle smoothness
- 8 Segments in a polyline curve

Display performance
- ☑ Apply solid fill
- ☐ Show text boundary frame only

Crosshair size
- 5

OK Cancel Apply Help

E. Select the ***Open and Save*** tab and change the settings on your screen to match the dialog box below.

2007

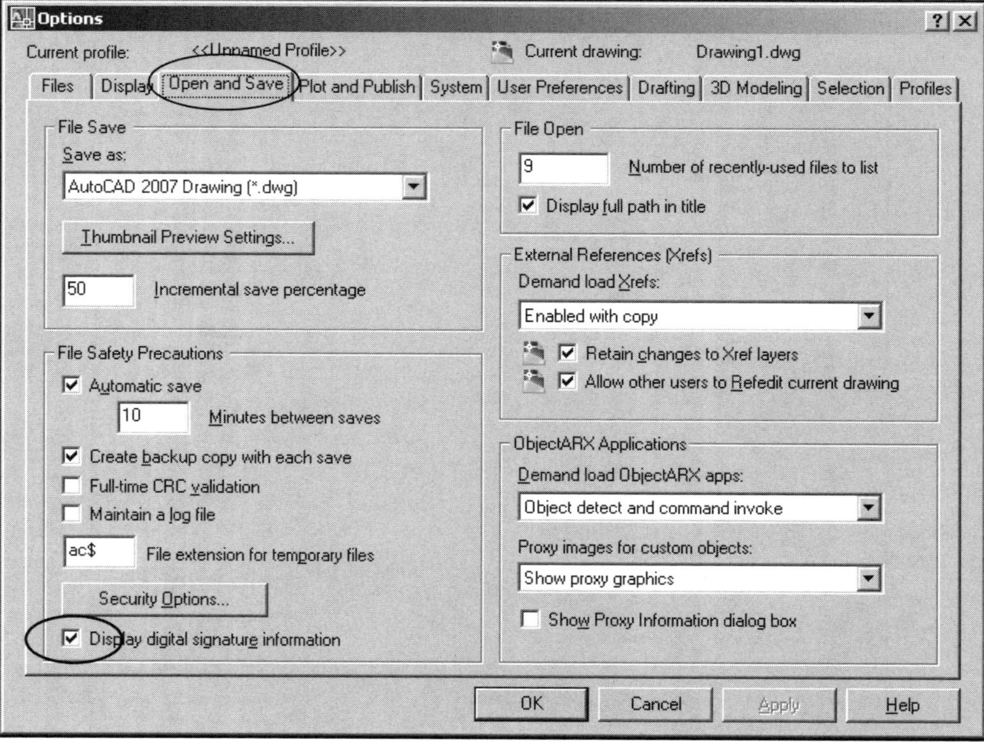

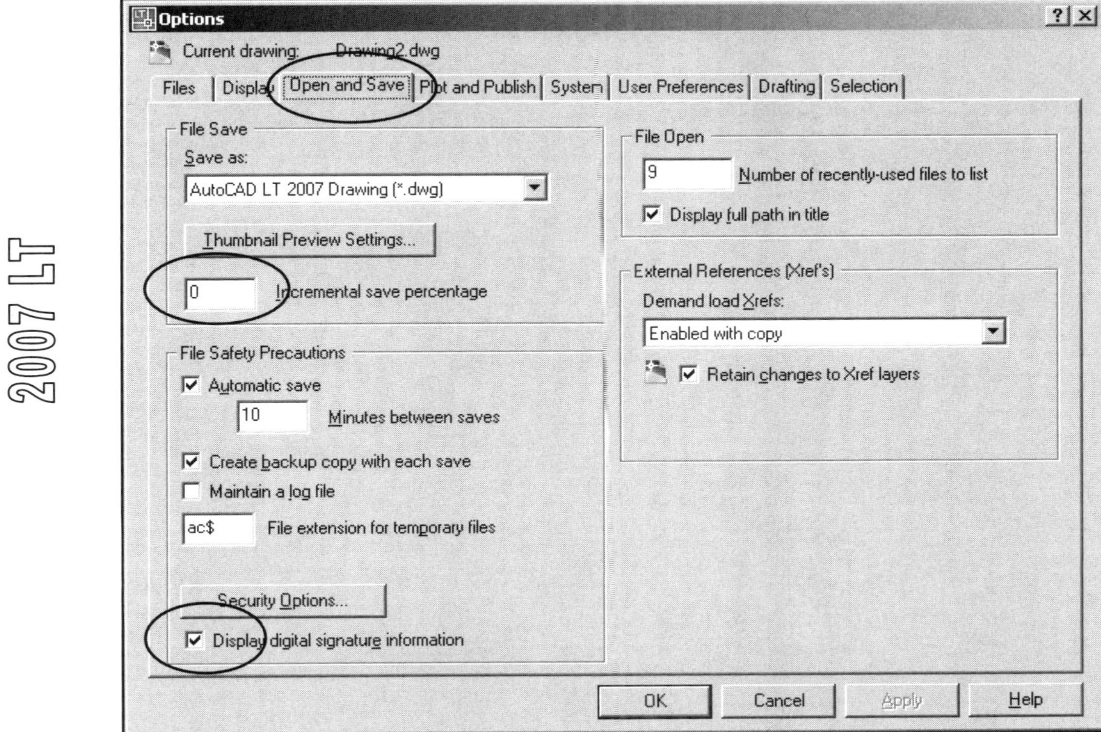

F. Select the ***Plot and Publish*** tab and change the settings on your screen to match the dialog box below.

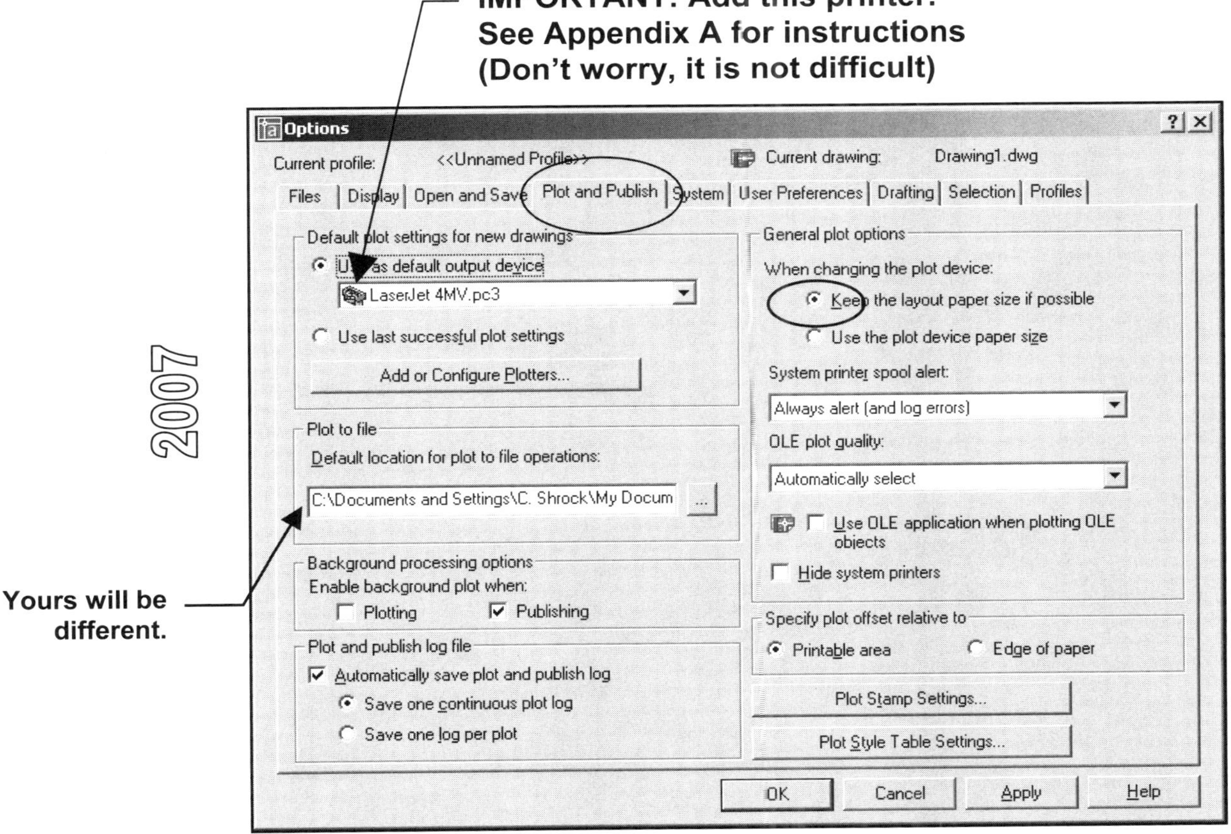

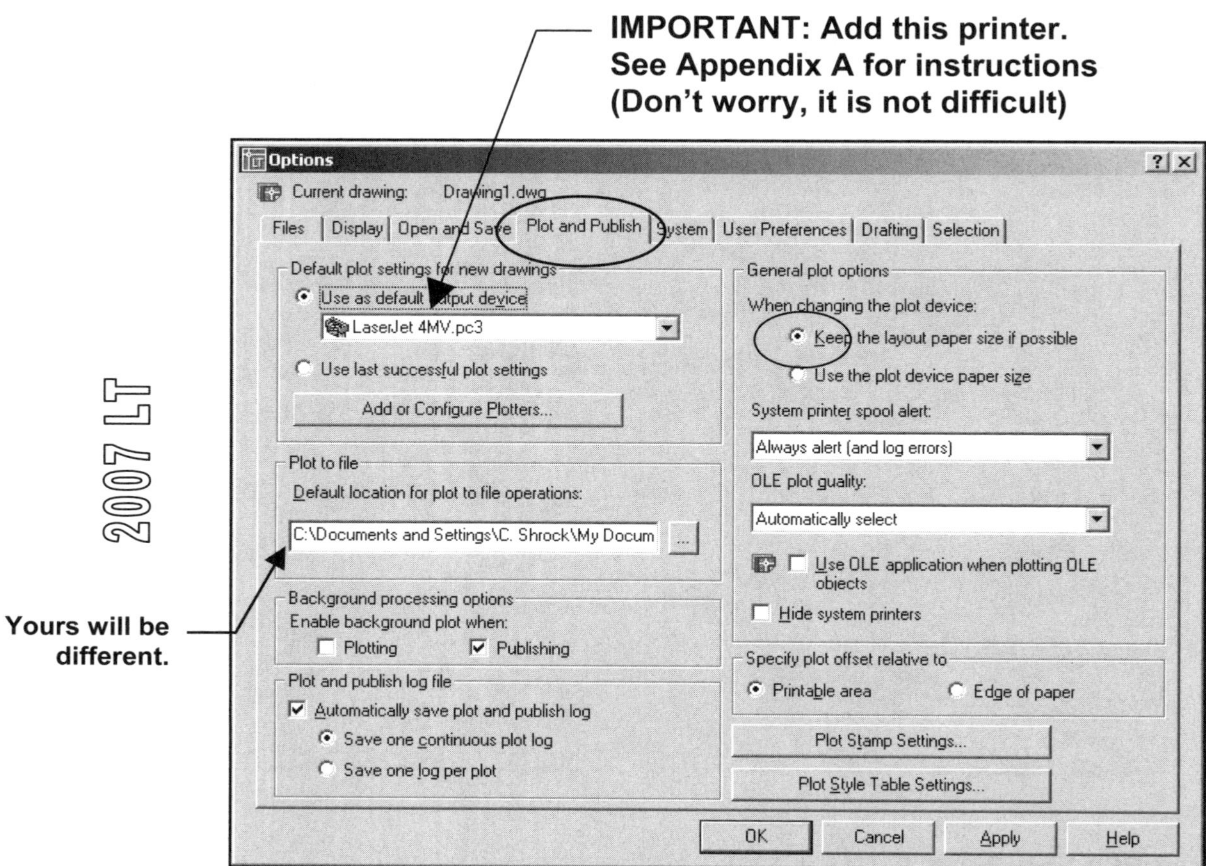

G. Select the ***System*** tab and change the settings on your screen to match the dialog box below.

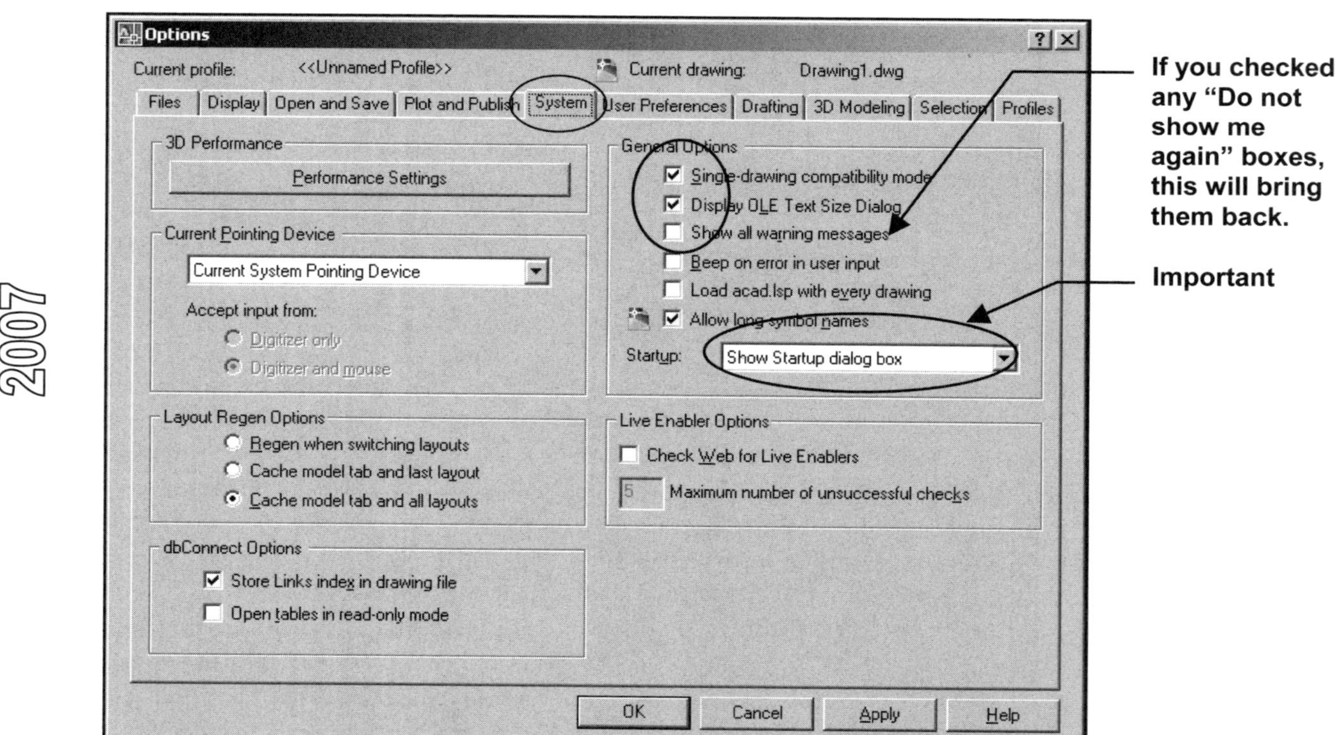

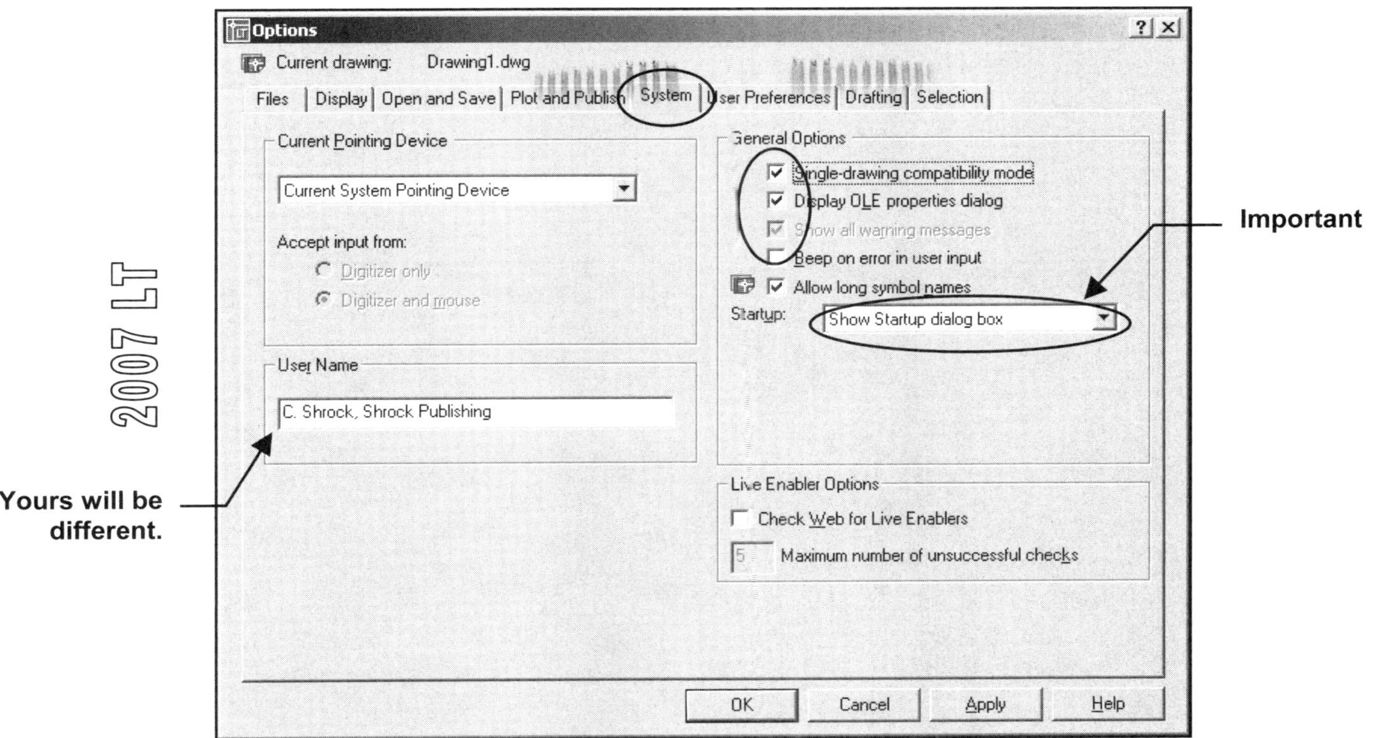

H. Select the ***User Preferences*** tab and change the settings on your screen to match the dialog box below.

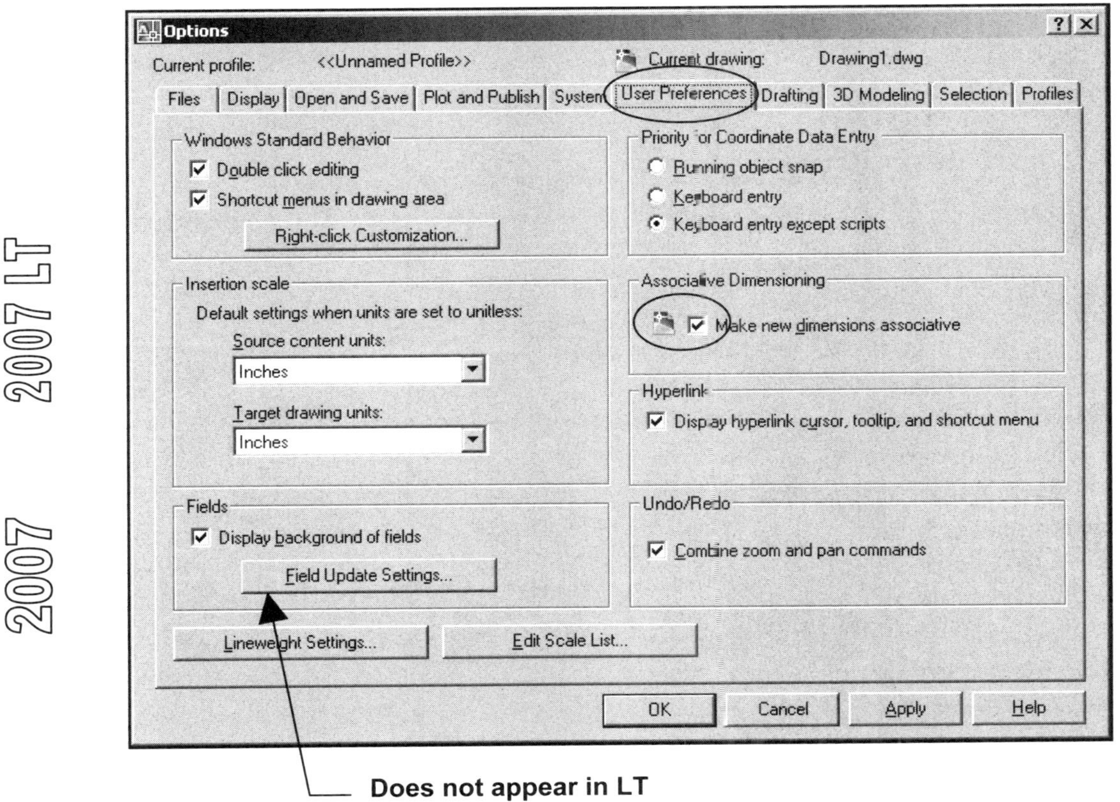

I. Select the ***Right-click Customization..*** box and change the settings on your screen to match the dialog box below.

2007 LT 2007

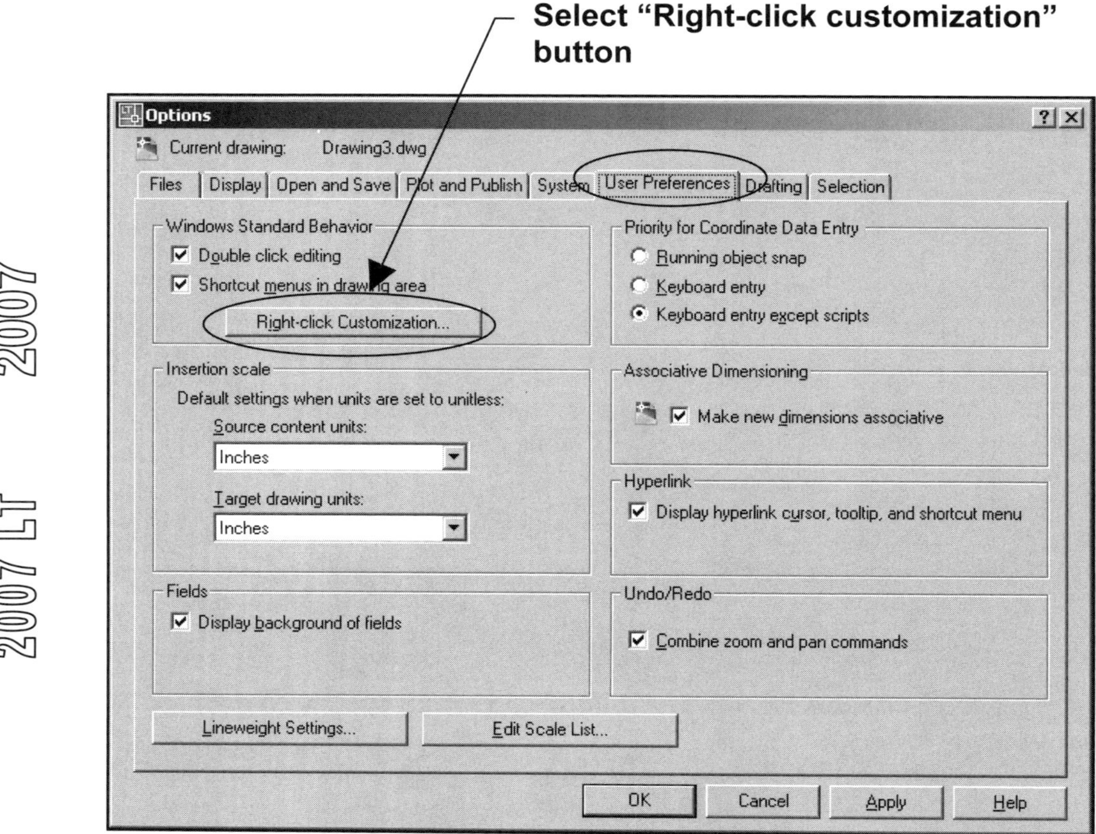

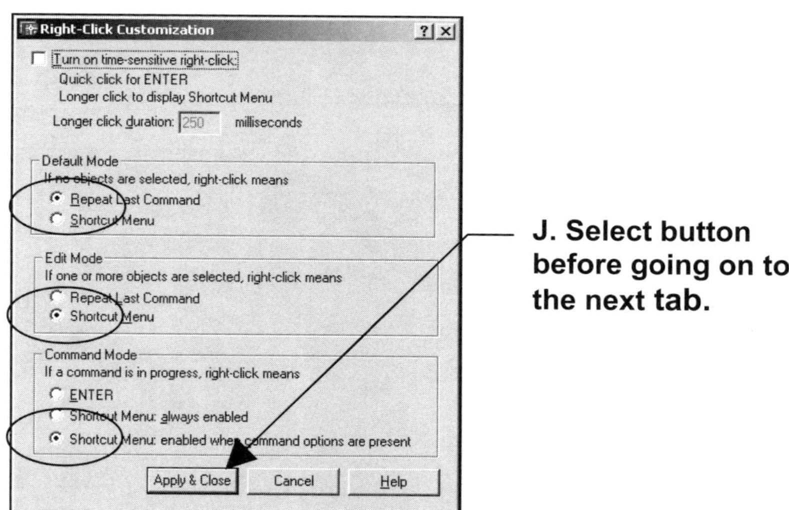

J. Select the ***Apply & Close*** button, shown above, before going on to the next tab.

K. Select the ***Drafting*** tab and change the settings on your screen to match the dialog box below.

2007

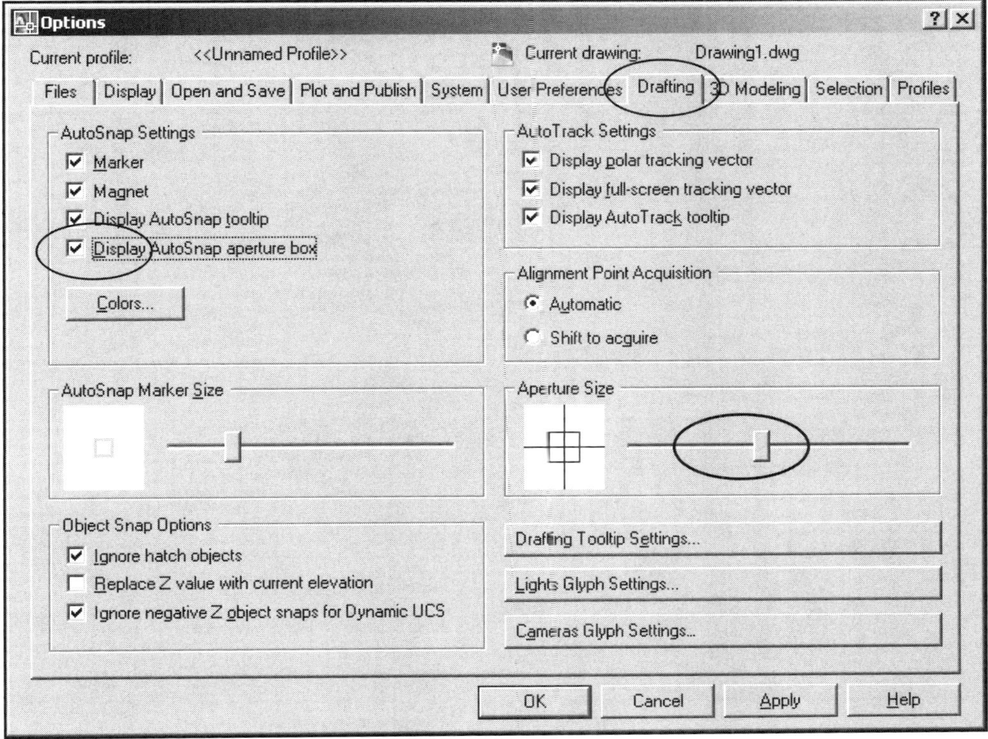

2007 LT

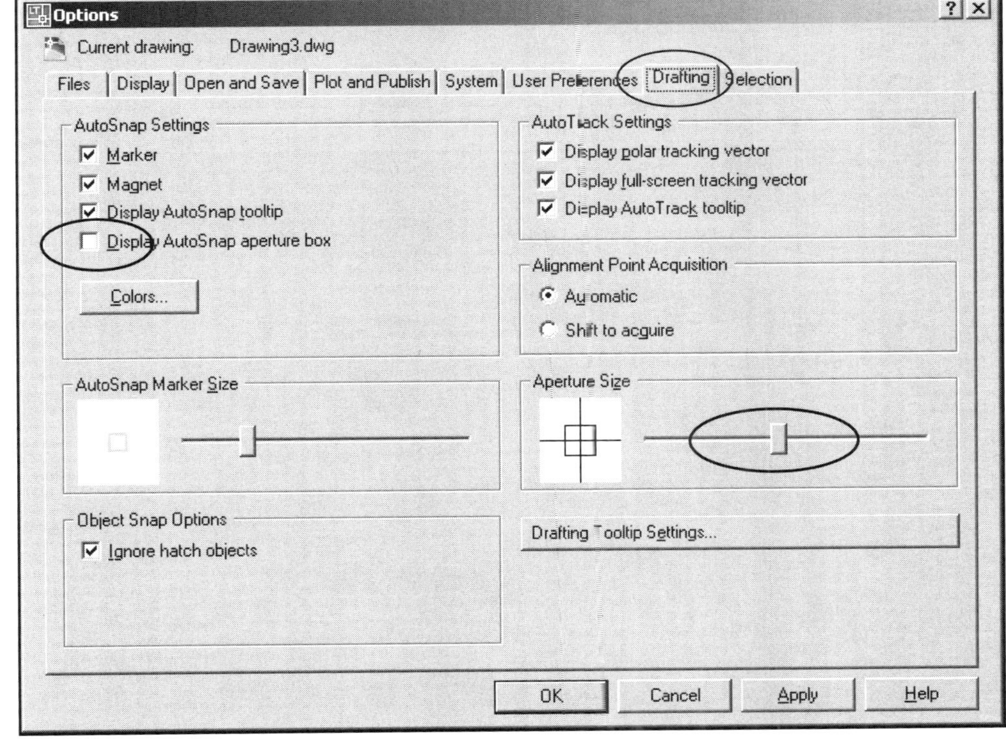

L. Select the ***Selection*** tab and change the settings on your screen to match the dialog box below.

2007

2007 LT

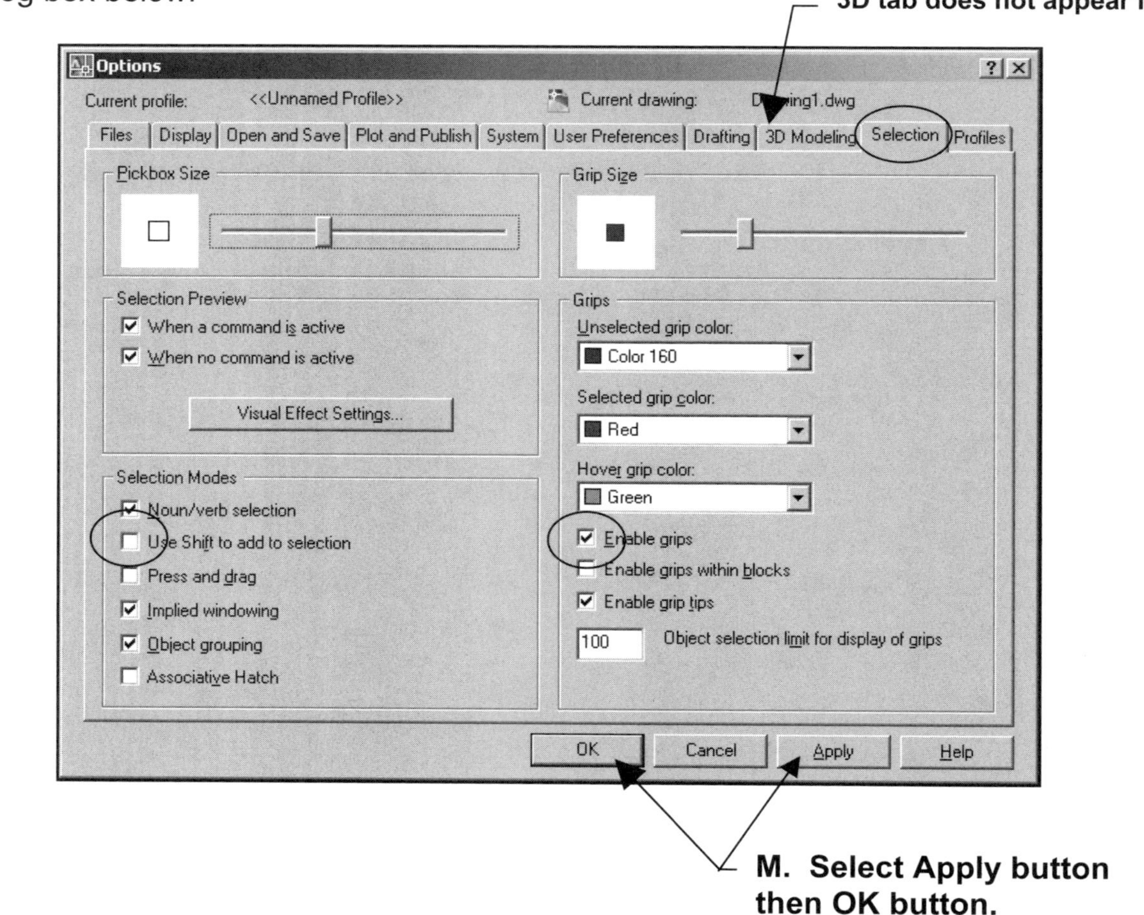

M. Select the ***Apply*** button then the ***OK*** button.

Customizing your Wheel Mouse

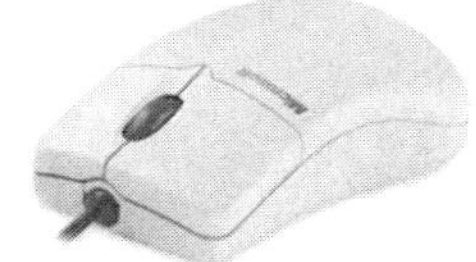

A Wheel mouse has two or more buttons and a small wheel between the two topside buttons. The default functions for the two top buttons are as follows:
Left Hand button is for **input**
Right Hand button is for **Enter** or the **shortcut menu**.

You will learn more about this later. But for now follow the instructions below.

Using a Wheel Mouse with AutoCAD®

To get the most out of your Wheel Mouse set the **MBUTTONPAN** setting to **"1"** as follows:

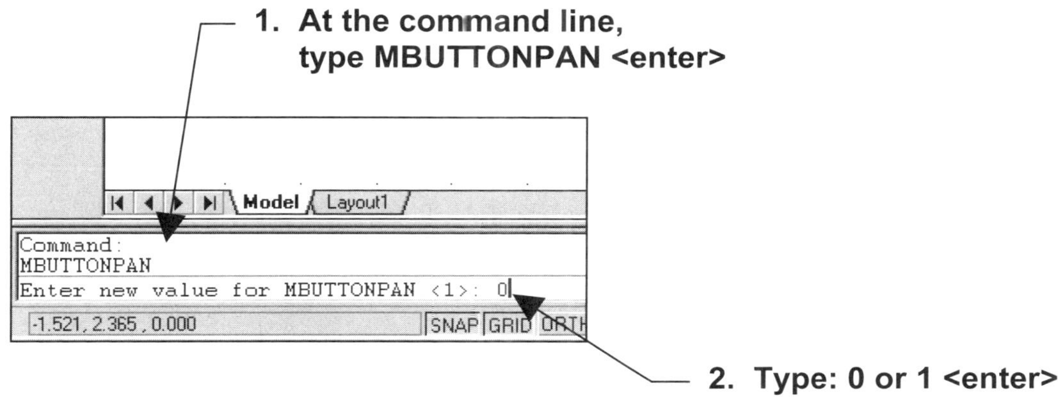

After you understand the function of the "Mbuttonpan" variable, you can decide whether you prefer the setting "0" or "1" as described below.

MBUTTONPAN setting 0:

ZOOM Rotate the wheel forward to zoom in
Rotate the wheel backward to zoom out

OBJECT SNAP Object Snap menu will appear when you press the wheel

MBUTTONPAN setting 1: (Factory setting)

ZOOM Rotate the wheel forward to zoom in
Rotate the wheel backward to zoom out

ZOOM EXTENTS Double click the wheel

PAN Press the wheel and drag

AutoCAD 2007 system requirements

Operating system:

Microsoft Windows XP Professional SP 1 or 2
Microsoft Windows XP Home SP 1 or 2
Microsoft Windows XP Tablet PC SP 2
Microsoft Windows 2000 SP 3 or 4 (SP4 recommended)

Browser

Microsoft Internet Explorer 6.0 SP 1or later

Hardware (required)

Pentium III or later with 800 MHz processor or better
Mouse, Track Ball or other pointing device
1024 x 768 video graphics display with true color*
CD-ROM drive for initial installation only

Hardware (optional)

Printer or Plotter
Serial or Parallel port (for peripheral devices)
Sound card with speakers
Disk drive (3-1/2 floppy, Zip or CDW) for saving files.

RAM and Hard Disk Space

512 MB of RAM minimum*
750MB of hard disk space
64MB of swap space (Recommended)
100MB free disk space in system folder (Recommended)

***Note: Increased RAM requirement**

The 3D visualization and rendering abilities of AutoCAD requires much more memory than ever before. It will run with 512MB but Autodesk recommends 2GB of RAM and a video card with at least 128MB of its own memory. Check Autodesk's website (www.autodesk.com) for the list of supported video cards.

LEARNING OBJECTIVES

Note:

This lesson should be used to determine whether or not you are ready for this level of instruction. If you have difficulty creating Exercises 1A, B and C, you should consider starting with the "Exercise Workbook for Beginning AutoCAD 2007".

LESSON 1

DISPLAY MULTIPLE DRAWINGS

AutoCAD has an option that allows you to have one or more drawings open at the same time. This option is **"Single-drawing compatibility mode".** If you have "Single-drawing compatibility mode" **ON**, only one drawing can be open on the screen. If you have "Single-drawing compatibility mode" **OFF**, multiple drawings may be open on the screen.

You may view the multiple drawings side by side (tiled) or as a full screen (cascade).
See examples of each on the next page.

How to configure AutoCAD to open *"Multiple drawings".*

1. Select **TOOLS / OPTIONS**
2. Select the **SYSTEM** tab.
3. Remove the **check mark** from "Single-drawing compatibility mode" box.
4. Select **APPLY** and the **OK** button.

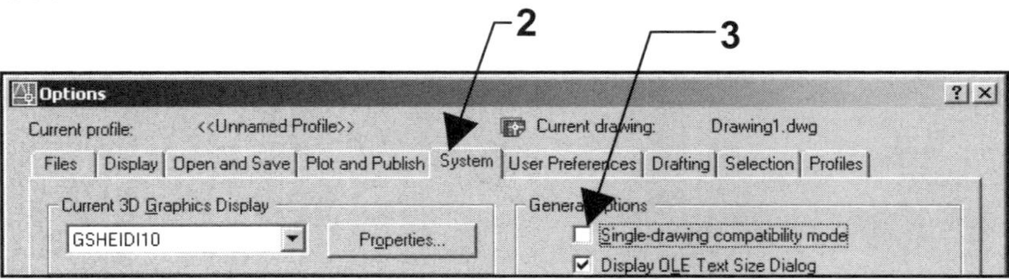

How to CLOSE multiple drawings.

1. Select the **Window** pulldown menu.
2. Select **Close** to close the active drawing or **Close All** to close all the drawings at once.

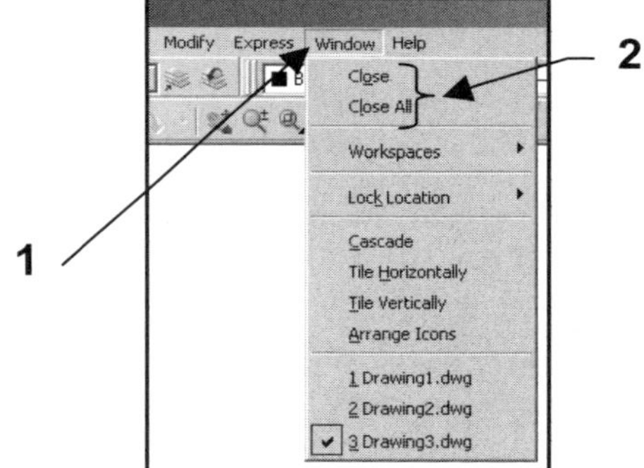

How to view multiple drawings.

1. Select the **WINDOW** pulldown menu.
2. Select Cascade, Tile Horizontally or Tile Vertically.

(Examples on next page)

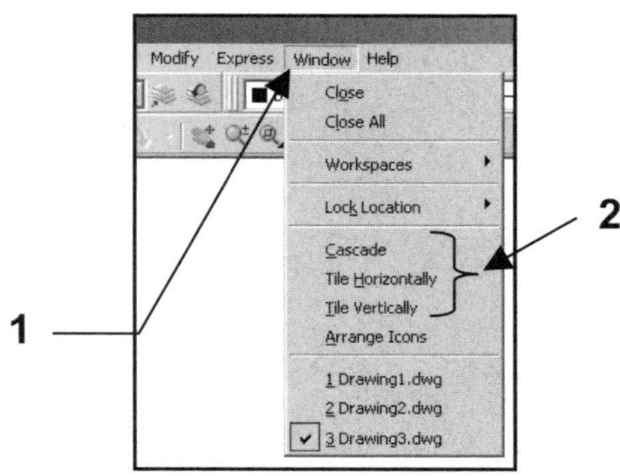

Example of "TILED"

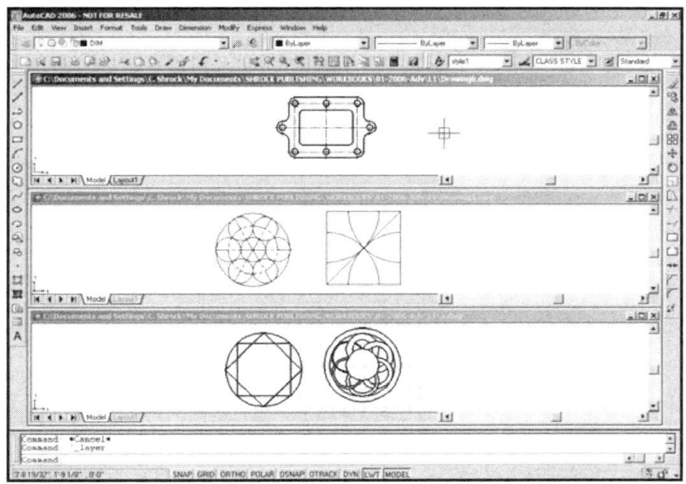

Tiled <u>Horizontally</u>

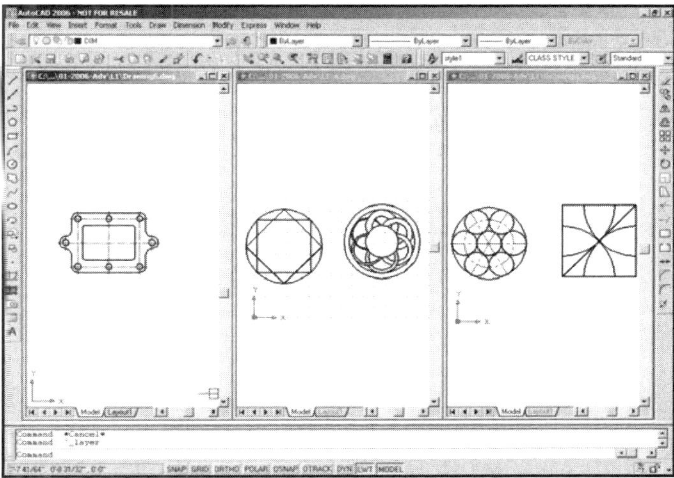

Tiled <u>Vertically</u>

Example of "CASCADE"

To view one of the multiple drawings, use one of the following:

1. Use **Crtl + tab** to toggle between drawings;

   or

2. Select **Window** pulldown menu, then select the drawing name from the list shown;

   or

3. Click on the **Title bar.**

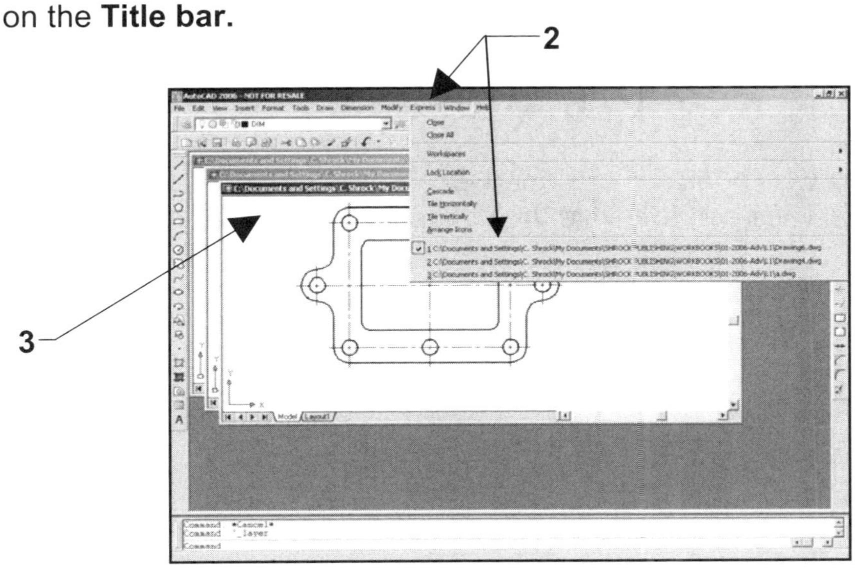

WARM UP DRAWINGS

The following drawings have been included for two purposes:

First purpose: To make sure you remember the commands taught in the Beginning Workbook and to get you prepared for the new commands in this Advanced workbook.

Second purpose: To confirm that you are ready for this level of instruction.

IMPORTANT, PLEASE READ THE FOLLOWING:

*This workbook assumes you already have enough basic AutoCAD 2007 knowledge to easily complete Exercises 1A, 1B and 1C. If you have difficulty with these exercises, you should consider reviewing "**Exercise Workbook for Beginning AutoCAD 2007**". If you try to continue without this knowledge you will probably get confused and frustrated. It is better to have a good solid understanding of the basics before going on.*

Note to Instructors

The Page Set ups in this workbook are designed for HP plotters, 4MV and 500. If your plotters are different, you will need to revise the plotting instructions throughout the lessons.

EXERCISE 1A

INSTRUCTIONS:
1. Start a **NEW** file and select **1workbook helper.dwt**. (Refer to Intro-2)
2. Draw the Objects below using Layers: Object and Center.
3. Do not Dimension.
4. Add your name anywhere on the drawing. Use Layer: Txt-Lit Style: Standard Height: .25
5. Save as: EX-1A
6. Plot from **"Model"** tab using the instructions on page 1-8.

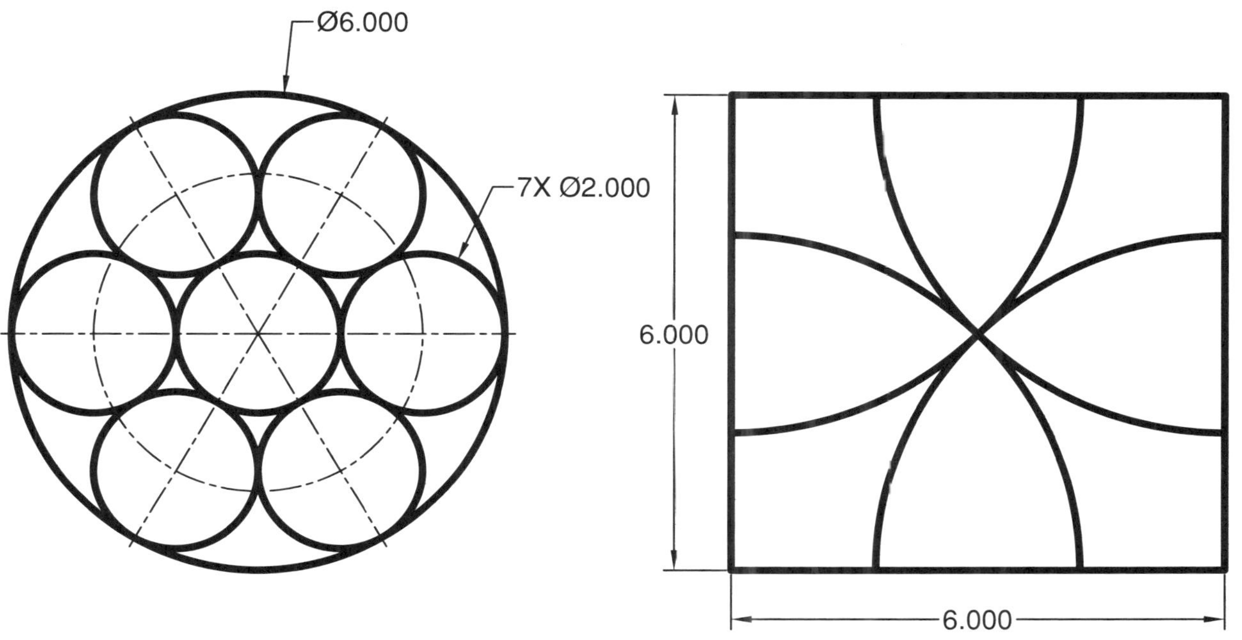

CONSTRUCTION HINTS

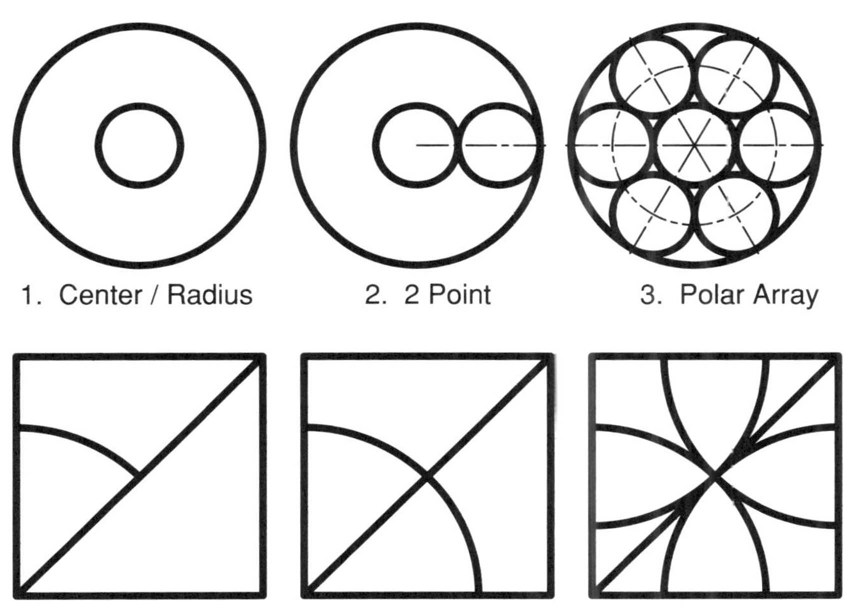

EXERCISE 1B

INSTRUCTIONS:
1. Start a **NEW** file and select **1workbook helper.dwt**.
2. Draw the Objects below using Layer: Object.
3. Do not Dimension.
4. Add your name anywhere on the drawing. Use Layer: Txt-Lit Style: Standard Height: .25
5. Save as: EX-1B.
6. Plot from **"Model"** tab using the instructions on page 1-8.

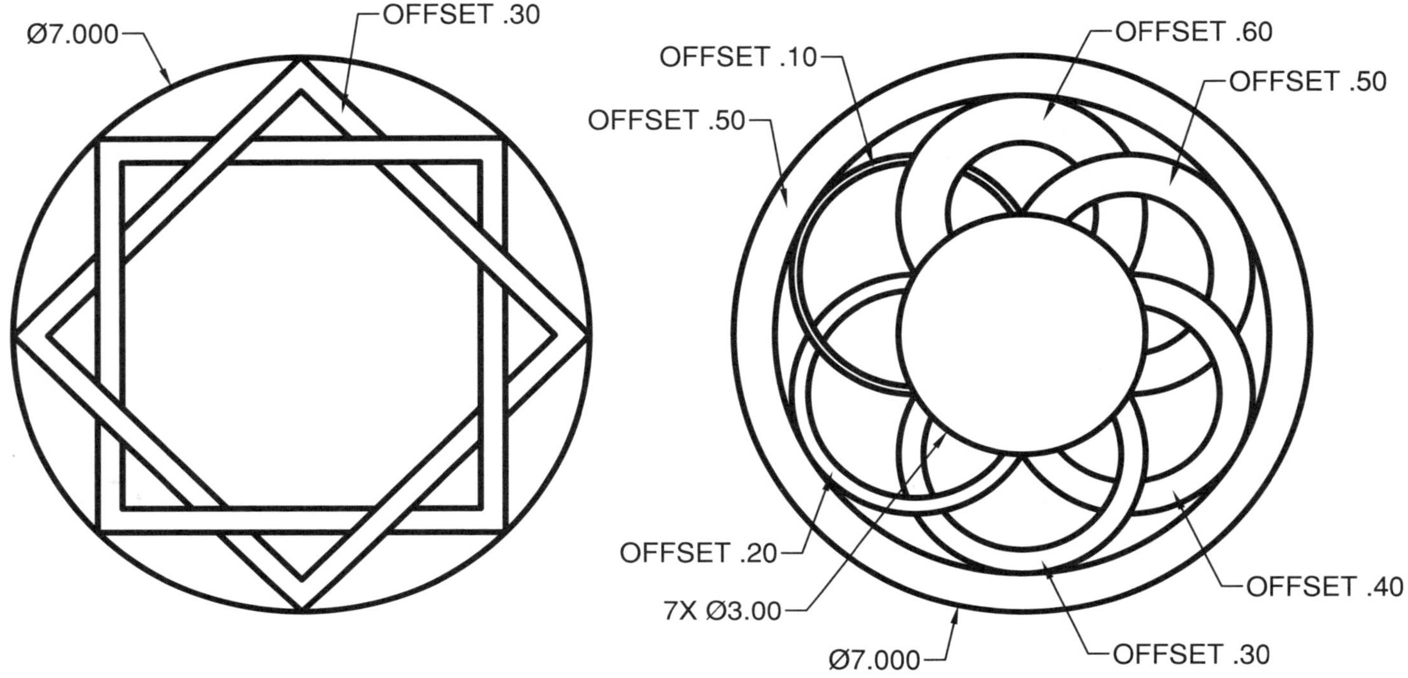

CONSTRUCTION HINTS

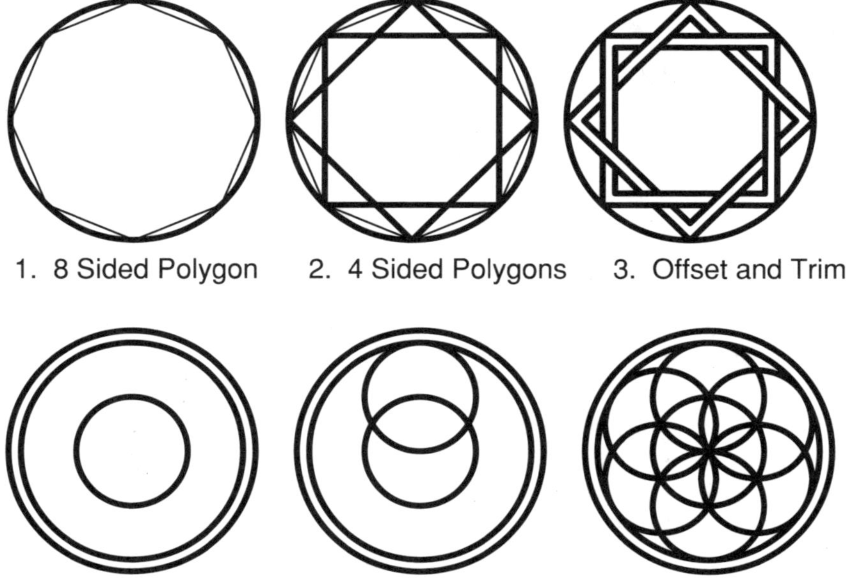

EXERCISE 1C

INSTRUCTIONS:

1. Start a **NEW** file and select **1workbook helper.dwt**.
2. Draw the Objects below using Layers: Object, Center and Dimension.
3. Dimension using Dimension Style **"Class Style"**.
4. Add your name anywhere on the drawing. Use Layer: Txt-Lit Style: Standard Height: .25
5. Save as: EX-1C
5. Plot from **"Model"** tab using the instructions on page 1-8.

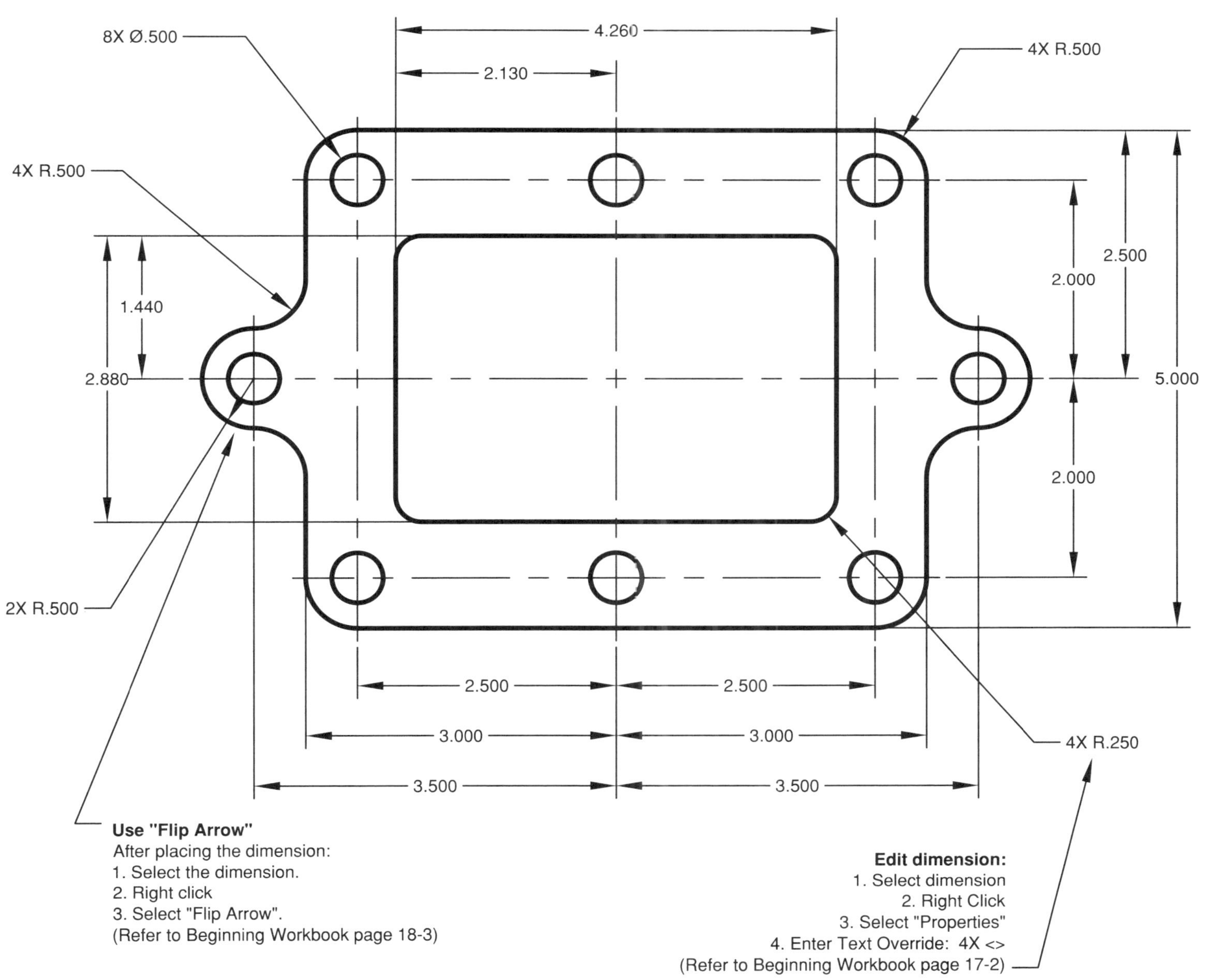

REVIEW PLOTTING FROM MODEL SPACE

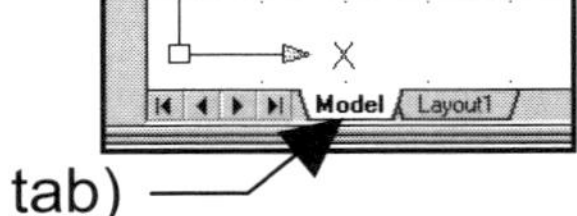

1. Open EX-1A (or a drawing you want to plot.)
2. Make sure that the Model tab is selected. (Model tab)
3. Select: View / Zoom / All
4. Select the **Plot** command by "right clicking" on the "Model" tab or using one of the following methods listed below:

Type = Print or Plot
Pulldown = File / Plot
Tool bar = Standard

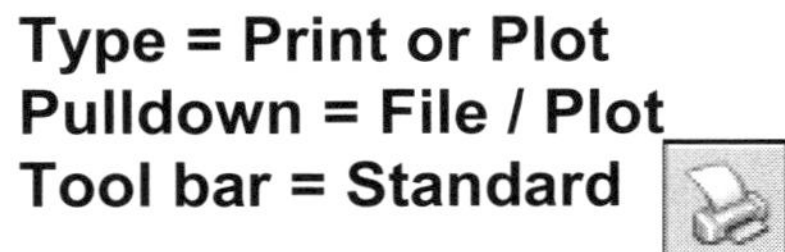

The Plot dialog box below should appear.

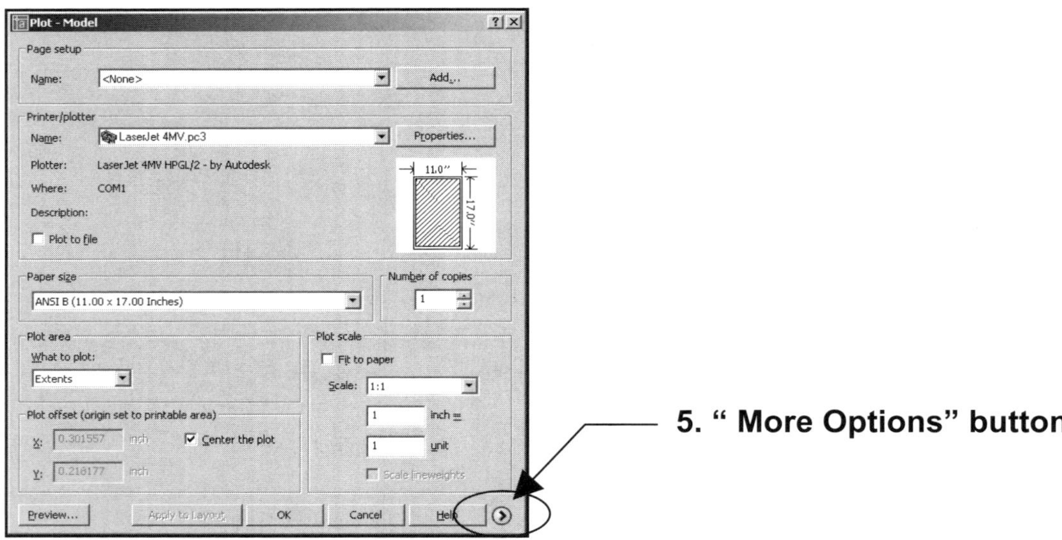

5. Select the "**More Options**" button to expand the dialog box if necessary.

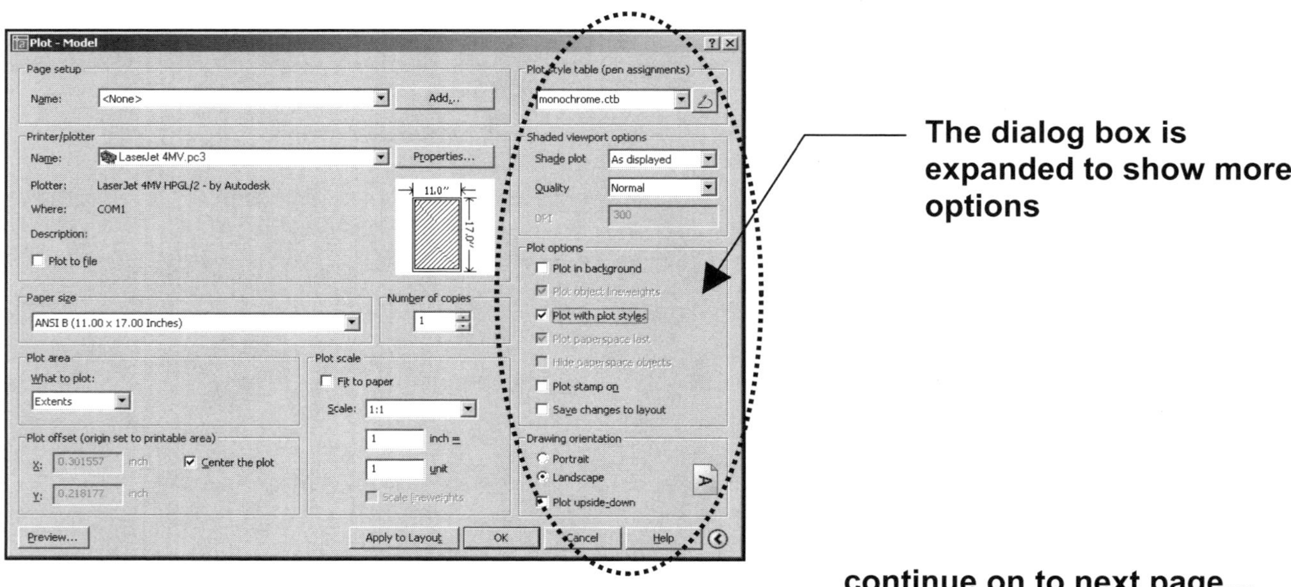

continue on to next page...

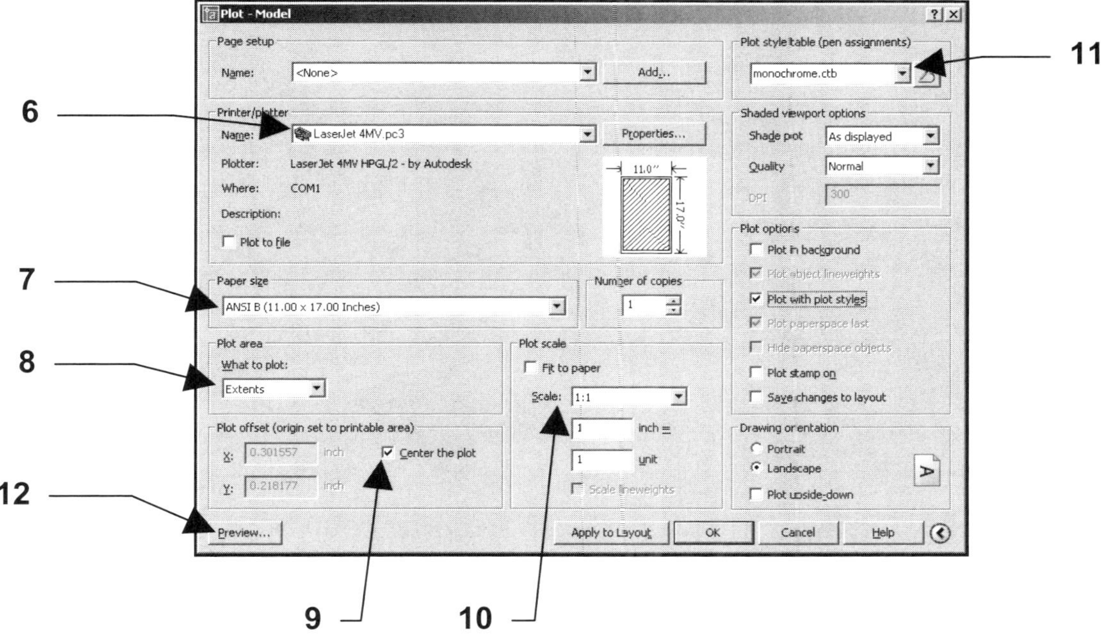

6. Select the "Printer / Plotter" **LaserJet 4MV** from the drop down list.
   Note: This Printer represents a size 17 X 11 for the exercises in this workbook. If it is not in the drop down list, you need to configure it on your system, even though you do not have this printer. Don't worry, this will not harm your computer. Refer to Appendix A for step by step instructions.
   (If you would like to use your own 8-1/2 X 11 printer, you may select it but refer to the note about scale selection in #10 below)

7. Select the Paper Size **ANSI B (11 X 17 inches)**

8. Select the Plot Area **EXTENTS**
   (Note: If Extents is listed, you have not drawn anything)

9. Select the Plot Offset **Center the plot**

10. Select Plot Scale **1 : 1**
    (Note: If you would like to print your drawing on a 8-1/2 X 11 printer, select the printer and select the "Fit to Paper" box and change the paper size.)

11. Select the Plot Style table **Monochrome.ctb**
    The following box will appear. Select **Yes**

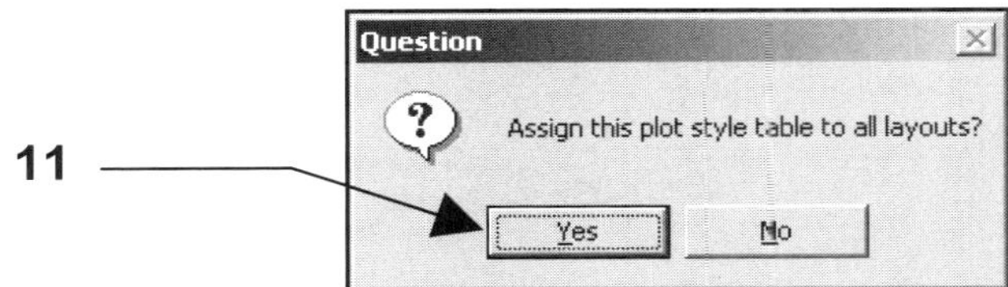

12. Select **Preview** button.

Continue on to next page...

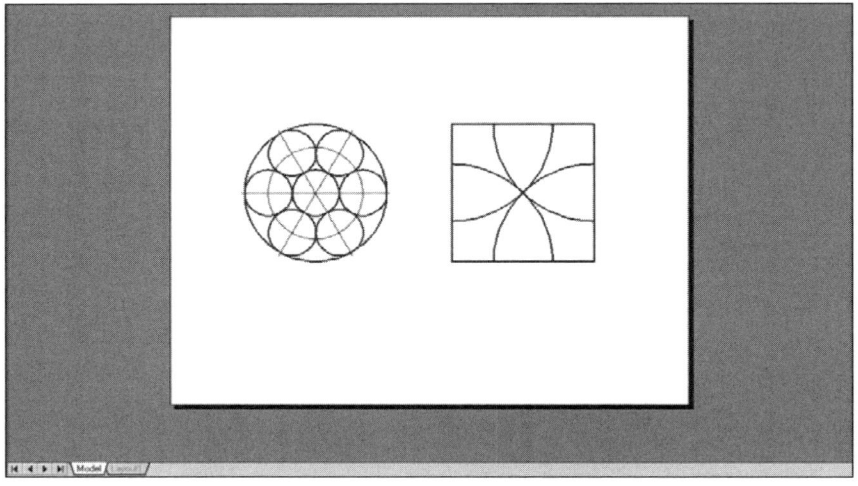

Does your display appear as shown above?
If yes, press <enter> and proceed to 13.
If no, recheck 1 through 11.

You have just created a **Page Setup**. All of the settings you have selected can now be saved so you do not have to select them all again when plotting another drawing.
To save the **Page Setup** you need to ADD it to the model tab within this drawing.

13. Select the **ADD** button.

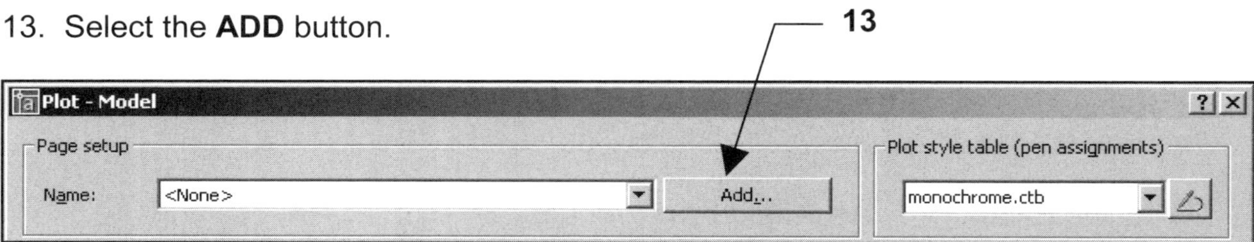

14. Type the new Page Setup name **Model-mono**
(This name identifies that you will use it when plotting the Model tab in monochrome.)

15. Select **OK** button

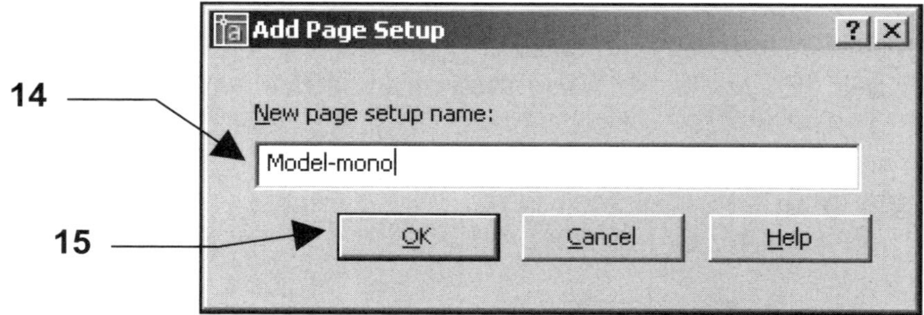

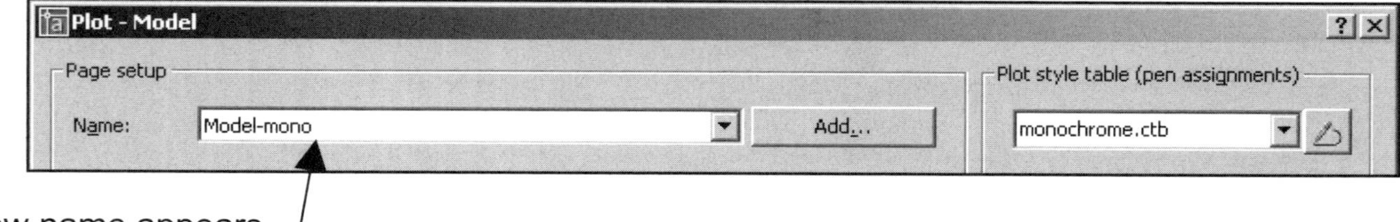

New name appears

continue on to next page....

16. Select **Apply to Layout** button.

17. Select the **OK** button to send the drawing to the printer or <u>select **Cancel** if you do not want to print the drawing at this time. The Page Setup will still be saved.</u>

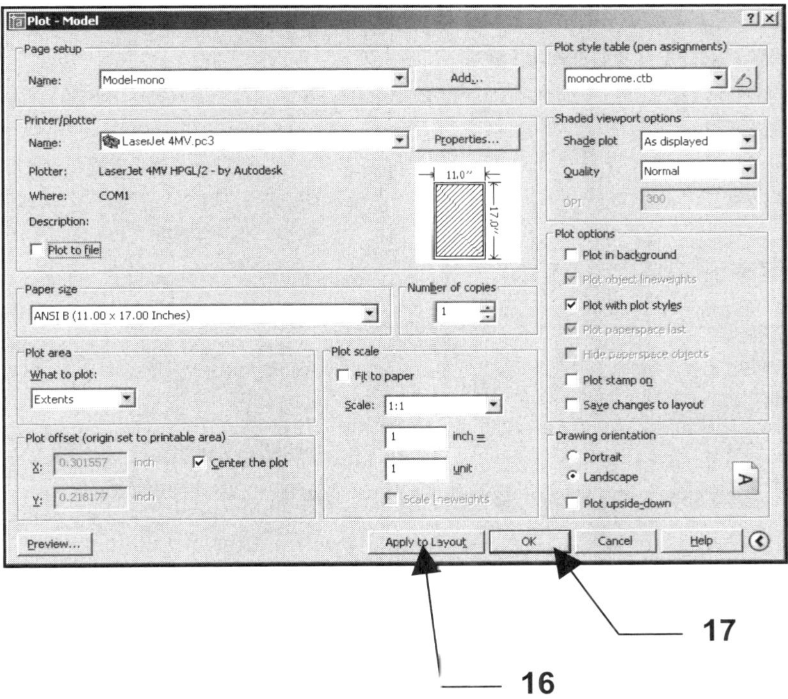

18. Save the entire drawing again. The Page Setup will be saved to the **Model tab** within this drawing and available whenever you wish to plot this drawing.

If you would like to use these setting to plot another drawing, such as EX-1B and EX-1C, you must repeat 1 through 18 or "Import" the page setup. To import the page setup refer to the next page.

Note: The instructions above are to be used when plotting while in "Model Space" ONLY.

IMPORT A PAGE SETUP

You may import a previously saved page setup from another drawing.

1. Open the drawing that you wish to plot.

2. Select the Model tab or a Layout tab.
(Note: You may only import Model page setups into Model and Layout page setups into Layout.)

3. Select **File / Plot**.

4. Select **Import** from the drop down list.

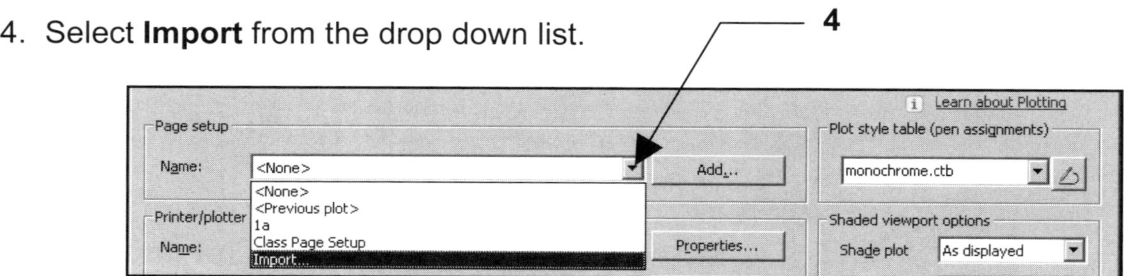

5. Find the drawing from which you wish to import a page setup.

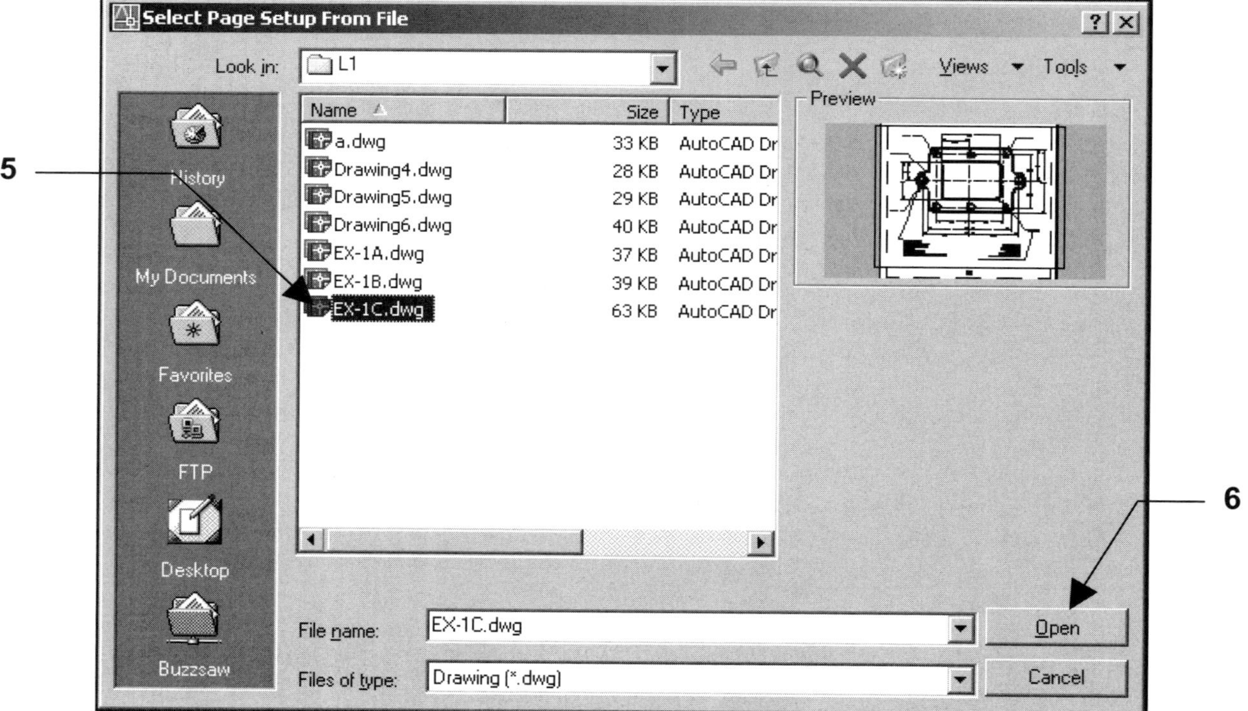

6. Select **Open**

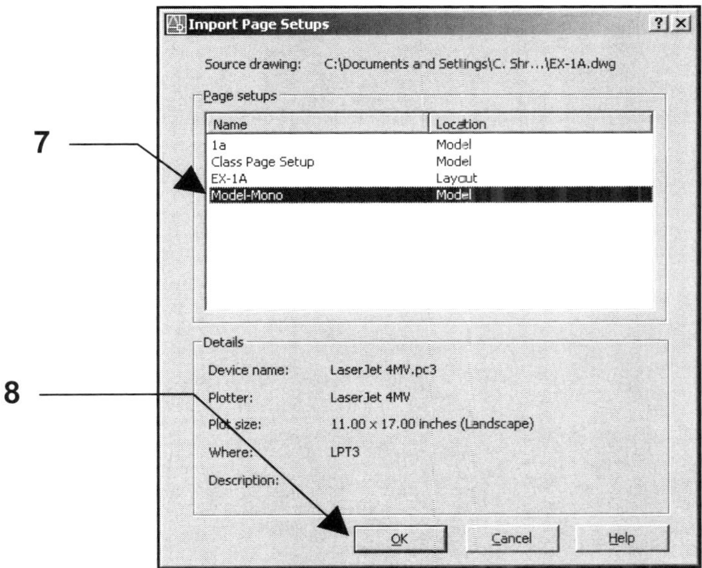

7. Select the page setup to import.

8. Select the **OK** button.

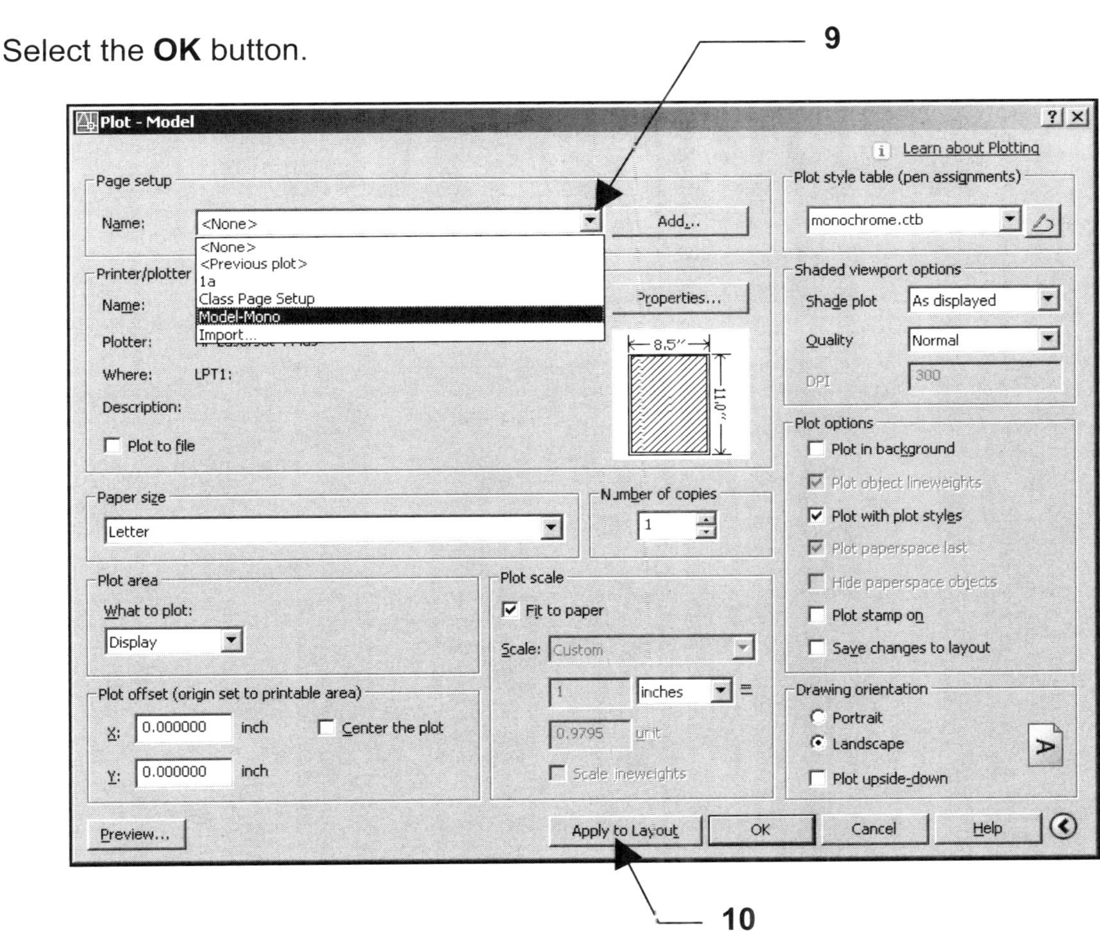

9. Select the imported page setup from the page setup name drop down list.

10. Select the **Apply to Layout** button.

Don't forget to save the drawing again if you wish to permanently attach this page setup.

NOTES:

LEARNING OBJECTIVES

After completing this lesson, you will be able to:

1. Customizing your Workspace.
2. Restore AutoCAD default workspace.
3. Export and Import a workspace.
4. Create a new Toolbar.
5. Delete a Toolbar command.
6. Add or delete a command on a pull-down menu.
7. Customize the Status Bar.
8. Import and Export a Profile
9. Understand the Redo command.

LESSON 2

CUSTOMIZING YOUR WORKSPACE

At this stage of your AutoCAD education you are probably not too interested in customizing AutoCAD. But I would like to at least introduce you to the new **Custom User Interface** to familiarize you with a few of the customizing options that are available. It is relatively easy to use and sometimes helpful. If you find it intriguing you may explore further using the AutoCAD "Help" system discussed in Lesson 1 in the Beginning workbook.

> Note: Those of you that are familiar with the customizing process of previous AutoCAD versions should note that the Custom User Interface file structure replaces the MNU/MNS file structure. If you are someone that enjoys customizing AutoCAD, you will find this tool will make customizing easier because all customizing is done within AutoCAD. You do not have to venture outside of AutoCAD.

Start by opening the **Customize User Interface** dialog box using one of the following:

TYPE = CUI
PULLDOWN = TOOLS / CUSTOMIZE / INTERFACE or VIEW / TOOLBARS
TOOLBAR = NONE

The following dialog box should appear.

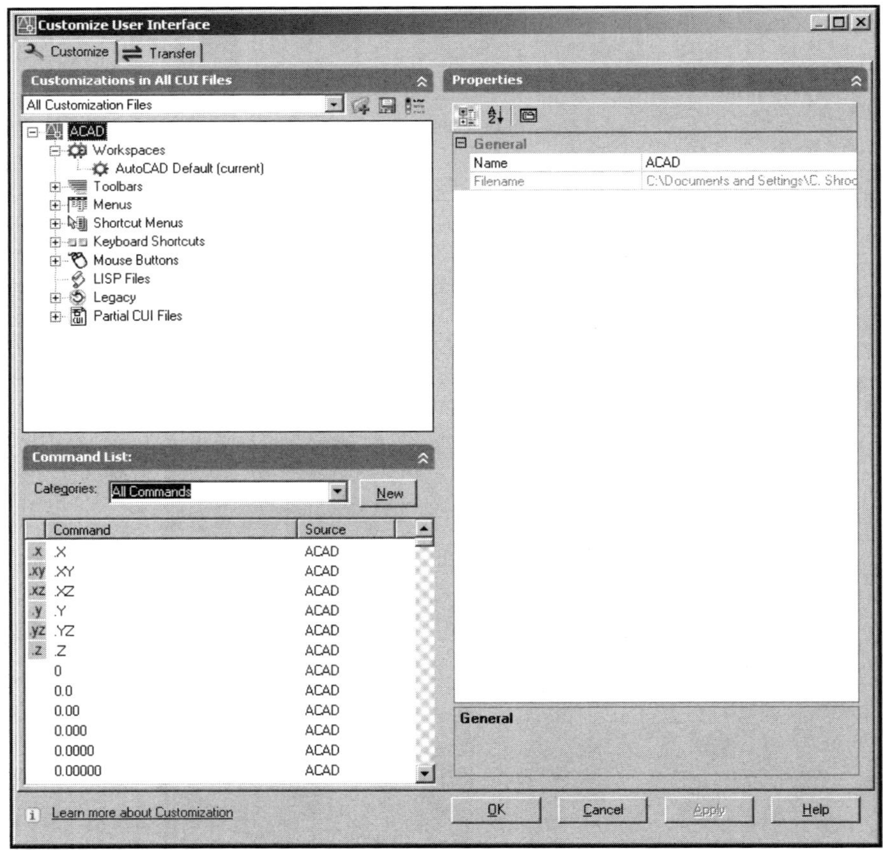

CREATING A WORKSPACE

Have you found yourself asking, "Do I really need all of these toolbars on the screen" or wishing that you could change the tools on a toolbar? Well now you can. You can create a "Workspace" all your own very easily.

Before you begin customizing your workspace it is a good idea to create a new one and then make the changes to that new one. This will allow you to always go back to the original AutoCAD default appearance.

Follow the instructions below to create a new workspace and make some changes. Don't be afraid that you will mess up your configuration. I will show you how to easily return to the AutoCAD default configuration. So don't worry.

1. Open the **Customize User Interface** dialog box as shown on the previous page.

2. Right click on **Workspaces.**

3. Select **New / Workspace**.

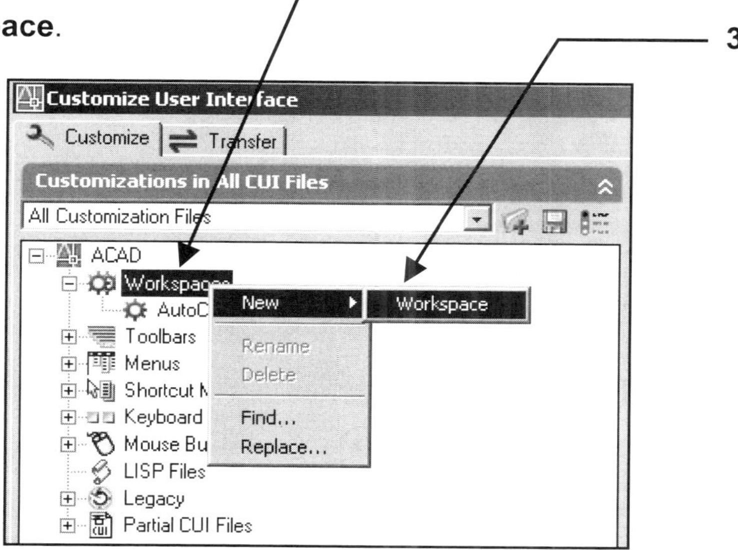

4. Enter the new workspace name **Class Workspace demo** <enter>.
   (Or whatever you would like to name the new workspace)

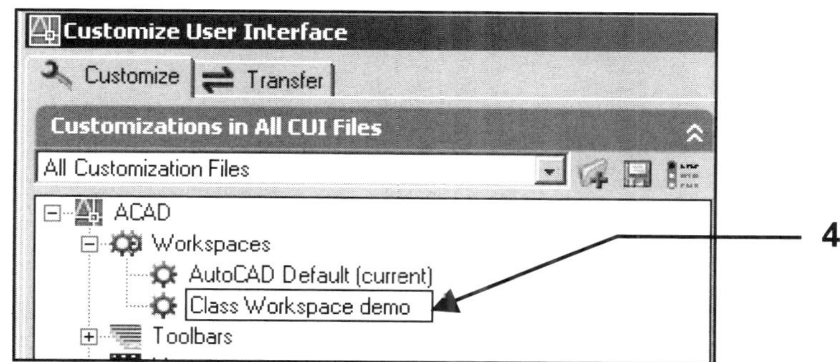

That's all there is to it. You have successfully created a new workspace.
The following pages will show you how to customize it.

Selecting the Toolbars to display

1. Open the Customize User Interface dialog box (if it isn't already open) and select the Workspace you wish to customize.

2. Right click on **Class Workspace demo** and select **Customize Workspace**. (Or select the **Customize Workspace** button in the **Workspace Contents** box.)

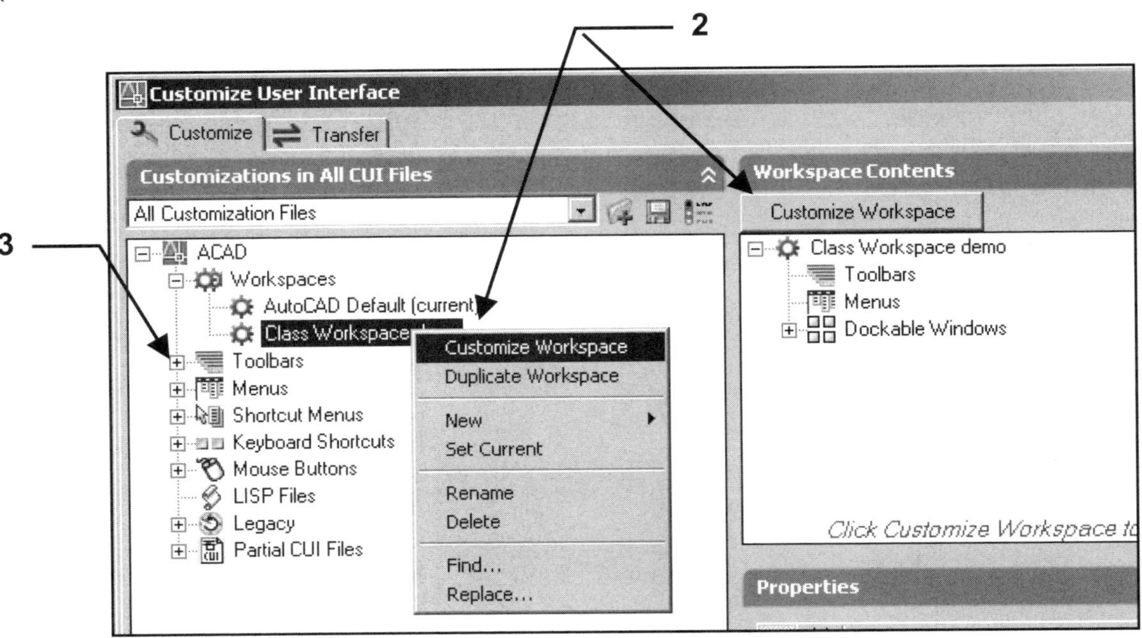

3. Select the **+** sign to the left of the **Toolbars** to display the toolbar list.

4. Select all of the Toolbars that you wish to be displayed on the screen. Notice that all of the toolbars are listed. Place a check in the box to the left of the toolbar name. The selected Toolbars will be listed in the **Workspace Contents** box, on the right.

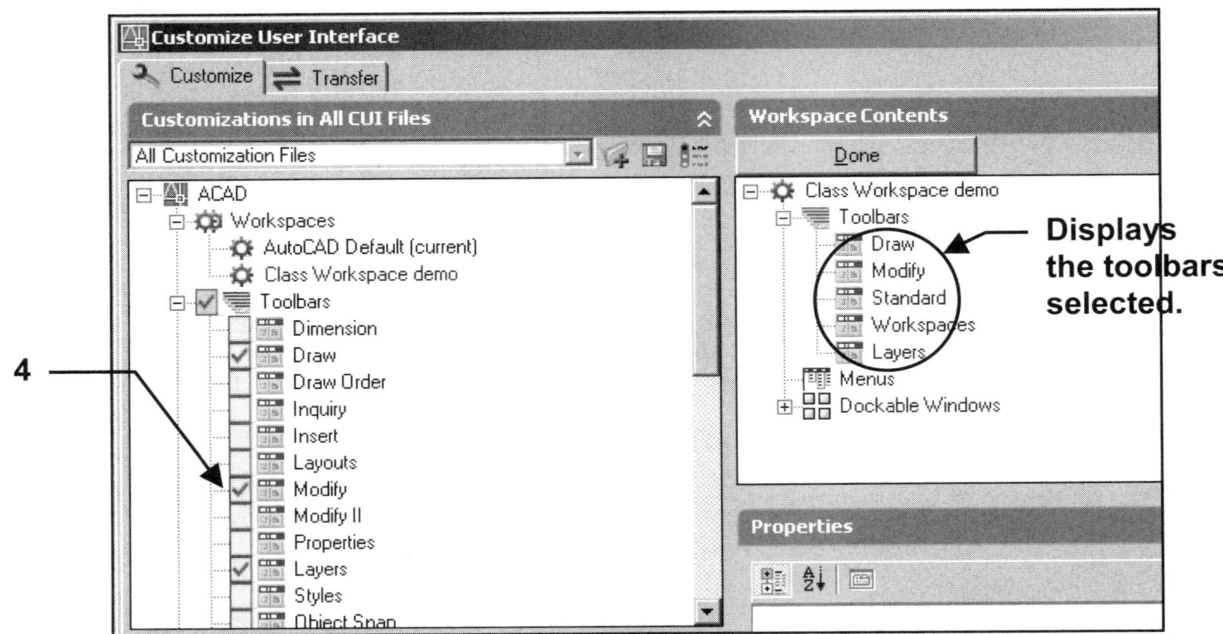

5. Now select the **+** sign beside the **Menus** and select the menus you wish to display.

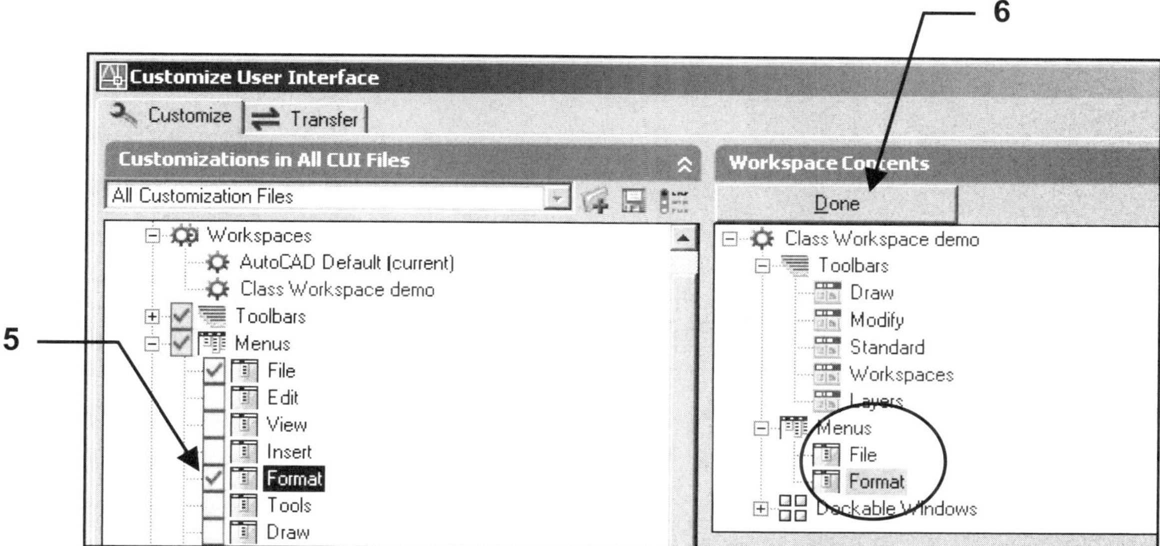

6. Select the **Done** box.

7. Right click on the Class Workspace demo and select Set Current.

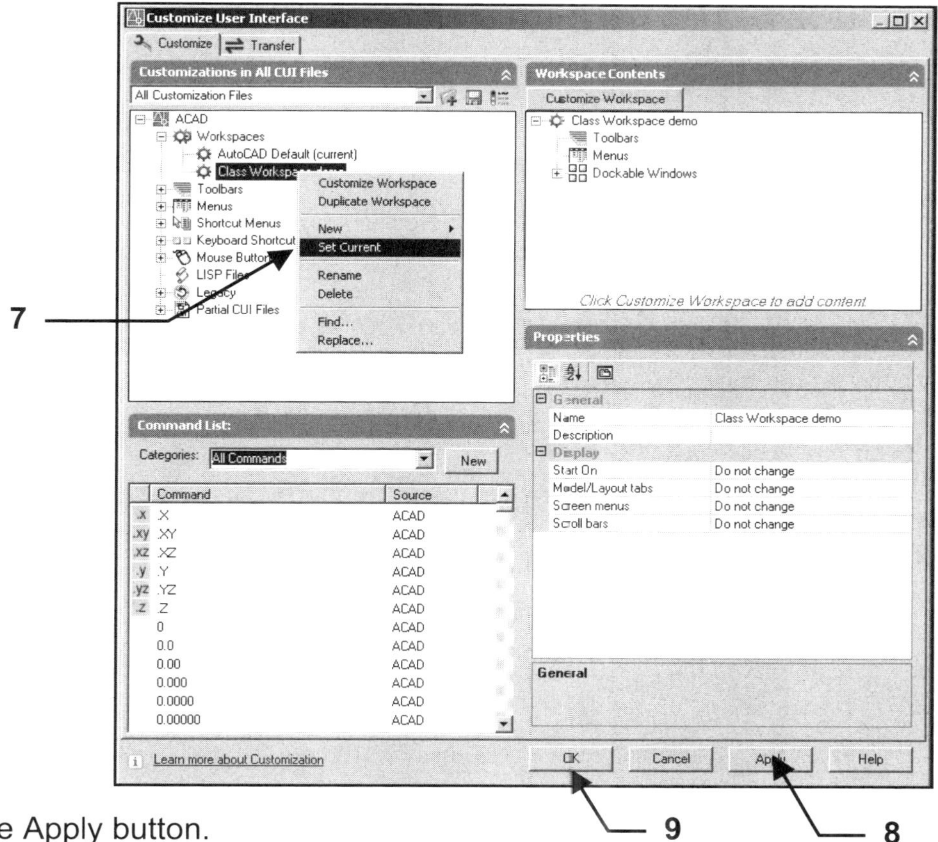

8. Select the Apply button.

9. Select the OK button. (Your workspace is automatically saved to this computer.)

Does your workspace appear different?

It should be displaying only the toolbars and menus that you selected.
Don't Panic, I told you I would show you how to return to the default workspace.
Turn to the next page.

How to return to the AutoCAD default workspace.

Notice the **Workspaces toolbar** located in the drawing area of the screen.
(If it is not there, open the Workspaces toolbar)

The **current** workspace is shown in the box.

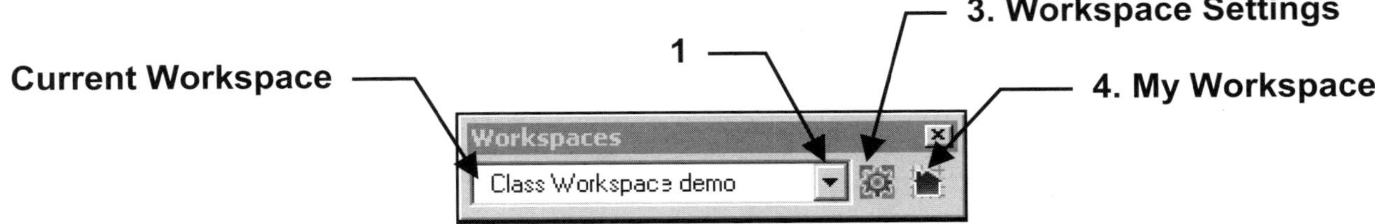

1. Select the **AutoCAD Classic** from the list using the drop down arrow.

The AutoCAD Classic Default workspace should now be displayed.

Workspaces toolbar options

3. Select the **Workspace Settings** tool. (See above)
or select **Tools / Workspaces / Workspace Settings**

4. Select which workspace to display when you click on the **My Workspace** tool shown above.

5. Lists all workspaces. "Un-Check" if you want one removed from the "My Workspace" list.

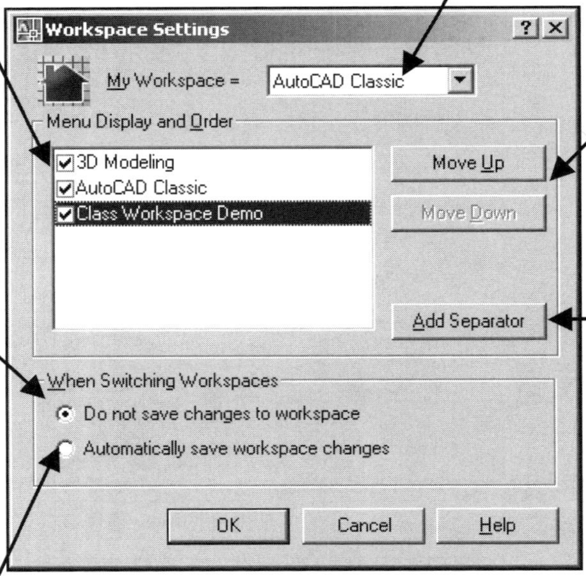

6. Move workspace up or down in the display order.

7. Add a separator line between workspace names in list.

8. Does not save changes you have made to a workspace when you switch to another workspace.

9. Saves changes you have made to a workspace when you switch to another workspace.

Workspace Settings
My Workspace = AutoCAD Classic
Menu Display and Order
3D Modeling
AutoCAD Classic
Class Workspace Demo
Move Up
Move Down
Add Separator
When Switching Workspaces
Do not save changes to workspace
Automatically save workspace changes
OK Cancel Help

EXPORT a WORKSPACE

Now that you have created your workspace you may wish to **Export** the workspace so you may **Import** the file onto another computer.

1. Select **TOOLS / CUSTOMIZE / EXPORT CUSTOMIZATIONS**
2. Select **Workspaces**.

Notice the **Transfer** tab is automatically selected.

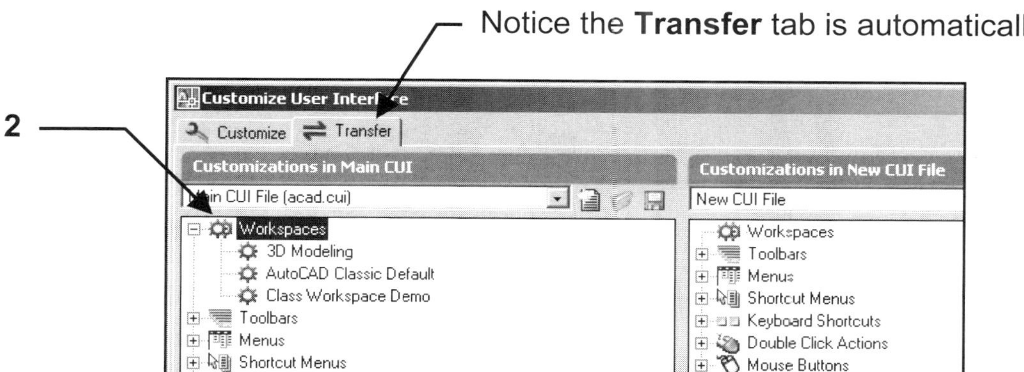

3. Select **Save As** from the drop down list.

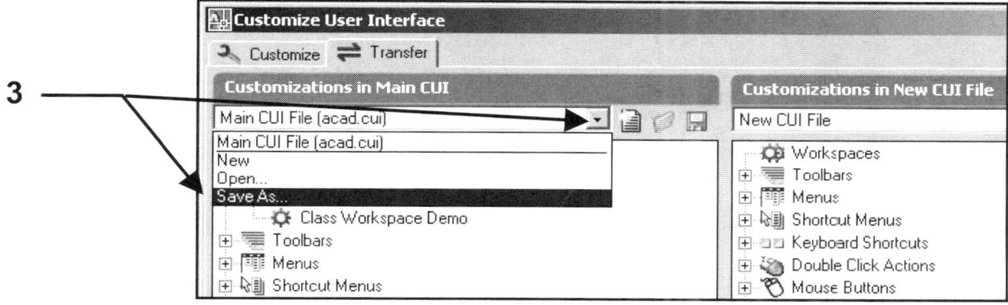

4. Locate where you wish to save the workspace file.
   (Note: This is a fairly large file and may not fit onto a 3-1/2" floppy disk.)

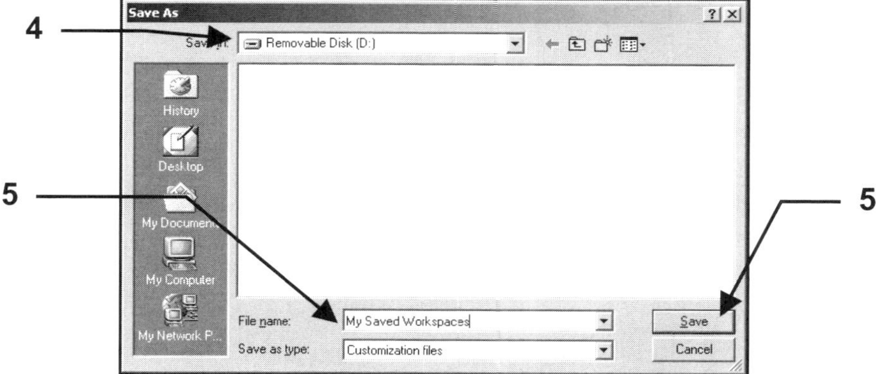

5. Enter a file name.
6. Select the Save button.

Your customized workspace is now saved and can be imported onto another computer, as shown on the next page.

IMPORT a WORKSPACE

You may **IMPORT** a previously exported workspace onto another computer. This will save you the effort of customizing the workspace to your preferences each time you use a different computer.

1. Select **TOOLS / CUSTOMIZE / IMPORT CUSTOMIZATIONS**
2. Select **OPEN** from the drop down list.

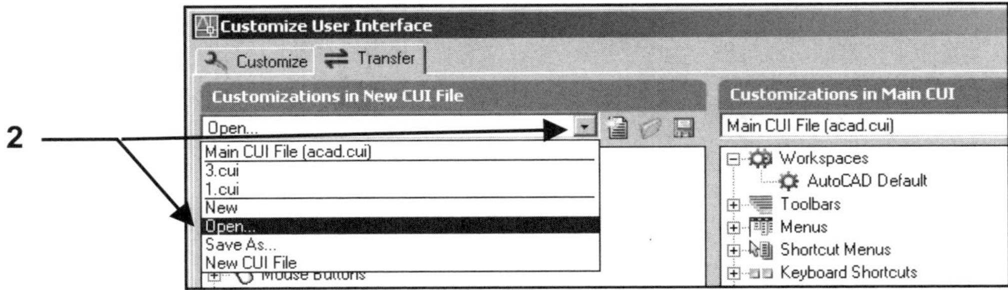

3. Locate the previously exported workspace file.
4. Select the file.
5. Select the **Open** button.

6. Drag and drop the workspace file, that you wish to import, from the left side to the word **Workspaces** on the right.

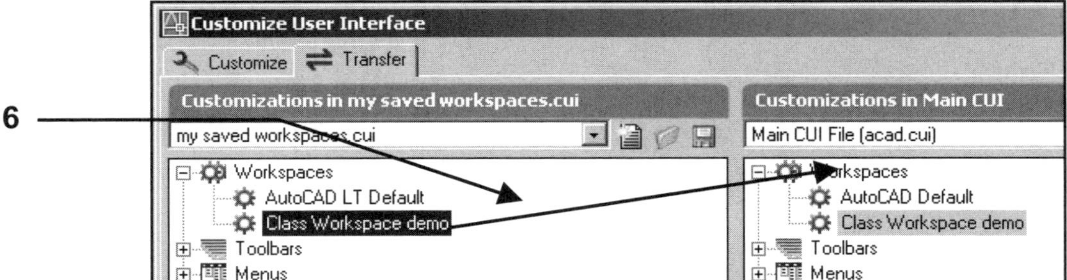

7. Select the **OK** button. *(Do not select the "X" in the upper right corner to close the dialog box. If you do, you will cancel the previous operations)*

HOW TO CREATE A NEW TOOLBAR

1. Select **TOOLS / CUSTOMIZE / INTERFACE**

2. Right click on **Toolbars** and select **New / Toolbar**.

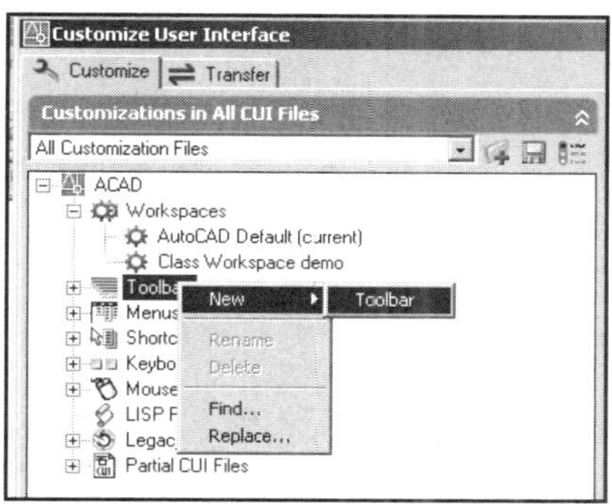

3. A new Toolbar (named Toolbar1) is placed at the bottom of the Toolbar list. Enter a name for the new Toolbar and press <enter>.

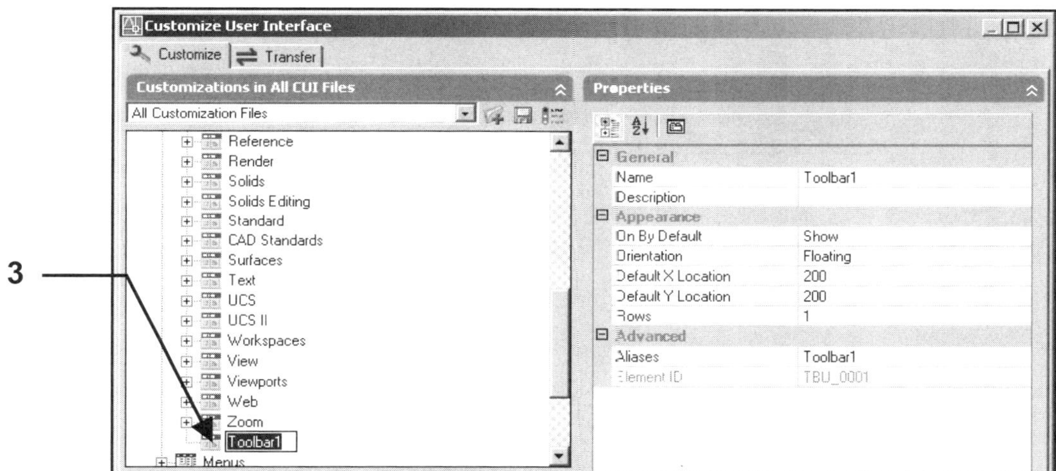

4. The new toolbar name will appear in the Properties pane on the right.

   Update the Properties:

   Description: Enter a description for the toolbar
   On By Default: Select Hide or Show.
   (If you choose Show, the toolbar will be displayed in all workspaces.)
   Orientation: Select Floating, Top, Bottom, Left or Right.
   Default X location: Enter a number for the X axis location for the toolbar.
   Default Y location: Enter a number for the Y axis location for the toolbar.
   Rows: Enter the number of rows for an undocked toolbar.
   Aliases: Enter an alias for the toolbar.

5. Drag and drop the command you want to add to the new toolbar.
   Drag from the Command List pane to a location just below the name of the toolbar.
   When a black line appears, drop the command.

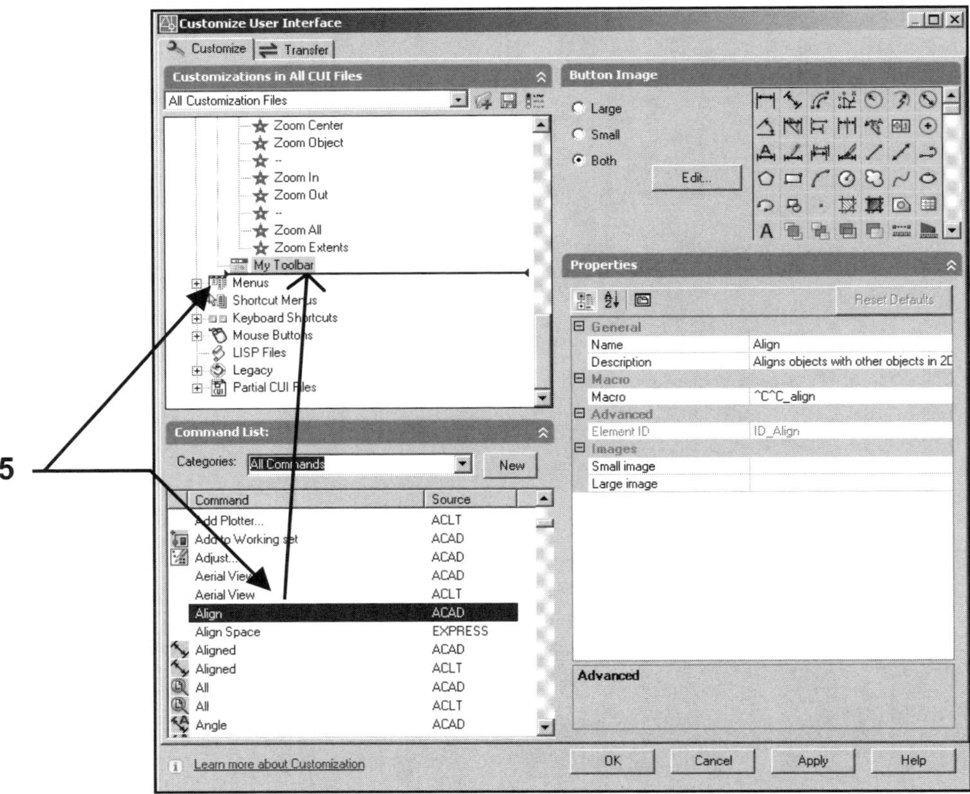

6. When you have finished adding commands to the new toolbar, click the **OK** button.

The Customize User Interface dialog box will close and your new toolbar should be on the screen.

HOW TO DELETE A COMMAND FROM A TOOLBAR

1. Select **TOOLS / CUSTOMIZE / INTERFACE**

2. Select the **+** plus sign beside **Toolbar** to display all of the toolbars.

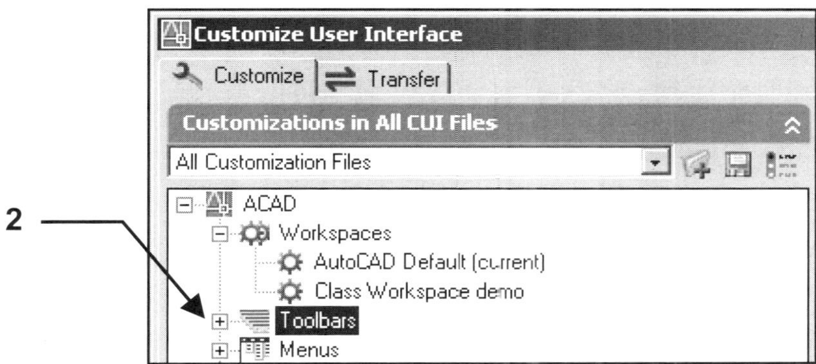

3. Select the toolbar whose command buttons you want to delete.
   Select the + plus sign beside the toolbar to display the command names.

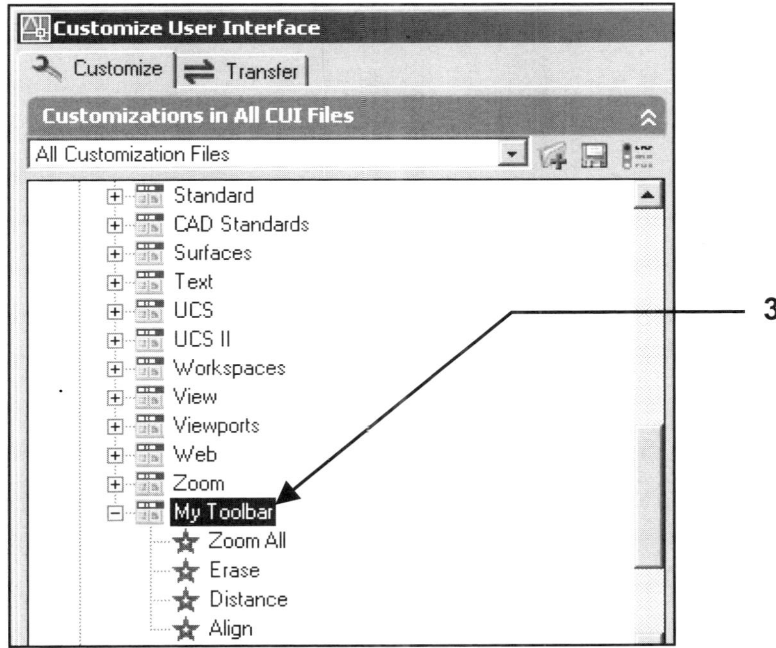

4. Right-click the name of the command you want to remove and select Delete.

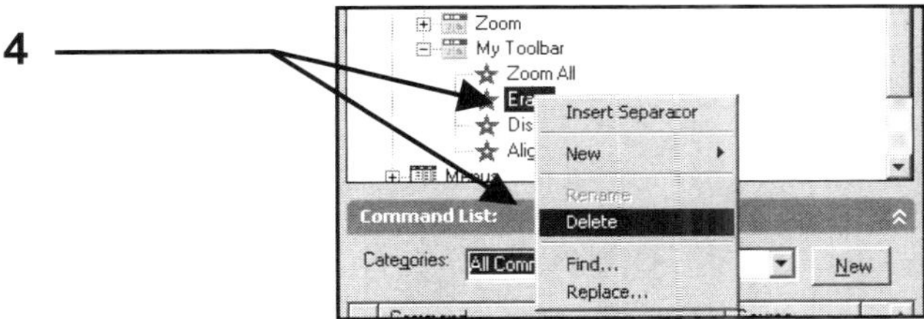

ADD A COMMAND TO A PULL-DOWN MENU

1. Select **TOOLS / CUSTOMIZE / INTERFACE**
2. Expand the Menus display by selecting the + plus sign.
3. Select the Menu, you wish to add to, from the list and expand it.
4. Select and drag the command from the command list to the Menu.

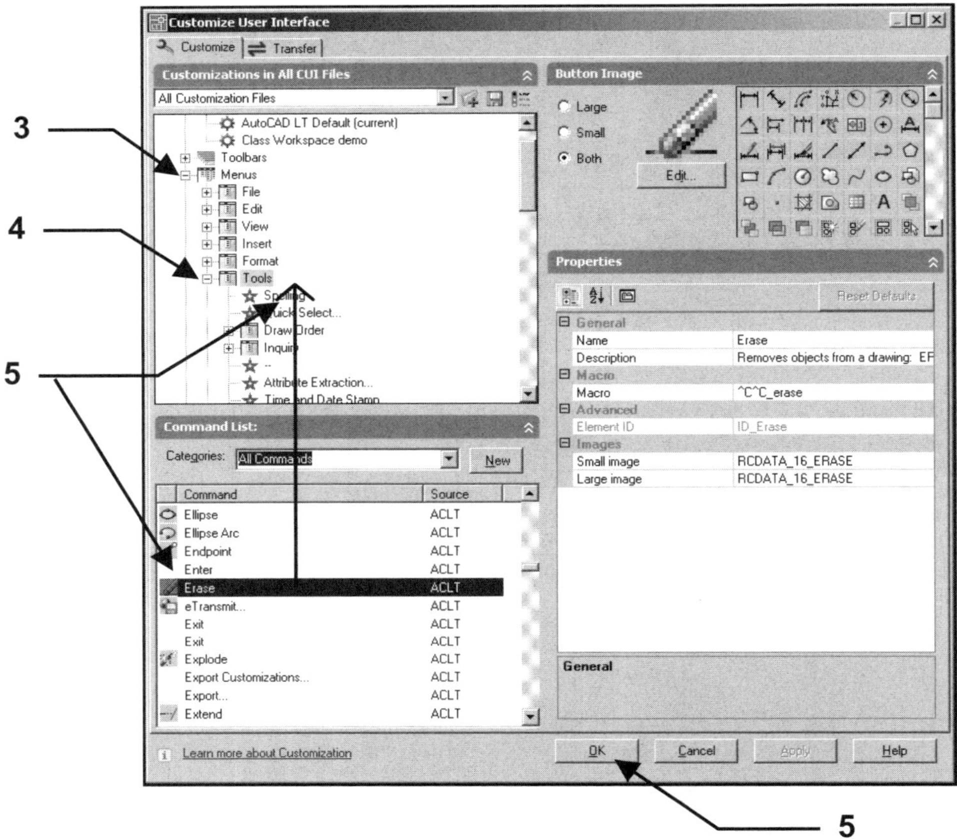

5. When you finish adding commands, select the **OK** button.

DELETE A COMMAND FROM A PULL-DOWN MENU

1. Right click on the command and select Delete from the list.

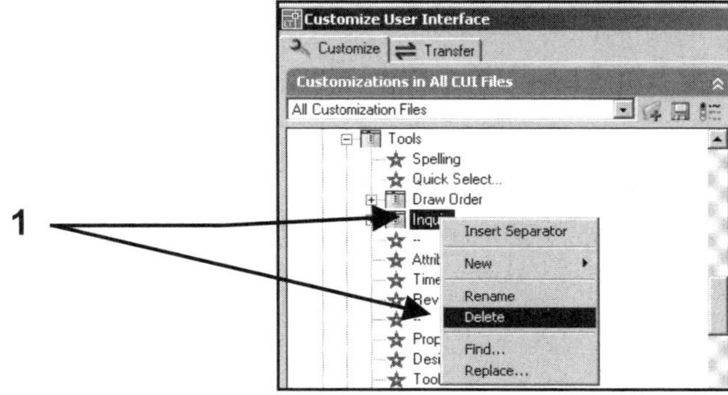

CREATE A WORKSPACE THE EASY WAY

Now that I showed you how the Customize User Interface is used, I will now show you the easy way to create your own workspace. But it was important for you to understand the complete process first.

1. Open AutoCAD.

2. Set up your drawing environment.
   Don't select the CUI. Just make the changes on the screen.
   Display toolbars, menus and palettes the way you want them.

3. Save your workspace.
   a. Click on the down arrow on the Workspaces toolbar.
   b. Select **Save Changes as**....

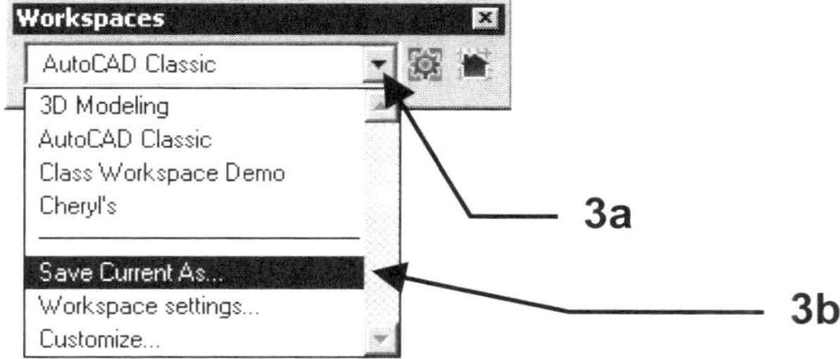

4. Type the name of your new workspace and select Save.

Note: If you want to <u>**modify an existing workspace**</u>, click on the down arrow and select the workspace and then select the Save button.

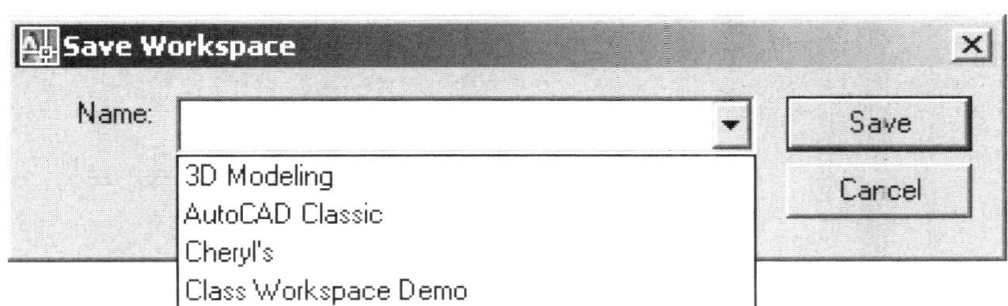

Set your Workspace as "My Workspace".

Now that you have created your Workspace you may designate it as the home workspace. This means that any time you select the **My Workspace** button, on the Workspaces toolbar, your workspace will appear. Of course this will only work if your workspace is on the computer you are currently working on. If it is not you will have to "Import" it. Refer to page 2-8.

1. Select the "**Workspace Settings**" button on the Workspaces toolbar.

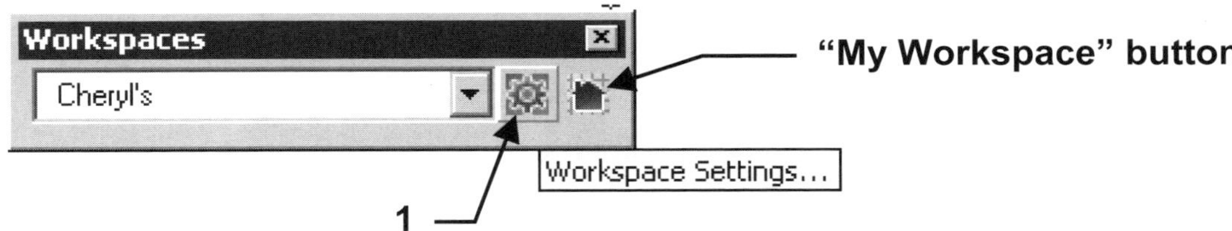

2. Click on the "**My Workspace" down arrow** and select the workspace from the list.

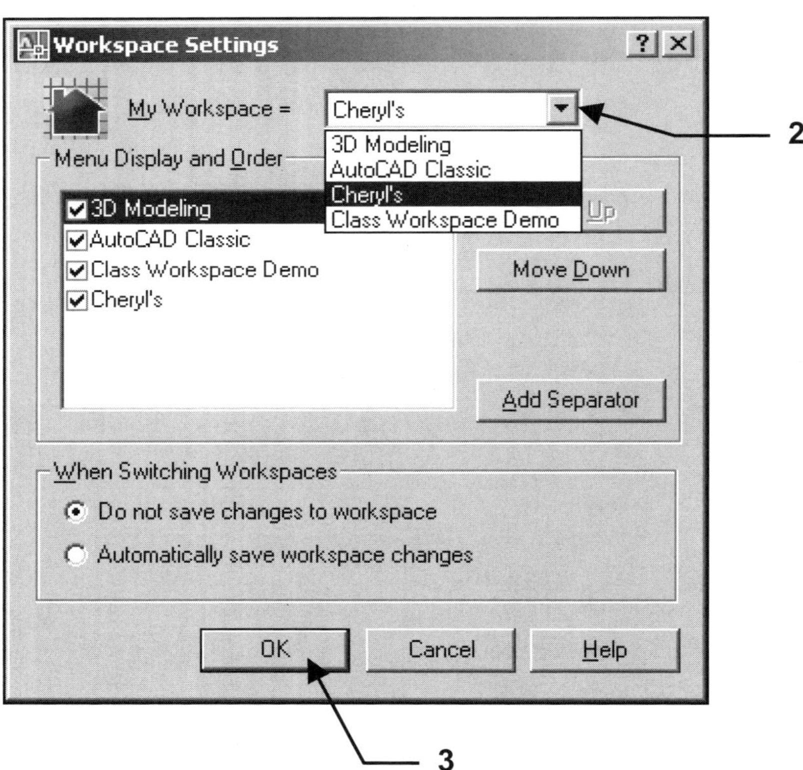

3. Select the **OK** button.

HOW TO CUSTOMIZE THE STATUS BAR

AutoCAD allows you to customize the Status Bar. For example, you may never use "OTRACK", so you could remove that button.

Personally, I see no reason to remove any buttons but you can make that decision.

1. Click the small down arrow at the right end of the status bar.

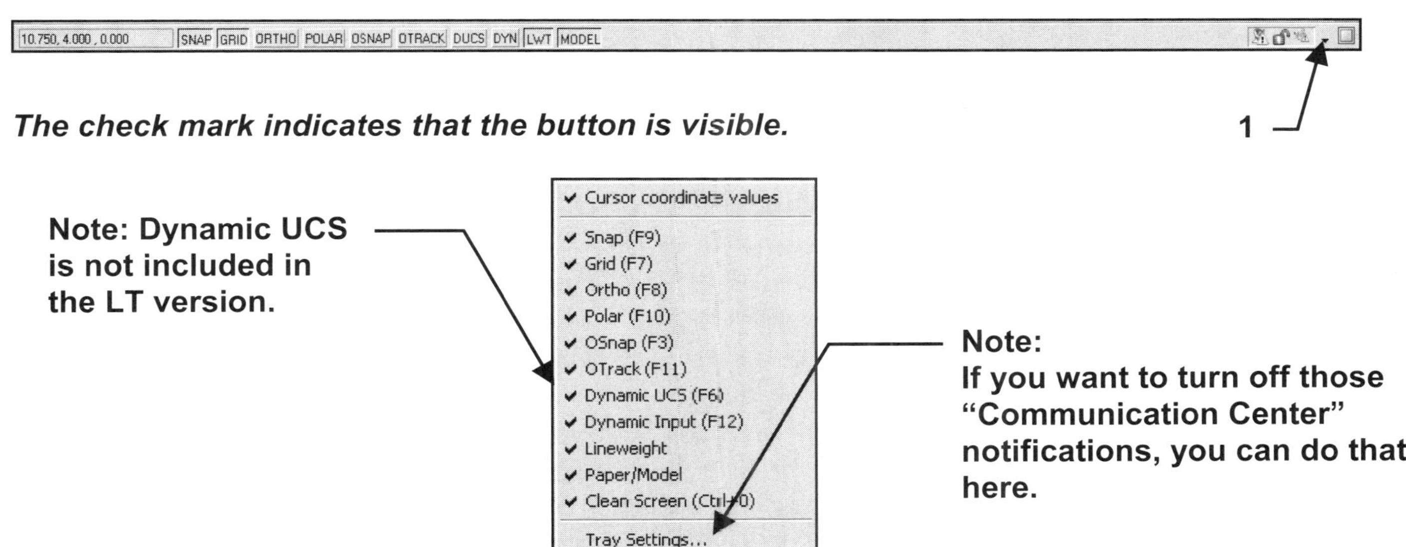

2. Click on the button name and the check mark will disappear and the button will disappear from the Status Bar.

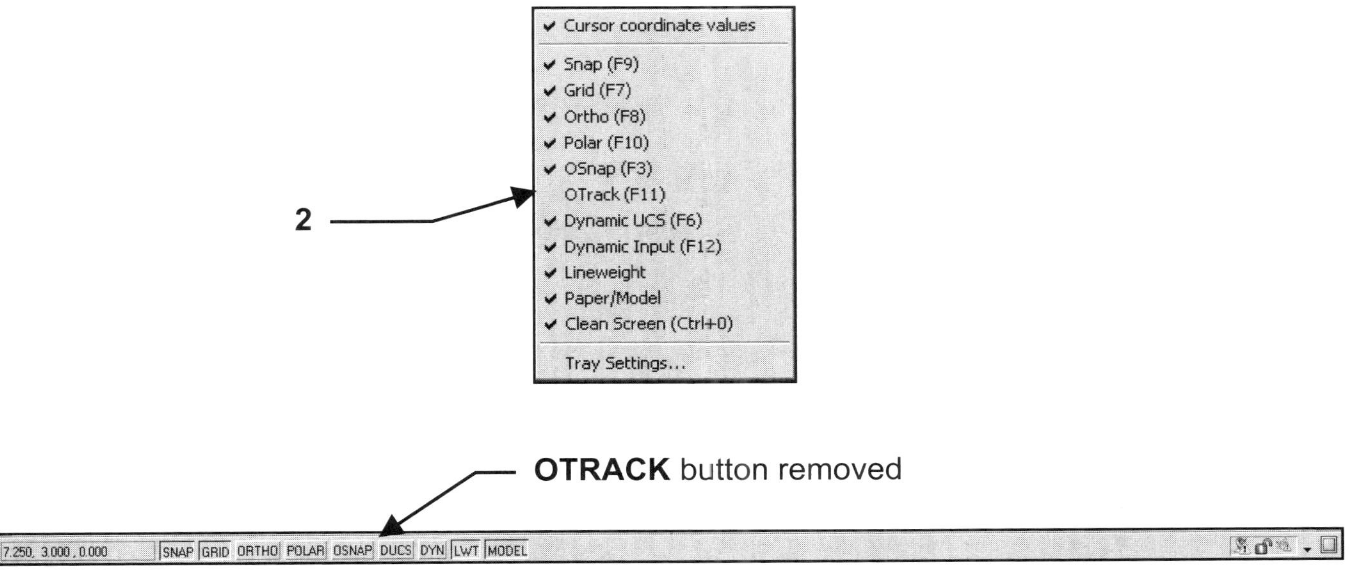

Note:

1. **To replace the button, click on the button name again, following steps 1 and 2 above.**
2. **Adding or removing a status line button is not saved in your Workspace or Profile.**

REDO multiple commands 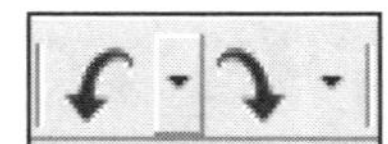

AutoCAD has created a new MREDO command to redo multiple UNDO commands.
You can redo recently undone commands from the Redo drop-down list.
Remember, redoing a command only works if you have Undone a command.

Note:
If the undo or redo arrows are grayed out, it means that there are no objects drawn and/or no undo commands have been issued.

How to redo multiple undo commands.

1. Start a new drawing.
2. Draw a line, rectangle and a circle.
3. Copy all 3 objects.

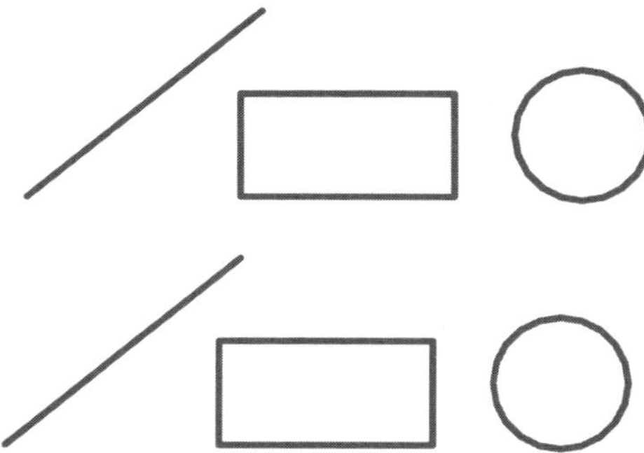

Your drawing should look approximately like this.

4. Click on the **UNDO** drop-down arrow. 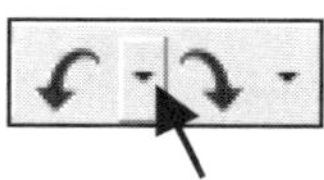

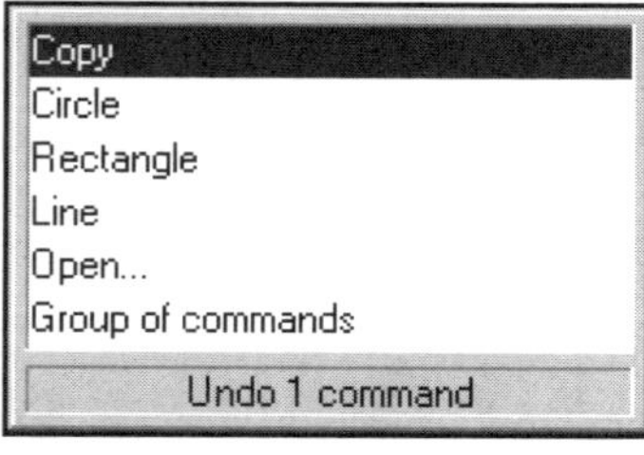

The UNDO list displays the creating and editing commands with the most recent at the top of the list.

5. Highlight the **Copy , Circle and Rectangle** command only, as shown below, and press the left mouse button. (To highlight, just drag the cursor down the list)

The Rectangles and Circles should have disappeared and only one Line remains.

Now let's "Redo" some of what we "Undo"ed.

6. Click on the **REDO** drop-down arrow. 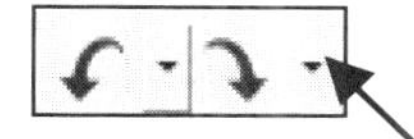

7. Highlight the Rectangle and Circle commands and press the left mouse button.

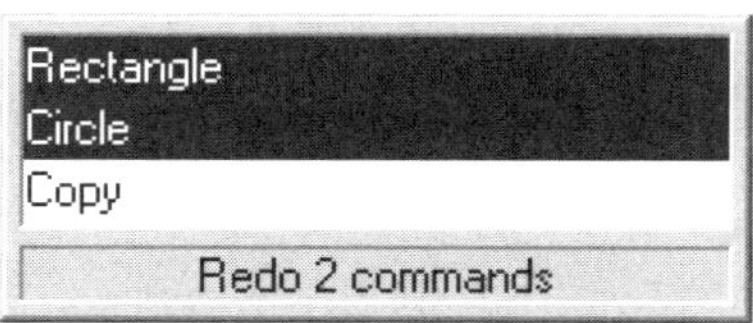

One Rectangle and one Circle should have reappeared.

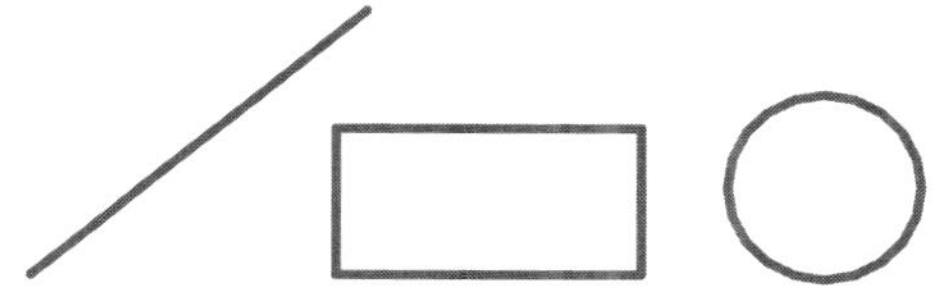

Notice that you may not "skip" commands. You must select the commands in order from the top down. This means if you have a command that is 5 commands down in the list, you may wish to use another method to make the change.

Understanding USER PROFILES

Note: This option not available to AutoCAD LT user's.

What is a User Profile?

Workspaces and Profiles change the display of your drawing environment, but they are not the same. You will probably want to create a customized workspace and profile.

Workspaces control the display of menus and toolbars in the drawing area. When you use or switch a workspace, you change the display of your drawing area. You can easily switch to another workspace within a drawing session. You manage your workspaces from the Customize User Interface dialog box.

Profiles control user options, drafting settings, paths, and values. Profiles are updated each time you make a change to an option, setting, or other value. You can manage most of your profiles from the Options dialog box.

Options that you may find useful to change are, Crosshair size, screen colors, Auto Snap marker size, Pick box size and many more.

When you make changes to the drawing display, the changes are stored in your profile and are displayed the next time you launch the program, regardless of your workspace settings.

The profile changes are not saved to a workspace.

In the following exercises, you will learn how to:

1. Create a profile.
2. Save (export) the profile for future use.
3. Retrieve (import) the profile.

EXERCISE 2A

CREATE A NEW PROFILE.

(Note: LT users do not have this option and cannot do this exercise.)

Although there are many settings and values you can change, we will limit the changes to the following:

1. Cross hair size
2. AutoCAD screen colors
3. AutoSnap marker size
4. Aperture size
5. Pick box size
6. Enable grips
7. Grip size

Note: If you are in a classroom, you should ask the instructor for approval before changing any settings other than the 8 listed above.

A. Select **Tools / Options…** or move the cursor to the command line and press the right mouse button then select **"OPTIONS"** from the short cut menu.

B. Select the **Profiles** tab.
   1. Select the **"Add to List"** button.

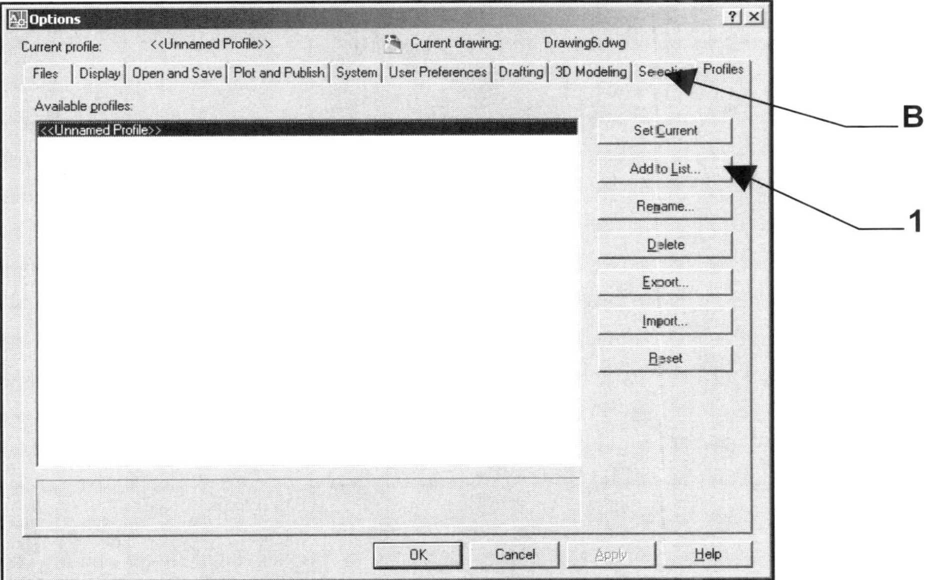

2. Type the profile name. ***(Type your name: Last-First) but*** <u>**DO NOT**</u> press <enter> Press the **TAB** key.

3. Type a description, such as: ***This is the profile (your name here) uses in class.***

4. Select the **"Apply and Close"** button.

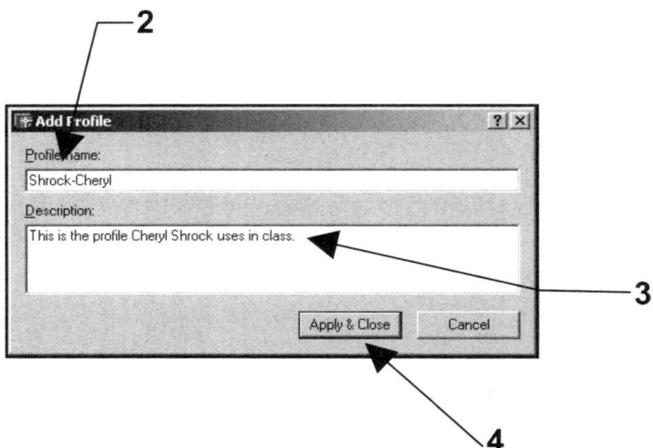

5. Select the Profile name you just created. (Click on it)

6. Select the **"Set Current"** button.

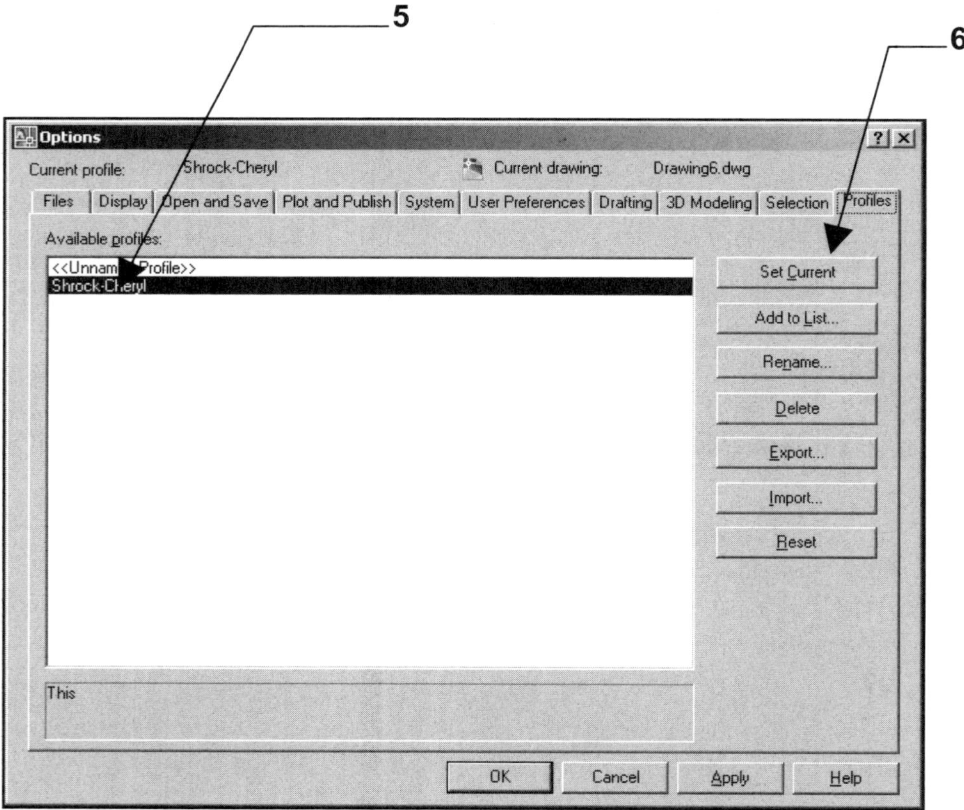

C. Select the **DISPLAY** tab.

1. Change the **"Crosshair Size"** to your preference from 5 to 100. The number represents a percentage of the screen size. 5 = 5% of the screen (default) 100 = 100% of the screen size

2. Select the **Colors...** button.

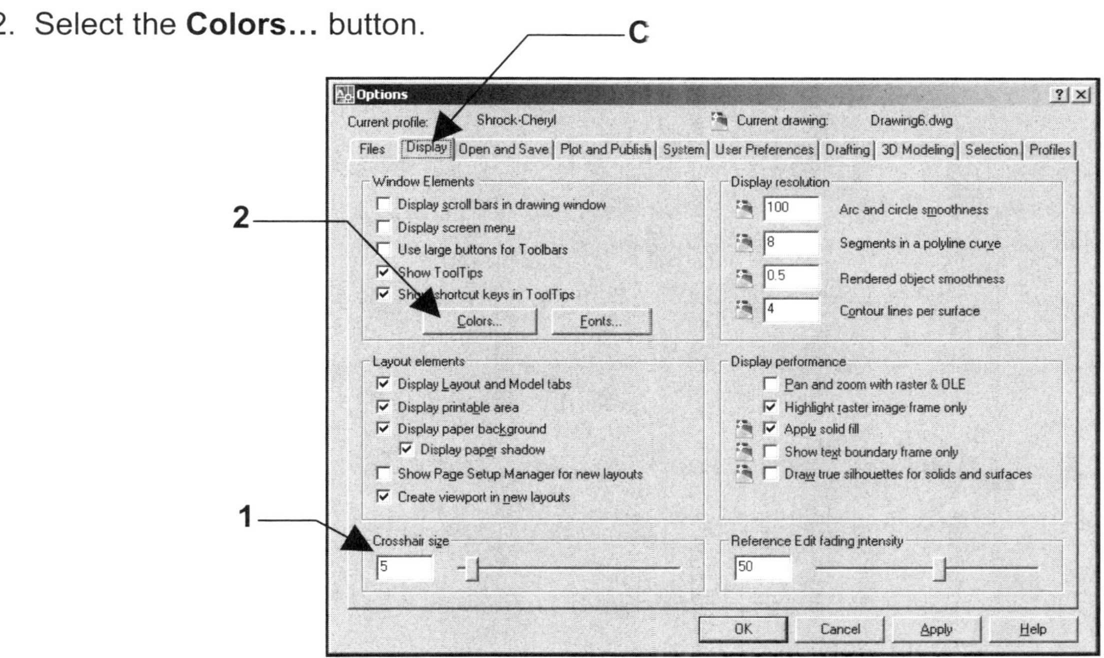

3. Select what you would like to change.

4. Select the color.

5. When complete, select the **"Apply & Close"** button.

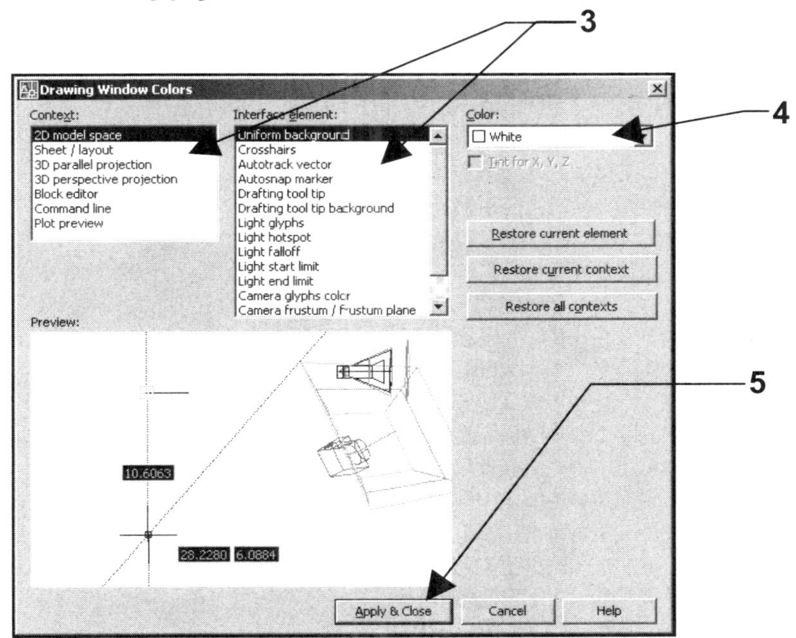

D. Select the **DRAFTING** tab.

1. Adjust the size of the "**AutoSnap Marker"** to your preference. Click and Drag the Slider Button to the left (min) or right (max). This is the marker that appears on objects at the preset object snap locations when using AutoSnap.

2. Adjust the size of the "**Aperture"** to your preference. Click and Drag the Slider Button to the left (min) or right (max). The aperture box appears when using **AutoTRACKING**.

3. When complete, select the **Apply Button**.

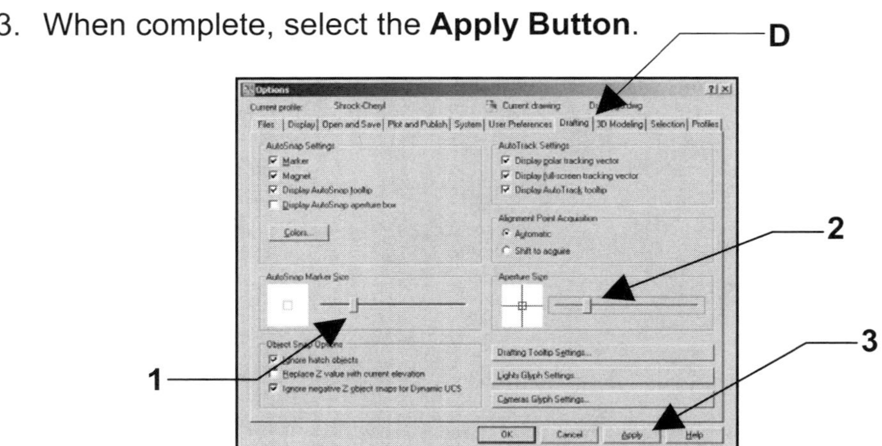

E. Select the **SELECTION** tab.

1. Adjust the **Pickbox Size** to your preference. The pickbox appears when *selecting objects. (The size will affect the selection precision)*

2. Enable (on) the display of grip boxes by placing a check mark in the box. Disable (off) by removing the check mark.

3. Adjust the size of the **Grip** boxes.

4. When complete, select the **Apply Button.**

5. Select the **OK** button to leave the **Options** dialog box.

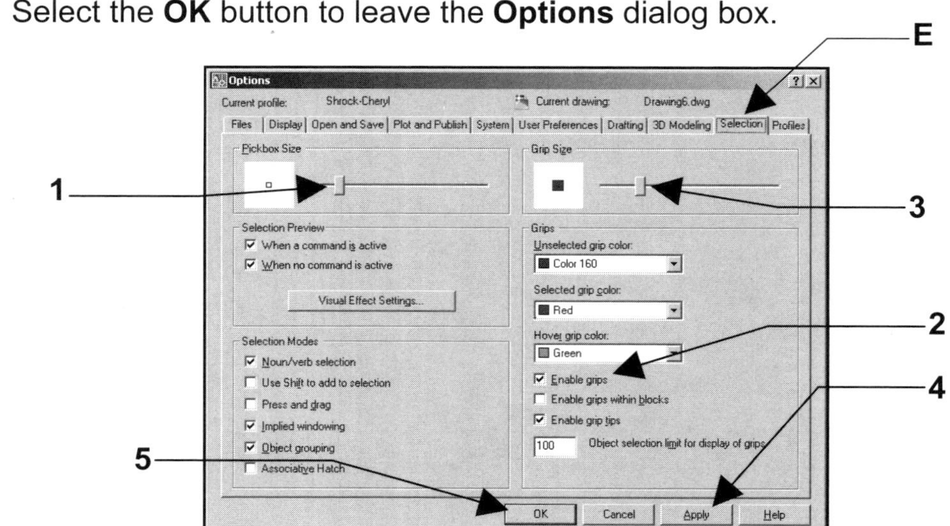

Now look at the screen. Is everything to your preference? If so, we will now save this Profile to a disk. This will allow you to transfer this profile to other computers.
(Note: the profile will automatically be saved to the computer you are working on right now. Exporting is only necessary if you want to take your Profile to another computer.)

F. EXPORTING A PROFILE.

1. Select **Tools / Options...** or move the cursor to the command line and press the right mouse button then select **"OPTIONS"** from the short cut menu.
2. Select the **PROFILES** tab.
3. Select the Profile that you wish to Export.
4. Select the **EXPORT** button.

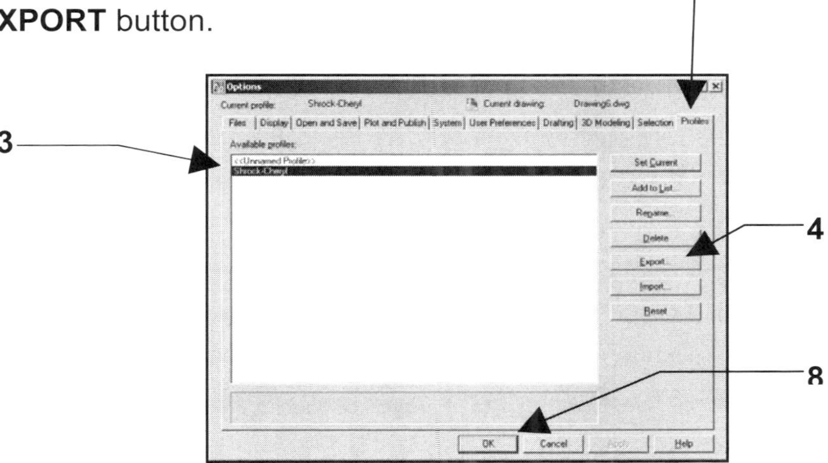

5. Select a "removable" disk, such as a 3-1/2 floppy, Zip disk or CD.
6. Type the name of the profile. *(Note: the file extension is **.ARG** for profiles)*
7. Select the **SAVE** button.
8. Now select the **OK** button.
   (See dialog box above)

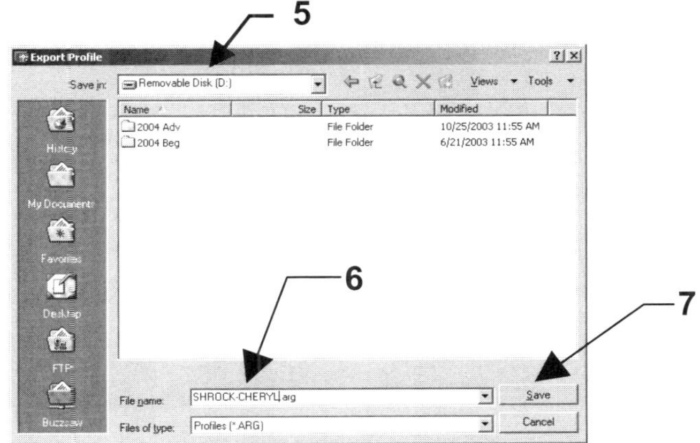

Note: If you make changes to your profile you must Export (save) the profile again as explained above.

If you want to move your profile to another computer, you must IMPORT. (Refer to the next page.)

EXERCISE 2B

IMPORTING A USER PROFILE.

Now that you have learned how to create a profile, you will want to use it.

<u>If the profile "does" exist on the computer</u>, you merely select it and then select **"SET CURRENT".**

<u>If the profile "does not" already exist on the computer you are working on</u>, you must **IMPORT** it and then "**SET CURRENT"**. (Follow the instructions below)

A. Select **Tools / Options…** or move the cursor to the command line and press the right mouse button then select **"OPTIONS"** from the short cut menu.

B. Select the **PROFILES** tab.
   1. Select the **IMPORT** button.

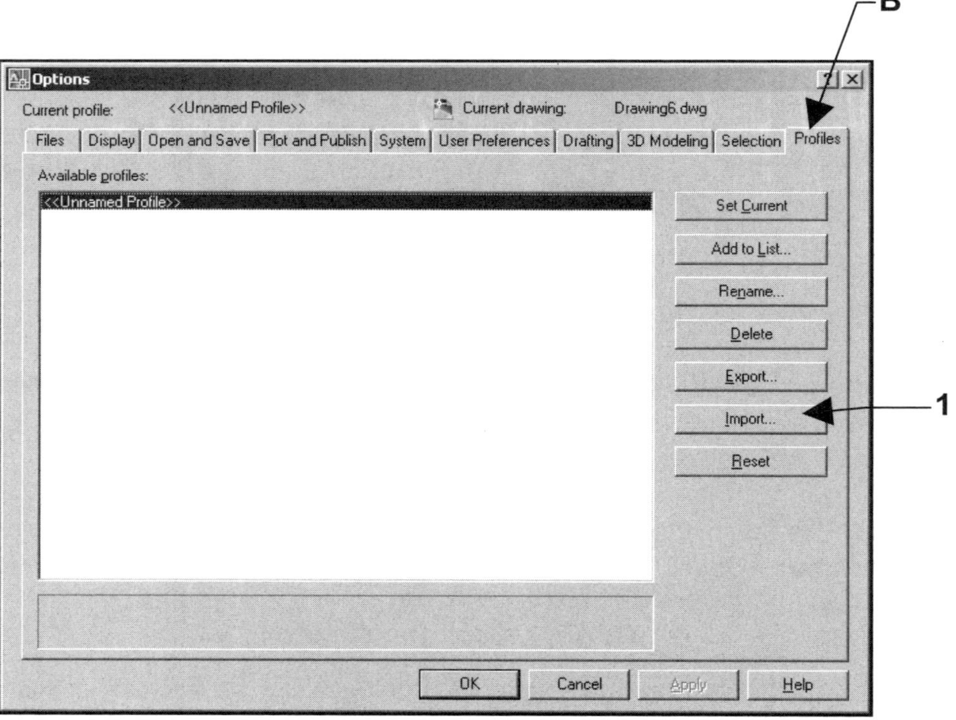

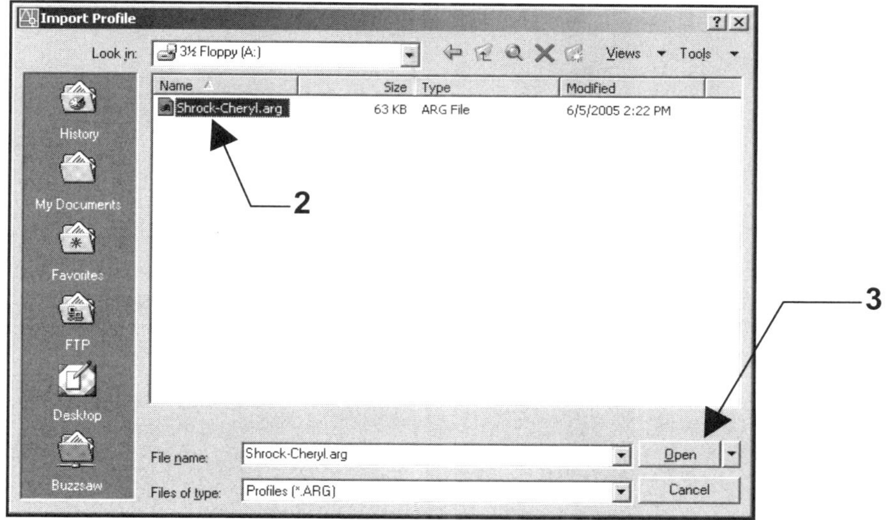

2. Find and select your previously ***exported profile*** file
3. Select the **OPEN** button.

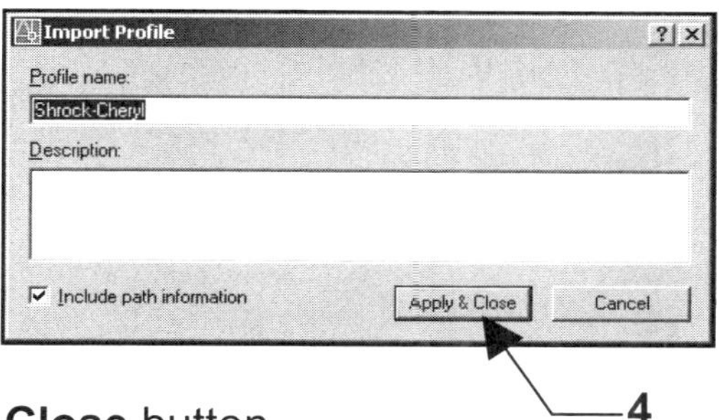

4. Select the **Apply & Close** button.

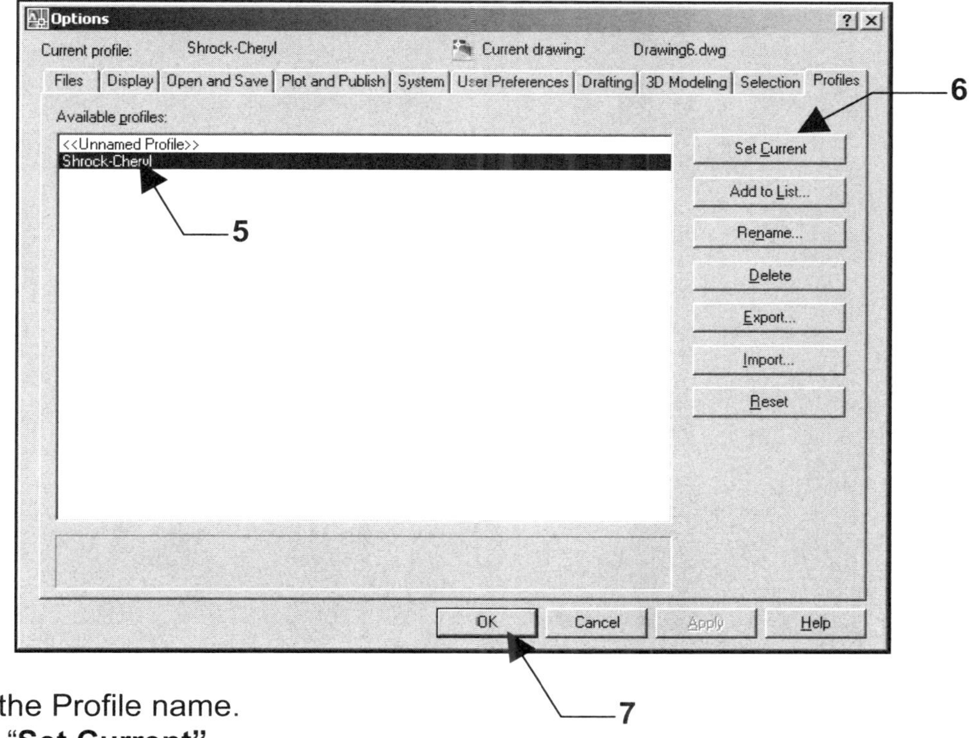

5. Select the Profile name.
6. Select **"Set Current"**
7. Select the **OK** button.

Notes:

LEARNING OBJECTIVES

After completing this lesson, you will be able to:

1. Create a Setup drawing.
2. Create Borders for 17 x 11, 24 x 18 and 36 x 24.
3. Lock a Viewport.
4. Create Page Setups for 17 x 11, 24 x 18 and 36 x 24.

LESSON 3

EXERCISE 3A

CREATE A MASTER DECIMAL SETUP DRAWING

The following instructions will guide you through creating a "Master" decimal setup drawing. This setup drawing is very similar to the setup drawing you may have made while using the Beginning workbook but create it anyway. In addition to giving you more practice, there are some minor differences in the settings, layers and styles.

NEW SETTINGS

A. Begin your drawing without a template as follows:

1. Select **"FILE / NEW".**
2. Select **"START FROM SCRATCH**" Box.
3. Select "**OK**".
4. Your screen should be blank, no grids and the current layer is 0.

Select Imperial or Metric

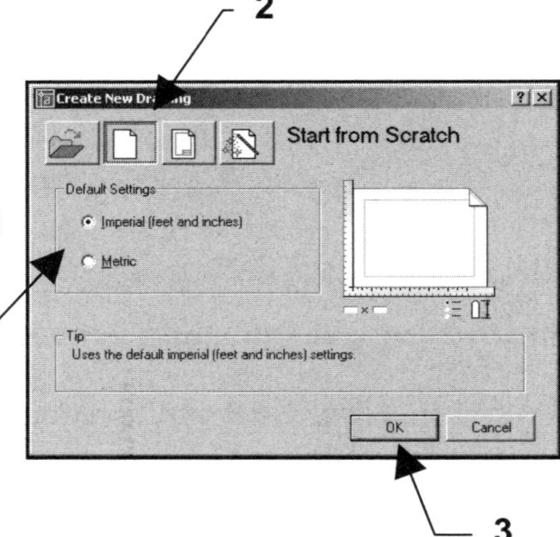

B. Set drawing specifications as follows:

1. Select **FORMAT / UNITS.**
   a. Change the "Type" and "Precision", then select **OK**.

Decimal

Precision

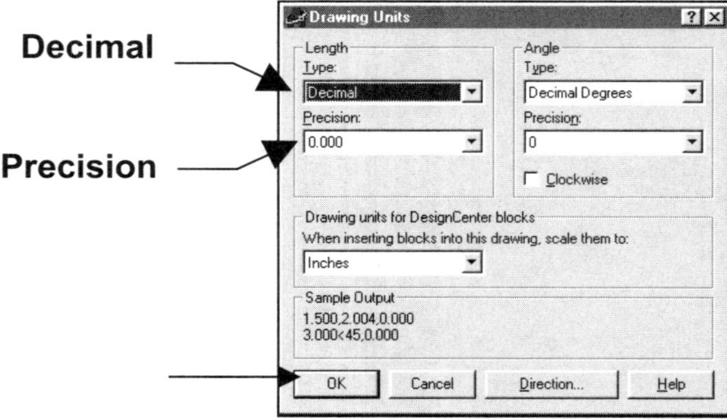

2. Select "**FORMAT / DRAWING LIMITS.**
   a. Lower left corner = 0.000,0.000
   b. Upper right corner = 17 , 11
   c. Use **"VIEW / ZOOM / ALL"** to generate the new limits.
   d. Set your **Grids** to **ON** to display the paper size.

3. Select **"TOOLS / DRAFTING SETTINGS / SNAP and GRID** tab.
   a. Change the settings as shown below and then select **OK**.

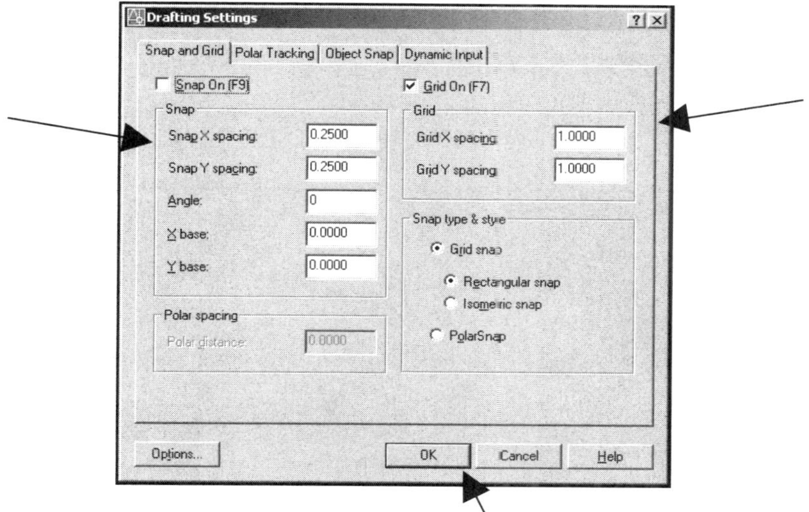

4. Set **"PICK BOX"** size (your preference).
   See page 2-18 E-1 for instructions.

NEW LAYERS

C. Create new layers.

1. First **Load** the linetypes listed below. (See page 26-4 Beg. Workbook)
   CENTER2
   HIDDEN
   PHANTOM2

2. Change the Lineweight settings to inches or metric and adjust the Display scale. (Refer to page 9-7 in the Beginning Workbook)

3. Assign names, colors, linetypes and plotability.
   (See page 26-3 Beg. Workbook)

NAME	COLOR	LINETYPE	LWT	PLOT
BORDER	RED	CONTINUOUS	.039 (1.00mm)	YES
CENTER	CYAN	CENTER2	Default	YES
CONSTRUCTION	WHITE	CONTINUOUS	Default	NO
DIMENSION	BLUE	CONTINUOUS	Default	YES
HATCH	GREEN	CONTINUOUS	Default	YES
HIDDEN	MAGENTA	HIDDEN	Default	YES
OBJECT	RED	CONTINUOUS	.024 (0.60mm)	YES
PHANTOM	MAGENTA	PHANTOM2	Default	YES
SECTION	WHITE	PHANTOM2	.031 (0.80mm)	YES
SYMBOL	RED	CONTINUOUS	.024 (0.60mm)	YES
TEXT HEAVY	WHITE	CONTINUOUS	Default	YES
TEXT LIGHT	BLUE	CONTINUOUS	Default	YES
THREADS	GREEN	CONTINUOUS	Default	YES
VIEWPORT	GREEN	CONTINUOUS	Default	NO
XREF	WHITE	CONTINUOUS	Default	YES

NEW TEXT STYLE

D. Create a text style.

1. Select "**FORMAT / TEXT STYLE".**
2. Create a NEW text style named CLASS TEXT.
3. Select the Font and Effects shown in the dialog box below.

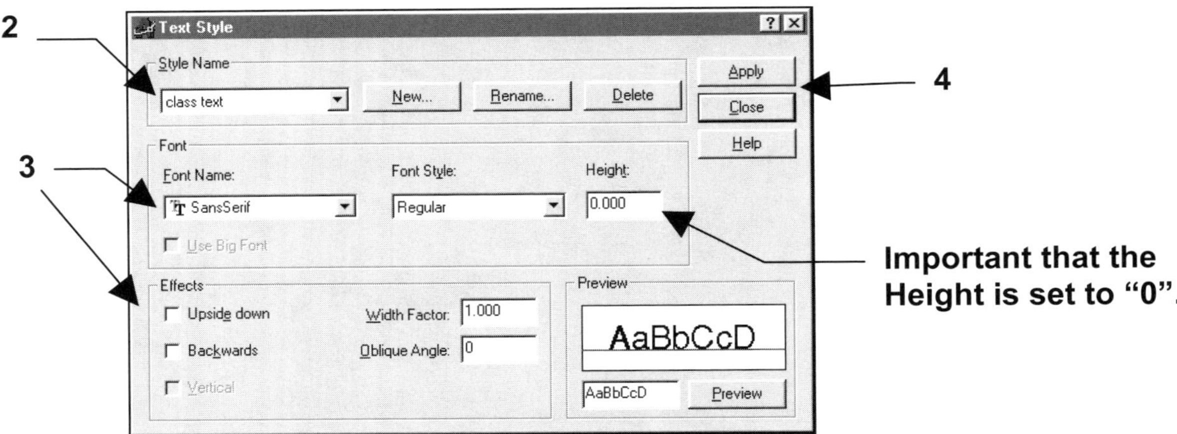

4. When complete select **APPLY,** then **CLOSE**.

NEW DIMENSION STYLE

E. Create a new Dimension Style named ***Class Style.***

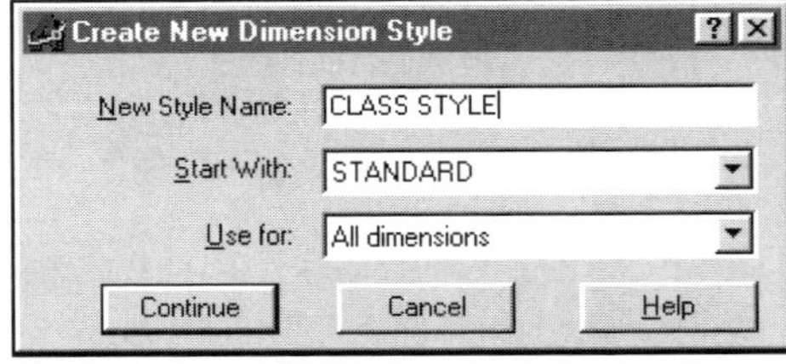

1. Select the **DIMENSION / STYLE** command.
2. Select the **NEW** button.
3. Make the changes to the following dialog boxes.

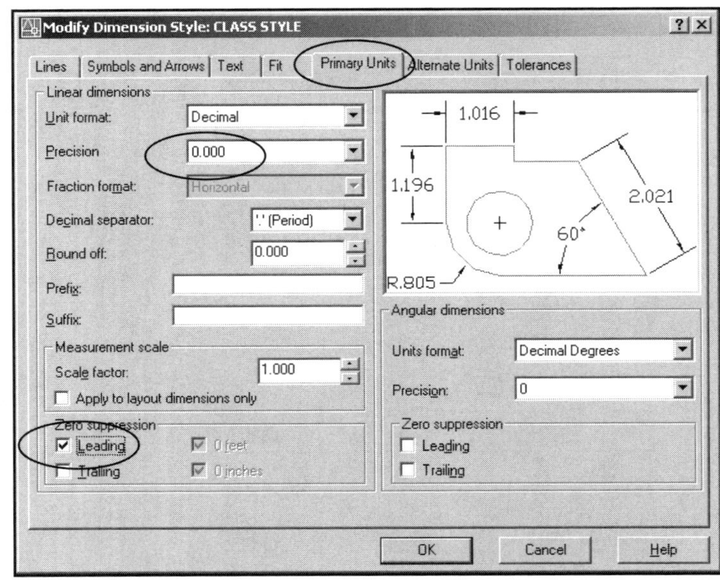

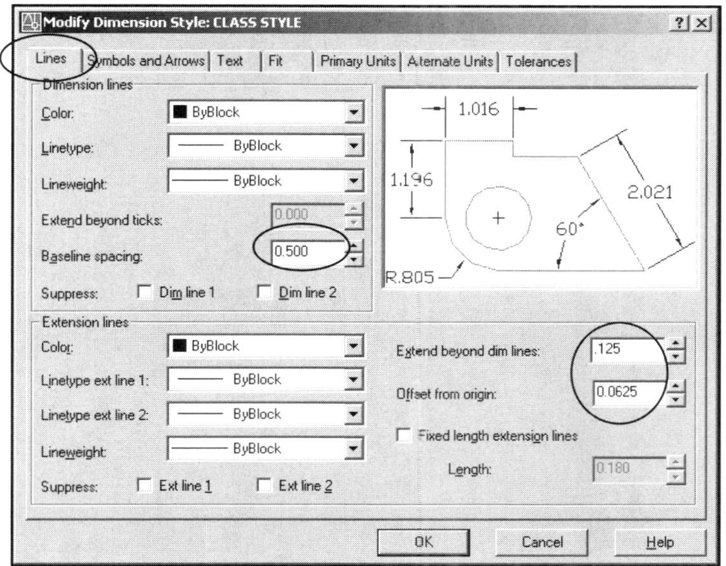
Modify Dimension Style: CLASS STYLE
Lines
Symbols and Arrows
Text
Fit
Primary Units
Alternate Units
Tolerances
Dimension lines
Color:
ByBlock
Linetype:
ByBlock
Lineweight:
ByBlock
Extend beyond ticks:
0.000
Baseline spacing:
0.500
Suppress:
Dim line 1
Dim line 2
1.016
1.196
2.021
60°
R.805
Extension lines
Color:
ByBlock
Linetype ext line 1:
ByBlock
Linetype ext line 2:
ByBlock
Lineweight:
ByBlock
Suppress:
Ext line 1
Ext line 2
Extend beyond dim lines:
.125
Offset from origin:
0.0625
Fixed length extension lines
Length:
0.180
OK
Cancel
Help

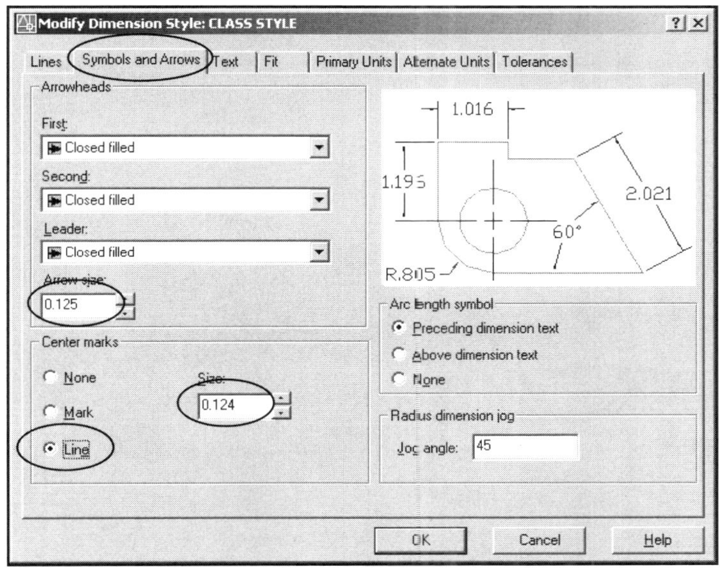
Modify Dimension Style: CLASS STYLE
Lines
Symbols and Arrows
Text
Fit
Primary Units
Alternate Units
Tolerances
Arrowheads
First:
Closed filled
Second:
Closed filled
Leader:
Closed filled
Arrow size:
0.125
1.016
1.196
2.021
60°
R.805
Arc length symbol
Preceding dimension text
Above dimension text
None
Center marks
None
Mark
Line
Size:
0.124
Radius dimension jog
Jog angle:
45
OK
Cancel
Help

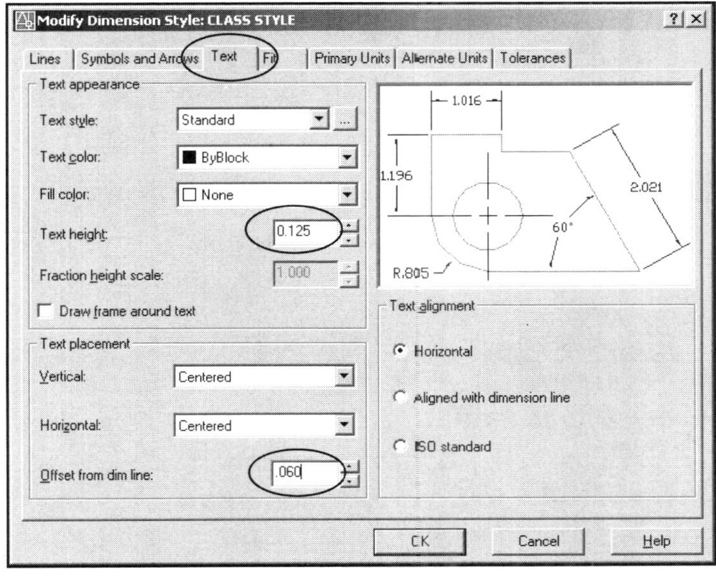
Modify Dimension Style: CLASS STYLE
Lines
Symbols and Arrows
Text
Fit
Primary Units
Alternate Units
Tolerances
Text appearance
Text style:
Standard
Text color:
ByBlock
Fill color:
None
Text height:
0.125
Fraction height scale:
1.000
Draw frame around text
1.016
1.196
2.021
60°
R.805
Text placement
Vertical:
Centered
Horizontal:
Centered
Offset from dim line:
.060
Text alignment
Horizontal
Aligned with dimension line
ISO standard
OK
Cancel
Help

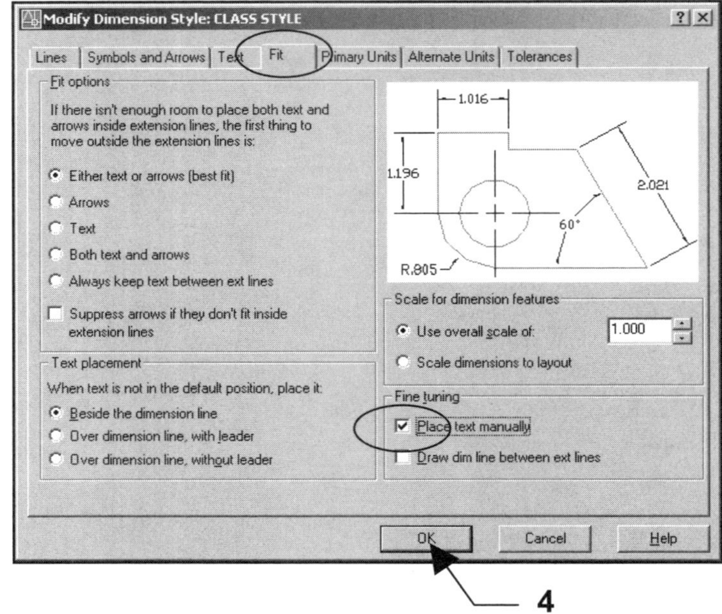

4. After changing the settings shown above, select the **OK** button.

5. Your new style **"Class Style"** should be listed.
   a. Select the "Class Style" Style.
   b. Select the **"Set Current"** button

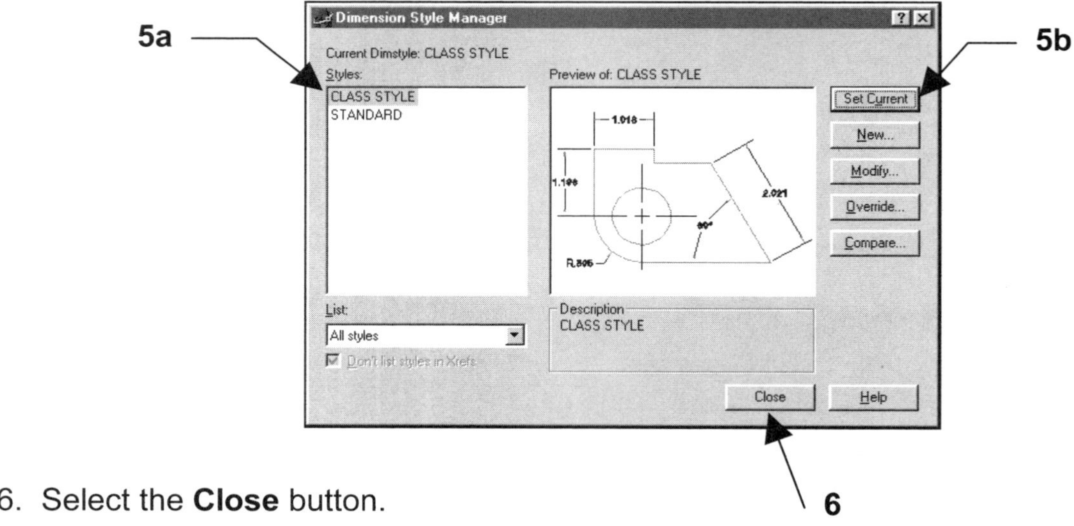

6. Select the **Close** button.

THIS NEXT STEP IS VERY IMPORTANT.

F. SAVE ALL THE SETTINGS YOU JUST CREATED.

1. Select **File / Save as**.
2. Save as: **My Decimal Setup.**

> ***NOTE: This was just step 1 to creating your setup drawing. Continue on through the following exercises, in this lesson, using this same file (My Decimal Setup). Each step adds more information to this setup drawing.***

EXERCISE 3B

CREATE A LAYOUT & BORDER FOR PLOTTING ON A 11 X 17 SHEET

The following instructions will guide you through creating a Border drawing that will be used in combination with "My Decimal Setup" when plotting. You will create a Layout and draw a border with a title block. All of this information will be saved and you will not have to do this again.

A. Open **My Decimal Setup**

B. Select a **LAYOUT** tab.

Note: If the "Page Setup Manager" dialog box shown below does not appear automatically, right click on the Layout tab, then select Page Setup Manager.

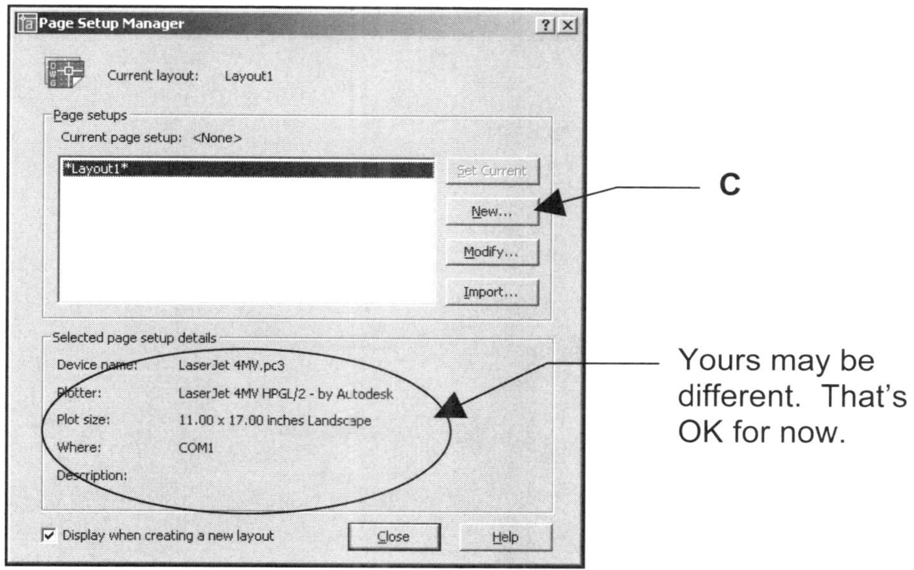

C. Select the **New** button. *(The "New Page Setup" dialog should appear)*

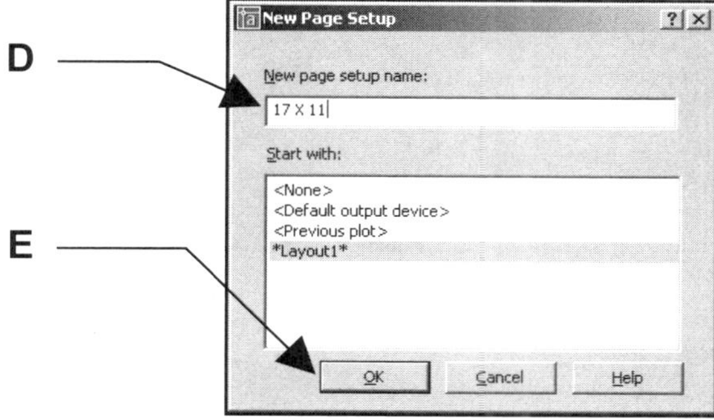

D. Type the new page setup name, **17 x 11**.
E. Select the **OK** button.

(The "Page Setup" dialog box should appear)

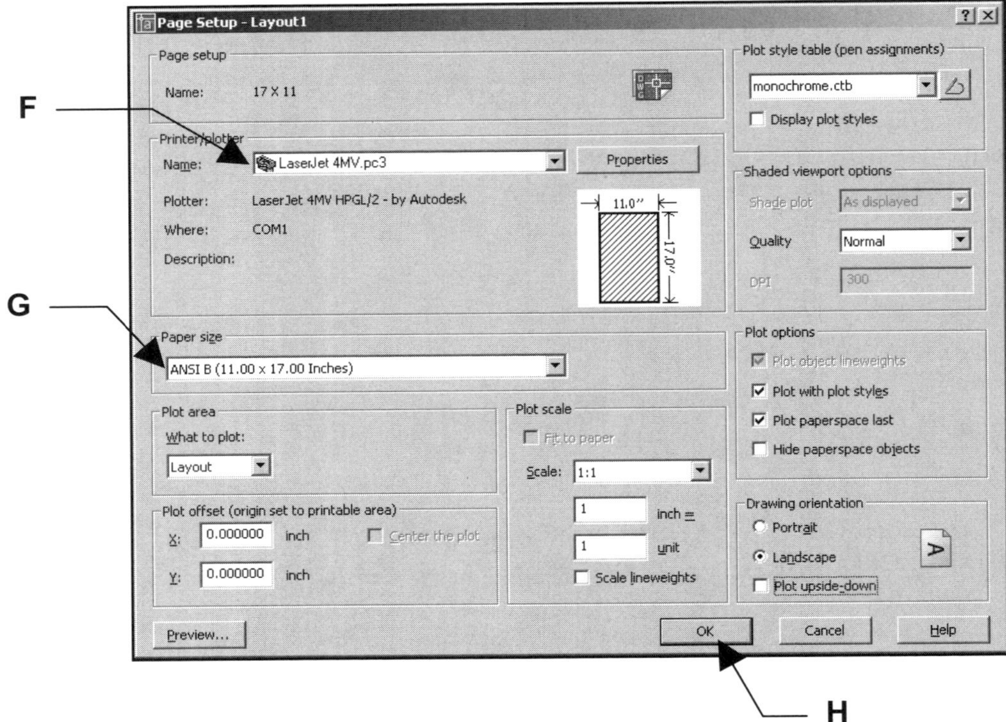

F. Select the **Printer / Plotter**.
G. Select the **Paper Size.**
H. Select the **OK** button.

The "Page Setup Manager" should have returned. Notice the new page setup name now appears in the list.

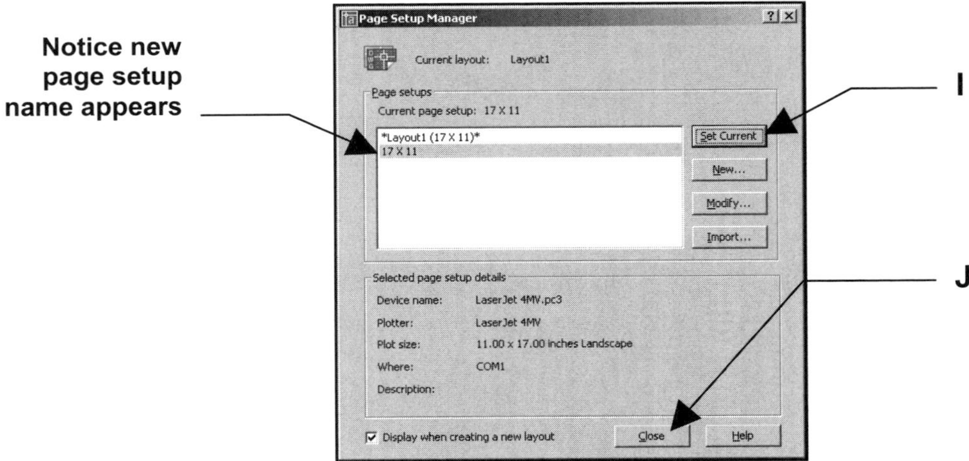

I. Select the **Set Current** button.
J. Select the **Close** button.

You should now have a sheet of paper displayed on the screen.
This sheet is the size you specified in the "Page Setup".
This sheet is in front of the drawing that is in Model Space.
The dashed line represents the printing limits for the device that you selected.

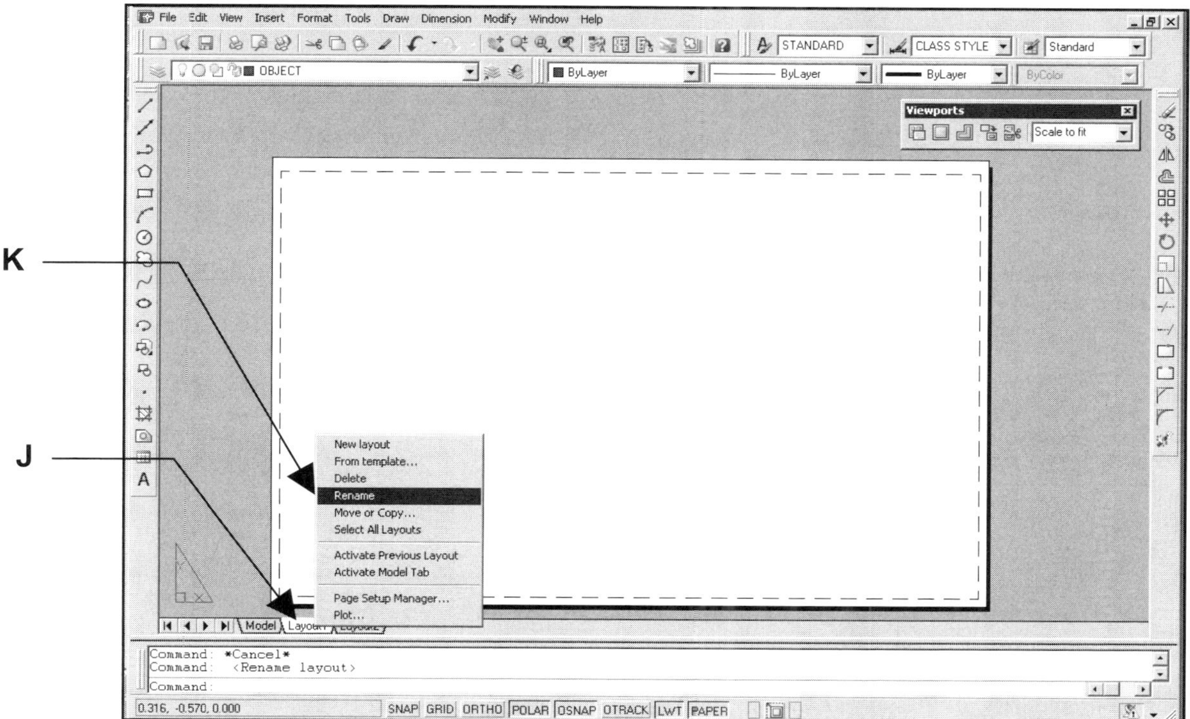

J. Right click on the Layout tab and select **Rename**.
K. Type the new name **11 X 17 (1 to 1)** then select **OK**.

Continue on to the next page....

L. Draw the Border with title block, shown below, on the sheet of paper shown on the screen.

M. When you have completed the Border, shown below:
   1. Select File / Save as
   2. Save as: **My Decimal Setup** (Again)

16.000
10.000
USE LAYER "BORDER" FOR ALL
BORDER LINES AND TITLE BLOCK
.25, .25
SCHOOL NAME
CITY, STATE
TITLE GOES HERE
TYPE YOUR NAME HERE
XX-XX-XX
1 = 1
EX-XXX

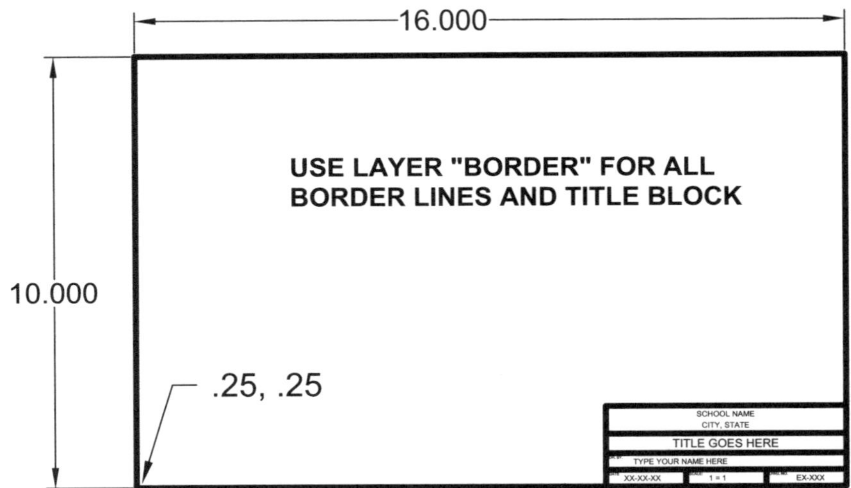

Important: Use "Single Line Text" to place the text in the Title Block below.

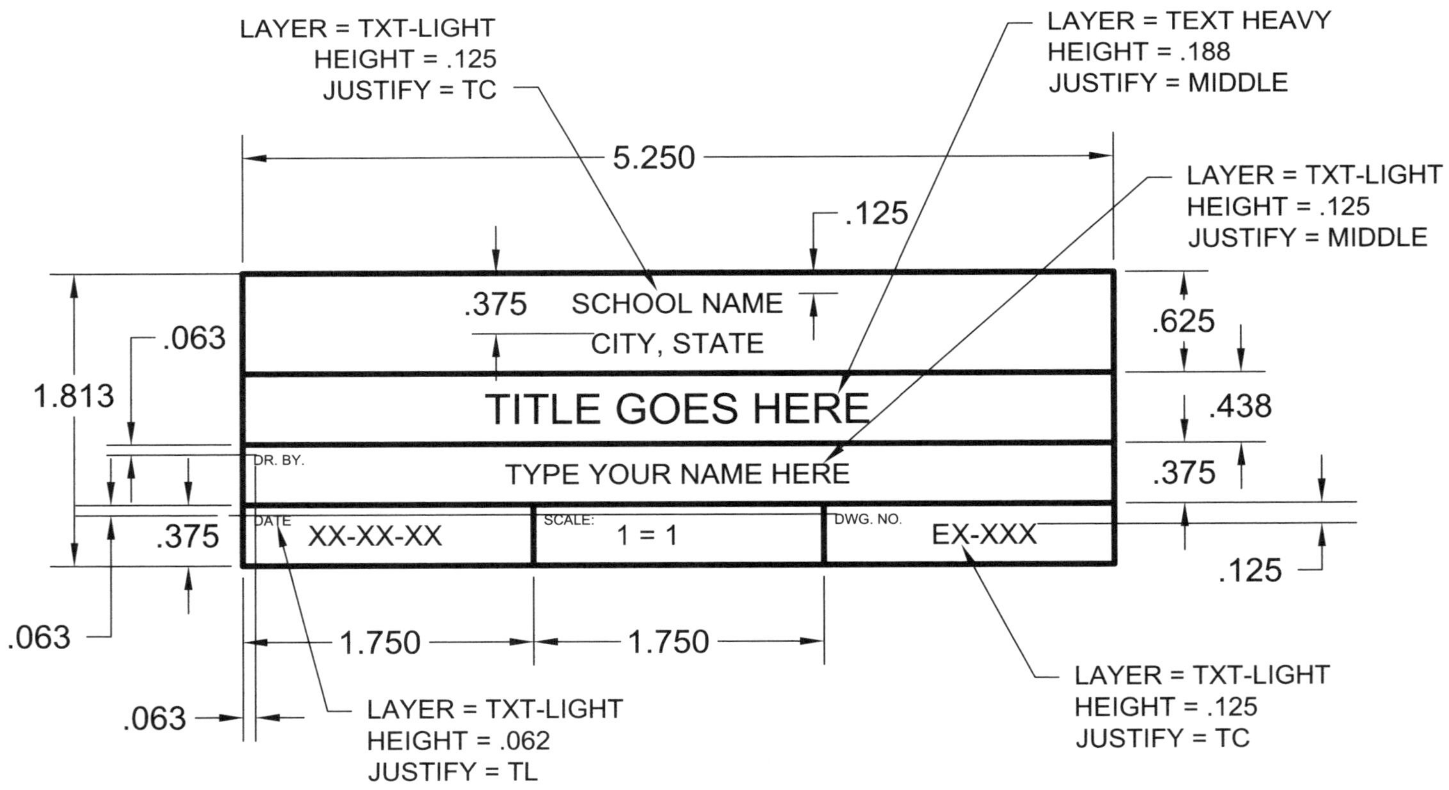

LAYER = TXT-LIGHT
HEIGHT = .125
JUSTIFY = TC
LAYER = TEXT HEAVY
HEIGHT = .188
JUSTIFY = MIDDLE
5.250
.125
LAYER = TXT-LIGHT
HEIGHT = .125
JUSTIFY = MIDDLE
.375
SCHOOL NAME
CITY, STATE
.063
.625
1.813
TITLE GOES HERE
.438
DR. BY.
TYPE YOUR NAME HERE
.375
DATE
XX-XX-XX
SCALE:
1 = 1
DWG. NO.
EX-XXX
.375
.125
.063
1.750
1.750
LAYER = TXT-LIGHT
HEIGHT = .125
JUSTIFY = TC
.063
LAYER = TXT-LIGHT
HEIGHT = .062
JUSTIFY = TL

EXERCISE 3C

CREATE A VIEWPORT

The following instructions will guide you through creating a VIEWPORT in the Border Layout sheet. Creating a viewport has the same effect as cutting a hole in the sheet of paper. You will be able to see through the viewport frame (hole) to Model.

A. Open **My Decimal Setup**

B. Select the **11 X 17 (1 to 1)** tab.

C. Select layer "Viewport"

D. Select the Single Viewport icon from the Viewports Toolbar or Type MV <enter>.

Single Viewport icon (LT's toolbar looks a little different but it works the same)

E. Draw a Single viewport approximately as shown.

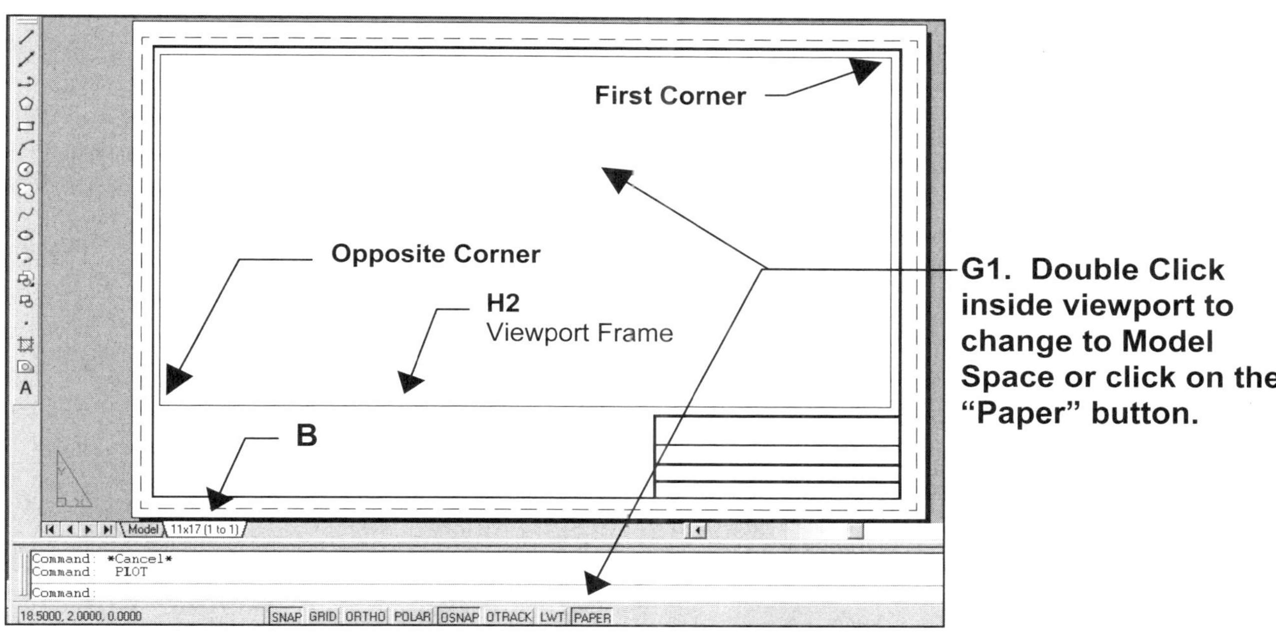

F. After successfully creating the Viewport, you should now be able to see through to Model. (Your grids should appear if they are ON.)

G. Adjust the Model space scale.

1. Select the **Paper** button so it changes to "Model" or double click inside the Viewport Frame.
2. Do **View / Zoom / All** before adjusting the scale.
3. Select 1:1 in the **VIEWPORT** toolbar.

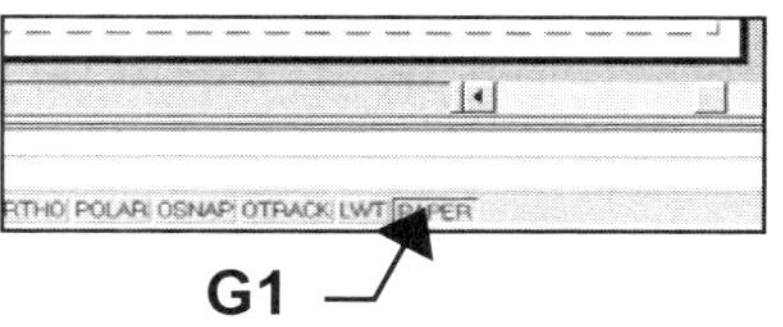

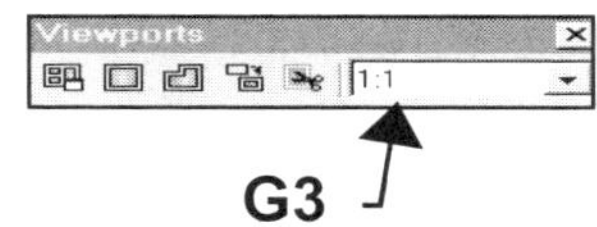

H. Lock the Viewport

This option will "lock" the adjusted scale within the selected viewport. If the viewport is "locked", the adjusted scale cannot be changed unless you "unlock" the viewport. When Zoom is used inside a locked viewport, the display of the entire layout is zoomed.

HOW TO LOCK A VIEWPORT

Method 1. a. Make sure you are in Paper Space.
b. Left click once on the Viewport frame.
c. Right Click (the short cut menu should appear).
d. Select "**Display Locked**" - Select Yes.

Method 2. a. Left click once on the Viewport frame.
b. Right click and select "Properties".
c. Select: **Misc / Displayed Locked / Yes**.

Method 3. a. Type **MV** at the command line and press <enter>.
b. Select "Lock" <enter>.
c. Type On or Off<enter>
d. left click on the Viewport frame.

J. Save as: **My Decimal Setup**

K. Continue on to 3-D.

EXERCISE 3D

CREATE A PAGE SETUP FOR 11 X 17 SHEET

The following instructions will guide you through the final steps for setting up the master file for plotting. These settings will stay with **My Decimal Setup** and you will be able to use it over and over again.

A. Open **My Decimal Setup** if it isn't already open.

B. Select the **11x17 (1 to 1)** layout tab.

You should be looking at your Border and Title Block now.

C. Select **File / Plot** or place the cursor on the **11x17 (1 to 1)** tab and press the right mouse button. Select **Plot** from the short cut menu.

The Plot dialog box shown below should appear. Select the "More Options" button if your dialog box does not appear the same as shown below

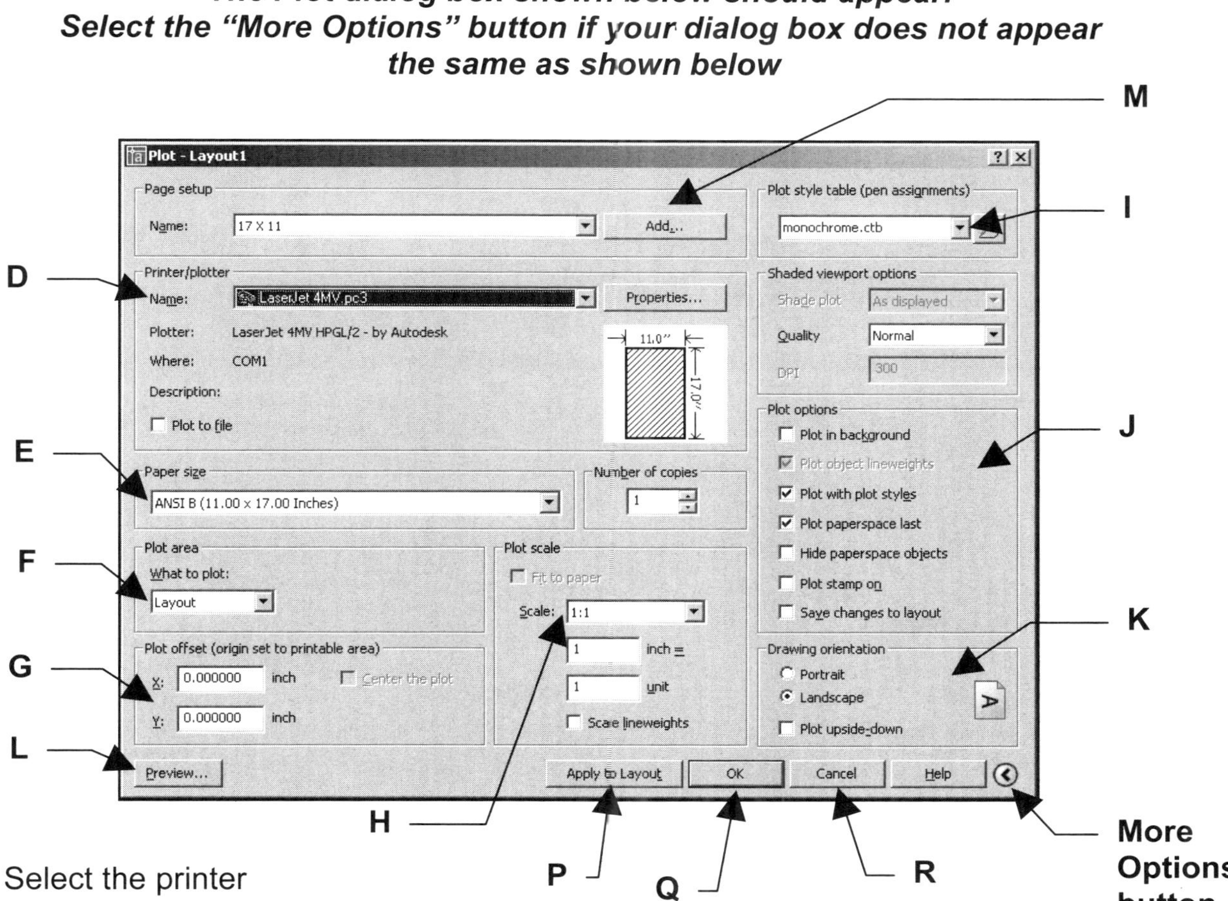

D. Select the printer

E. Select the Paper Size

F. Select the Plot Area

G. Plot offset should be 0 for both X and Y.

H. Select scale 1 : 1

I. Select the Plot Style Table "Monochrome.ctb"

J. Select the Plot options shown

K. Select Drawing Orientation: Landscape

L. Select **Preview** button.

1. If the drawing is centered on the sheet, press the Esc key and continue on to **M**.

2. If the drawing does not look correct, press the Esc key and check all your settings, then preview again.

M. Select the **ADD** button.

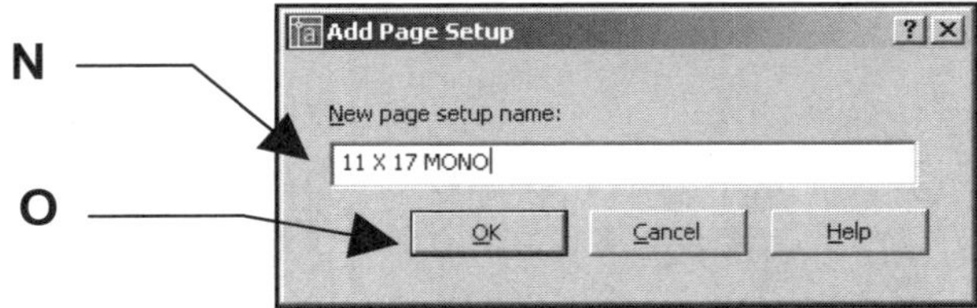

N. Type the new page setup name: 11 X 17 Mono

O. Select OK button.

The settings are now saved for future use.

P. Select the "**Apply to Layout**" button.

Q. If your computer **is** connected to the plotter / printer selected, select the **OK** button to plot, then proceed to **S**.

R. If your computer is not connected to the plotter / printer selected, select the Cancel button to close the Plot dialog box and proceed to **S**. Note: Selecting Cancel will cancel your selected setting if you did not save the page setup as specified in **M**.

S. Save the drawing
   1. Select **File / Save As**
   2. Save as: **My Decimal Setup**

You have now completed the Page Setup "11 X 17 MONO" for layout tab 11X17(1to 1) within the master drawing " ***My Decimal Setup"****. When you use this tab, use this page setup. You will understand better after you finish EX-3H.*

EXERCISE 3E

CREATE A LAYOUT & BORDER FOR PLOTTING ON A 24 X 18 SHEET

The following instructions will guide you through creating a new Layout and a new border to be used for plotting drawings on a 24 x 18 sheet. All of this information will be saved and you will not have to do this again.

A. Open **My Decimal Setup**

B. Select a **LAYOUT** tab.

Note: If the "Page Setup Manager" dialog box shown below does not appear automatically, right click on the Layout tab, then select Page Setup Manager.

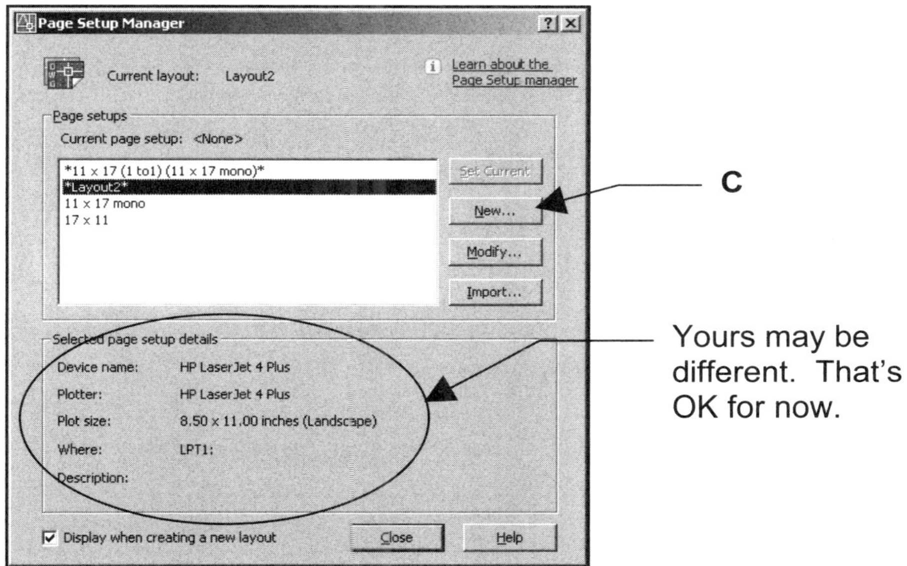

C. Select the **New** button. (*The "New Page Setup" dialog should appear)*

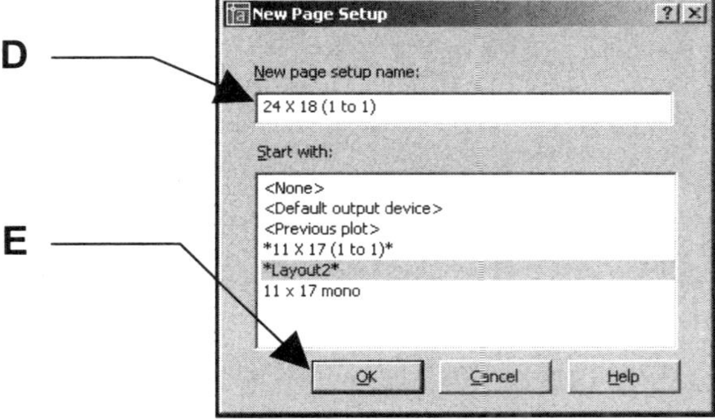

D. Type the new page setup name, **24 x 18 (1 to 1)**.
E. Select the **OK** button.

The "Page Setup" dialog box should appear.

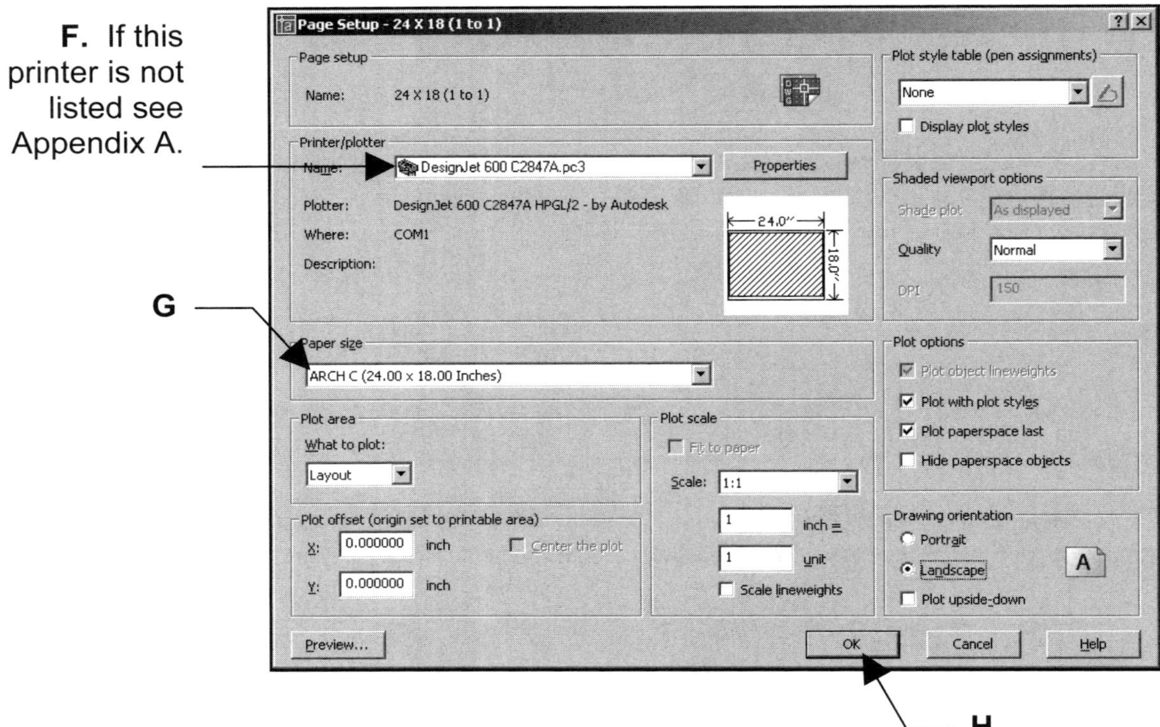

F. Select the **Printer / Plotter**.
G. Select the **Paper Size.**
H. Select the **OK** button.

The "Page Setup Manager" should have returned. Notice the new page setup name now appears in the list.

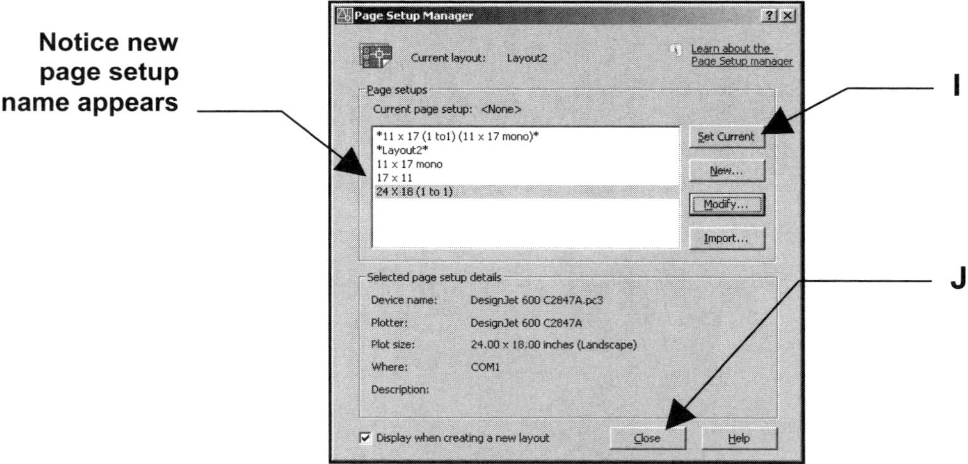

I. Select the **Set Current** button. (Don't forget this step, it is very important.)
J. Select the **Close** button.

You should now have a sheet of paper displayed on the screen.
This sheet is the size you specified in the "Page Setup" (24 X 18).
This sheet is in front of the drawing that is in Model Space.
The dashed line represents the printing limits for the device that you selected.

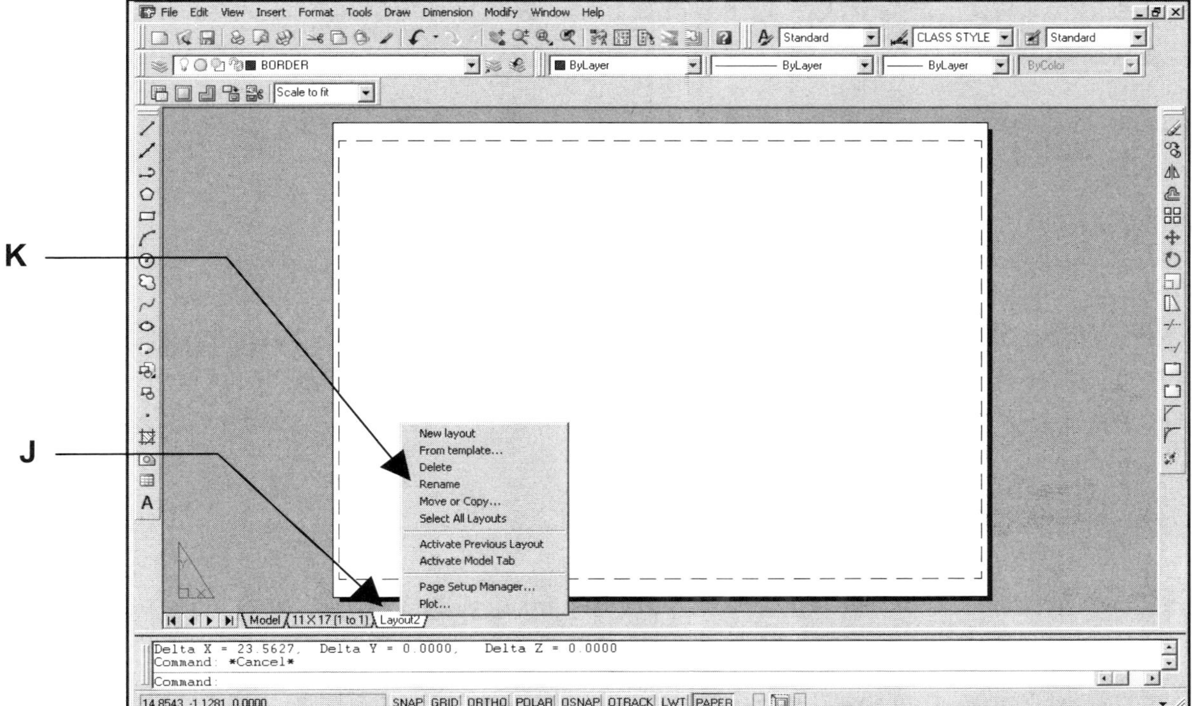

J. Right click on the Layout2 tab and select **Rename**.
K. Type the new name **24 X 18 (1 to 1)** then select **OK**.

L. Now we need to draw a new larger RECTANGLE, shown below, to fit on the larger sheet of paper shown on the screen. (Use Layer = Border)

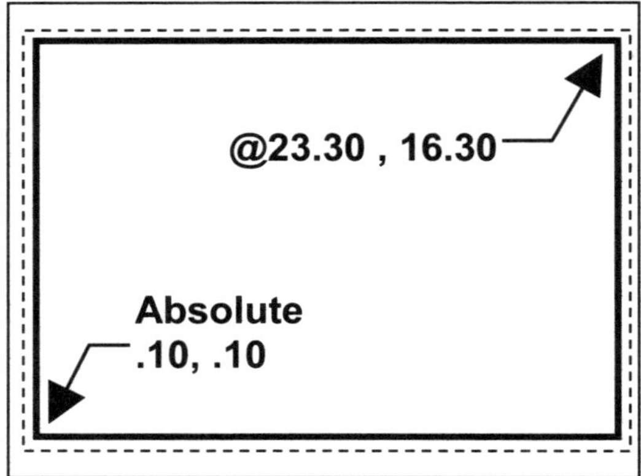

Now, you do not want to draw the EX-3B title block all over again. The next steps are to create a block from the EX-3B title block and then insert it into this layout. This is how thinking ahead saves you time.

M. Select the **"11 X 17 (1 to 1)"** Layout tab.

1. You need to be in "Paper Space".
2. Select "Draw / Block / Make" command. (Refer to page 8-2 for "Block" instructions.)

   a. Make a BLOCK of the "Title Block" ONLY. Do not include the Border Rectangle. Notice where the "Basepoint" is selected below.

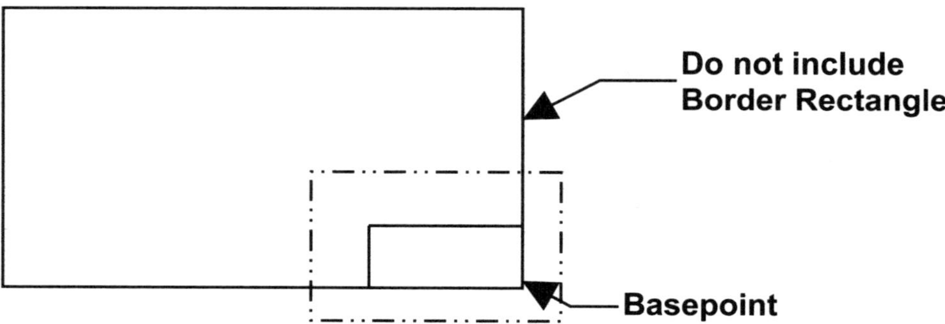

N. Select the **24 X 18 (1 to 1)** Layout tab.

1. **Very important:** Change to Layer = Border.
2. Select **INSERT / BLOCK.** (Refer to page 8-6 for "Insert" instructions.)
3. Select the "Title block" block then select OK.
4. Insertion Point should be the lower right corner of the Border Rectangle on the screen.

Do not scale the inserted Block. It may look smaller because the border rectangle is larger than the previous border, but the title block is the same size it was in 11 X 17 (1 to 1) layout.

O. Create a Viewport. (Refer to Exercise 3C for step by step instructions.)

1. Select Layer = Viewport
2. Select "Single Viewport" icon.
3. Draw a Viewport **approximately** 1/4" inside (smaller) the border lines. (Approximately as shown.)

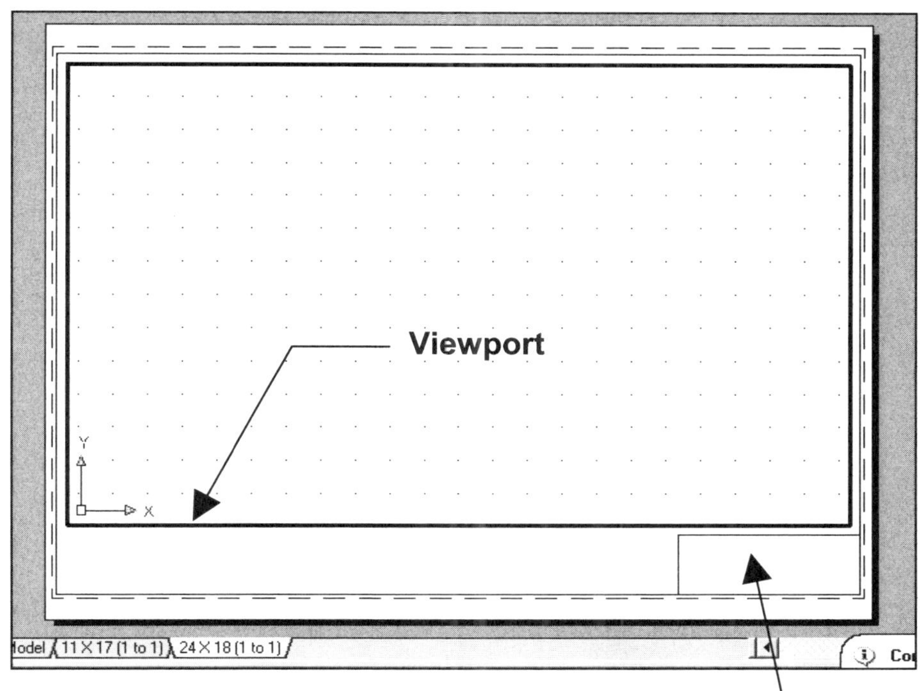

"EXPLODE" this Title block before editing.

P. IMPORTANT: Change to **Model Space** before going on to **Q**. (Double click inside the viewport or use the status line.)

Q. Change the **DRAWING LIMITS.**
   a. Lower left corner = 0 , 0
   b. Upper right corner = 24 , 18

R. (Very important) Select **VIEW / ZOOM / ALL.**

S. Adjust the Model Space scale to 1 : 1 using the **VIEWPORT toolbar.**

S

T. Lock the Viewport.
(Refer to page 3-12.)

Now the Viewport scale will not change when you zoom in or out.

U. Save as: **My Decimal Setup (again).**

Now you have 2 master layout borders. One to be used when plotting on a sheet size 17 x 11 and one to be used when plotting on a sheet size 24 x 18 at a scale of 1 : 1. But we are not quite done yet. Continue on to EX-3F.

EXERCISE 3F

CREATE A PAGE SETUP FOR 24 X 18 SHEET

The following instructions will guide you through creating a **PAGE SETUP** named: **"24 X 18 Mono".** You need this Page Setup to plot the **24 X18 (1 to 1)** layout.

A. Open **My Decimal Setup** if it isn't already open.

B. Select the **24 X 18 (1 to 1)** layout tab.

You should be looking at your Border and Title Block now.

C. Select **File / Plot** or place the cursor on the **24 X 18 (1 to 1)** tab and press the right mouse button. Select **Plot** from the short cut menu.

The Plot dialog box shown below should appear. Select the "More Options" button if your dialog box does not appear the same as shown below.

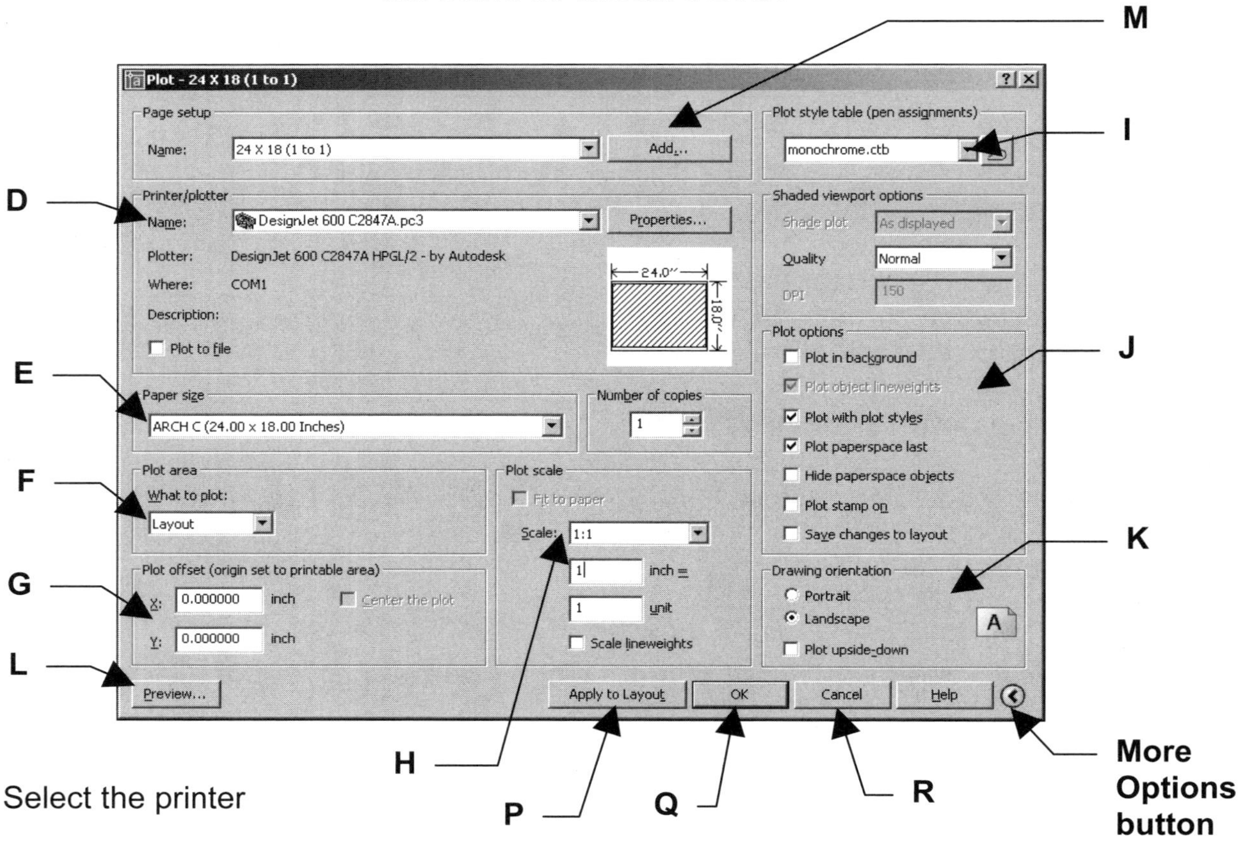

D. Select the printer

E. Select the Paper Size

F. Select the Plot Area

G. Plot offset should be 0 for both X and Y.

H. Select scale 1 : 1

I. Select the Plot Style Table "Monochrome.ctb"

J. Select the Plot options shown

K. Select Drawing Orientation: Landscape

L. Select **Preview** button.

1. If the drawing is centered on the sheet, press the Esc key and continue on to **M**.

2. If the drawing does not look correct, press the Esc key and check all your settings, then preview again.

M. Select the **ADD** button.

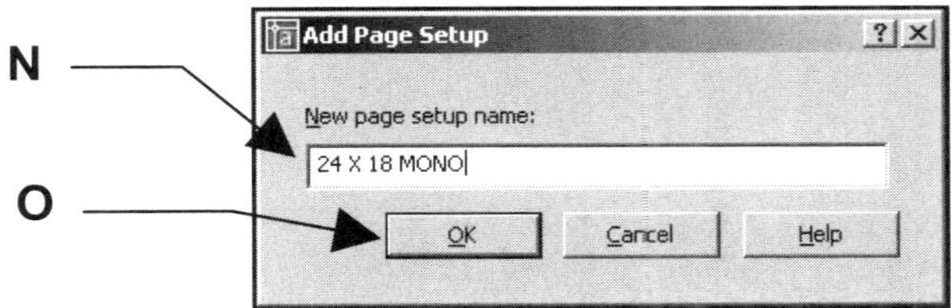

N. Type the new page setup name: 24 X 18 Mono

O. Select OK button.

The settings are now saved for future use.

P. Select the "Apply to Layout" button.

Q. If your computer <u>**is**</u> connected to the plotter / printer selected, select the **OK** button to plot, then proceed to **S**.

R. If your computer is <u>not</u> connected to the plotter / printer selected, select the Cancel button to close the Plot dialog box and proceed to **S**. Note: Selecting Cancel <u>will cancel</u> your selected setting if you did not save the page setup as specified in **M**.

S. Save the drawing
1. Select File / Save As
2. Save as: **My Decimal Setup**

You have now completed the Page Setup "24 X 18 MONO" for layout tab 24 X 18 (1 to1) within the master drawing " ***My Decimal Setup"****. When you use this tab, use this page setup. You will understand better after you finish EX-3H.*

EXERCISE 3G

CREATE A LAYOUT & BORDER FOR PLOTTING ON A 24 X 36 SHEET

The following instructions will guide you through creating a new Layout and a new border to be used for plotting drawings on a 24 x 36 sheet. All of this information will be saved and you will not have to do this again.

A. Open **My Decimal Setup**

B. Right Click on the 24 X 18 (1 to 1) layout tab.
   1. Select "New Layout" from the menu. (A new Layout1 tab will appear.)

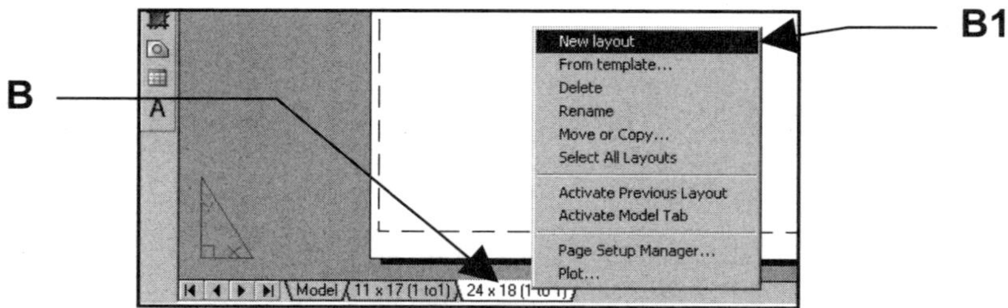

C. Select the **LAYOUT1** tab.

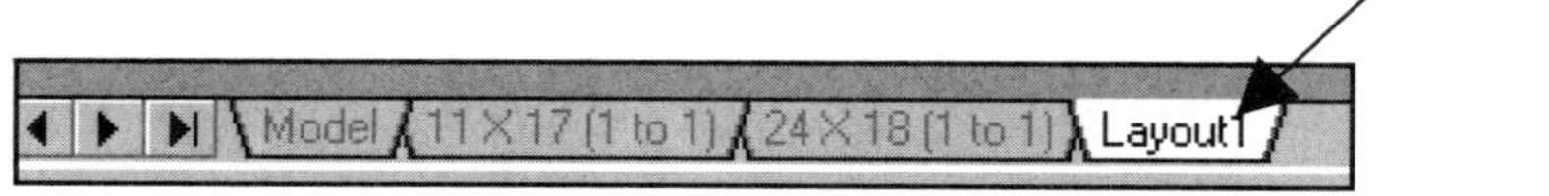

Note: If the "Page Setup Manager" dialog box, shown on the right, does not appear automatically, right click on the Layout1 tab, then select Page Setup Manager.

D. Select the **New** button.
   (*The "New Page Setup" dialog should appear)*

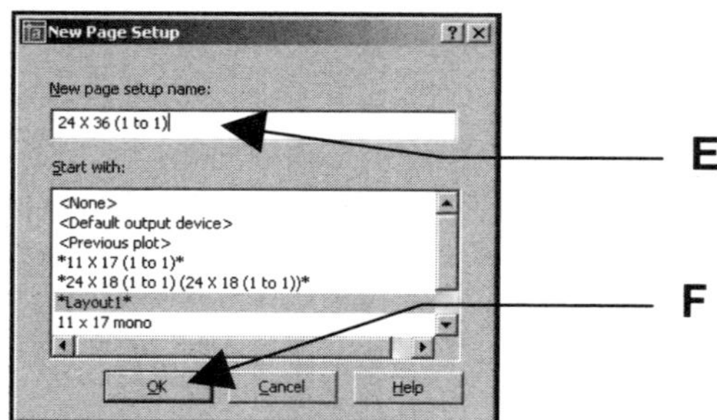

E. Type the new page setup name, **24 x 36 (1 to 1)**.

F. Select the **OK** button.

The "Page Setup" dialog box should appear.

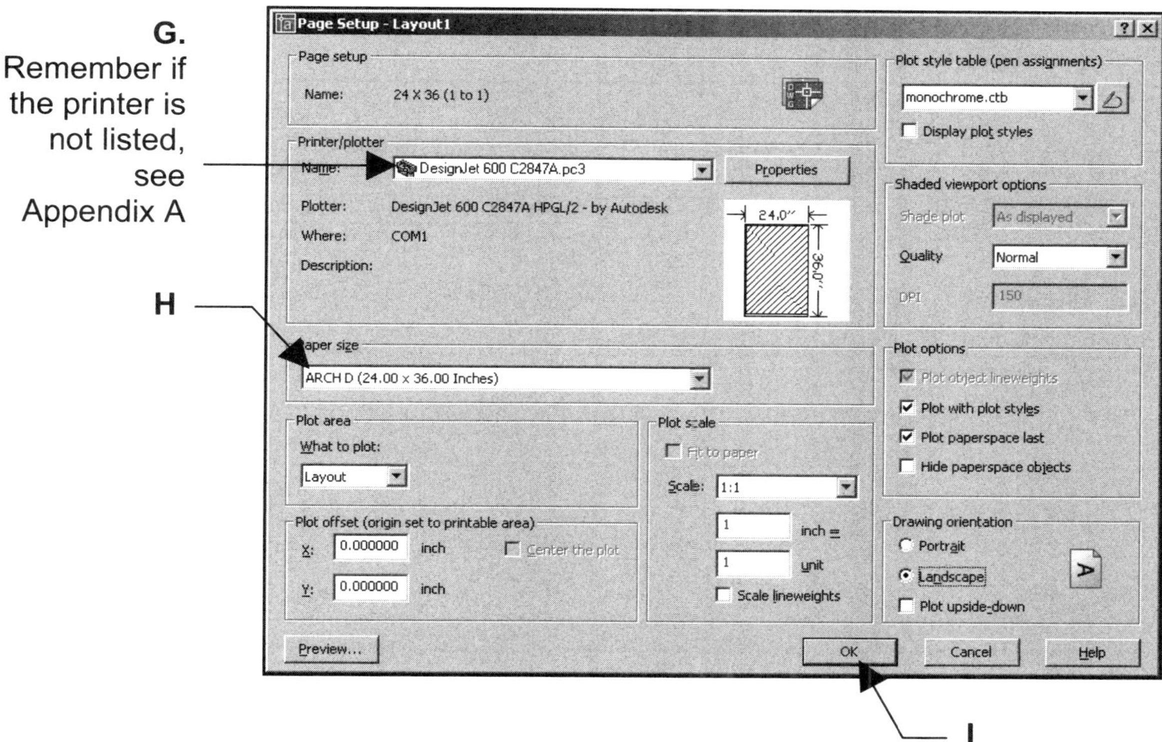

G. Select the **Printer / Plotter**.
H. Select the **Paper Size.**
I. Select the **OK** button.

The "Page Setup Manager" should have returned. Notice the new page setup name now appears in the list.

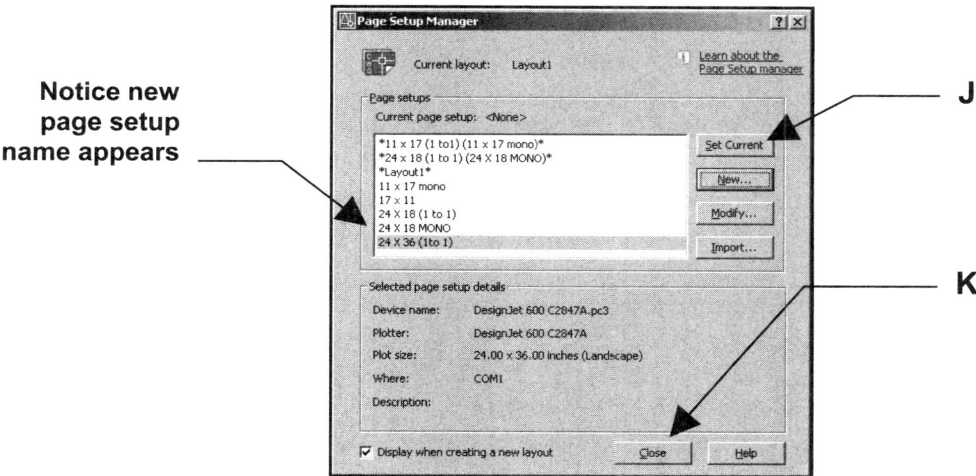

J. Select the **Set Current** button. (Don't forget this step, it is very important).

K. Select the **Close** button.

You should now have a sheet of paper displayed on the screen.
This sheet is the size you specified in the "Page Setup" (24 X 36).
This sheet is in front of the drawing that is in Model Space.
The dashed line represents the printing limits for the device that you selected.

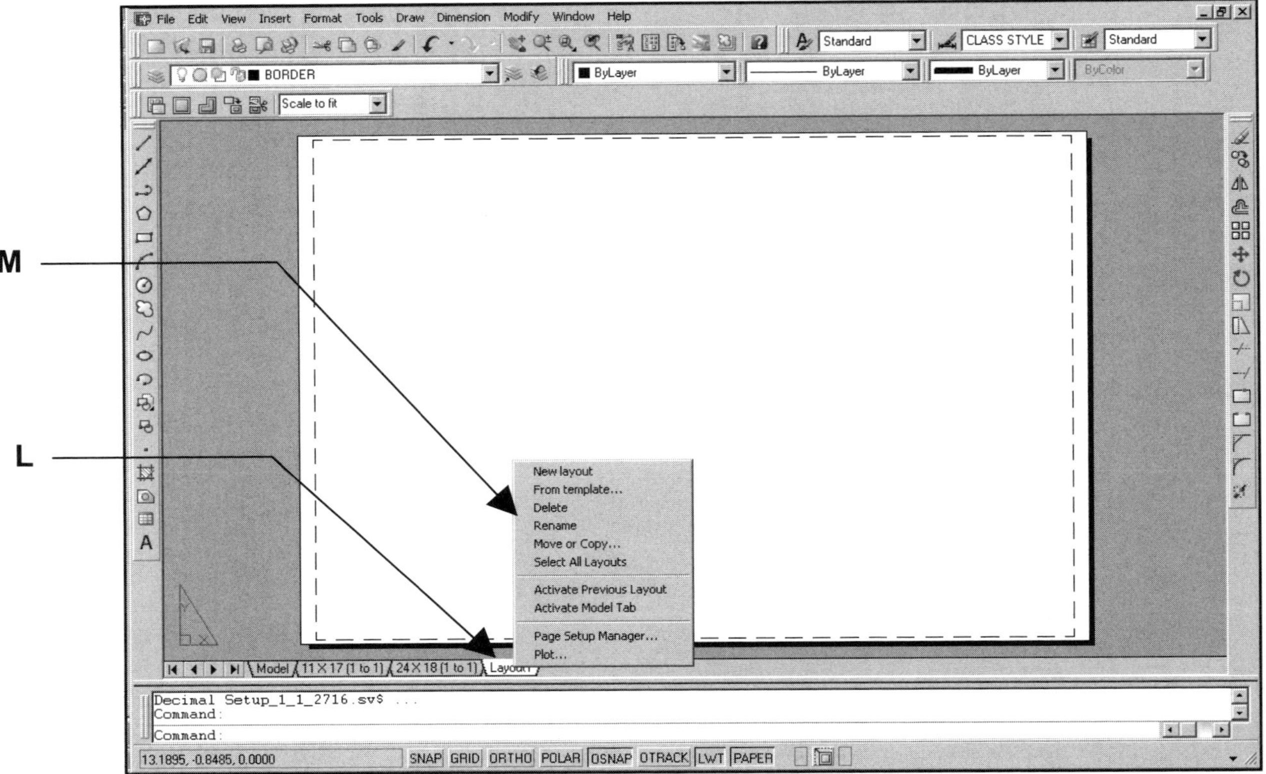

L. Right click on the Layout1 tab and select **Rename**.

M. Type the new name **24 X 36 (1 to 1)** then select **OK**.

N. Now we need to draw a new even larger RECTANGLE, shown below, to fit on the even larger sheet of paper shown on the screen. (Use Layer = Border)

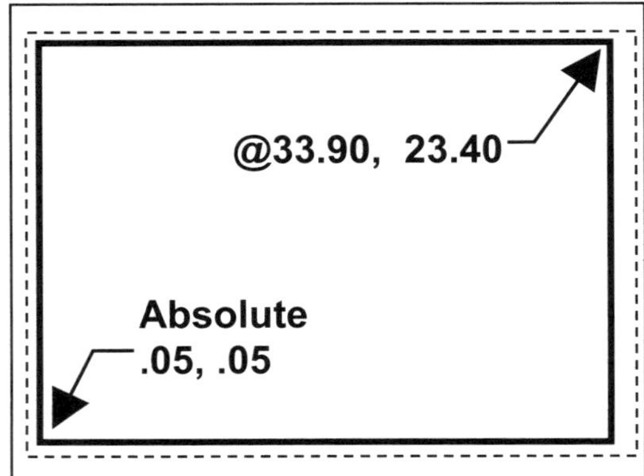

To save time again, you will insert the block, "Title Block", that you created previously.

O. Insert the block, "Title Block".
   1. **Very important:** Change to Layer = Border
   2. Select **INSERT / BLOCK** (Refer to page 8-6 for "Insert" instructions)
   3. Select the "Title block" block then select OK.
   4. Insertion Point should be the lower right corner of the Border Rectangle on the screen.

Do not scale the inserted Block. It may look smaller, because the border rectangle is even larger than the previous border, but the title block is the same size it was in 11 X 17 (1 to 1) layout.

P. Create a Viewport (Refer to Exercise 3C for step by step instructions if necessary)

   1. Select Layer = Viewport
   2. Select "Single Viewport" icon.
   3. Draw a Viewport <u>**approximately**</u> 1/4" inside (smaller) the border lines. (Approximately as shown)

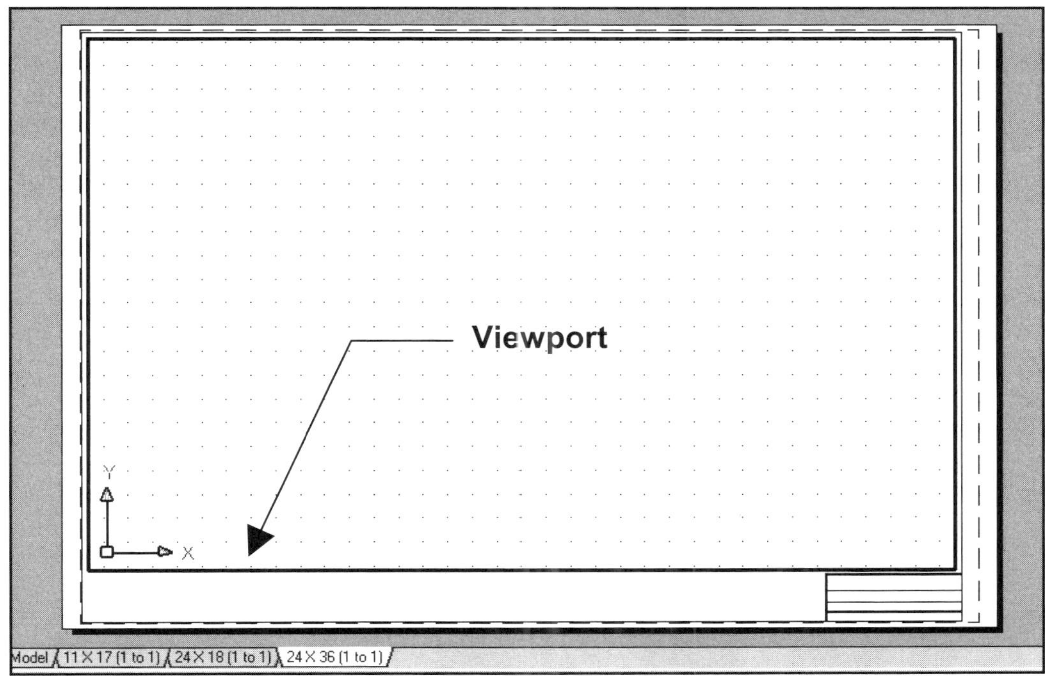

Q. IMPORTANT: Change to **Model Space** before going on to R. (Double click inside the viewport or use the status line.)

R. (Very important) Change the **DRAWING LIMITS.**
   a. Lower left corner = 0 , 0
   b. Upper right corner = 36, 24

S. Select **VIEW / ZOOM / ALL.**

T. Adjust the Model Space scale to 1 : 1 using the **VIEWPORT toolbar.**

U. Lock the Viewport .
(Refer to page 3-12)

Now the Viewport scale will not change when you zoom in or out.

V. Save as: **My Decimal Setup (again).**

Now you have 3 master layout borders.
One to be used when plotting on a sheet size 17 x 11 at a scale of 1 : 1.
One to be used when plotting on a sheet size 24 x 18 at a scale of 1 : 1.
One to be used when plotting on a sheet size 24 X 36 at a scale of 1 :1.

You are almost done. Just one more step to complete the "My Decimal Setup" master setup drawing. Continue on to EX-3H.

EXERCISE 3H

CREATE A PAGE SETUP FOR 24 X 36 SHEET

The following instructions will guide you through creating a **PAGE SETUP** named: **"24 X 36 Mono".** You need this Page Setup to plot the **24 X 36 (1 to 1)** layout.

A. Open **My Decimal Setup** if it isn't already open.

B. Select the **24 X 36 (1 to 1)** layout tab.

You should be looking at your Border and Title Block now.

C. Select **File / Plot** or place the cursor on the **24 X 36 (1 to 1)** tab and press the right mouse button. Select **Plot** from the short cut menu.

The Plot dialog box shown below should appear. Select the "More Options" button if your dialog box does not appear the same as shown below

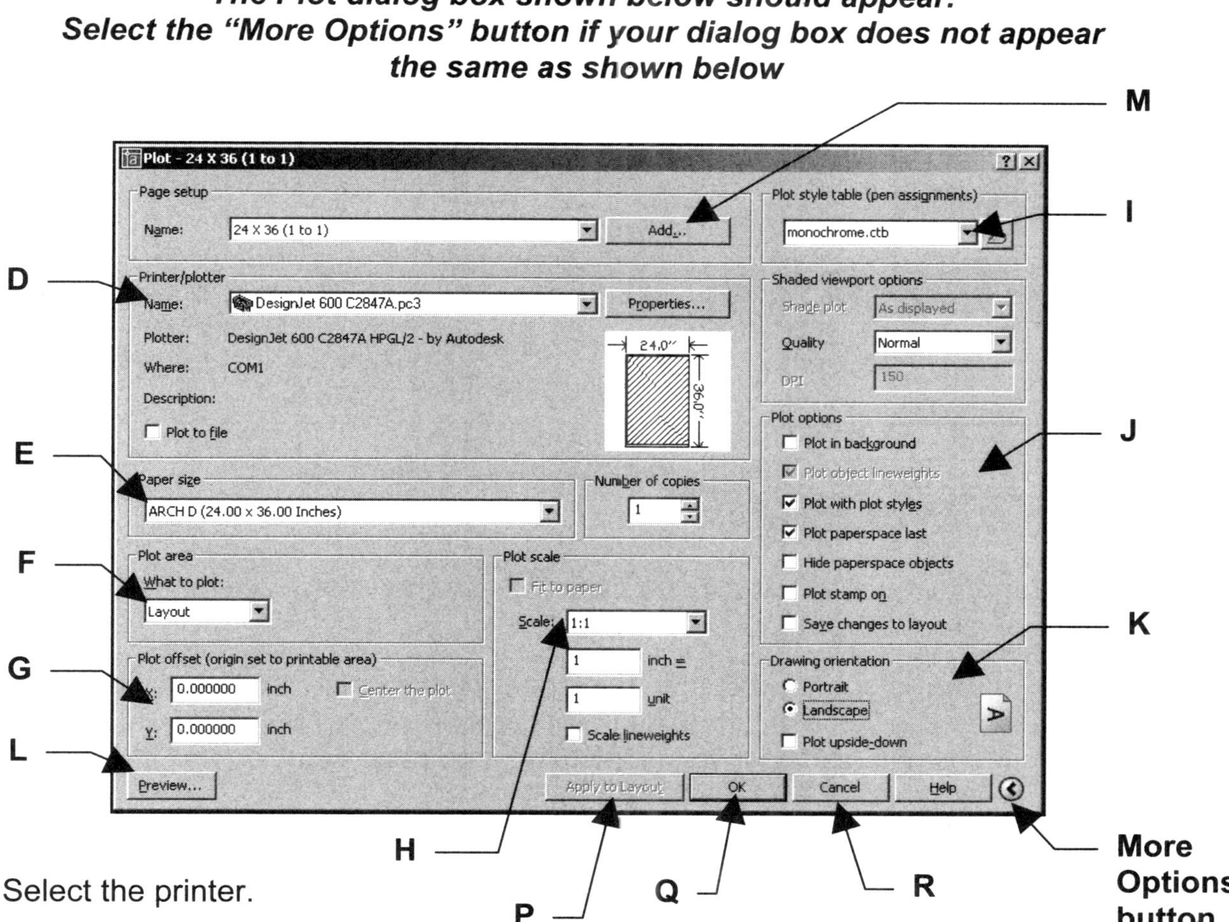

D. Select the printer.

E. Select the Paper Size .

F. Select the Plot Area.

G. Plot offset should be 0 for both X and Y.

H. Select scale 1 : 1.

I. Select the Plot Style Table "Monochrome.ctb".

J. Select the Plot options shown.

K. Select Drawing Orientation: Landscape.

L. Select **Preview** button.

1. If the entire drawing is displayed on the sheet, press the Esc key and continue on to **M**. (Note: It will not be centered using this selected printer.)

2. If the drawing does not look correct, press the Esc key and check all your settings, then preview again.

M. Select the **ADD** button.

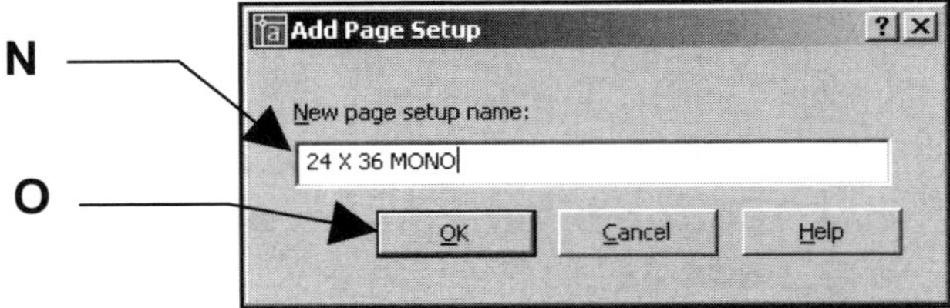

N. Type the new page setup name: 24 X 36 Mono

O. Select OK button.

The settings are now saved for future use.

P. Select the "Apply to Layout" button.

Q. If your computer **is** connected to the plotter / printer selected, select the **OK** button to plot, then proceed to **R**.

R. If your computer is not connected to the plotter / printer selected, select the Cancel button to close the Plot dialog box and proceed to **R**. Note: Selecting Cancel will cancel your selected setting if you did not save the page setup as specified in **M**.

S. Save the drawing.
   1. Select File / Save As.
   2. Save as: **My Decimal Setup.**

You have now completed the Page Setup "24 X 36 MONO" for layout tab 24 X 36(1to 1) within the master drawing " ***My Decimal Setup".*** *When you use this tab, use this page setup.*

Summary

Now let's think about what you have achieved in this lesson.

EX-3A Select Preferences to draw with.

1. Started a NEW drawing file.
2. Selected your preferences such as: Units, Drawing Limits, Snap and Grid.
3. Created New Layers, Text Styles and a Dimension Style.

EX-3B Select Printer and Paper size to plot on.

1. Selected the Printer / Plotter.
2. Selected the "Size" of sheet you want to plot on.
3. Drew a master border with title block.

EX-3C Cut a hole in the sheet to see model space and adjust Model Space scale.

1. Created a Viewport.
2. Adjusted the scale of the Viewport.
3. Locked the Viewport.

EX-3D Set up Plotting specifications and saved them.

1. Selected Printer / Plotter.
2. Selected Paper size.
3. Selected what area to plot.
4. Selected the scale for paperspace.
5. Selected Plot Style Table.
6. Selected Plot Options
7. Selected Drawing Orientation (Portrait or Landscape)
8. Previewed to make sure all specifications are correct.
8. Added the specifications to your file.

EX-3E Repeat 3B and 3C for a larger size of paper.

EX-3F Repeat 3D with different specifications.

EX-3G Repeat 3B and 3C for an even larger size of paper.

EX-3H Repeat 3D with different specifications.

Do you see a pattern forming?

<u>The result is a Master setup drawing file with 3 layout tabs predefined for 3 sizes of paper and 3 page setups (plotting specifications) for each layout.</u>

Would you like to print a drawing on your letter size Printer?

Anytime you want to print a quick draft on your letter size printer do the following:

1. Open the drawing on to the screen.

2. Select **File / Plot**.

3. Select your printer. (If your printer is not listed refer to Appendix A.)

4. The "AutoCAD Warning" box, shown below, will appear. Select the **OK** button.

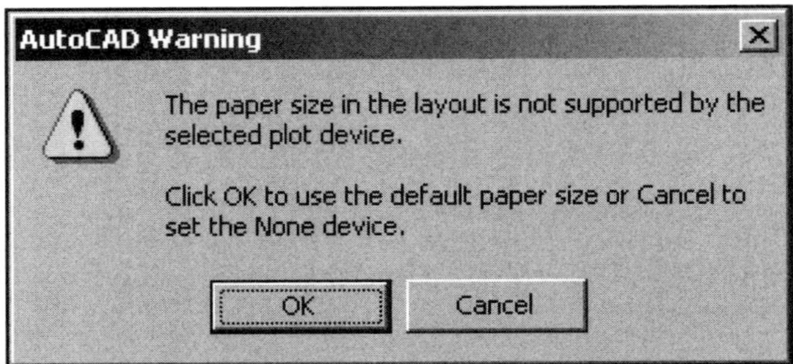

5. Select "**Paper Size**".
   Note: The paper size probably already changed to the default size for the printer.

6. Select "**Extents**" for the Plot Area.

7. Select "**Center** the Plot" for Plot Offset.

8. Select "**Fit to Paper**" for Plot Scale.

9. Select the "**Plot Style Table**".
   a. Black only = Monochrome.
   b. Color = None. (Of course your printer must be capable of printing in color.)

10. **Preview**
    a. If it looks correct press enter.
    b. If it doesn't look correct recheck 3 through 9 above.

11. Consider "Adding" this as a page setup so you don't have to select all the settings next time. Hmmmmm...think about it. (Refer to EX-3D, 3F or 3H)

LEARNING OBJECTIVES

After completing this lesson, you will be able to:

1. Create a “Feet and inches” Setup drawing.
2. Create another Border for 24 x 18
3. Create Page Setups for 1/4” = 1’ and 1/8” = 1’

LESSON 4

EXERCISE 4A

CREATE AN MASTER FEET-INCHES SETUP DRAWING

The following instructions will guide you through creating a "Master" feet-inches setup drawing. You will follow the same steps you did in Lesson 3 with a few additional scaling issues.

NEW SETTINGS

A. Begin your drawing without a template as follows:

1. Select **"FILE / NEW".**
2. Select **"START FROM SCRATCH"** Box.
3. Select "**OK**".
4. Your screen should be blank, no grids and the current layer is 0.

Select Imperial or Metric

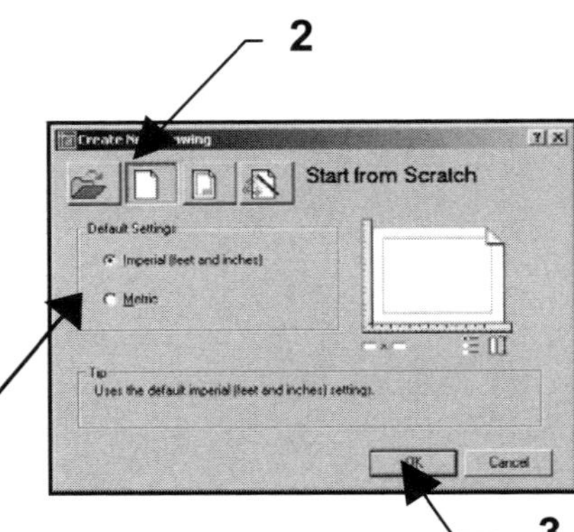

B. Set drawing specifications as follows:

1. Select **FORMAT / UNITS**
   a. Change the "Type" and "Precision" then select **OK**

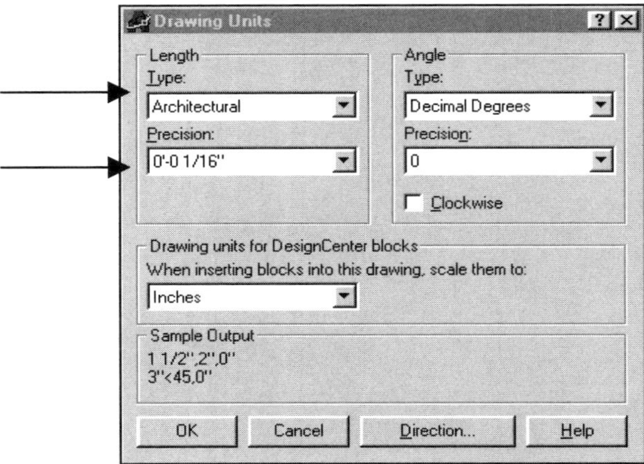

2. Select **FORMAT / DRAWING LIMITS**
   a. Lower left corner = 0'-0", 0'-0"
   b. Upper right corner = 24, 18. (Notice this is inches, not feet.)
   c. Use **"VIEW / ZOOM / ALL"** to generate the new limits.

3. Select **TOOLS / DRAFTING SETTINGS / SNAP and GRID** tab.
   a. Change the settings as shown below and then select **OK**.
   b. Set your grids to **ON** to display the paper size.

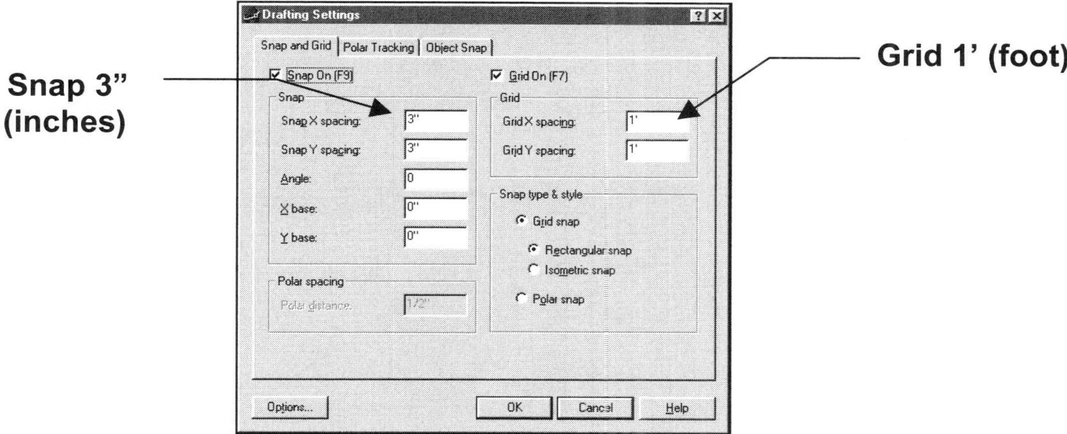

4. Set **"PICK BOX"** size (your preference).
   Use "Tools / Options / Selection" tab.

NEW LAYERS

C. Create new layers.
   1. **Load** linetype **DASHED2** and **PHANTOM**
   2. Change the Lineweight settings to inches or metric and adjust the Display scale.
   3. Assign names, colors, linetypes and plotability. (Use Format / Layer)

NAME	COLOR	LINETYPE	LWT	PLOT
BORDER	RED	CONTINUOUS	.039 (1.00mm)	YES
CABINETS	CYAN	CONTINUOUS	.016 (0.40mm)	YES
CONSTRUCTION	WHITE	CONTINUOUS	Default	**NO**
DIMENSION	BLUE	CONTINUOUS	Default	YES
DOORS	GREEN	CONTINUOUS	.016 (0.40mm)	YES
ELECTRICAL	CYAN	CONTINUOUS	.024 (0.60mm)	YES
FURNITURE	MAGENTA	CONTINUOUS	.016 (0.40mm)	YES
HARDSCAPE	WHITE	CONTINUOUS	.024 (0.60mm)	YES
HATCH	GREEN	CONTINUOUS	Default	YES
HIDDEN	MAGENTA	DASHED2	Default	YES
MISC	CYAN	CONTINUOUS	.016 (0.40mm)	YES
OBJECT	RED	CONTINUOUS	.024 (0.60mm)	YES
PLUMBING	9	CONTINUOUS	.016 (0.40mm)	YES
PROPERTY LINE	WHITE	PHANTOM	.024 (0.60mm)	YES
SYMBOLS	GREEN	CONTINUOUS	.024 (0.60mm)	YES
TEXT HEAVY	WHITE	CONTINUOUS	Default	YES
TEXT LIGHT	BLUE	CONTINUOUS	Default	YES
VIEWPORT	GREEN	CONTINUOUS	Default	**NO**
WALLS	RED	CONTINUOUS	.024 (0.60mm)	YES
WINDOWS	GREEN	CONTINUOUS	.016 (0.40mm)	YES
WIRING	CYAN	DASHED2	.024 (0.60mm)	YES
XREF	WHITE	CONTINUOUS	Default	YES

NEW TEXT STYLE

D. Create 2 text styles named ***CLASS TEXT*** and ***ARCH TEXT***

1. Select **"FORMAT / TEXT STYLE".**
2. Make the changes shown in the dialog boxes below.

CLASS TEXT

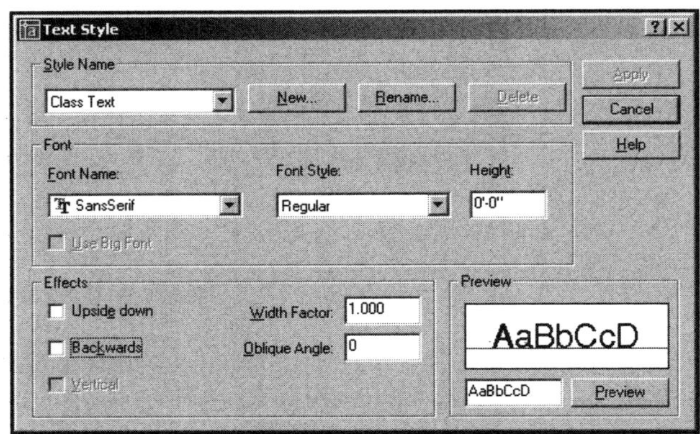

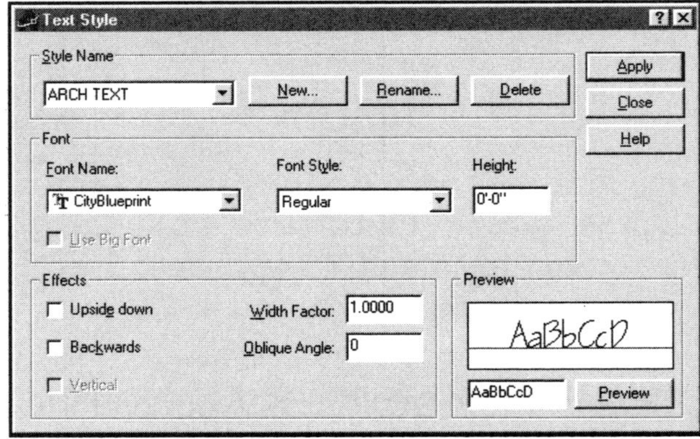

3. When complete, select **APPLY** then **CLOSE**.

NEW DIMENSION STYLE

E. Create a new Dimension Style named ***Arch Dim***

1. Select **FORMAT / DIMENSION STYLE**
2. Select the **NEW** button.
3. Enter the **ARCH DIM** name.
4. Select the **Continue** button.
5. Make the changes to the following dialog boxes.

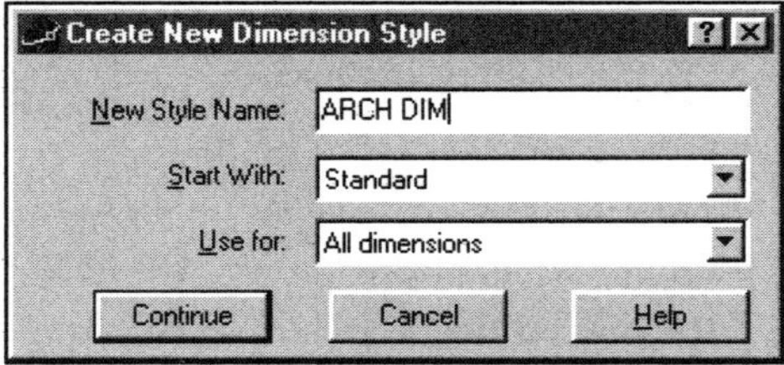

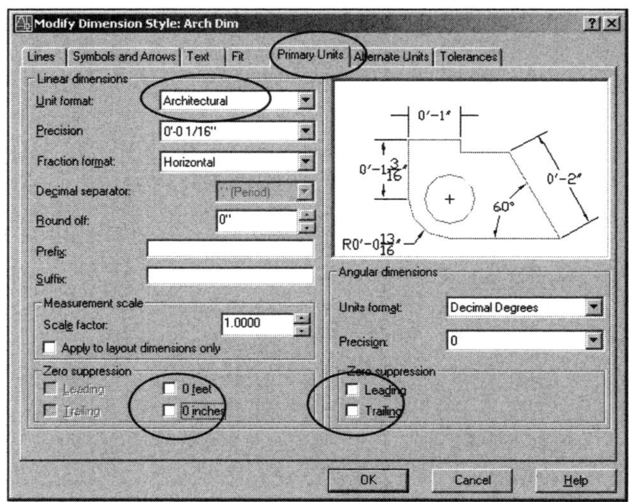

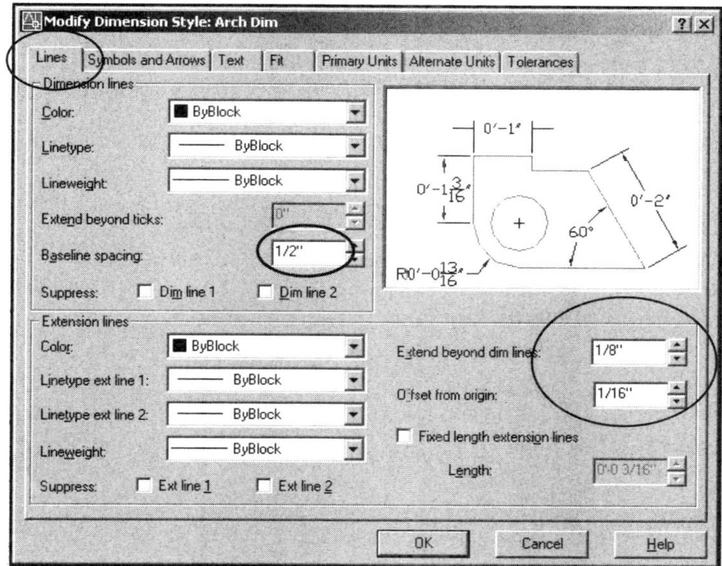
Modify Dimension Style: Arch Dim
Lines
Symbols and Arrows
Text
Fit
Primary Units
Alternate Units
Tolerances
Dimension lines
Color:
ByBlock
Linetype:
ByBlock
Lineweight:
ByBlock
Extend beyond ticks:
0"
Baseline spacing:
1/2"
Suppress:
Dim line 1
Dim line 2
Extension lines
Color:
ByBlock
Linetype ext line 1:
ByBlock
Linetype ext line 2:
ByBlock
Lineweight:
ByBlock
Suppress:
Ext line 1
Ext line 2
Extend beyond dim lines:
1/8"
Offset from origin:
1/16"
Fixed length extension lines
Length:
0'-0 3/16"
OK
Cancel
Help

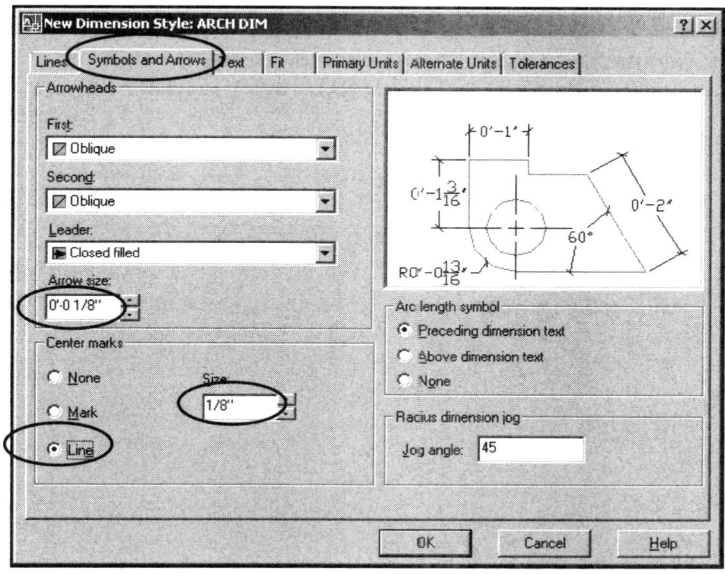
New Dimension Style: ARCH DIM
Lines
Symbols and Arrows
Text
Fit
Primary Units
Alternate Units
Tolerances
Arrowheads
First:
Oblique
Second:
Oblique
Leader:
Closed filled
Arrow size:
0'-0 1/8"
Center marks
None
Mark
Line
Size:
1/8"
Arc length symbol
Preceding dimension text
Above dimension text
None
Radius dimension jog
Jog angle:
45
OK
Cancel
Help

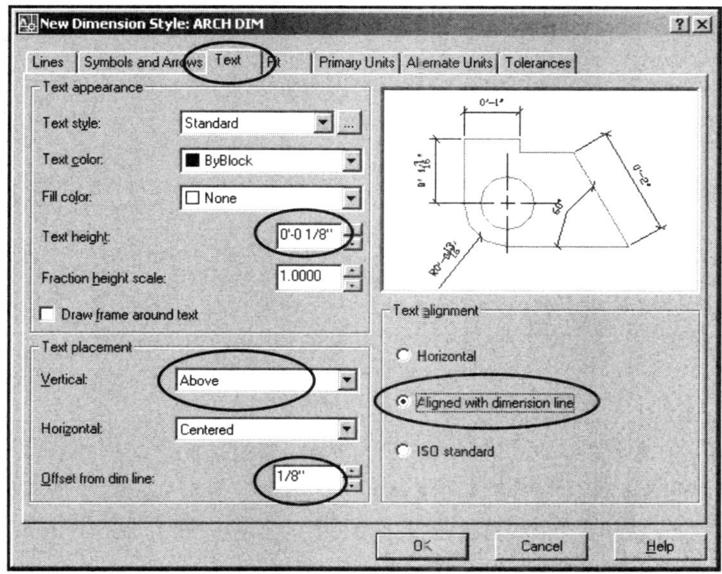
New Dimension Style: ARCH DIM
Lines
Symbols and Arrows
Text
Fit
Primary Units
Alternate Units
Tolerances
Text appearance
Text style:
Standard
Text color:
ByBlock
Fill color:
None
Text height:
0'-0 1/8"
Fraction height scale:
1.0000
Draw frame around text
Text placement
Vertical:
Above
Horizontal:
Centered
Offset from dim line:
1/8"
Text alignment
Horizontal
Aligned with dimension line
ISO standard
OK
Cancel
Help

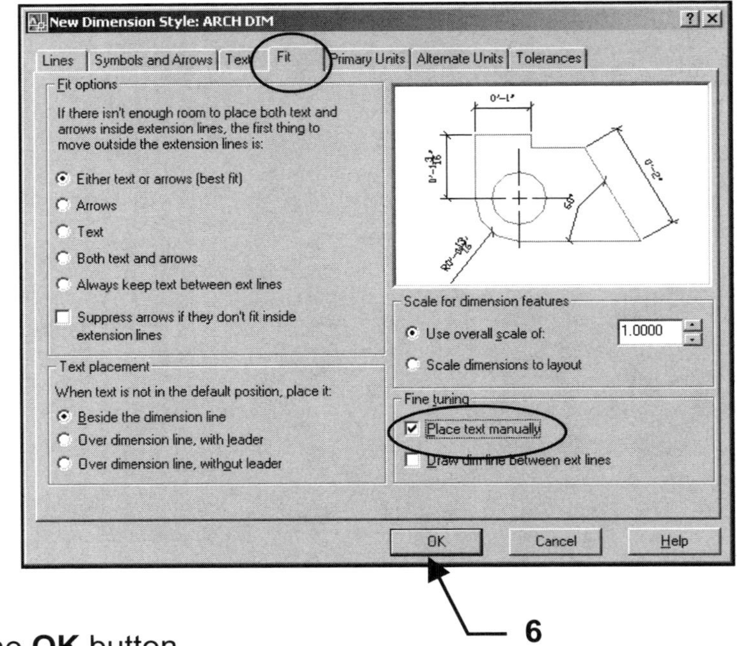

6. **NOW** select the **OK** button.

7. *Select the new style "**ARCH DIM**".*
   a. Select the **"Set Current"** button.

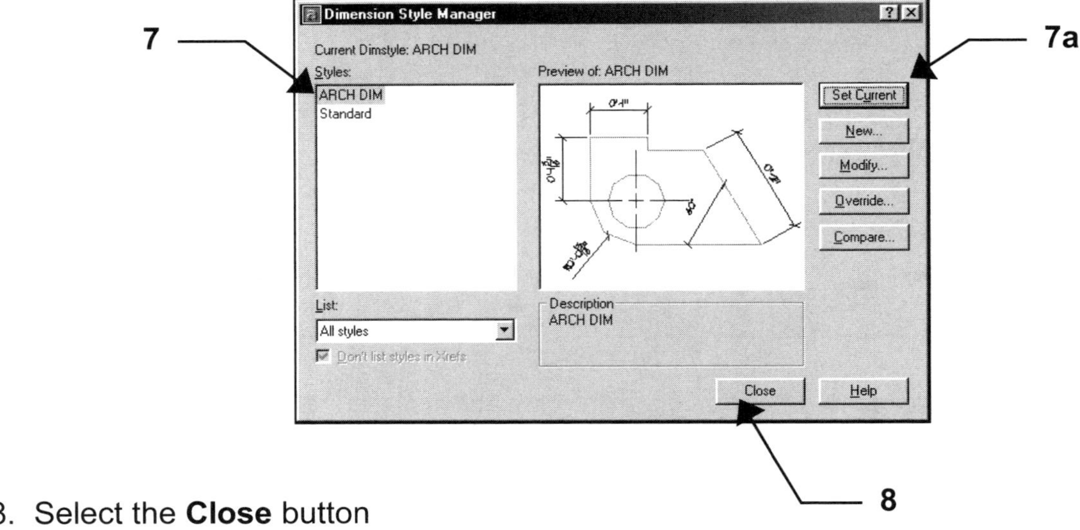

8. Select the **Close** button

THIS NEXT STEP IS VERY IMPORTANT.

F. SAVE ALL THE SETTINGS YOU JUST CREATED

1. Select   File / Save as
2. Save as: **My Feet-Inches Setup**

NOTE: This was just step 1 to creating your setup drawing. Continue on through the following exercises, in this lesson, using this same file (My Feet-Inches Setup). Each step adds more information to this drawing.

EXERCISE 4B

CREATE AN ARCHITECTURAL BORDER FOR PLOTTING

The following instructions will guide you through creating a Border drawing that will be used in combination with "My Feet-Inches Setup" when plotting. You will create a Layout and draw a border with a title block. All of this information will be saved and you will not have to do this again.

A. Open **My Feet-Inches Setup**

B. Select a **LAYOUT** tab.

Note: If the "Page Setup Manager" dialog box shown below does not appear automatically, right click on the Layout tab, then select Page Setup Manager.

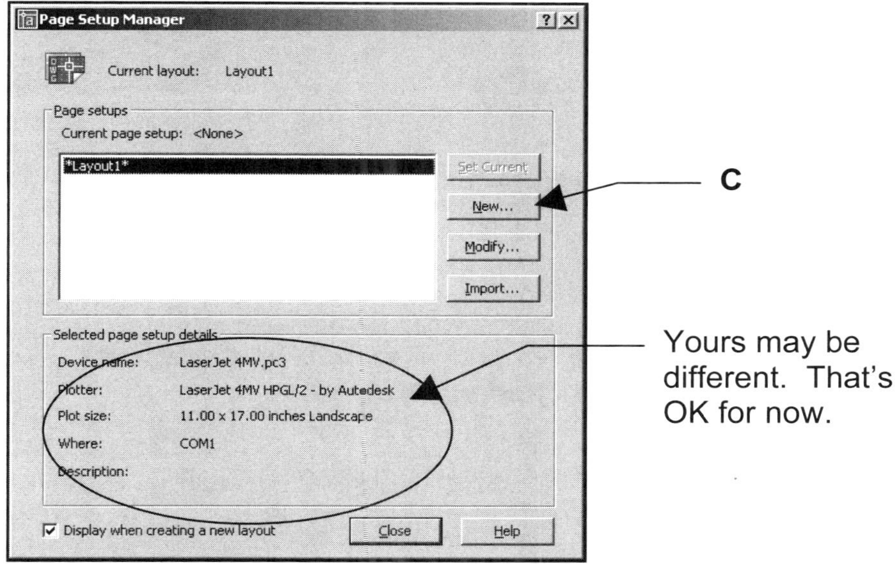

C. Select the **New** button. *The "New Page Setup" dialog should appear.*

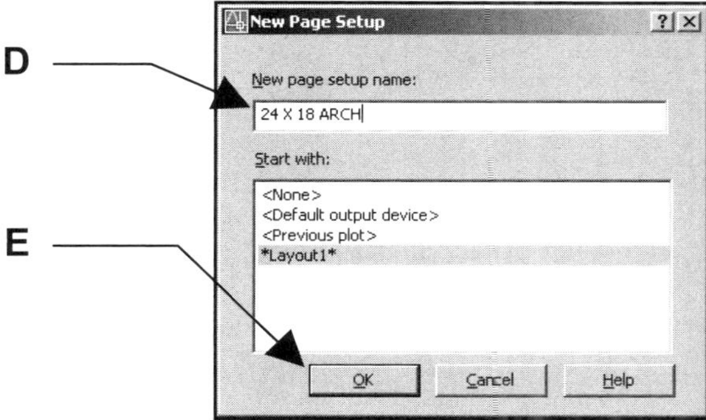

D. Type the new page setup name, **24 X 18 Arch**.
E. Select the **OK** button.

The "Page Setup" dialog box should appear.

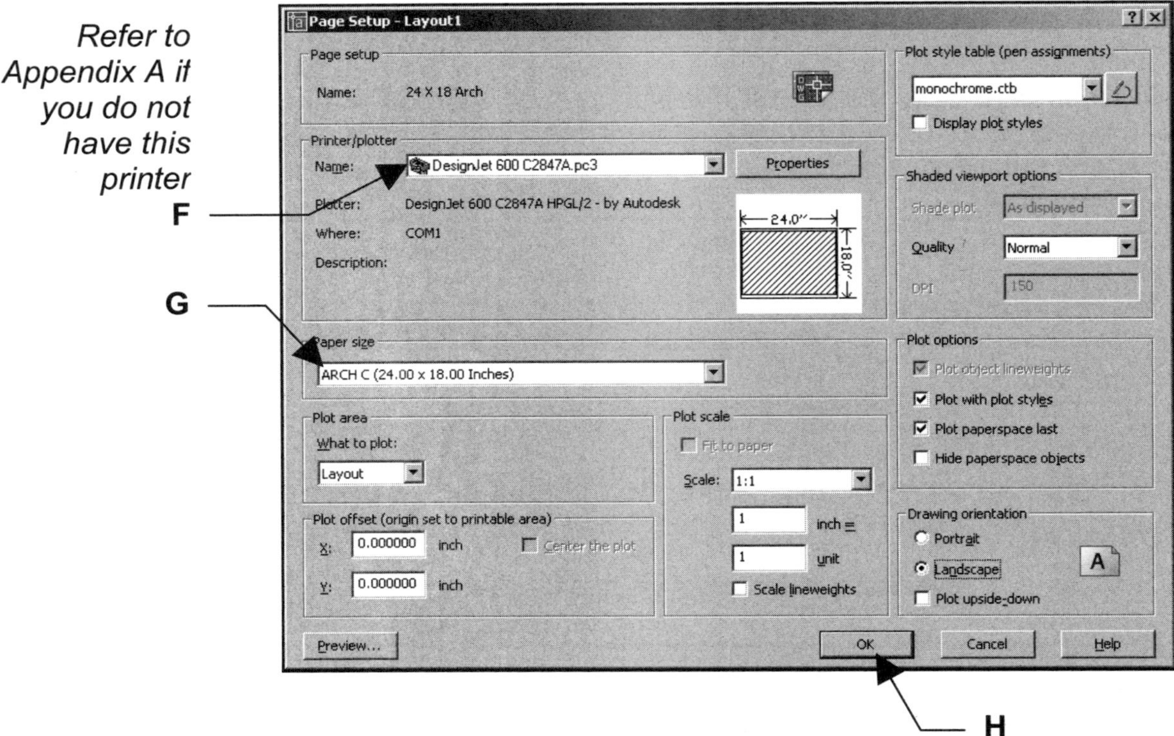

F. Select the **Printer / Plotter** shown.
G. Select the **Paper Size** shown.
H. Select the **OK** button.

The "Page Setup Manager" should have returned. Notice the new page setup name now appears in the list.

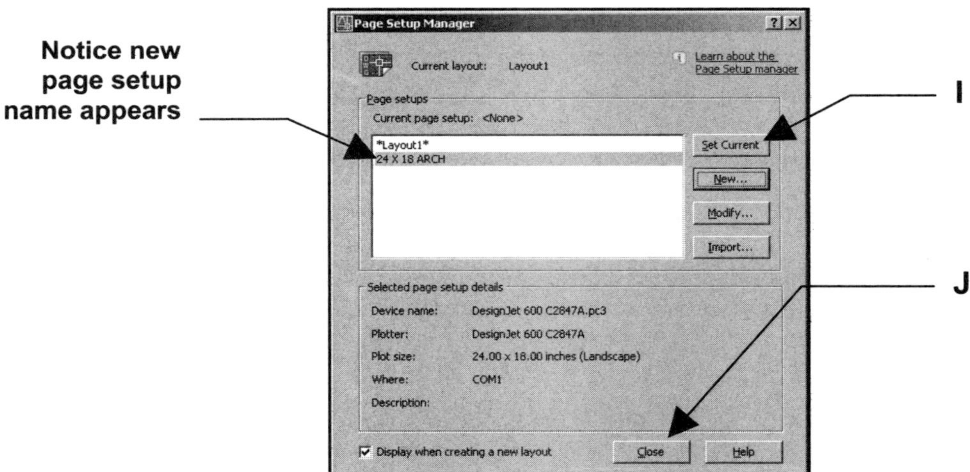

I. Select the **Set Current** button.
J. Select the **Close** button.

You should now have a sheet of paper displayed on the screen.
This sheet is the size you specified in the “Page Setup”.
This sheet is in front of the drawing that is in Model Space.
The dashed line represents the printing limits for the device that you selected.

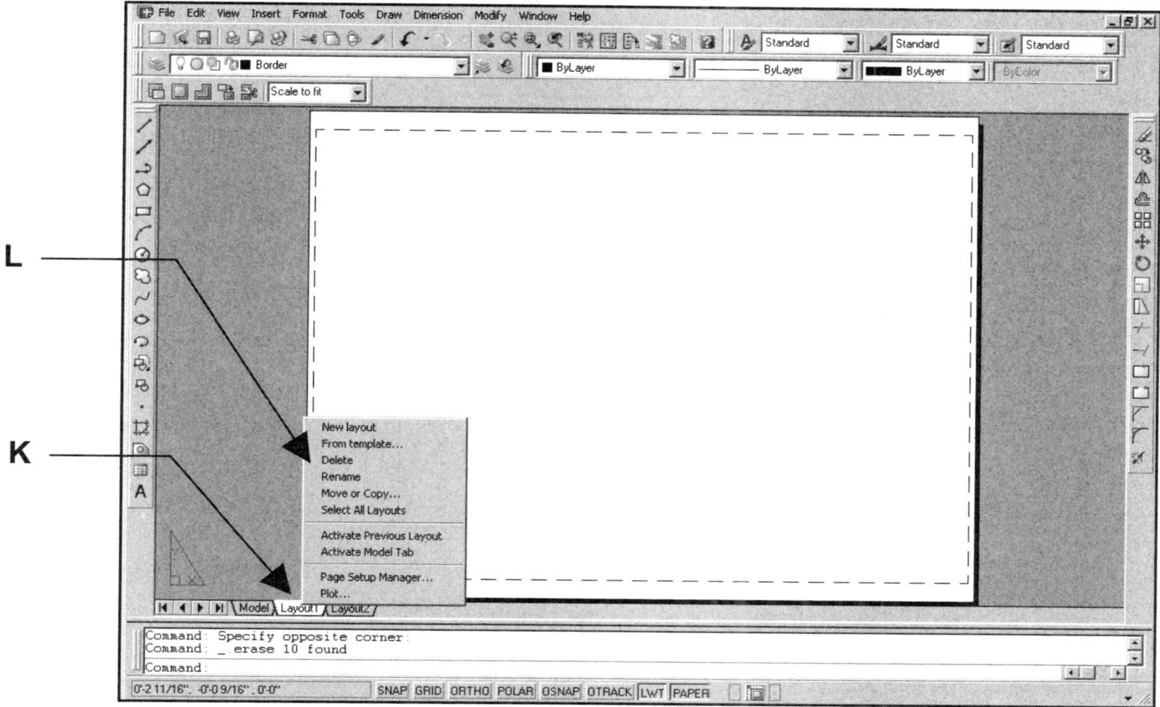

K. Right click on the Layout tab and select **Rename**.
L. Type the new name **24 X 18 (1 to 1)** then select **OK**.

M. Draw the Border with title block, shown below, on the sheet of paper shown on the screen.

N. When you have completed the Border, shown below:

1. Select File / Save as
2. Save as: **My Feet-Inches Setup** (Again)

23-3/8"
16-3/8"
1/8", 1/8"
SCHOOL NAME
CITY, STATE
TITLE GOES HERE
NAME
X-XX-XX
X=X
XXX

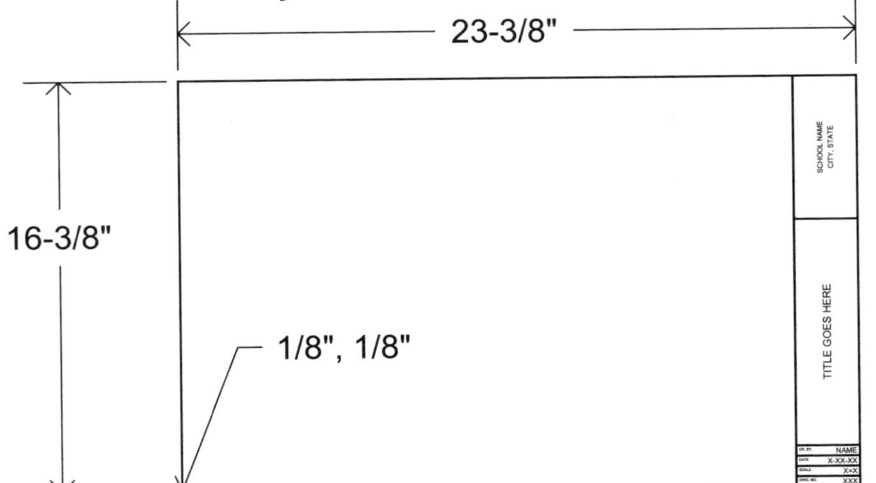

Use Layer Border for border lines and Title block lines.

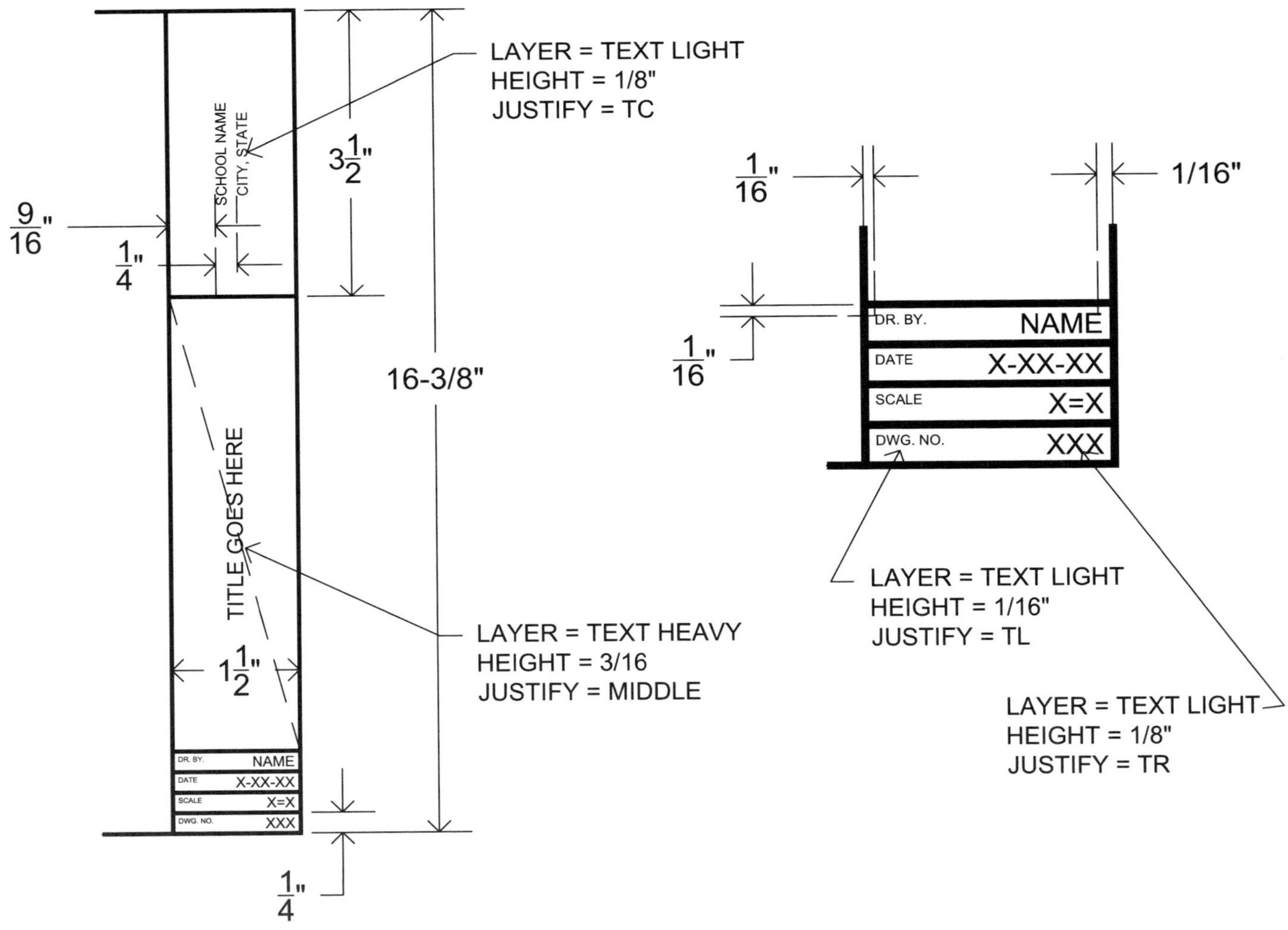

LAYER = TEXT LIGHT
HEIGHT = 1/8"
JUSTIFY = TC
SCHOOL NAME
CITY, STATE
3 1/2"
9/16"
1/4"
16-3/8"
TITLE GOES HERE
1 1/2"
LAYER = TEXT HEAVY
HEIGHT = 3/16
JUSTIFY = MIDDLE
DR. BY.
NAME
DATE
X-XX-XX
SCALE
X=X
DWG. NO.
XXX
1/4"
1/16"
1/16"
1/16"
LAYER = TEXT LIGHT
HEIGHT = 1/16"
JUSTIFY = TL
LAYER = TEXT LIGHT
HEIGHT = 1/8"
JUSTIFY = TR

EXERCISE 4C

CREATE A VIEWPORT & ADJUST SCALE

The following instructions will guide you through creating a VIEWPORT in the Border Layout sheet. Creating a viewport has the same effect as cutting a hole in the sheet of paper. You will be able to see through the hole to Model Space. Then adjust the Model Space scale to 1: 1.

A. Open **My Feet-Inches Setup**

B. Select the **24 X 18 (1 to 1)** tab.

C. Select layer **"Viewport".**

D. Select the **Single Viewport icon** from the Viewports Toolbar.

Single Viewport icon (LT's toolbar looks a little different.)

E. Draw a single viewport **<u>approximately</u>** 1/8" inside (smaller) the border lines, as shown.

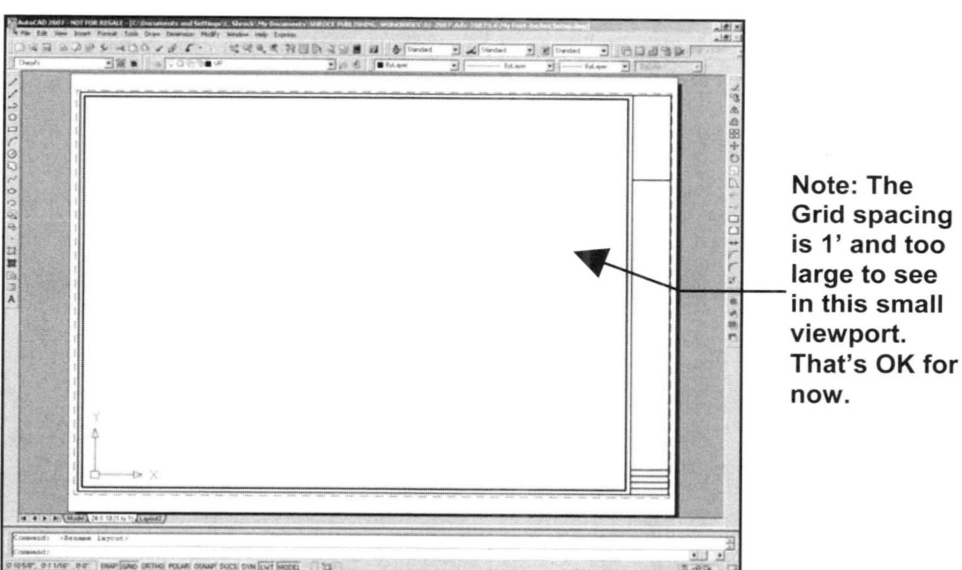

F. Change the **Paper button** to **Model** or double click inside the Viewport frame.

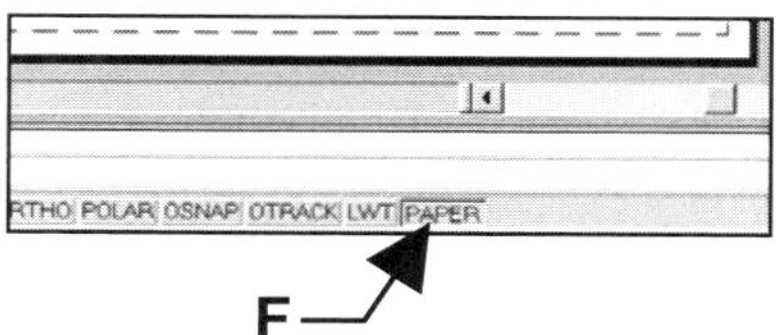

G. Select **View / Zoom / All**

H. Adjust the Model space scale.
   1. Select 1:1 in the **VIEWPORT** toolbar.

I. **Lock** the Viewport. (Refer to 3-12 if necessary.)

J. File / Save as: **My Feet-Inches Setup** (again).

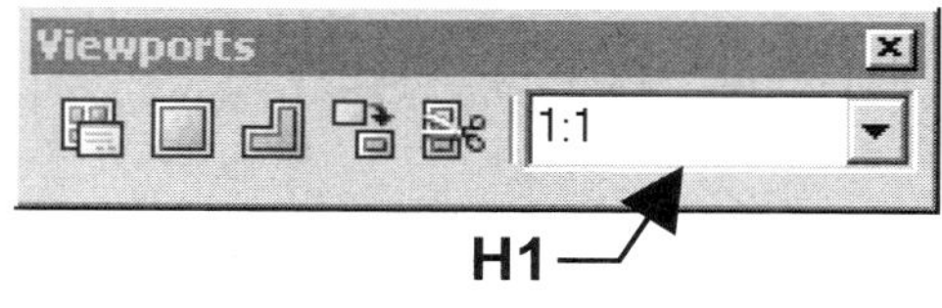

EXERCISE 4D

CREATE A PAGE SETUP FOR 24 X 18 SHEET

The following instructions will guide you through creating a **PAGE SETUP** named: **"24 X 18 ARCH".** You need this Page Setup to plot the **24 X18 (1 to 1)** layout.

A. Open **My Feet-Inches Setup** if it isn't already open.

B. Select the **24 X 18 (1 to 1)** layout tab.

You should be looking at your Border and Title Block now.

C. Select **File / Plot** or place the cursor on the **24 X 18 (1 to 1)** tab and press the right mouse button. Select **Plot** from the short cut menu.

The Plot dialog box shown below should appear. Select the "More Options" button if your dialog box does not appear the same as shown below.

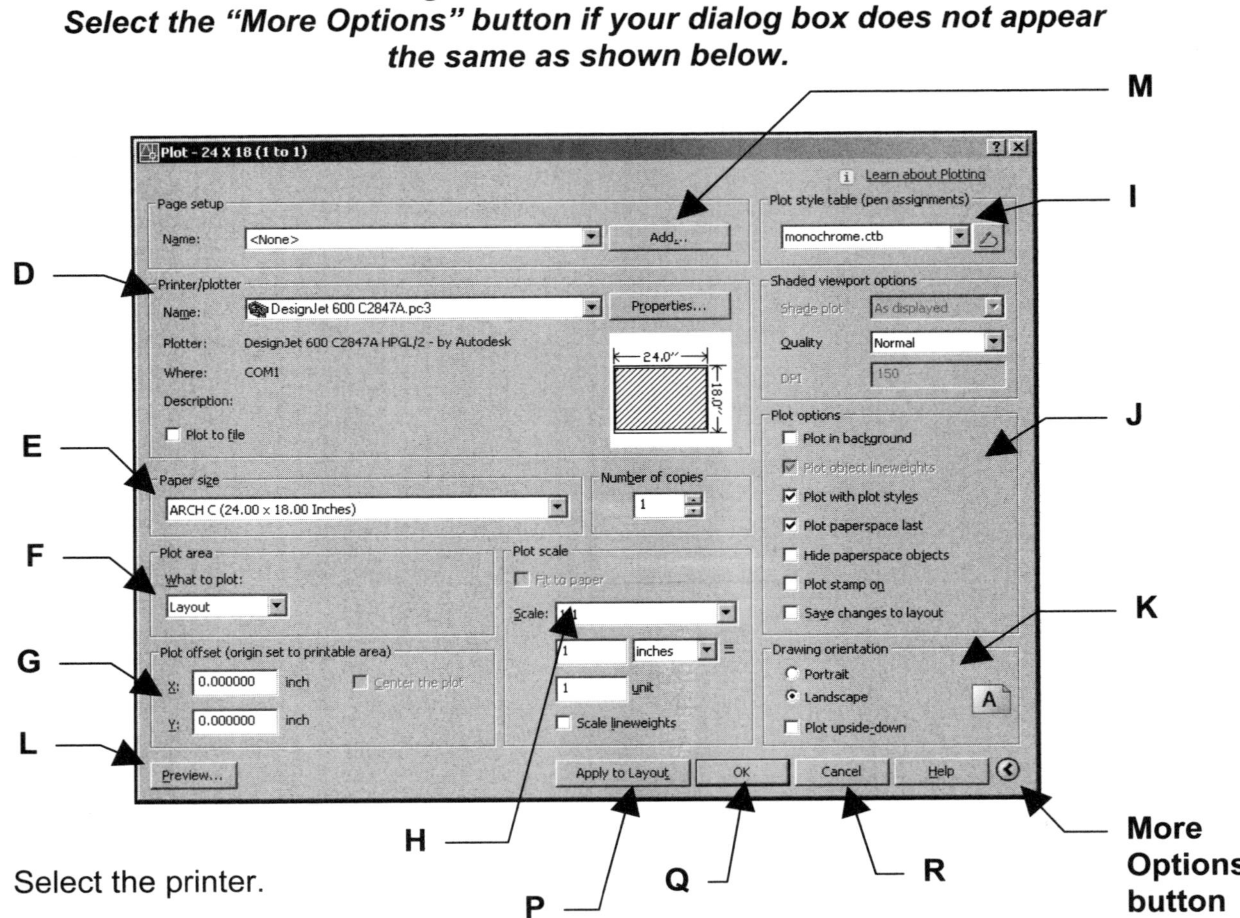

D. Select the printer.

E. Select the Paper Size.

F. Select the Plot Area.

G. Plot offset should be 0 for both X and Y.

H. Select scale 1 : 1.

I. Select the Plot Style Table "Monochrome.ctb".

J. Select the Plot options shown.

K. Select Drawing Orientation: Landscape.

L. Select **Preview** button.

1. If the drawing is centered on the sheet, press the Esc key and continue on to **M**.

2. If the drawing does not look correct, press the Esc key and check all your settings, then preview again.

M. Select the **ADD** button.

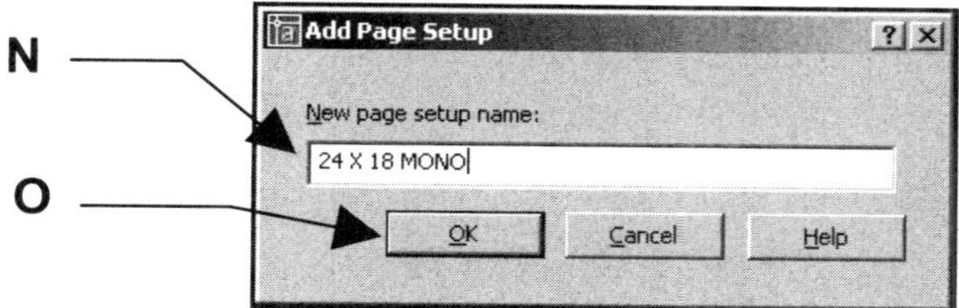

N. Type the new page setup name: **24 X 18 MONO**

O. Select OK button.

The settings are now saved for future use.

P. Select the **"Apply to Layout"** button.

Q. If your computer **is** connected to the plotter / printer selected, select the **OK** button to plot, then proceed to **S**.

R. If your computer is not connected to the plotter / printer selected, select the Cancel button to close the Plot dialog box and proceed to **S**. Note: Selecting Cancel will cancel your selected setting if you did not save the page setup as specified in **M**.

S. Save the drawing.
   1. Select **File / Save As**.
   2. Save as: **My Feet-Inches Setup.**

You have now completed the Page Setup "24 X 18 MONO" for layout tab 24 X 18(1to 1) within the master drawing " ***My Feet-inches Setup".*** *When you use this tab, use this page setup.*

EXERCISE 4E

CREATE A LAYOUT FOR 1/4" = 1'

In this exercise you are going to create a new layout by copying an existing layout. Then you will adjust the scale of the Model Space to 1/4" = 1' and lock the Viewport. This will be very easy.

A. Open **My Feet-Inches Setup** (If not already open).
B. Select the **24 X 18 (1 to 1)** tab.
C. Right Click on the **24 X 18 (1 to 1)** tab and select **Move or Copy..** from the menu.

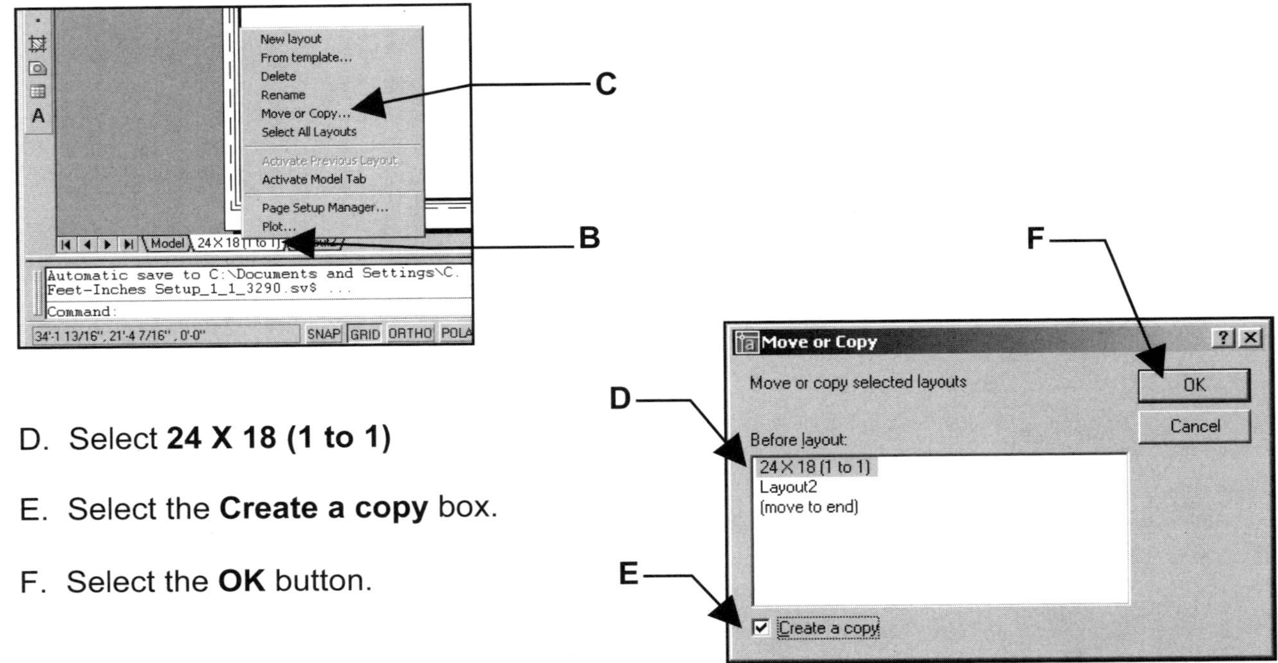

D. Select **24 X 18 (1 to 1)**

E. Select the **Create a copy** box.

F. Select the **OK** button.

You should now have a duplicate of the "24 X 18 (1 to 1)" layout tab.

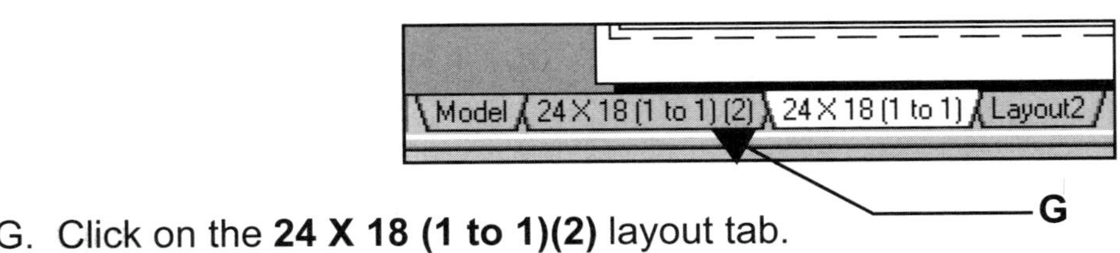

G. Click on the **24 X 18 (1 to 1)(2)** layout tab.

Note: This layout is identical to the 24 X 18 (1 to 1) tab in every way. It even includes the Page Setup that specifies what printer, paper size and plot style table to use. What a time saver, and wasn't that easy?

Now let's make some changes to this new layout so you can use it to plot drawings at a scale of 1/4" = 1'.

H. Unlock the Viewport.

I. Change to Model Space. **(Very Important)**

J. Change the "**Drawing Limits**" of Model Space.
   1. Lower Left corner = **0'- 0", 0'- 0"**
   2. Upper right corner = **96', 72'** (This is feet, not inches.)

K. Select **View / Zoom / All.**

L. Adjust the scale of Model space.
   1. Make sure you are in Model Space.
   2. Select 1/4" = 1' in the **VIEWPORT** toolbar.

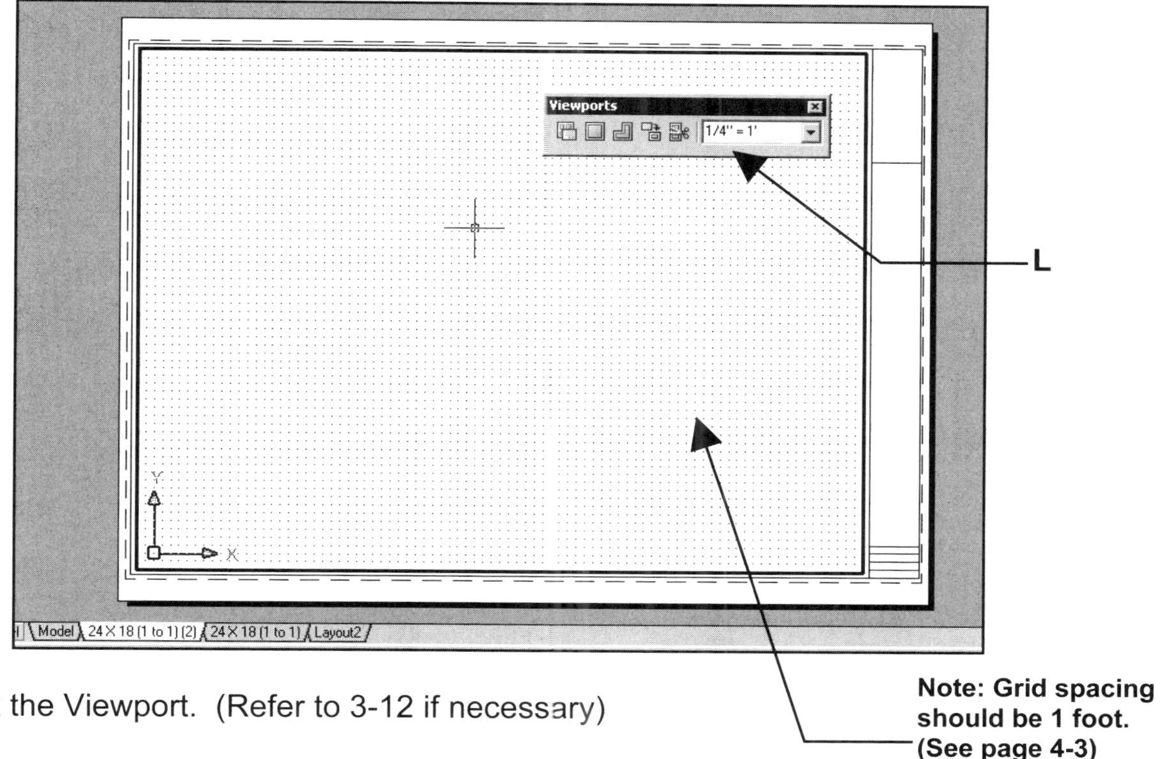

M. **Lock** the Viewport. (Refer to 3-12 if necessary)

N. Rename the Layout tab to: **24 X 18 Qtr-Ft.**

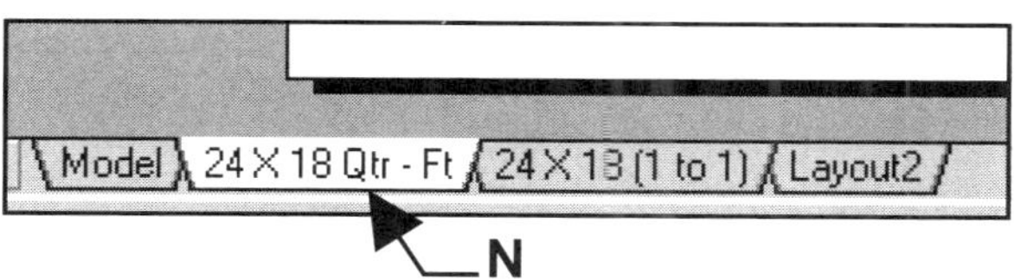

O. File / Save as: **My Feet-Inches Setup (again).**

Note: Use page set up **24 X 18 MONO** when plotting this layout.

You now have created a layout to be used when you have a drawing that requires plotting at a scale of 1/4" = 1'. A big drawing, such as a house.

Be patient, it will all be more clear at the end of this Lesson.

EXERCISE 4F

CREATE A LAYOUT FOR 1/8" = 1'

In this exercise you are going to create another new layout by copying an existing layout. Then you will adjust the scale of the Model Space to 1/8" = 1' and lock the Viewport. Again, this will be very easy.

A. Open **My Feet-Inches Setup** (If not already open).
B. Select the **24 X 18 (1 to 1)** tab.
C. Right Click on the **24 X 18 (1 to 1)** tab and select **Move or Copy..** from the menu.

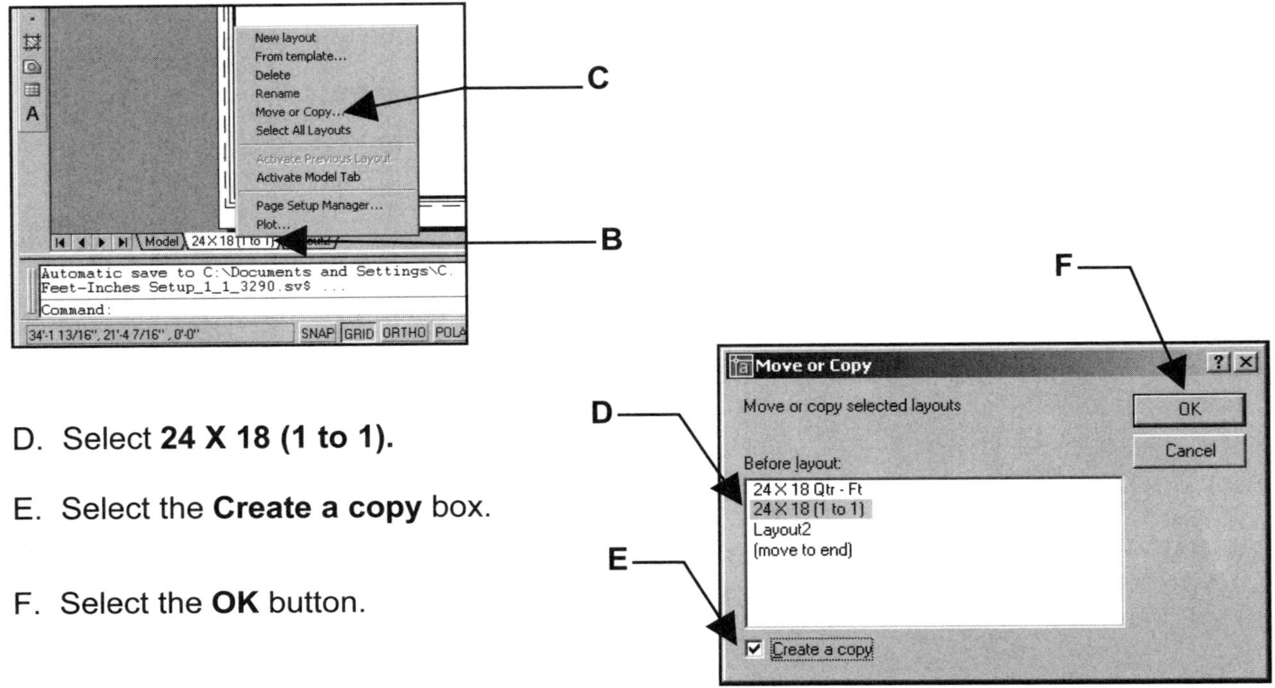

D. Select **24 X 18 (1 to 1).**

E. Select the **Create a copy** box.

F. Select the **OK** button.

You should now have a duplicate of the "24 X 18 (1 to 1)" layout tab.

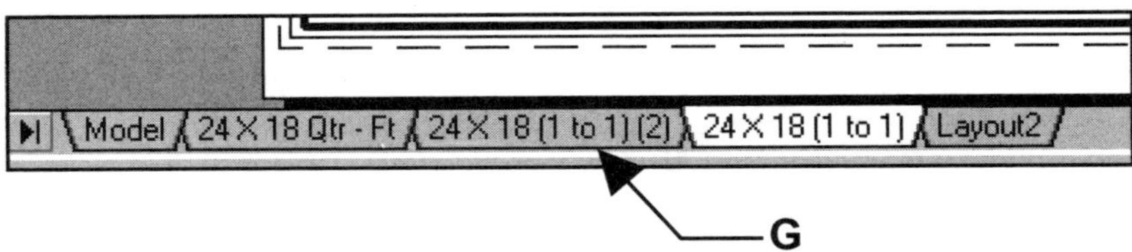

G. Click on the **24 X 18 (1 to 1)(2)** layout tab.

Now let's make some changes to this new layout so you can use it to plot drawings at a scale of 1/8" = 1'.

H. Unlock the Viewport.

I. Change to Model Space. **(Very important)**

J. Change the "**Drawing Limits**" of Model Space.
  1. Lower Left corner = **0'- 0", 0'- 0"**
  2. Upper right corner = **192', 144'** (This is feet, not inches.)

K. Select **View / Zoom / All**

L. Adjust the scale of Model space.
  1. Make sure you are in Model Space.
  2. Select 1/8" = 1' in the **VIEWPORT** toolbar.

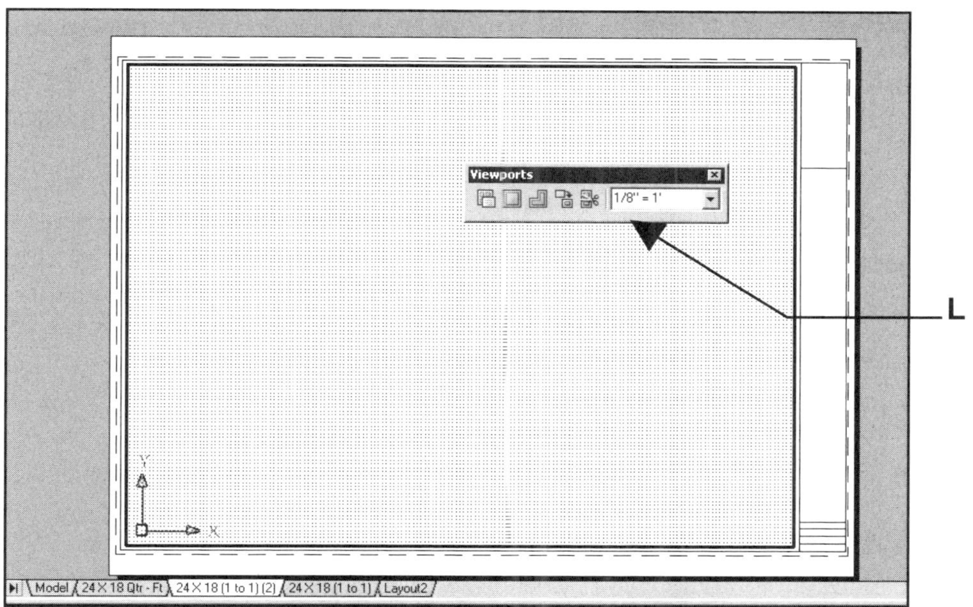

M. **Lock** the Viewport. (Refer to 3-12 if necessary.)

N. Rename the Layout tab to: **24 X 18 Eighth-Ft.**

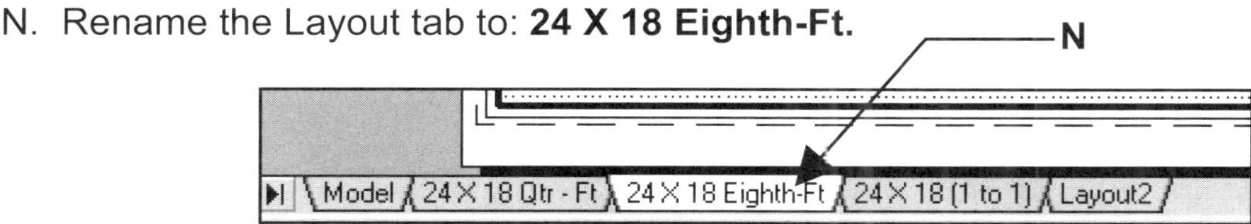

O. File / Save as: **My Feet-Inches Setup (again).**

Note: Use page set up **24 X 18 MONO** when plotting this layout.

You now have created another layout to be used when you have a drawing that requires plotting at a scale of 1/8" = 1'. A much bigger drawing.

You will understand more when you finish this Lesson.

EXERCISE 4G

MOVE & DELETE LAYOUT TABS

In an effort to keep the drawing organized you now need to move the layouts tabs into a more practical order. The logical order would be to start with the smallest on the left and progress larger to the right.

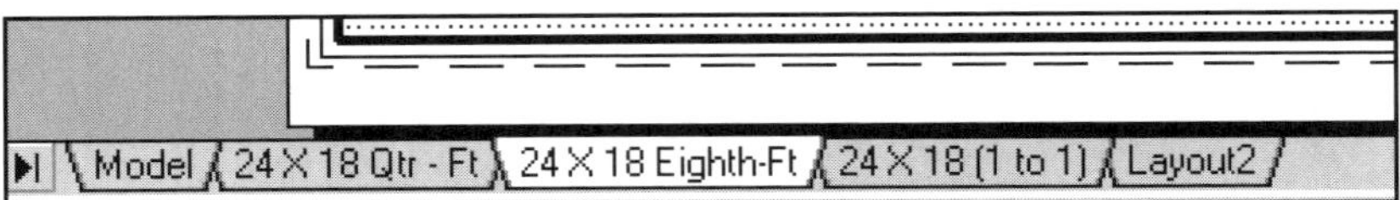

1. Select the **24 X 18 Qtr-Ft** layout tab.
2. Right Click on this layout tab and select **Move or Copy…** from the menu.
3. Select (**Move to the End).**

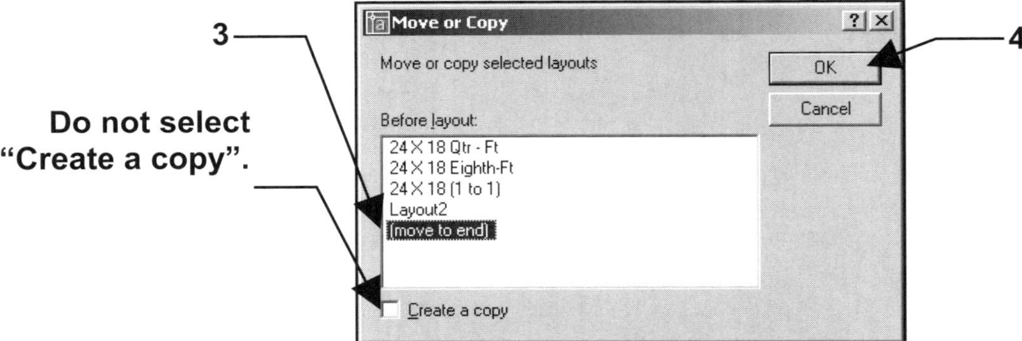

4. Select the **OK** button.

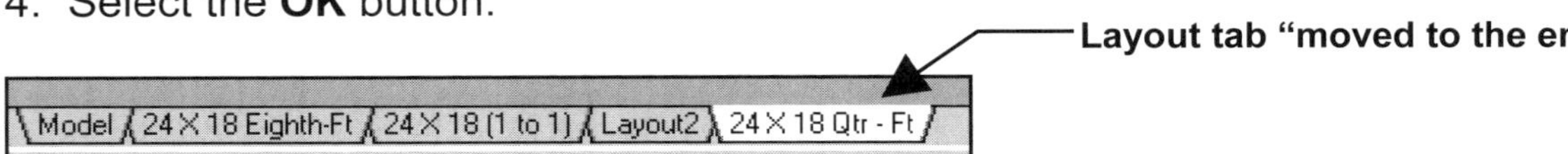

5. Now do the same to the **24 X Eighth-Ft** layout tab.

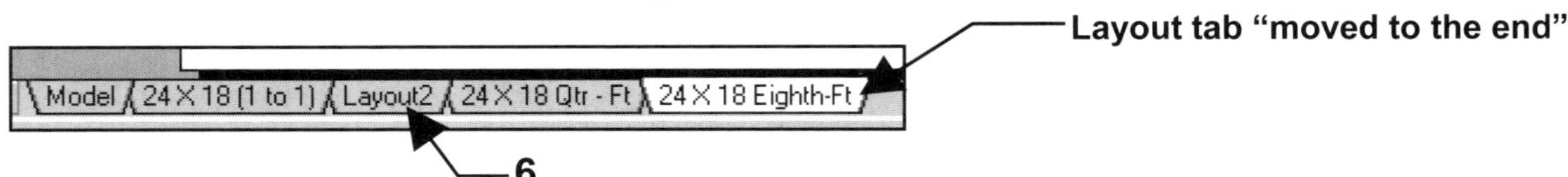

6. Select the **Layout2** tab. (If the Page Setup Manager appears select **Close.**)

7. Right Click and select **Delete** from the menu.

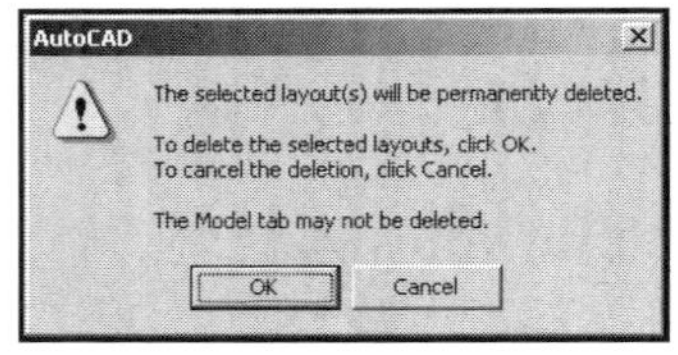

A warning will appear.
Select OK.
You can always add a Layout tab if you need one.

Your layout tabs should now be organized in a logical order as shown below.

Model | 24 X 18 (1 to 1) | 24 X 18 Qtr - Ft | 24 X 18 Eighth-Ft

8. Save again: File / Save as: **My Feet-Inches Setup.**

EXERCISE 4H

EXPERIMENTING WITH THE LAYOUTS

After you have finished this Exercise you should be able to determine the difference between the 3 layouts and why they are useful.

1. Open **My Feet-Inches Setup**. (If it isn't already open.)

2. Select the "**24 X 18 (1 to 1)**" tab.

3. Make sure you are in Model Space. Using Layer Object, draw a **Rectangle** just slightly smaller than the viewport frame.

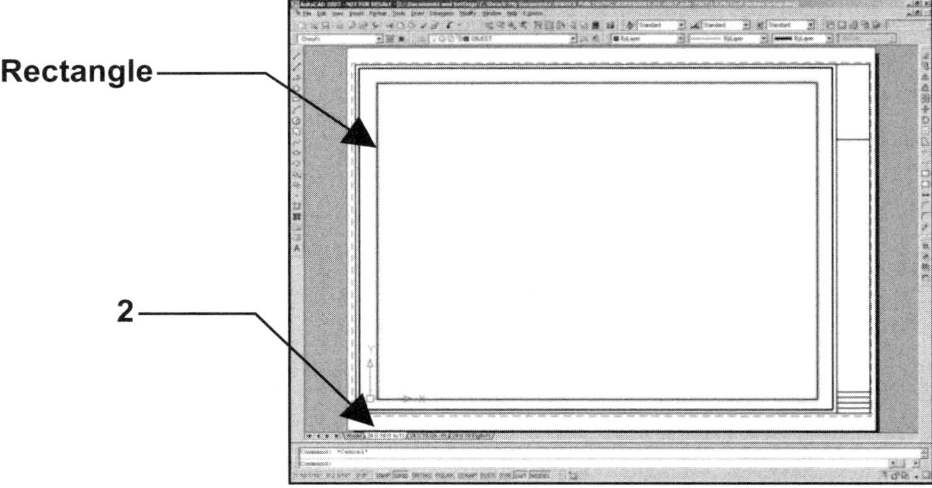

4. Select the "**24 X 18 Qtr-Ft**" tab.

5. Make sure you are in Model Space. Using Layer Object, draw another **Rectangle** just slightly smaller than the viewport frame.

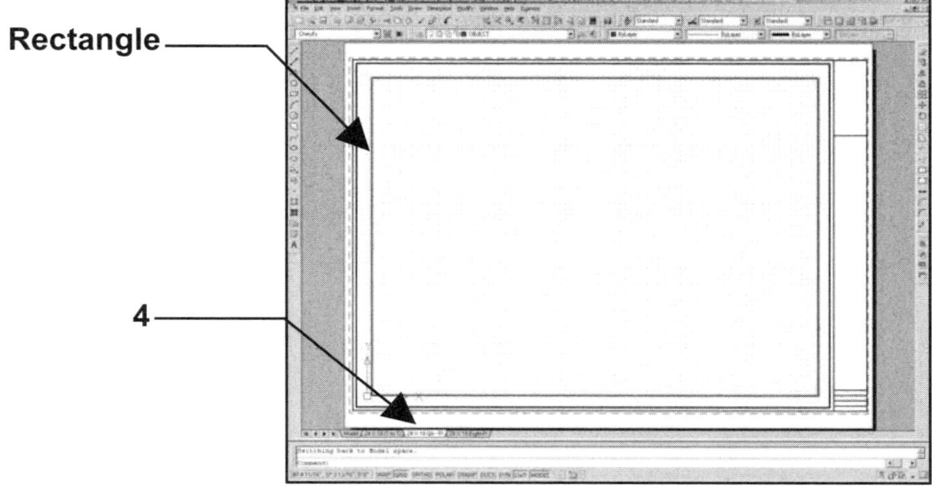

6. Select the "**24 X 18 Eighth-Ft**" tab.

7. Make sure you are in Model Space. Using Layer Object, draw another **Rectangle** just slightly smaller than the viewport frame.

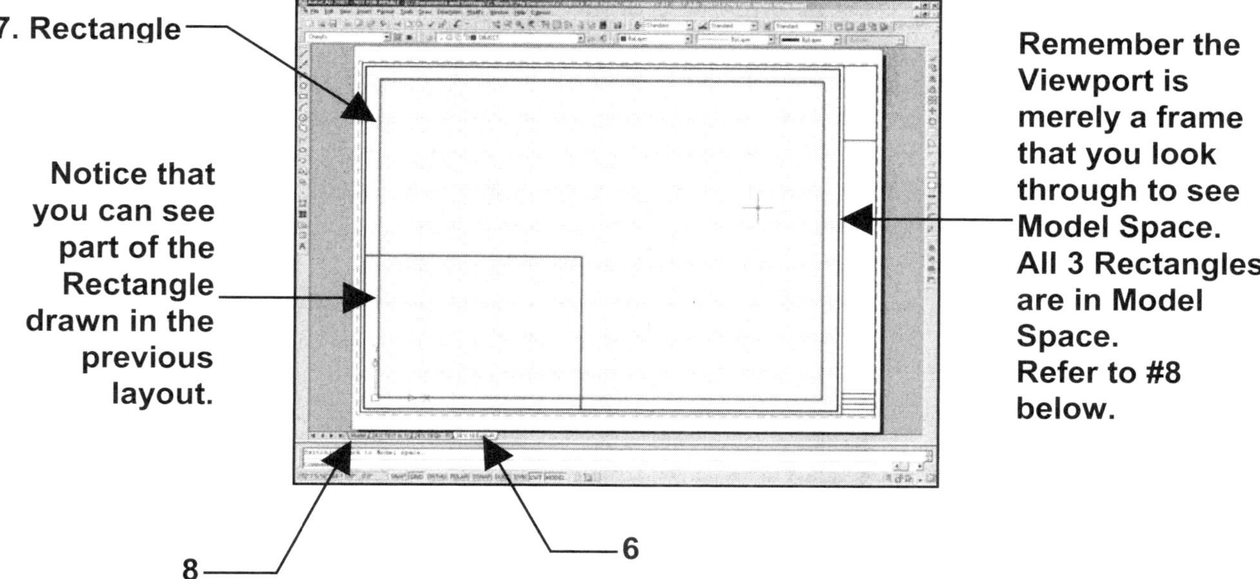

8. Now select the **"Model"** tab.

9. Select **View / Zoom / All.**

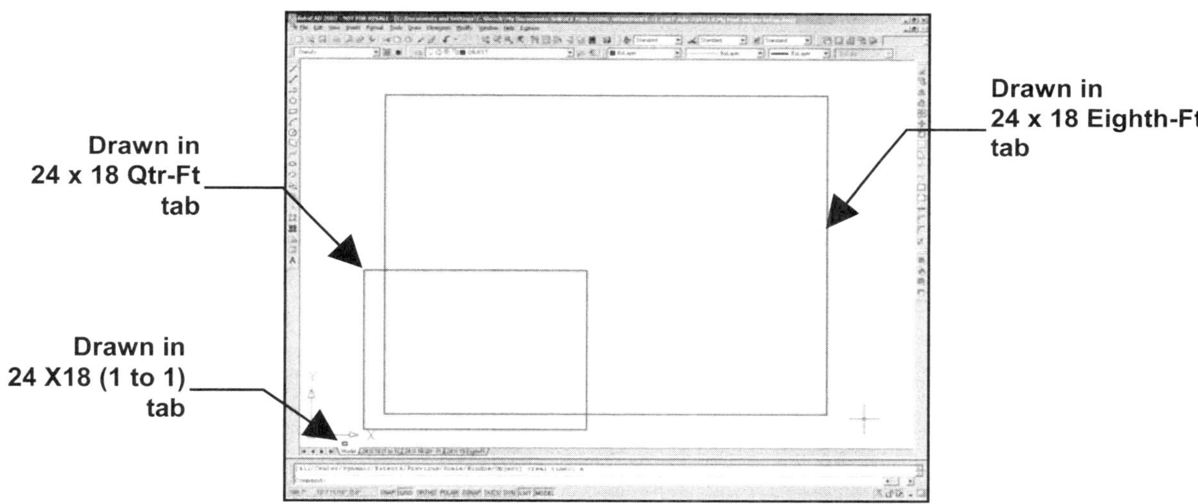

Each layout allows you to see more of the model space area.
If you were drawing a floor plan you would probably start with the **Qtr-ft** layout. If the drawing started getting too large for that drawing area simply select the **Eighth-Ft** tab and you will have a larger drawing area.
But all three layouts will be printed on a 24 X 18 sheet.

Think about it. This may seem a little confusing right now but it will become clearer as you use the layouts in the coming exercises.

Summary

Let's review what you achieved in this lesson.

4A. Select Preferences to draw with.

1. Specified Units, Drawing Limits, Snap and Grid.
2. Created New Layers, Text Styles and a Dimension Style.

4B. Select Paper Size 24 X 18 to plot on and drew a Border and Title Block.

4C. Created a Viewport, Adjusted the Scale of Model Space and Locked the Viewport.

4D. Set up Plotting specs (Page Setup) and saved them. (24 X 18 All Black)
This Page Setup can be used to plot any Layout tab in this Master Setup drawing, as long as it is plotted on 24 X 18 sheet size.

4E/F. Create new Layouts.
It is good drawing management to setup Layouts already prepared to use when plotting with different scales.
4C created a layout for plotting model space at a scale of 1 : 1.
4E created a layout for plotting model space at a scale of 1/4" = 1'
4F created a layout for plotting model space at a scale of 1/8" = 1'

4G. Manipulate the Layout tabs.
You learned how to move or delete layout tabs to organize your setups.

<u>Important to understand</u>

Layouts are merely a process of preparing your drawing for printing. You may have up to 16 Layouts. In this lesson you have been shown how to create a master setup drawing to use when you draw in Architectural units. The Layouts already have the scale adjusted for plotting 1 : 1, 1/4" = 1' or 1/8" = 1'.

Notice that you only need one page setup (24 X 18 MONO) to plot any of the 3 Layouts. But in lesson 3 you needed 3 different page setups. Why? Because you need a page setup for each <u>sheet size</u>. In lesson 3, we are using <u>3 different sheet sizes</u>. 17 X 11, 24 X 18 and 24 X 36. In this lesson, we are only using one sheet size, 18 X 24.

LEARNING OBJECTIVES

After completing this lesson, you will be able to:

AutoCAD 2007 Only

1. Create a Multiline Style.
2. Draw a multiline.
3. Edit a Multiline.
4. Use Freeze and Thaw to create multiple drawings.

LT users only

5. Double Line
6. Use Freeze and Thaw to create multiple drawings.

LESSON 5

MULTILINE (Not available in LT, refer to page 5-13)

The **MULTILINE** is a set of parallel lines called elements. The set can have up to 16 parallel lines.

These multilines can basically be drawn in the same manner as a line, with first and second endpoints. You can even use the CLOSE option.

The drawn multiline set is one object. Sort of like Polylines or Blocks.

The walls on a Floor plan would be an example of an Architectural application.
The belts on a conveyor would be an example of a Mechanical application.
Multiple circuit lines on a schematic would be an example of an Electrical application.
Roads and Intersection would be an example of a Civil application.

When you draw a multiline, you can use the **STANDARD** style, which has two lines, or create your own. Instructions for "Creating a Multiline Style" are on page 5-4.

First we will discuss how to draw with the MULTILINE command using the AutoCAD default Multiline Style "Standard".

1. Select the **MULTILINE** command using one of the following commands:

    TYPE = ML
    PULLDOWN = DRAW / MULTILINE
    TOOLBAR = NONE

    Command: _mline
    Current settings: Justification = Top, Scale = 1.00, Style = STANDARD

2. Specify start point or [Justification/Scale/STyle]: ***select the start point (P1)***
3. Specify next point: ***select the next point (P2)***
4. Specify next point or [Undo]: ***select the next point (P3)***
5. Specify next point or [Close/Undo]: ***select the next point (P4)***
6. Specify next point or [Close/Undo]: ***Close <enter>***

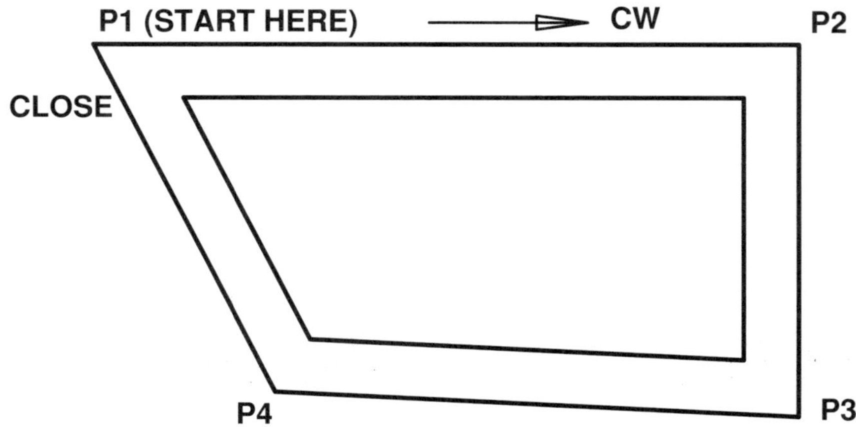

JUSTIFICATION

Because there are multiple lines, you must specify the justification of the multiline. Basically, this means: *What is your cursor hanging on to as you draw the multiline*? Is it the top, middle or bottom of the set, while you specify the endpoints. Your choices are **Top, Zero and Bottom.** These justifications are defined for drawing the multiline in a **Clockwise** direction.

1. Select the Multiline command
2. Select Justification option
3. Select Top, Zero or Bottom

Examples of Justification (drawn clockwise):

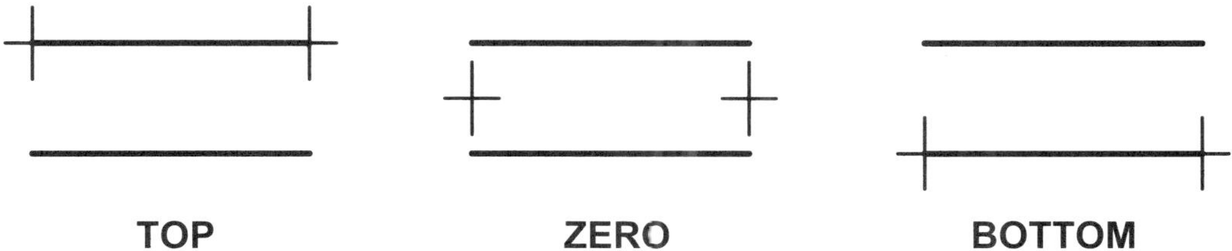

TOP **ZERO** **BOTTOM**

SCALE

The SCALE option is a **scale factor** of the width. It will increase or decrease the original specified width by multipling the original width by the specified scale.

1. Select the Multiline command
2. Select the Scale option
3. Enter the multiline scale

STYLE

The STYLE option allows you to select a previously created multiline style from the command line rather than having to select the FORMAT / MULTILINE pulldown menu.

1. Select the Multiline command
2. Select the Style option
3. Enter the Multiline Style name.

CREATING A MULTILINE STYLE

The following instructions will guide you through creating a new multiline style.

1. Select the Multiline Style command using one of the following commands:

    TYPE = mlstyle
    PULLDOWN = FORMAT / MULTILINE STYLE
    TOOLBAR = NONE

The following dialog box will appear.

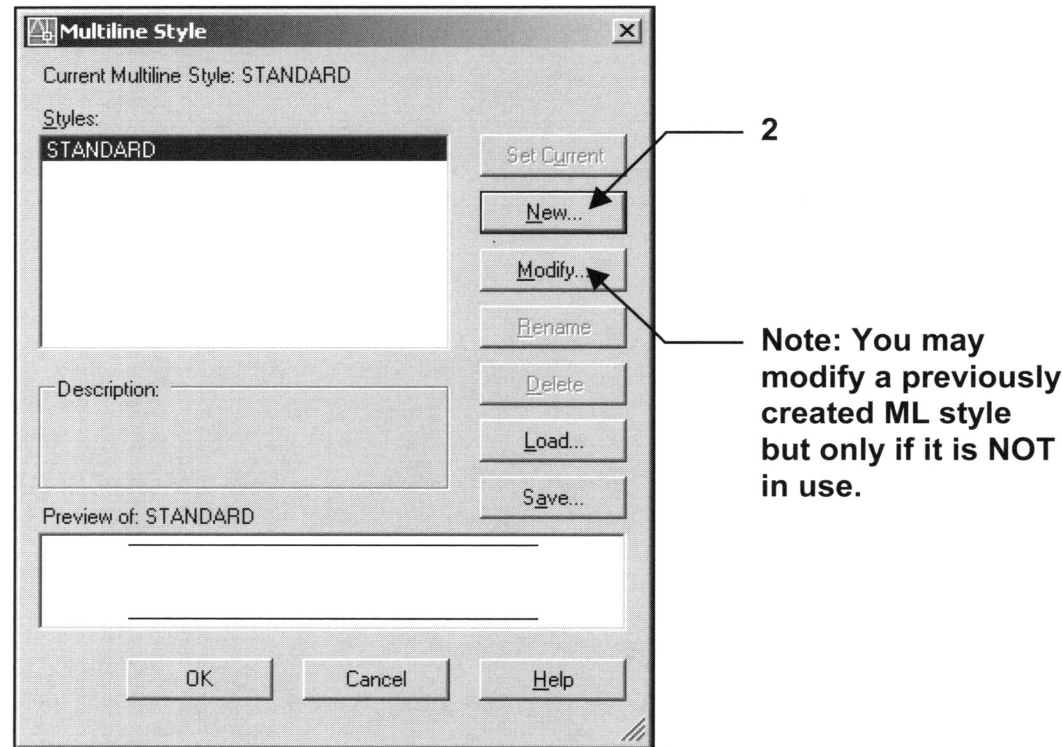

2. Select the **New** button.

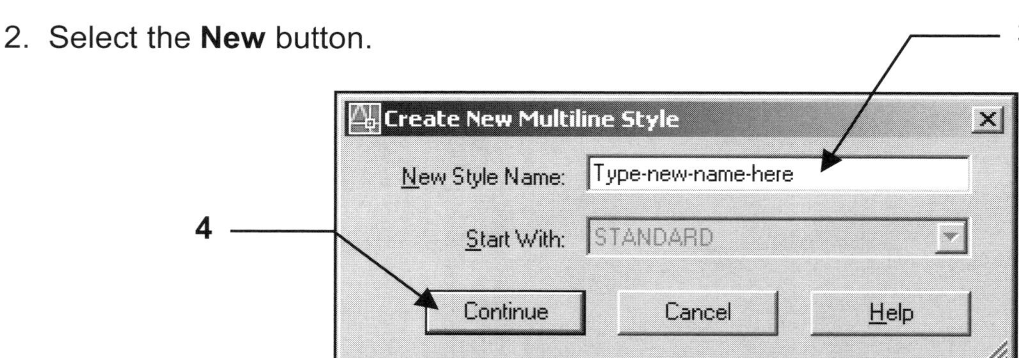

3. Type the new multiline style name. (Blank spaces are not allowed)

4. Select the **Continue** button.

The following dialog box will appear.

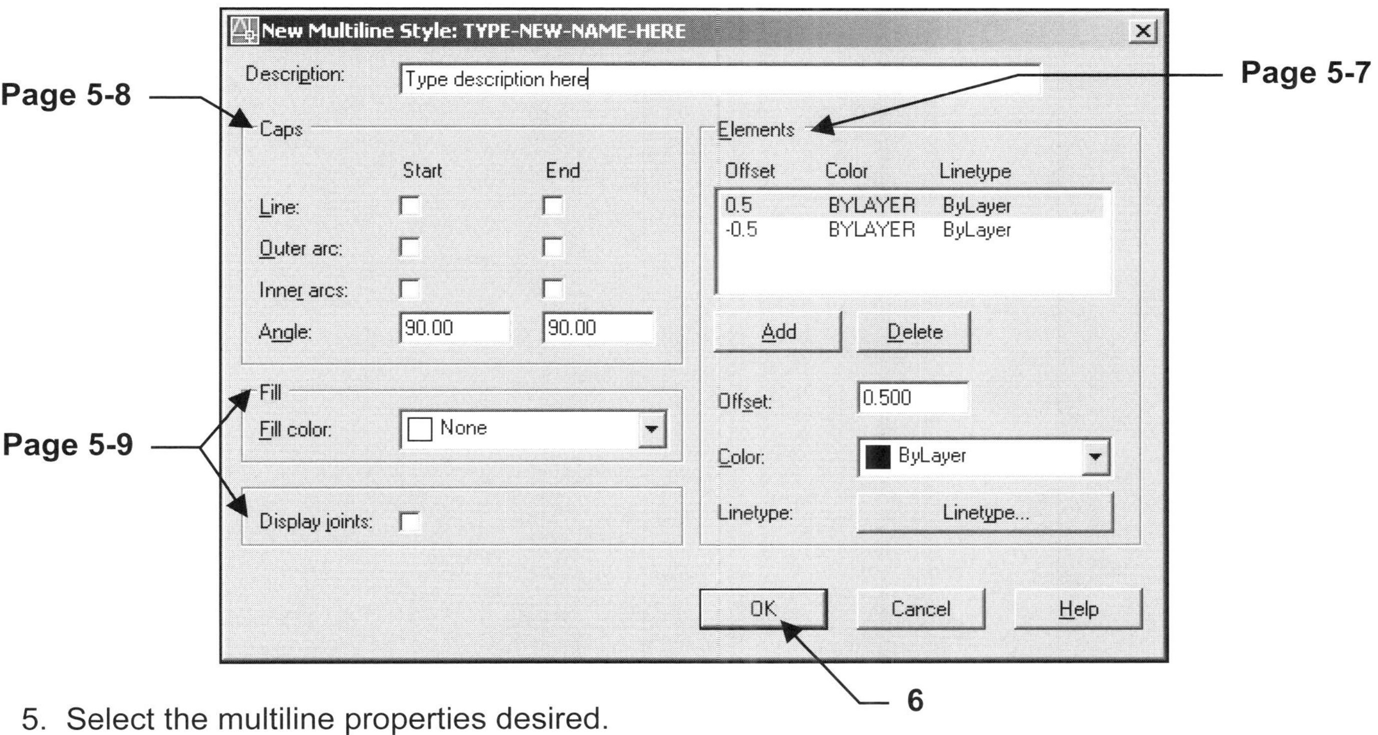

5. Select the multiline properties desired.
   (A detailed description of each follows on page 5-7 through 5-9.)

6. Select the **OK** button.

The following dialog box will appear.

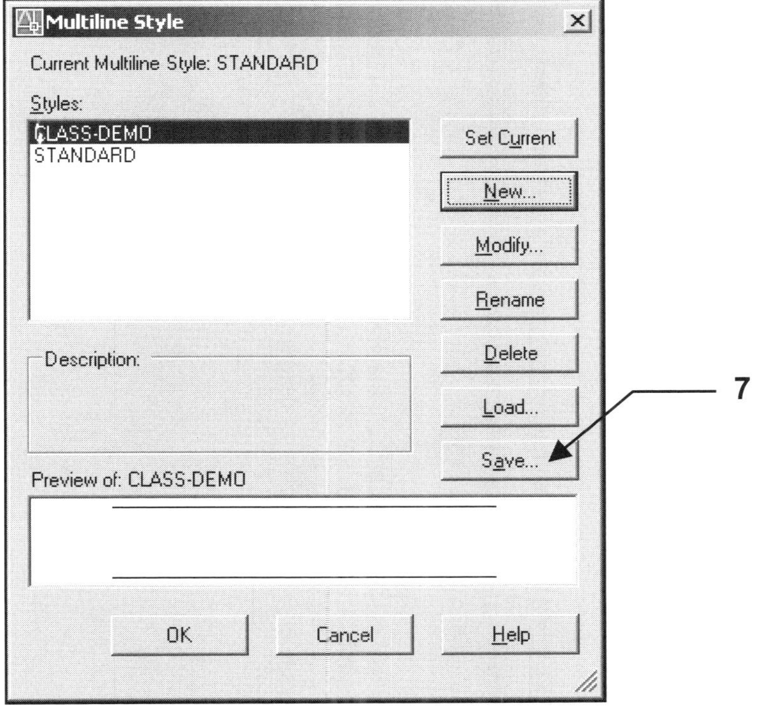

7. Select the **Save** button.

The following dialog box will appear.

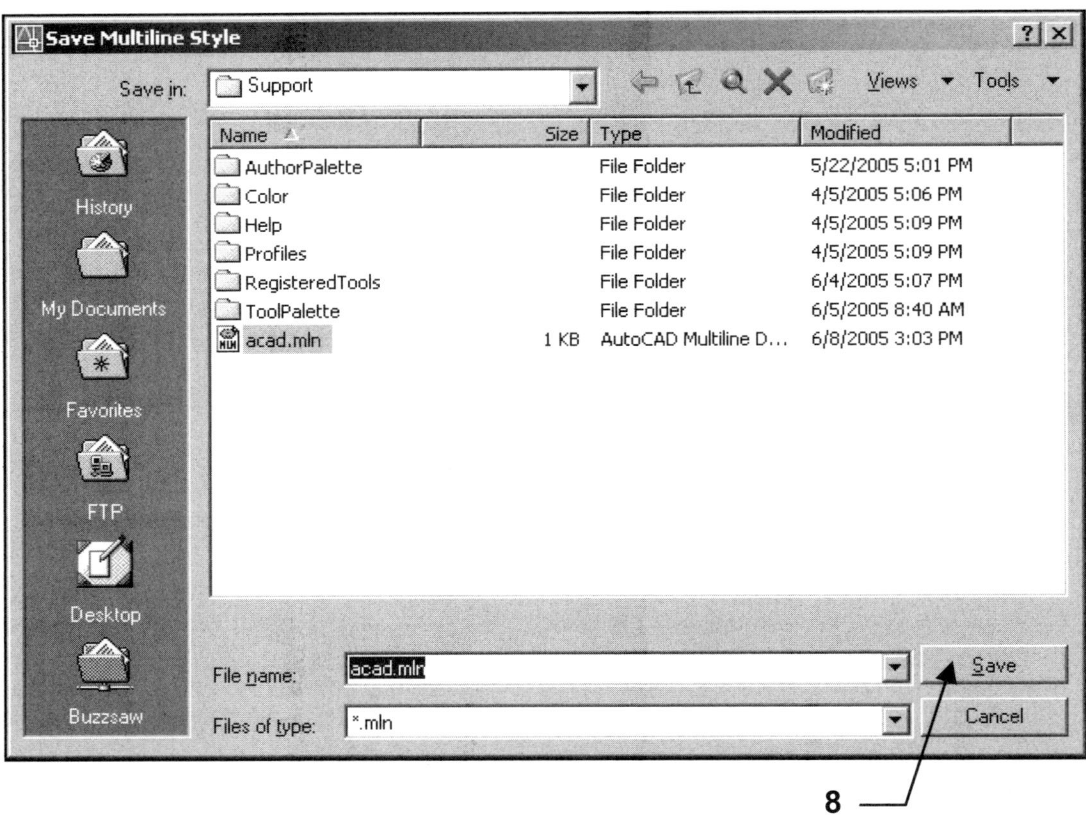

8. Select the **Save** button.
   (Note: You may change the "Save in" location, such as a removable disk, but <u>do not change the file name, **acad.mln**</u>. Many multiline styles can be saved within the **acad.mln** file. (You will understand better after you read the instructions for <u>Loading multilines on page 5-10</u>.)

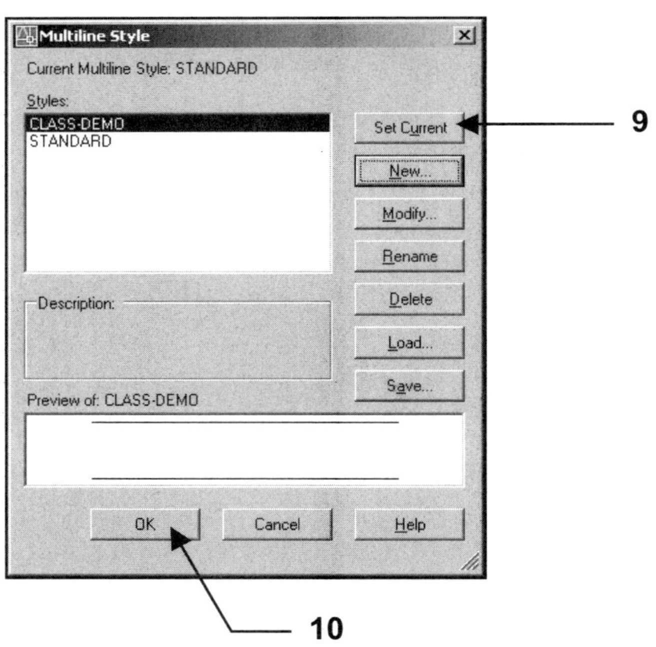

9. Select **Set Current**.

10. Select the **OK** button.

MULTILINE PROPERTIES

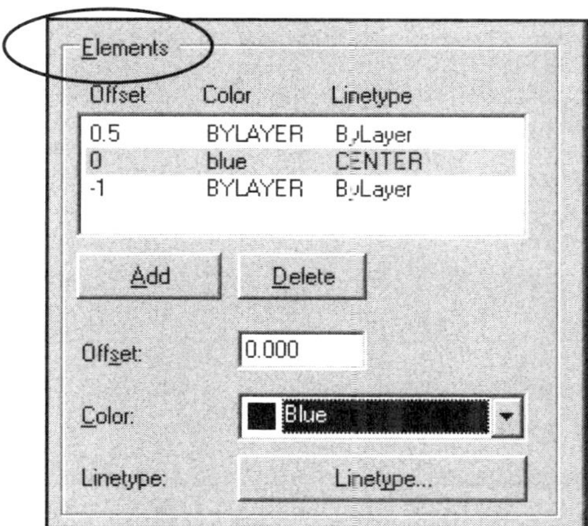

ELEMENTS:

Each individual line within the multiline set has its own **Offset, Color and Linetype.** These are called line elements.

The **offset** is the distance from the middle of the multiline set.
A positive offset is above the middle. A negative offset is below the middle. "0" offset is on the middle. The default offset is .50 above the middle and .50 below the middle.
The positive and negative offsets do not have to be equal. You may have an offset of positive .50 and negative 1.00.

Example: The offset values shown above would appear like this:

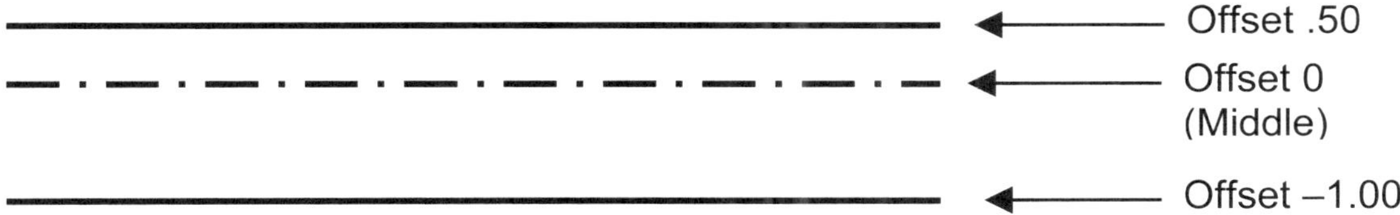

Notice that you can also define the color and linetype for each line.

To define a different offset:
1. Highlight one of the existing lines.
2. Change the offset in the offset box.

To delete a line:
1. Highlight one of the existing lines.
2. Select the Delete button.

To add a line:
1. Select the Add button. (Note: An line will be added as "0" offset, Bylayer Bylayer)

To define a color or linetype:
1. Highlight one of the existing lines.
2. Select the color and linetype.

MULTILINE PROPERTIES continued…

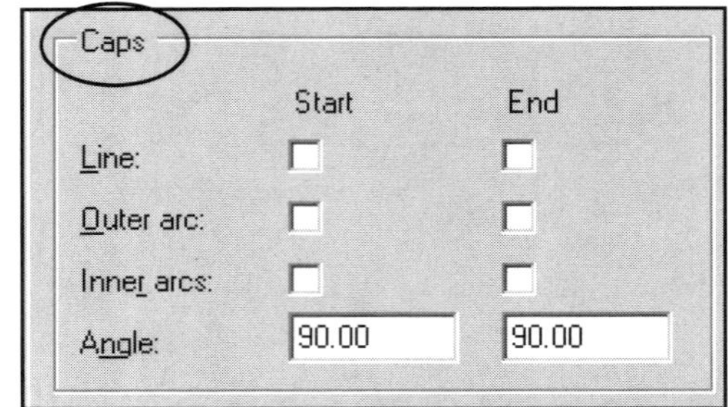

CAPS

Controls the appearance of the Start and End caps of the multiline.
You may

Examples shown below.

LINE
A line will appear only at the Start and the End of the Multiline.

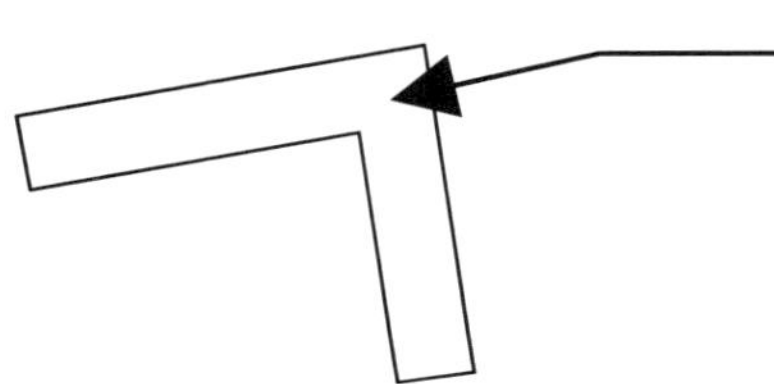

No Line here unless you select "Display Joints". See page 5-9.

OUTER ARC
An Arc will automatically be calculated and drawn to the 2 outer most lines.

INNER ARC
An Arc will automatically be caluculated and drawn to the 2 inner most lines.

ANGLE
The Start and End may have an angle between 10 to 170 degrees. A line will appear if you have selected LINE (above) for Start or End.

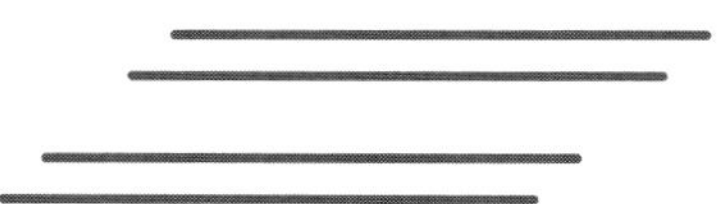

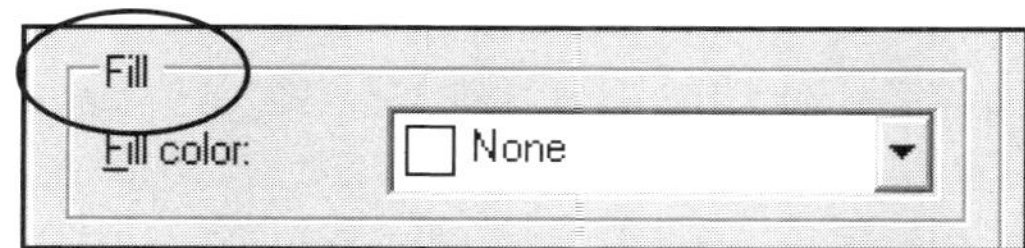

FILL

Sets the background fill color of the multiline.
Select one of the basic colors from the drop down list or "Select Color" to display the Select Color dialog box.

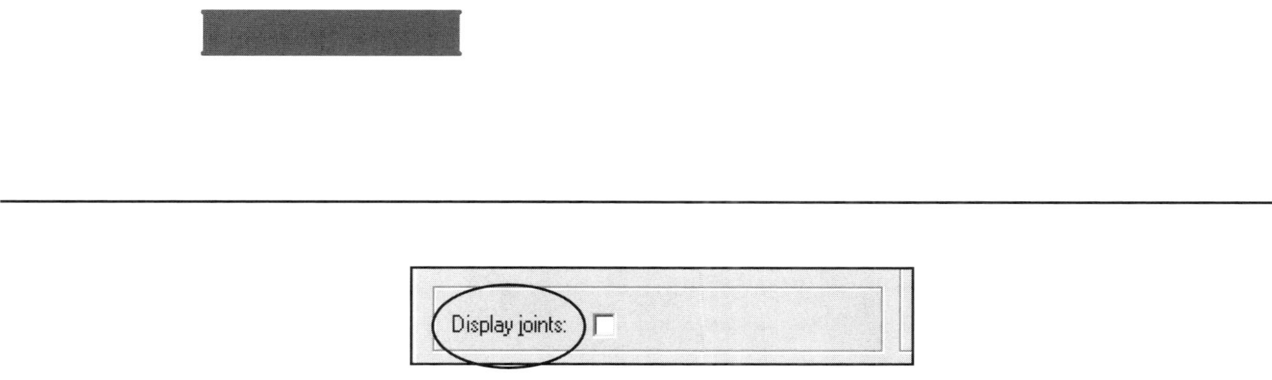

DISPLAY JOINTS

Controls the display of the joints at the vertices of each multiline segment. A joint is also known as a miter. Check the box to see joints.

DESCRIPTION

Descriptions are optional and can be up to 255 characters, including spaces

LOADING A MULTILINE STYLE

The following instructions will guide you through how to LOAD an existing multiline style from one drawing to another.

1. Select the Multiline Style command using one of the following commands:

   TYPE = mlstyle
   PULLDOWN = FORMAT / MULTILINE STYLE
   TOOLBAR = NONE

   The following dialog box will appear.

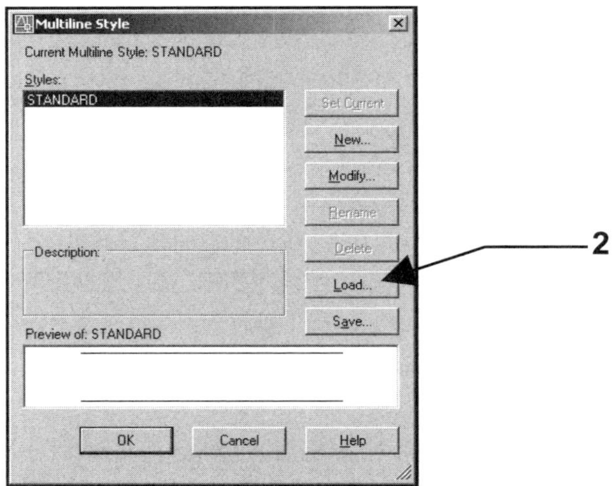

2. Select the **LOAD** button.

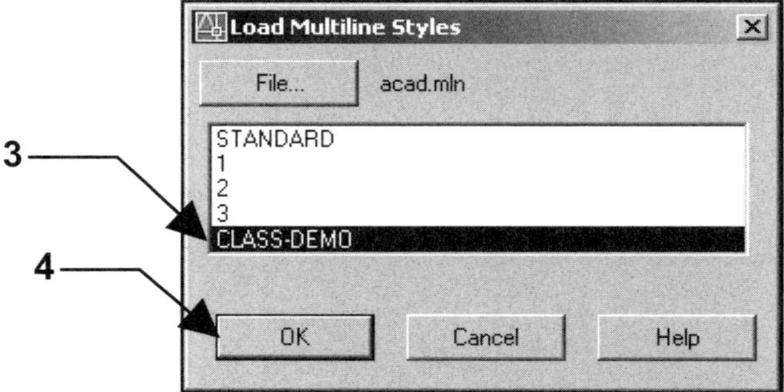

3. Select the Multiline Style from the list.
   (Note: If the Multiline Style is not listed select the **FILE** button. Locate the Multiline Style that you want to Load and select OPEN.)

4. Select the **OK** button.

NOTE: *If you insert a drawing file and it has a multiline style with the same name as an existing multiline file in the existing drawing on the screen, AutoCAD will change the incoming multiline style name by adding a "1" to the name. Example: Walls will become Walls1.*

EDITING MULTILINES

<u>There are 3 ways to edit a multiline.</u>
1. Multilines Edit Tools. (shown below)
2. Most common editing commands. (page 5-12)
3. Explode and treat as normal lines.

1. Multiline Edit Tools

If you have two multilines in a drawing, you can control the way they intersect.
Multilines can intersect in a <u>cross</u> or a <u>Tee</u> shape.
The shapes can be closed, open, or merged.
You may also add or delete a Vertex and "Cut" an opening and "Weld" it back together.

To open the **Multilines Edit Tools** dialog box, double click on a multiline.

The following dialog will appear.

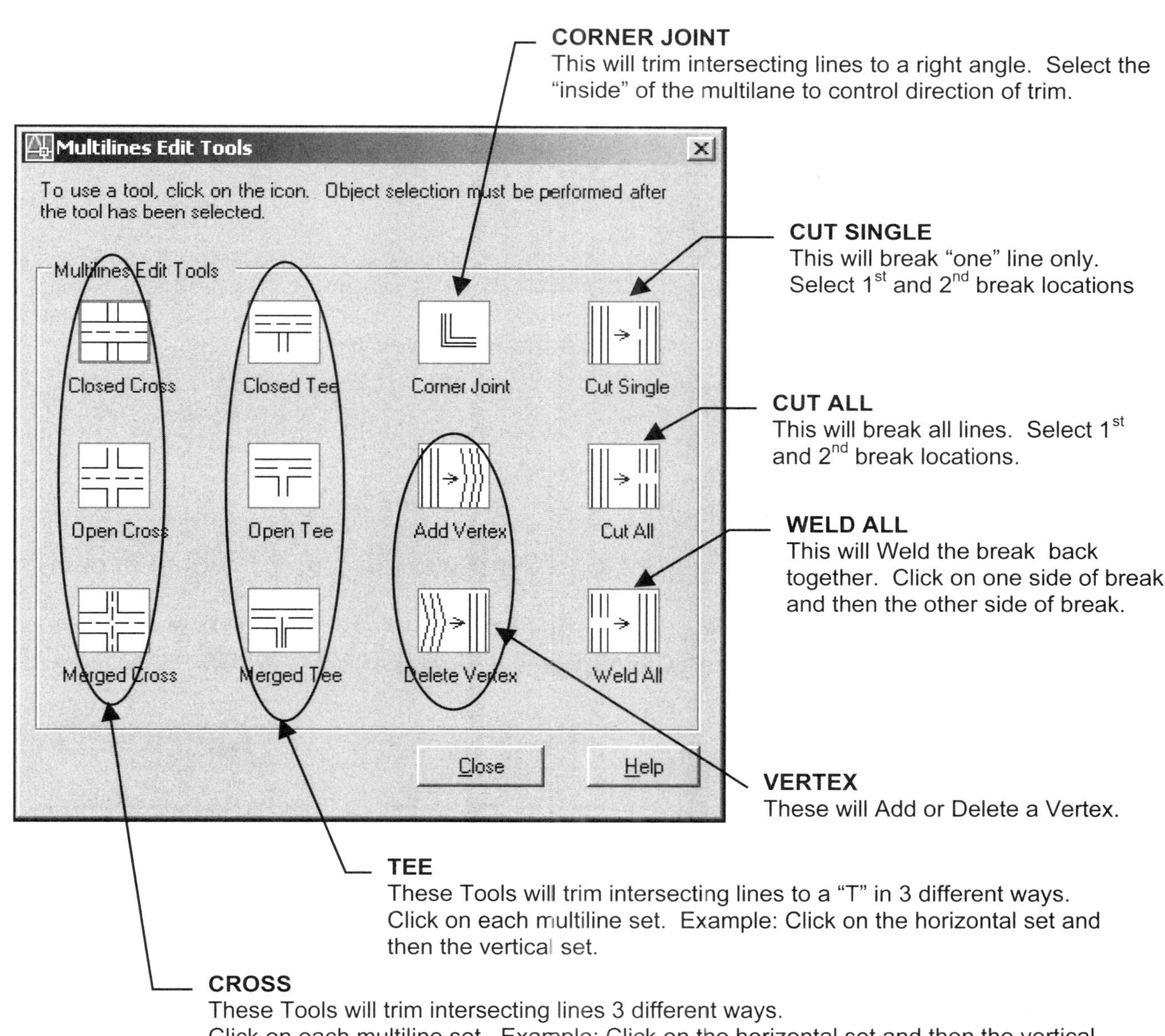

2. Most Common Editing Commands

You can use most of the common editing commands on multilines except, Break, Chamfer, Fillet, Lengthen and Offset.

But you have to read the command line carefully.

Example:
Use the “trim” command instead of the “multilines edit tools”.

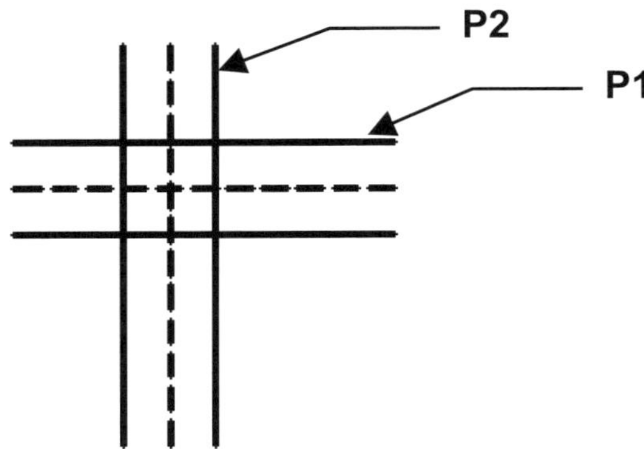

1. Select the Trim command

2. Command:
   TRIM
   Current settings: Projection=UCS, Edge=None
   Select cutting edges ...
   Select objects or <select all>: ***select the cutting edges (P1)***

3. Select objects: ***<enter>***

4. Select object to trim or shift-select to extend or
   [Fence/Crossing/Project/Edge/eRase/Undo]: ***select the multiline to trim (P2)***

5. Enter mline junction option [Closed/Open/Merged] <Merged>: ***select one of the options***

6. Select object to trim or shift-select to extend or
   [Fence/Crossing/Project/Edge/eRase/Undo]: ***<enter>***

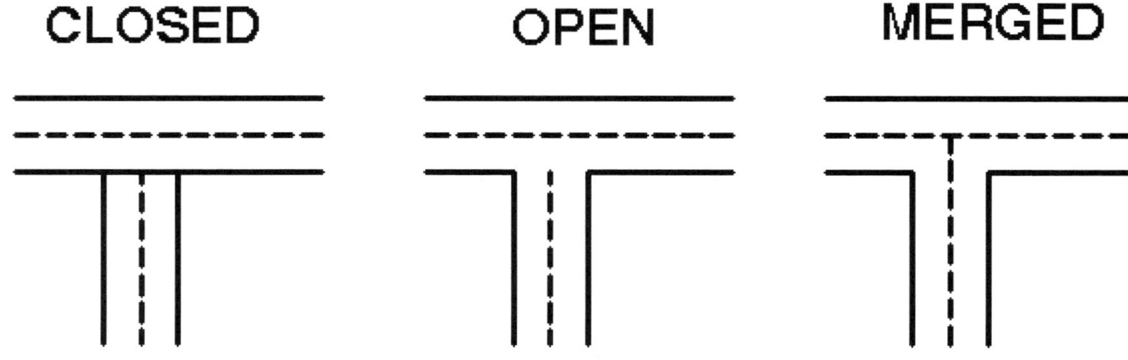

DOUBLE LINES (LT users only)

The Double Line command allows you to draw two parrallel lines at the same time. This is a very easy and useful command. You specify the distance between the two lines (width), specify whether the two lines are open or closed at the ends (Caps) and a few other options explained below.

Note: Even though you draw both lines at the same time, each line is a separate entity and can be edited individually.

Walls on a floorplan would be an example of an architectural application.
The thickness of a metal plate would be an example of a mechanical application.

1. Select the Double Line command using one of the following:
   TYPE = DL
   PULLDOWN = DRAW / DOUBLE LINE
   TOOLBAR = None

   Command: _dl
   DLINE
2. Specify start point or [Break/Caps/Dragline/Snap/Width]: ***place the first endpoint or select an option***
3. Specify next point or [Arc/Break/CAps/CLose/Dragline/Snap/Undo/Width]: ***place the second endpt***
4. Specify next point or [Arc/Break/CAps/CLose/Dragline/Snap/Undo/Width]: ***place the next endpt or <enter> to stop.***

OPTIONS

BREAK Creates a gap where a double line starts or ends. You can turn this ON or OFF. The default is ON.

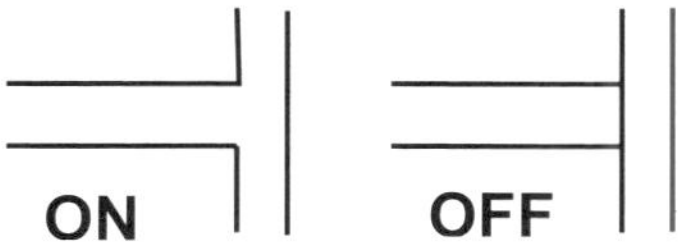

ON **OFF**

CAPS To set the ends of the double lines to be open or closed. The options are: **Both** – the start and the end, **Endpoint** – the end only, **Start point** – the first endpoint only, **None** – both ends open, **Auto** – close all ends that are not joined with another double line.

Both **End** **Start** **None**

DRAGLINE Basically this means "what is your cursor hanging on to?" Because there are two lines, you have to decide which line the cursor is attached to, or do you want the cursor in the center of the two lines. Your choices:

Right **Center** **Left**

OPTIONS CONTINUED

SNAP Automatically snaps the new double line to an object close to the new endpoint of the double line. The default is ON. If you have Break and Snap ON, the new double line will automatically attach the new double line to the object and break a gap in the object between the double lines. Also you have the option of determining the **size** of the snapping area around the crosshairs. This area is called the ***pixel search area***.

UNDO You undo the last double line segment. Must be done before you close or end the command.

WIDTH The distance between the two double lines.

DRAWING DOUBLE ARCS

1. Select the Double line command.
2. Place the first endpoint.
3. Arc/Break/CAps/CLose/Dragline/Snap/Undo/Width/<next point>: ***A*** <enter**>.**
4. Break/CAps/CEnter/CLose/Dragline/Endpoint/Line/Snap/Undo/Width/<second point>: ***Place the second point for the arc's circumference or CE for the Arc's center point.***
5. Endpoint: ***Pick the second endpoint.***

The Double Line below uses CE option to place the Arc's center point. Caps are "ON" also.

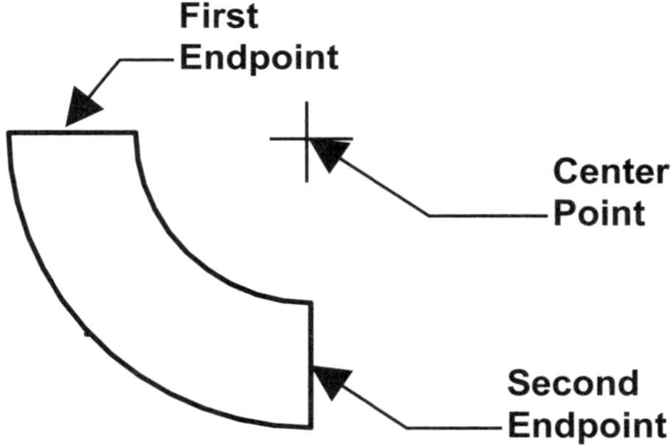

EXERCISE 5A

Create a MULTILINE STYLE

Sorry LT users, you can't do this lesson. Please skip to EX-5E.

1. Open **My Decimal Setup**
2. Select Layout tab **11 X 17 (1 to 1)**
3. Create the two Multiline Styles shown below.
   Use the instructions "Creating a Multiline Style" on page 5-4.
   Note: Use Linetypes Continuous, Center and Hidden.
4. NAMES = **5A1 and 5A2**
5. Save this drawing as **EX-5A.**

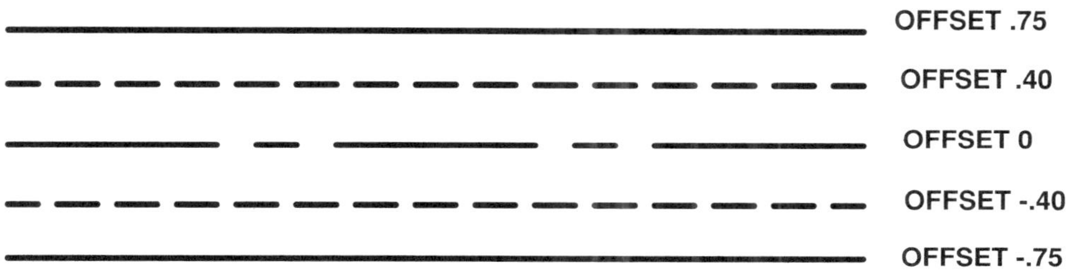

OFFSET .75
OFFSET .40
OFFSET 0
OFFSET -.40
OFFSET -.75

5A1

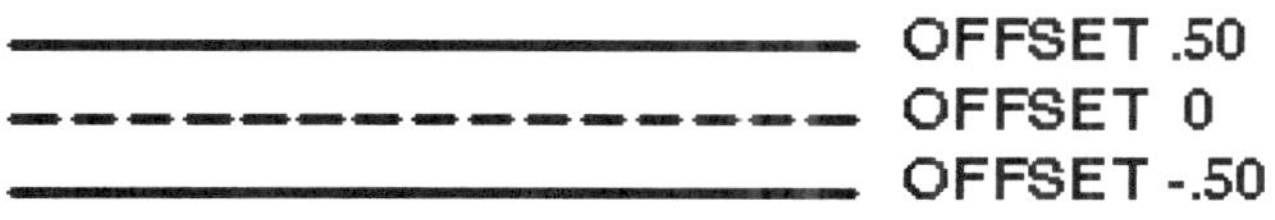

OFFSET .50
OFFSET 0
OFFSET -.50

5A2

EXERCISE 5B

Drawing a MULTILINE

Sorry LT users, you can't do this lesson. Please skip to EX-5E.

1. Open **My Decimal Setup**
2. Select Layout tab **11 X17 (1 to 1)**
3. Using the default multiline style **STANDARD**, draw the two objects below.
4. You must select the correct justification for each object. Pay close attention to where the dimensions are located on the object. (Inside or outside.)
5. Save this drawing as **EX-5B.**

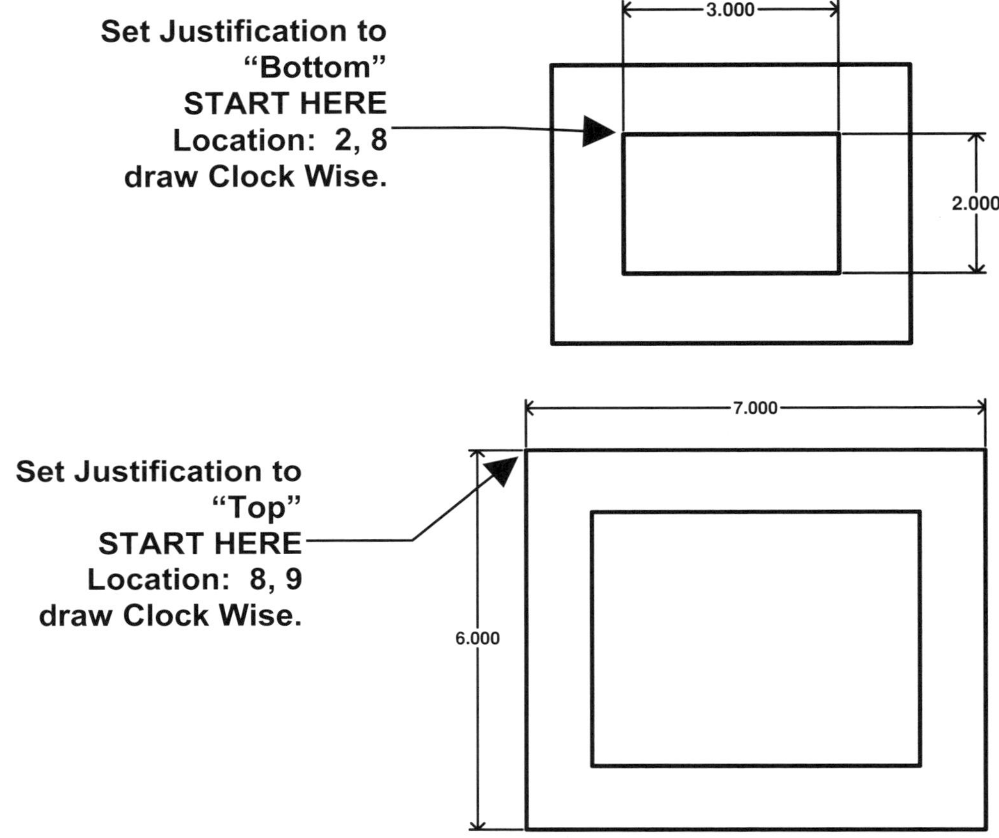

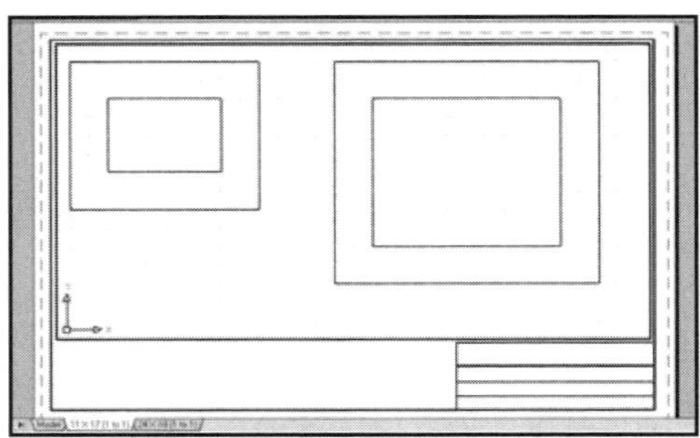

Try using DDE with Ortho on. That is the easiest way to draw these objects using the Multiline command.

If you used the correct justification, started at the correct location and drew in the clockwise direction, your drawing should appear like the example shown here.

EXERCISE 5C

TRIMMING A MULTILINE

Sorry LT users, you can't do this lesson. Please skip to EX-5E.

1. Open **My Decimal Setup**
2. Select Layout tab **11 X 17 (1 to 1)**
3. Using the multiline **5A2**, draw the 8 sets of crossing multilines below. Use Justification zero.
4. Using **MULTILINE EDIT TOOLS,** edit the multilines.
5. Save this drawing as **EX-5C.**

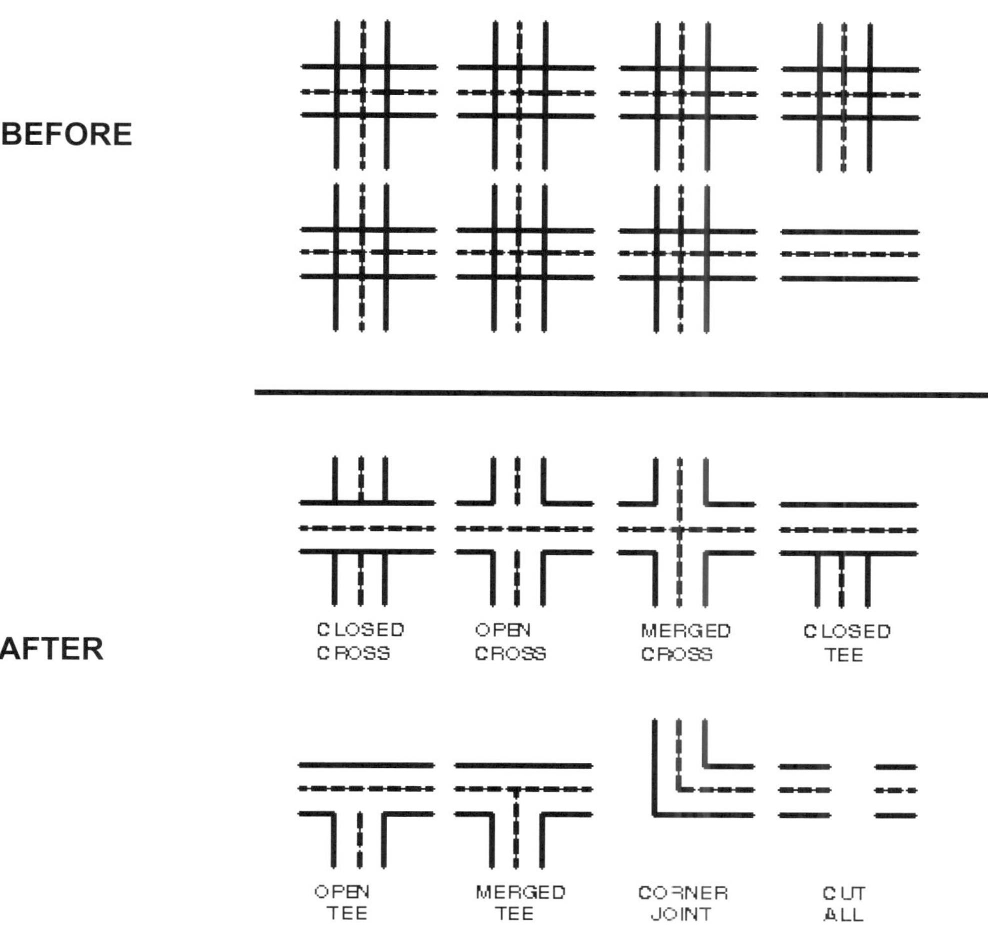

Modify the first two rows to appear like the lower two rows.

EXERCISE 5D

CREATE A FLOOR PLAN USING MULTILINE

LT users refer to Exercise 5E

A. Open **My Feet-Inches Setup**

B. Select the **24 x 18 (Qtr-Ft)** tab.

C. Create 3 new MULTILINE STYLES.

1. Name = WALLS6	1. Name = WALLS4	1. Name = WINDOW
2. Offset = 3 and -3	2. Offset = 2 and -2	2. Offset = 3, -3 & 0
3. Color = Bylayer	3. Color = Bylayer	3. Color = Bylayer
4. Linetype = Bylayer	4. Linetype = Bylayer	4. Linetype=Bylayer

D. Draw the Floor Plan (refer to the drawing on page 5-20.)
   1. Use "WALLS6" for the exterior walls
   2. Use "WALLS4" for the interior walls
   3. Use Layer Walls
   4. Don't forget to set the JUSTIFICATION to BOTTOM and draw CW.
   5. Leave spaces for the Windows and Doors OR use Cut all.

E. DOORS
   1. Use Layer DOORS.
   2. Door Size = 30" wide X 2" thick (4" space behind door)

F. FURNITURE
   1. Use Layer FURNITURE.
   2. Place approximately as shown.
   3. Sizes:

Desk = 3' x 5' (3' x 15" return)	Credenzas = 18" x 5'	Chairs = 2' Sq.
Sofa = 3' x 6'	Lamp Table = 2' Sq.	
File Cabinets = 15" x 24"	Copier = 2' x 3'	

G. ELECTRICAL
   1. Use Layer ELECTRICAL for switches and fixtures.
   2. Use Layer **WIRING** for the wire from switches to fixtures.
   3. Sizes:

   Wall Outlet = 8" dia. Overhead Lights = 16" dia. Switches = Text (ht = 6")

H. AREA TITLES (Office, Lobby and Reception)
   1. Use Layer = Text Heavy Text Style = Arch Text Height = 9" in model space

I. DIMENSION in model space. (Trans-spatial does not work well with Multilines)

J. Save as: Ex-5D

K. Plot all 4 drawings listed below using "Freeze and Thaw". (See examples on pg. 5-21)
   1. Complete drawing (all layers thawed)
   2. Floor Plan only (Freeze layers: Furniture, Electrical and Wiring)
   3. Floor Plan & Electrical (Thaw layers: Electrical and Wiring)
   4. Floor Plan & Furniture (Freeze layers: Elect. & Wiring. Thaw layer: Furniture)

EXERCISE 5E

CREATE A FLOOR PLAN USING DOUBLE LINE

This exercise is for LT users only.

A. Open **My Feet-Inches Setup**

B. Select the **24 X 18 (Qtr-Ft)** tab.

C. Draw the Floor Plan (refer to the drawing on page 5-20.)
1. The exterior walls are 6" thick
2. The interior walls are 4" thick
3. Use Layer Walls
4. Don't forget to set the Dragline.
5. Leave spaces for the Windows and Doors OR use trim or break.

E. DOORS
1. Use Layer DOORS
2. Door Size = 30" wide X 2" thick (4" space behind door)

F. FURNITURE
1. Use Layer FURNITURE
2. Place approximately as shown.
3. Sizes:

Desk = 3' x 5' (3' x 15" return) | Credenza's = 18" x 5' | Chairs = 2' Sq.
Sofa = 3' x 6' | Lamp Table = 2' Sq.
File Cabinets = 15" x 24" | Copier = 2' x 3'

G. ELECTRICAL
1. Use Layer ELECTRICAL for switches and fixtures.
2. Use Layer **WIRING** for the wire from switches to fixtures.
3. Sizes:

Wall Outlet = 8" dia. | Overhead Lights = 16" dia. | Switches = Text (ht = 6")

H. AREA TITLES (Office, Lobby and Reception) in PAPERSPACE.
1. Use Layer = Text Heavy Text Style = Arch Text Height = 3/16" in paper space.

I. DIMENSION in Paperspace.

J. Save as: Ex-5E

K. **Plot all 4 drawings listed below**: (See examples on page 5-21)
1. Complete drawing (all layers thawed)
2. Floor Plan only (Freeze layers: Furniture, Electrical and Wiring)
3. Floor Plan & Electrical (Thaw layers: Electrical and Wiring)
4. Floor Plan & Furniture (Freeze layers: Elect. & Wiring. Thaw layer: Furniture)

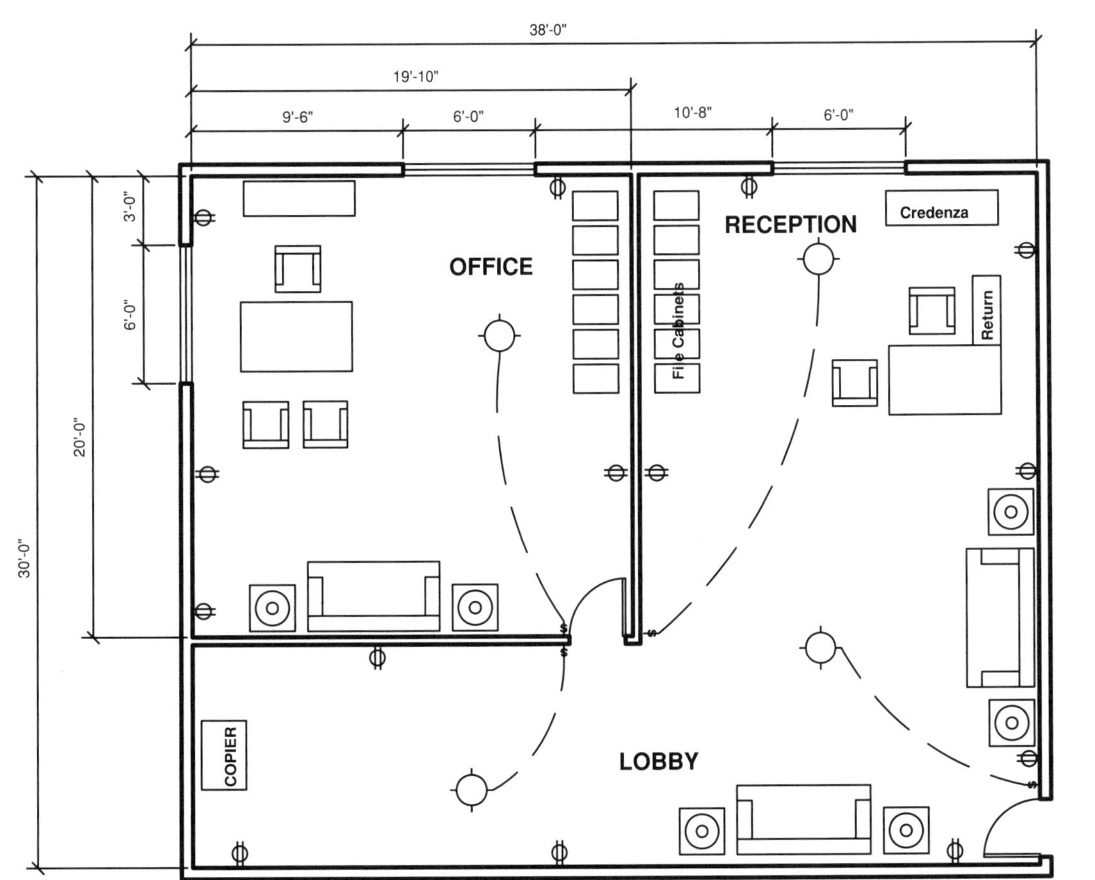

THE OFFICE

IF YOU USED MULTILINES:
Dimension in Model Space.
Trans-spatial dimensioning <u>does not work</u> well with multilines.

IF YOU USED DOUBLE LINES:
Dimension in Paper space.
Trans-spatial dimensioning <u>works well</u> with double lines.

EXERCISE 5D OR 5E

FREEZE AND THAW EXAMPLES

1. **Complete drawing**
   all layers thawed

2. **Floor Plan only**
   Freeze layers:
   Furniture, Electrical & Wiring

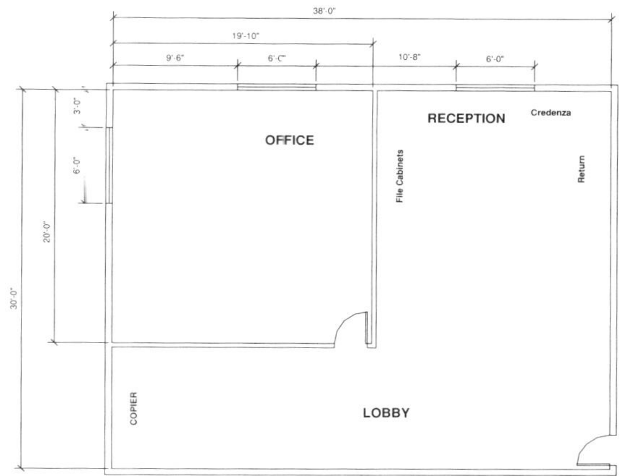

3. **Floor Plan & Furniture**
   Freeze layers:
   Electrical and Wiring
   Thaw layer: Furniture

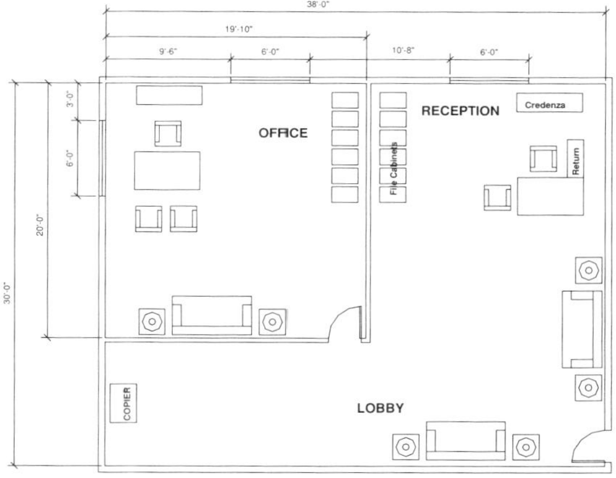

4. **Floor Plan & Electrical**
   Thaw layers:
   Electrical & Wiring
   Freeze layer: Furniture

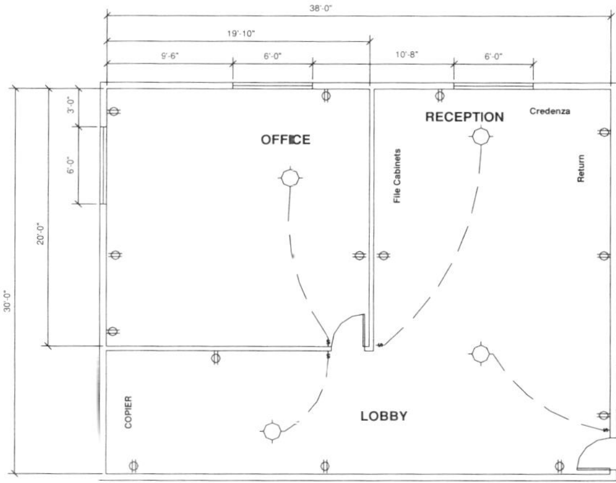

Note: You made 4 separate plots from one master drawing. Think about it.

NOTES:

LEARNING OBJECTIVES

After completing this lesson, you will be able to:

1. Create an Isometric object.
2. Create an Ellipse on an Isometric plane.

LESSON 6

ISOMETRIC DRAWINGS

An ISOMETRIC drawing is a pictorial drawing. It is primarily used to aid in visualizing an object. There are 3 faces shown in one view. They are: Top, Right and Left. AutoCAD provides you with some help constructing an isometric drawing but you will be doing most of the work. The following are instructions on the isometric aids AutoCAD provides. **Note: This is not 3D. Refer the "Introduction to 3D" section in this workbook.**

ISOMETRIC SNAP and GRID

1. Select **TOOLS / DRAFTING SETTINGS** or right click on the Grid button then select **Settings** from the short cut menu.

   The following dialog box should appear.

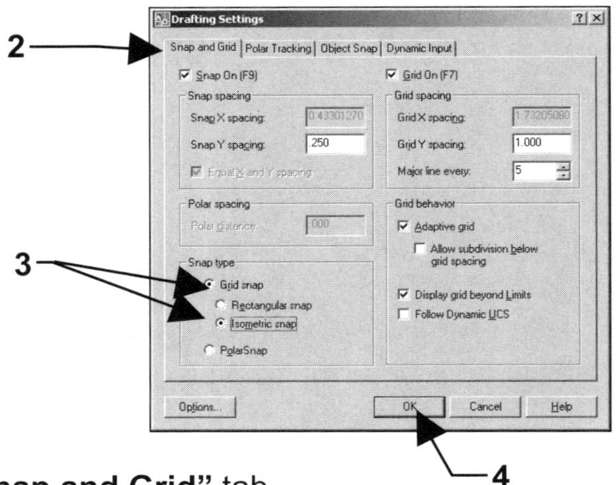

2. Select the "**Snap and Grid"** tab.
3. Select **"Grid snap"** and **"Isometric snap".**
4. Select **OK.**

NOTE: The Grid pattern and Cursor have changed. The grids are on a 30 degree angle. The cursor's horizontal crosshair is now on a 30 degree angle also. This indicates that the LEFT ISOPLANE is displayed.

ISOPLANES

There are 3 isoplanes, Left, Top and Right. You can toggle to each one by pressing the **F5** key. When drawing the left side of an object you should display the Left Isoplane. When drawing the Top of an object you should display the Top Isoplane. When drawing the Right side of an object you should display the Right Isoplane.

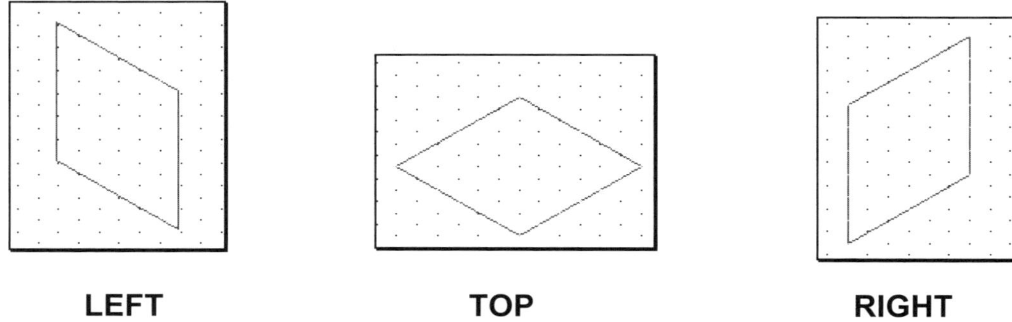

5. **Grid, Snap and Ortho should be ON**.

6. Now try drawing the 3 x 3 x 3 cube below using "Line".
   a. Start with the Left Isoplane. (You may use **DIRECT DISTANCE ENTRY**)
   b. Next change to the Top Isoplane (Press F5 once) and draw the top.
   c. Next change to the Right Isoplane (Press F5 once) and draw the Right side.

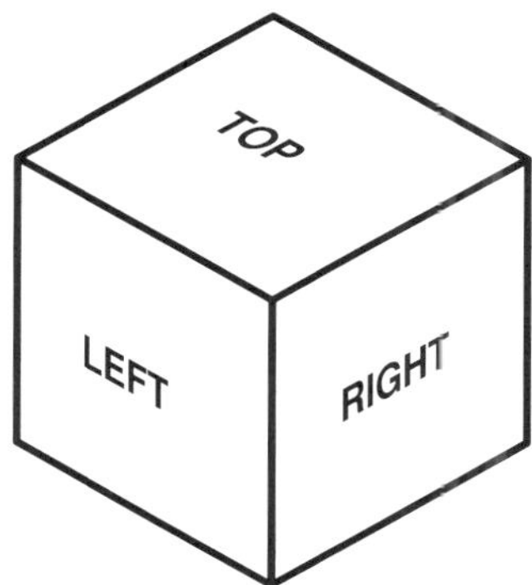

ISOMETRIC ELLIPSE

An isometric ellipse is a very helpful option when drawing an isometric drawing. AutoCAD calls an isometric ellipse an ISOCIRCLE. This option is located in the Ellipse command. ***IMPORTANT: the ISOCIRCLE option can only be selected when you have the Isometric Grid and Snap ON. This option will not appear if <u>Isometric Grid and Snap</u> is turned OFF.***

1. Change to the Left Isoplane. (Isometric Grid and Snap must be on.)
2. Select the **ELLIPSE** command. ***(Note: If you use the Pull-down Menu, select "Axis,End" method. <u>Do not select "Center")</u>***

   Command: _ellipse
   Specify axis endpoint of ellipse or [Arc/Center/Isocircle]: ***type "I" <enter>***
   Specify center of isocircle: ***specify the center location for the new isocircle***
   Specify radius of isocircle or [Diameter]: ***type the radius <enter>***

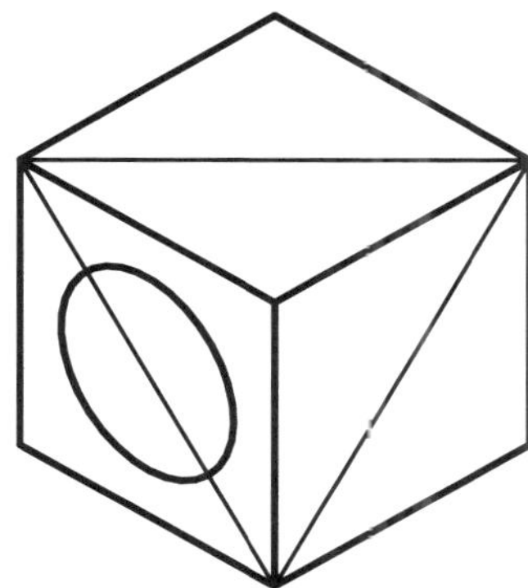

3. Now try drawing an Isocircle on the top and right, using the method above. Don't forget to change the "Isoplane" depending on where you draw the Isocircle.

EXERCISE 6A

ISOMETRIC ASSEMBLY

1. Open "MY DECIMAL SETUP"
2. Select the "24-18 (1 to 1)" tab.
3. Change to "MODEL SPACE" (Not Model tab)
4. Make sure the viewport scale is set to 1:1.
5 Lock the viewport, if is not already.
6 Set the "Snap type & style" to isometric snap (Refer to page 6-2 if necessary)
7. Change to isoplane "top".
8. Draw the objects below. Do not dimension.
9. Save as: EX-6A.

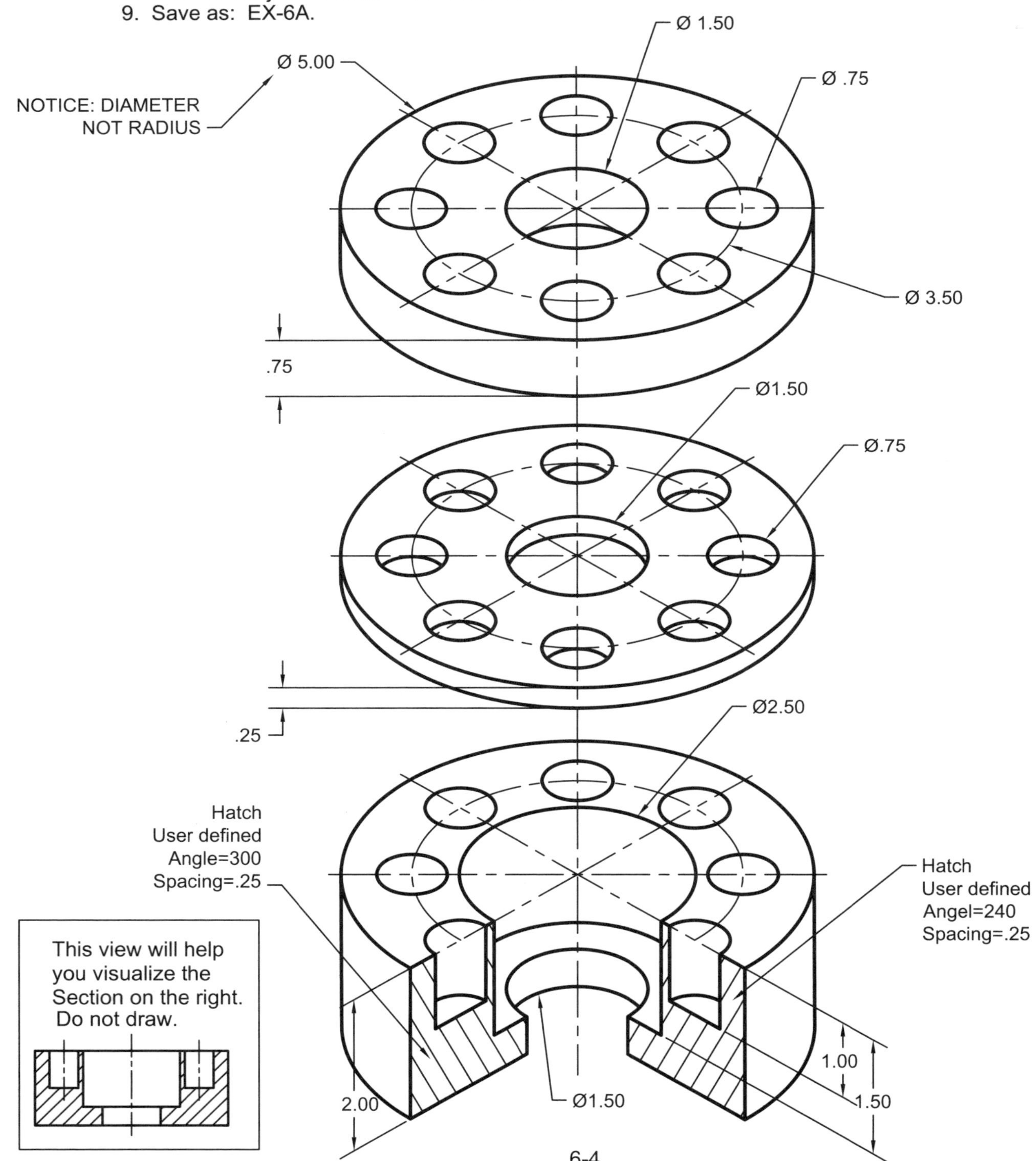

EXERCISE 6B

ISOMETRIC OBJECT

1. Open "MY DECIMAL SETUP"
2. Select the "24-18 (1 to 1)" Layout tab.
3. Change to "MODEL SPACE". (Not Model tab.)
4. Make sure viewport scale is set to 1:1 and locked, if not already.
5 Set the "snap type & style" to isometric snap (Refer to page 6-2 if necessary)
6. Change the isoplane with F5 when necessary.
7. Draw the objects below. DO NOT DIMENSION. (Dimensioning will be discussed in Lesson 7.)
8. Save as: EX-6B

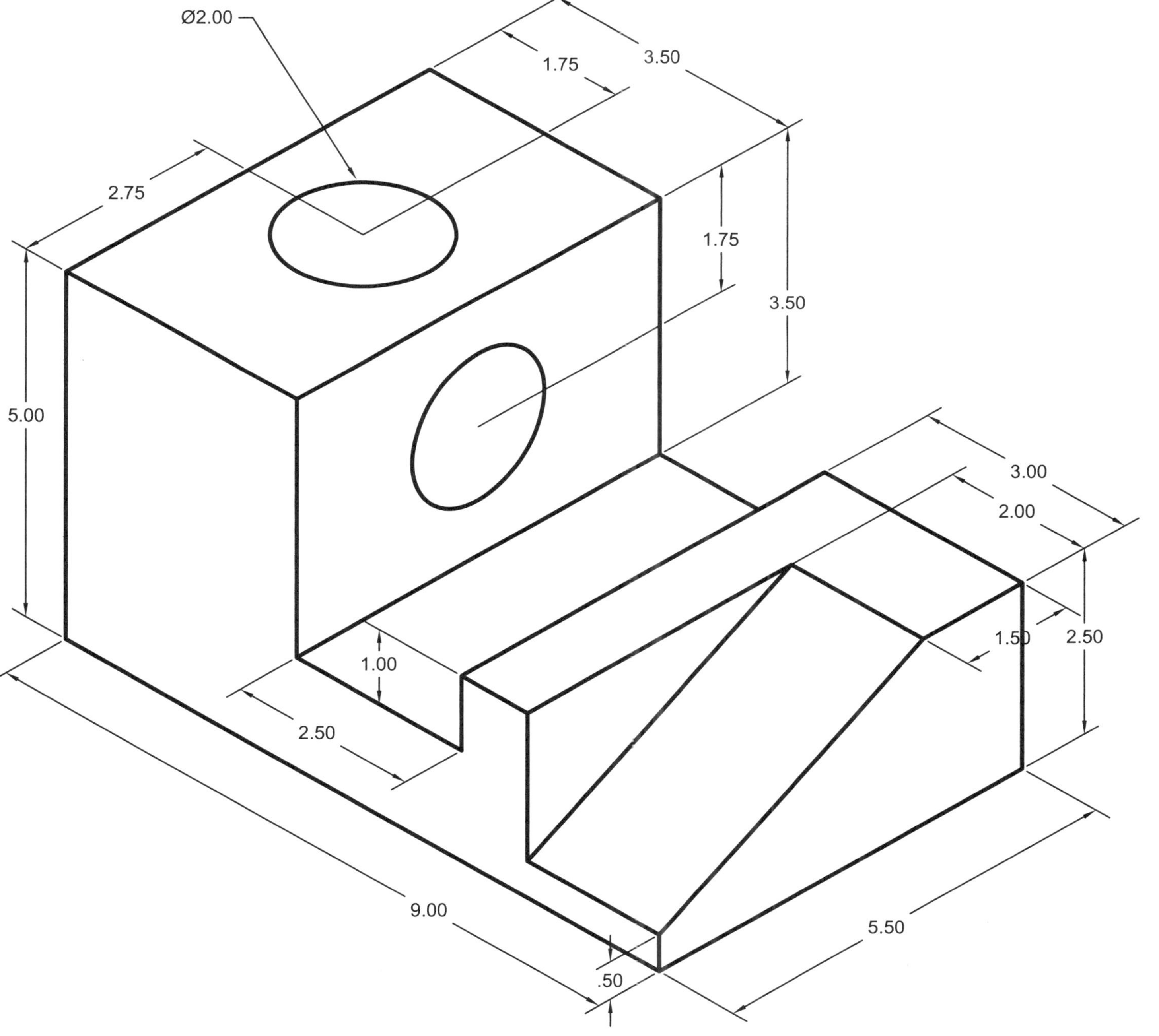

EXERCISE 6C
ABSTRACT HOUSE

1. Open "My Feet-Inches Setup"
2. Select the "24 x 18 Qtr-Ft" tab.
3. Change to Model Space. (Not Model tab.)
4. Make sure your viewport scale is 1/4" = 1' and Locked, if not already.
5. Draw the abstract house below. Do not dimension.
6. Save as Ex-6C.

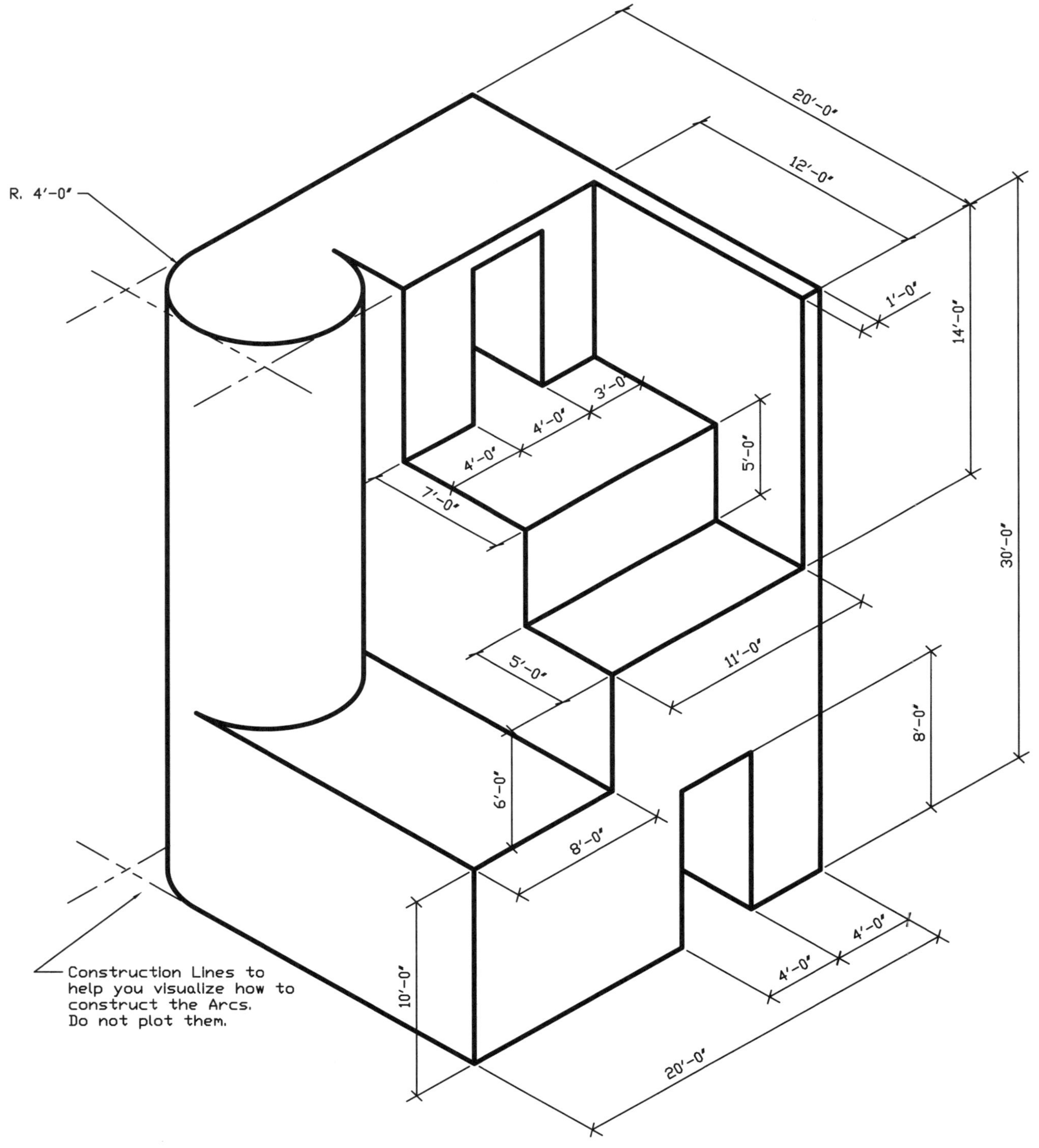

LEARNING OBJECTIVES

After completing this lesson, you will be able to:

1. Cut and Paste from one document to another.
2. Dimension an Isometric object.
3. Use Grips
4. Create Isometric Text.

LESSON 7

COPY, CUT and PASTE   

You can't use the "Copy" command to copy objects from one drawing to another drawing. You must use **EDIT / COPY** and **EDIT / PASTE**.

AutoCAD allows you to select objects in a drawing **(EDIT / COPY**) and then copy those objects to another drawing **(EDIT / PASTE).** Or you may remove the objects from the original drawing **(EDIT / CUT)** and then copy **(EDIT / PASTE)** them to another drawing.

You can also copy, cut and paste between AutoCAD and other documents such as Microsoft Word, Excel or any other Windows application.

Copy, Cut and Paste between AutoCAD drawings. (Simplified)

1. Open two AutoCAD drawings and Tile Vertically. (Example below)
2. Activate the drawing that contains the objects you want to copy. (Click in the drawing area)
3. Select the **COPY** or **CUT** command using one of the following:

TYPE = Copyclip
PULLDOWN = EDIT / COPY
TOOLBAR = STANDARD

TYPE = Cutclip
PULLDOWN = EDIT / CUT
TOOLBAR = STANDARD

TYPE = Pasteclip
PULLDOWN = EDIT/PASTE
TOOLBAR = STANDARD

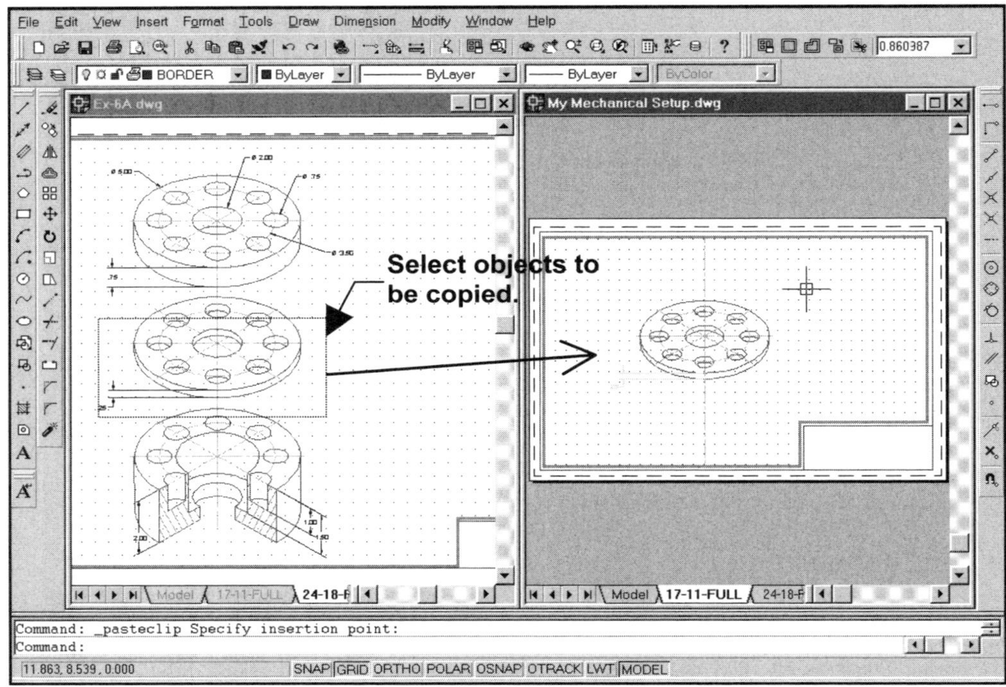

4. Select the objects to copy then <enter>. (Use any AutoCAD selection methods) If you selected **CUT**, the objects selected will disappear.
5. Activate the drawing in which the objects will be pasted.
6. Select the **PASTE** command.
7. Place the objects with the cursor by dragging the objects to the correct location and click.

Note: In the example above the base point for insertion was not requested. If you wish to designate a specific basepoint see **COPY with a Base Point** on the next page.

Where do the objects go when you select Copy or Cut?

Copies of the objects are temporarily stored on the Windows Clipboard. The Clipboard retains only the most recent information that was copied. When another object is copied, the newly copied object replaces the last objects stored within the Clipboard's memory.

ADDITIONAL OPTIONS

COPY with a BASE POINT or COPYBASE
This option is the same as copy except it allows you to specify a base point for the objects.

COPY LINK
Copies **all** objects currently on the screen and automatically selects the ORIGIN as the basepoint. You will not be prompted to select objects.

PASTE as BLOCK or PASTEBLOCK
Copies objects into the same or different drawing as a BLOCK.

PASTE to ORIGINAL COORDINATES or PASTEORIG
Pastes objects into the new drawing at the same coordinate position as the original drawing. You will not be prompted for an insertion point.

PASTE SPECIAL or PASTESPEC
Use when pasting from other applications into AutoCAD.

Copy, Cut and Paste between other applications and AutoCAD. (Simplified)

How to copy and paste text from a **word processing** program to an **AutoCAD drawing**.

1. Open the Text document.
2. Highlight the text that you wish to copy.
3. Select the COPY command.
4. Open the AutoCAD drawing.
5. Select the PASTE command.

Once the text is in the drawing you may use the "Properties" palette to make changes to the appearance. If you want to change the wording, double click on the text and you will be temporarily returned to the software that originated the text. Make the changes and then select **File / Close and return**.

DRAW ORDER

The **DRAWORDER** command changes the display order of objects. You may easily display an object in front or behind another object.

1. Left click on the object to select it.

2. Right click and select **Draw Order** from the menu.

3. Select one of the following:

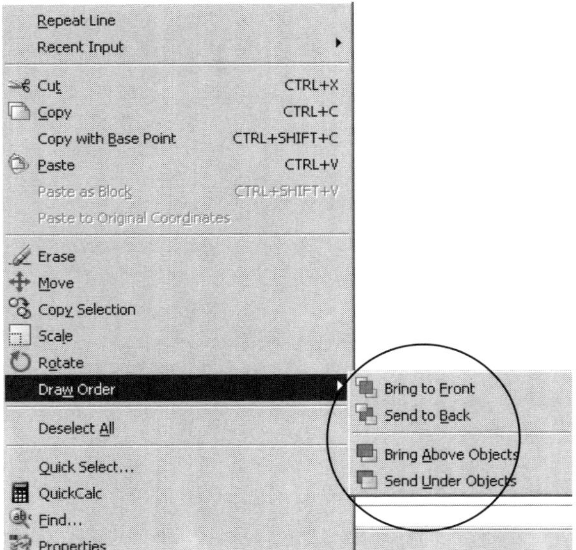

DIMENSIONING AN ISOMETRIC DRAWING

Dimensioning an isometric drawing in AutoCAD is a two step process. First you dimension it with the dimension command **ALIGNED**. Then you adjust the angle of the extension line with the dimension command **OBLIQUE**.

STEP 1. Dimension the object using the dimension command **ALIGNED**. 

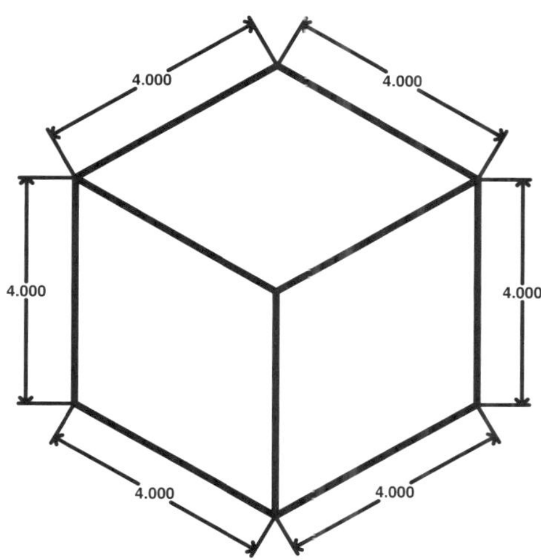

STEP 2. Adjust the angle of the extension lines using the dimension command **OBLIQUE.**
(Do each dimension individually)

1. Select **DIMENSION / OBLIQUE** or *Type: **DIMEDIT** and select **O** for Oblique* or 
2. Select the dimension you wish to adjust. (click on it) <enter>
3. Type the Obliquing angle for the **extension line**. (30, 150, 210 or 330)

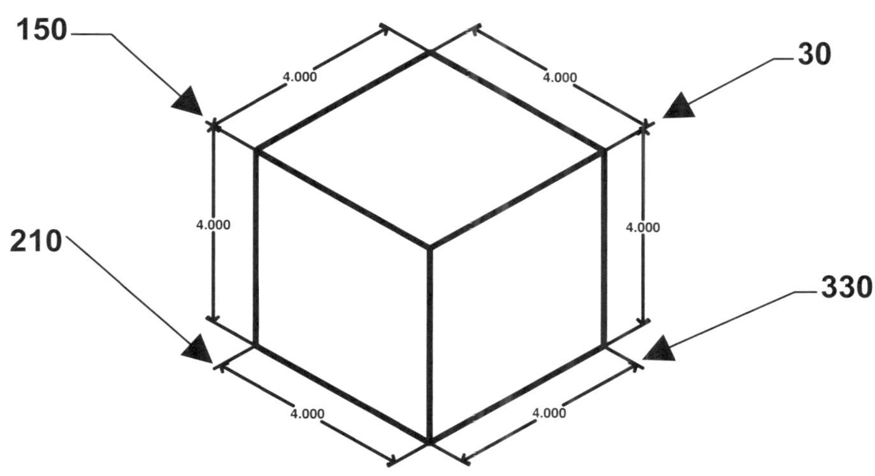

ISOMETRIC TEXT

AutoCAD doesn't really have an Isometric Text. But by using both **Rotation** and **Oblique angle**, we can make text appear to be laying on the surface of an Isometric object.

1. First create two new text styles. One with an oblique angle of **30** and the other with an oblique angle of **minus 30**.

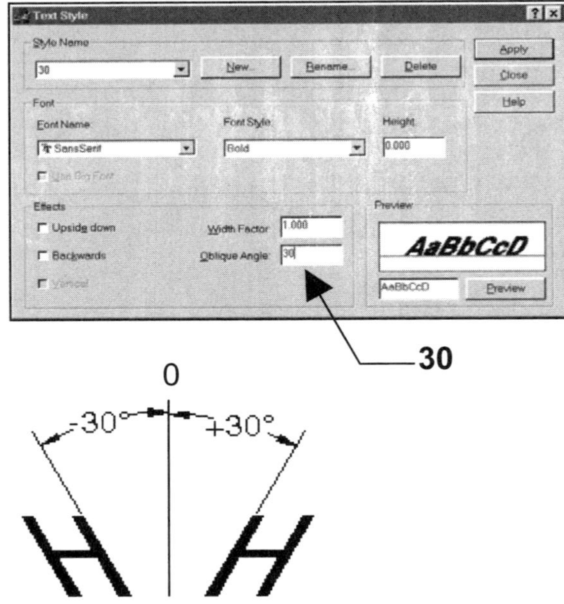

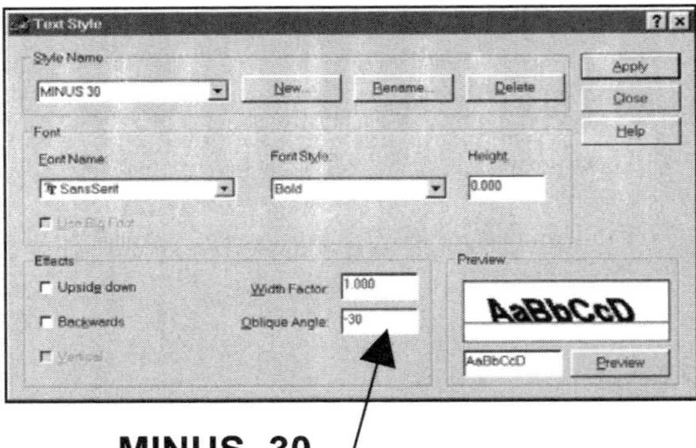

2. Next select the appropriate text style and rotation as follows:

   a. Select the text style with an obliquing angle of 30 or minus 30.
   b. Select **DRAW / TEXT / SINGLE LINE**
   c. Place the START POINT or Justify.
   d. Type the Height.
   e. Type the Rotation angle.
   f. Type the text.

This is an example of what you can achieve. The text appears to be on the top and side surfaces.

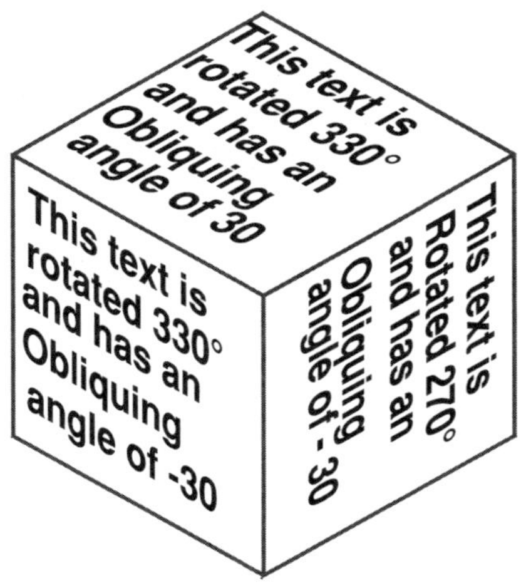

EXERCISE 7A

Copy and Paste between two drawings.

1. Confirm that "Single-drawing Compatibility mode" is OFF (Refer to page 1-2)
2. Open **My Decimal Setup** and select the **11 X 17 (1 to 1)** tab.
3. Also, open **Ex-6A**.
4. Tile Vertically. (Example below)
5. Activate the Ex-6A drawing (Click in the drawing area)
6. Select the **COPY** command using one of the following:

TYPE = Copyclip or CTRL + C
PULLDOWN = EDIT / COPY
TOOLBAR = STANDARD

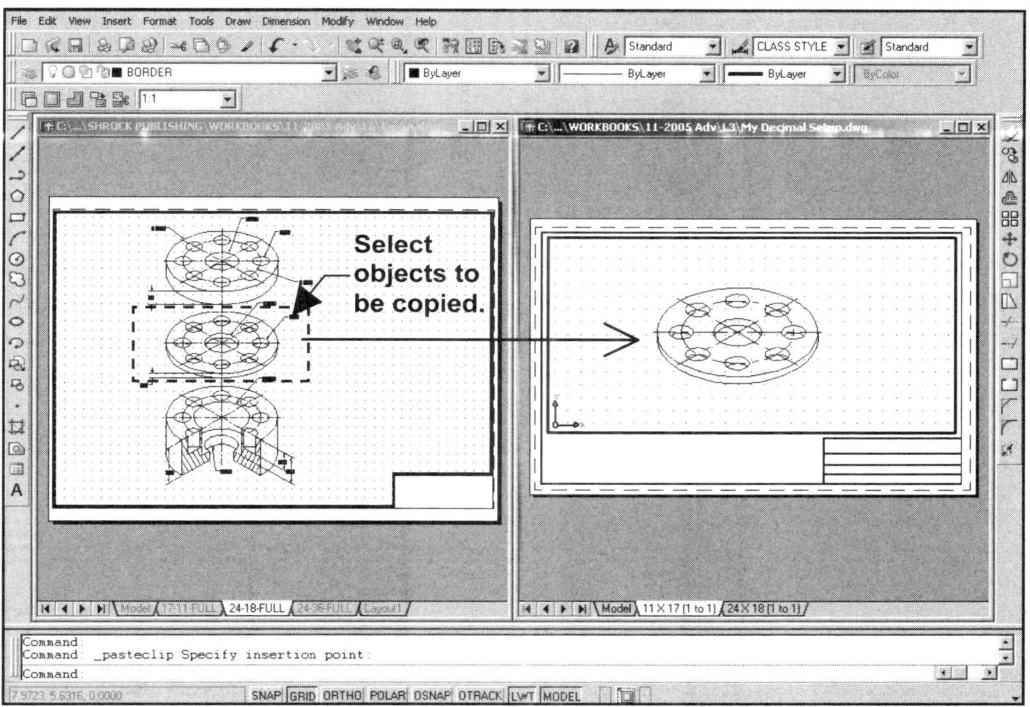

7. Select the objects to copy with a crossing window then <enter>.
8. Activate the **My Decimal Setup** drawing.
9. Select Model Space. (Do not paste into Paper Space.)
10 Select the **PASTE** command using one of the following methods.

TYPE = Pasteclip or CTRL + V
PULLDOWN = EDIT / PASTE
TOOLBAR = STANDARD

11. Place the objects with the cursor in the drawing area and left click. (As shown above.)

12. **Do not save** this drawing. It is for practice only.

EXERCISE 7B

OBLIQUE DIMENSIONING

1. Open drawing EX-6B.
2. Dimension the isometric object shown below. (Refer to page 7-5 if necessary)
3. Use "DIMENSION / ALIGNED" first.
4. Then "DIMENSION / OBLIQUE.
   (Remember, the oblique angle is the desired angle for the extension line.)
5. Use grips to move the dimensions if necessary.
6. Save as: EX-7B

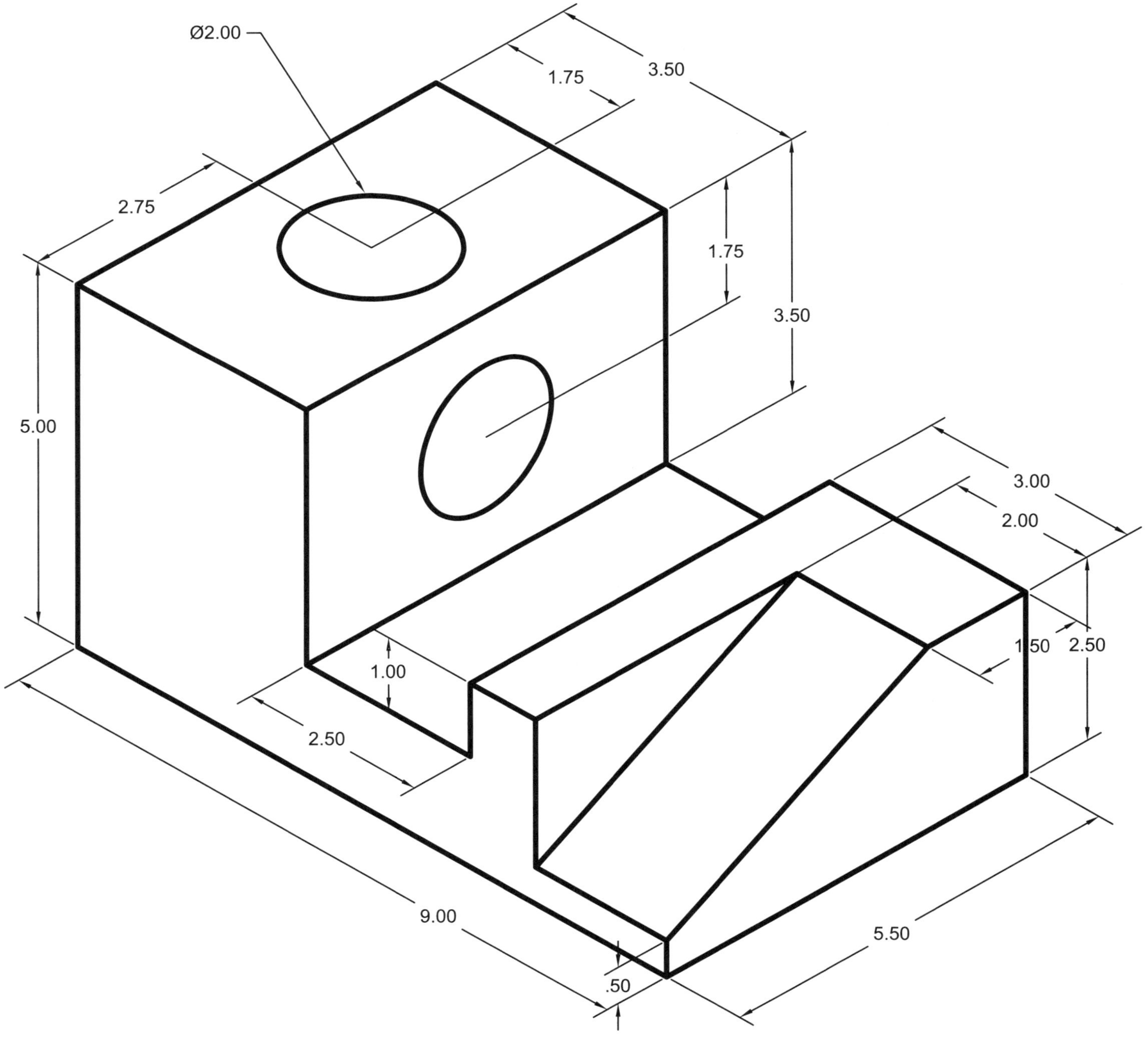

EXERCISE 7C

OBLIQUE DIMENSIONING

1. Open drawing EX-6C.
2. Dimension the abstract house.
3. Use "Dimension / Aligned" first. (dimension in paperspace)
4. Now use "Dimension / Oblique"
5. Use grips to move the dimensions if necessary.
6. Save as EX-7C.

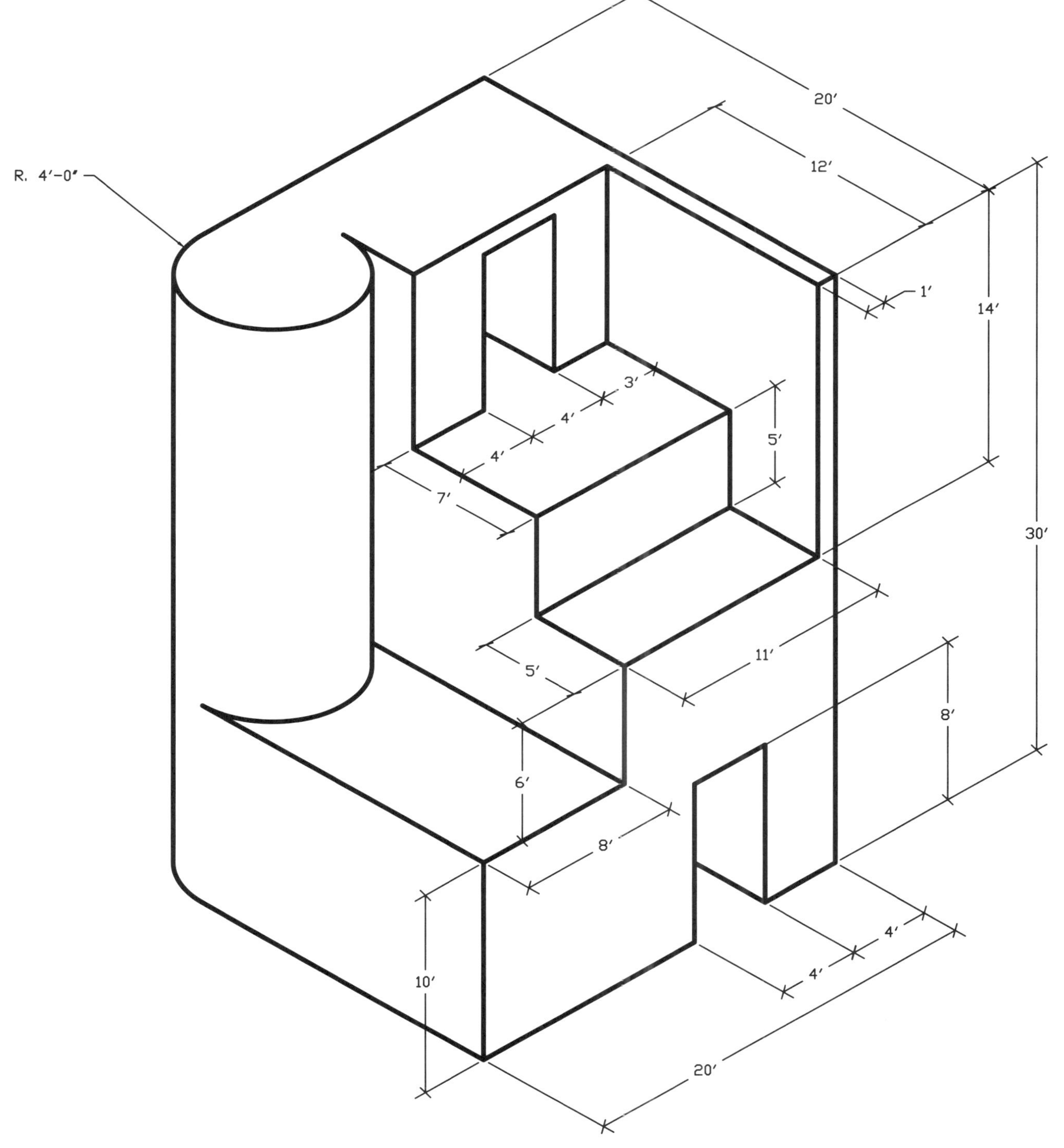

EXERCISE 7D

ISOMETRIC TEXT

1. Open My Decimal Setup.
2. Select the 11 X 17 (1 to 1) tab.
3. First create two new text styles. One with an oblique angle of 30 and the other with an oblique angle of minus 30.

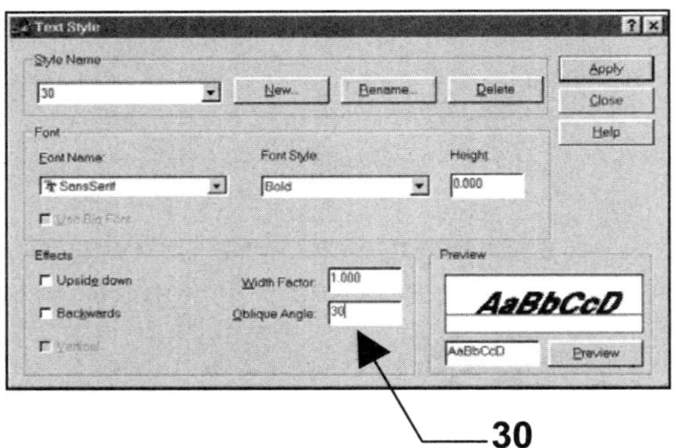

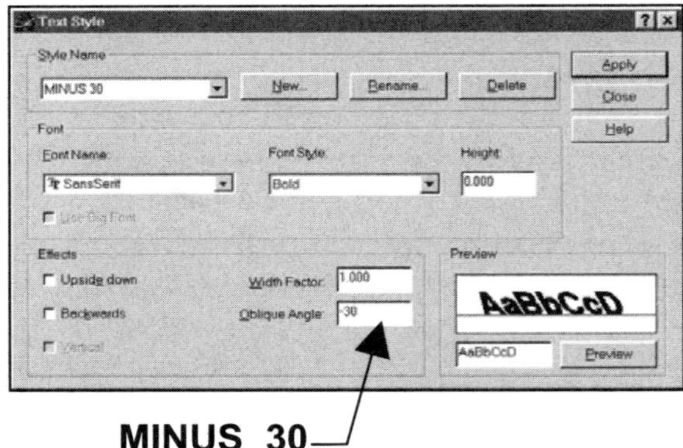

4. Draw a 4 inch cube.

5. Next select the appropriate text style and rotation to create the text on the cube shown below.

   a. Select the text style with an oblique angle of 30 or minus 30.
   b. Select **DRAW / TEXT / SINGLE LINE**
   c. Place the START POINT approximately as shown.
   d. Type the Height = .25
   e. Type the Rotation angle.
   f. Type the text.

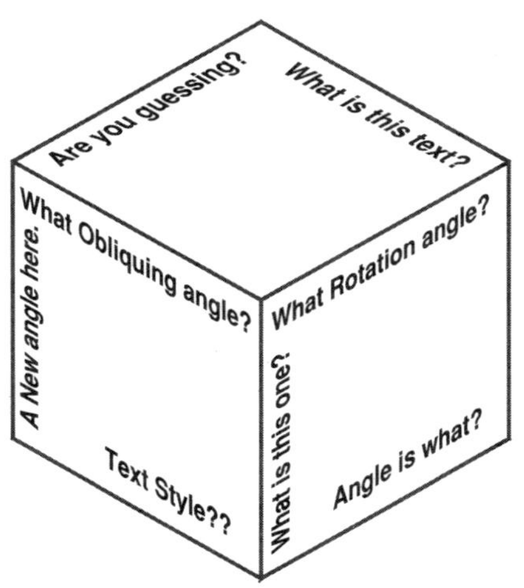

LEARNING OBJECTIVES

After completing this lesson, you will be able to:

1. Create a Block (Review)
2. Insert a Block (Review)
3. Assign and use Attributes
4. Define a Dynamic Block

LESSON 8

Let's review "How to create a block"

You learned how to create Blocks, in Lesson 28, in the Exercise Workbook for Beginning AutoCAD 2006. Because Blocks are so useful, let's review "How to create a block" one more time. Then we will learn how to add "Attributes" to those blocks and create Dynamic Blocks to make them even more useful.

CREATING A BLOCK

1. First draw the objects that will be converted into a Block.

   For this example a circle and 2 lines are drawn.

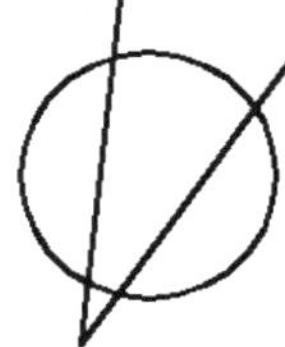

2. Select the **BMAKE** command using one of the following:

   TYPE = B
   PULLDOWN = DRAW / BLOCK / MAKE
   TOOLBAR =DRAW

   (The Block Definition dialog box will appear.)

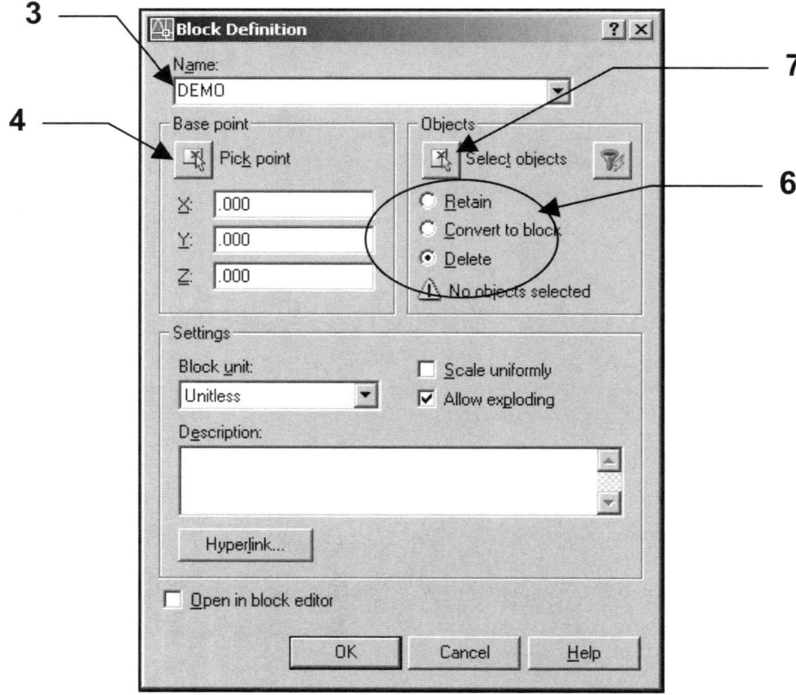

3. Enter the New Block name in the **Name** box.

4. Select the **Pick Point** button. (Or you may type the X, Y and Z coordinates.)
The Block Definition box will disappear and you will return temporarily to the drawing.

5. Select the location where you would like the insertion point for the Block.
Later when you insert this block, the block will appear on the screen attached to the cursor at this insertion point. Usually this point is the CENTER, MIDPOINT or ENDPOINT of an object.

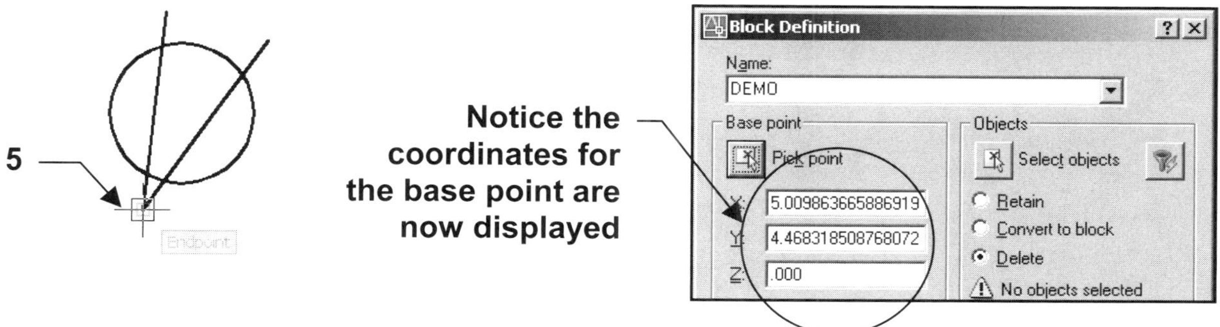

6. Select an option.

   It is important that you select one and understand the options below.

   Retain
   If this option is selected, the original objects will stay visible on the screen after the block has been created.

   Convert to block
   If this option is selected, the original objects will disappear after the block has been created, but will immediately reappear as a block. It happens so fast you won't even notice the original objects disappeared.

   Delete
   If this option is selected, the original objects will disappear from the screen after the block has been created.

7. Select the **Select Objects** button.

   The Block Definition box will disappear and you will return temporarily to the drawing.

8. Select the objects you want in the block, then press <enter>.

The Block Definition box will reappear and the objects you selected should be illustrated in the Preview Icon area.

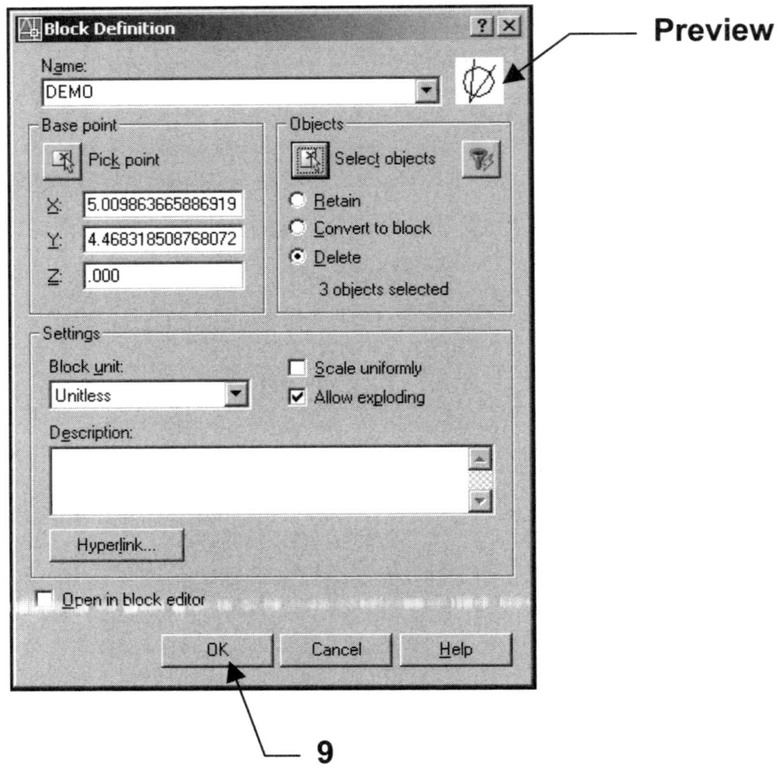

9. Select the **OK** button.
   The new block is now stored in the drawing's block definition table.

10. To verify the creation of this Block, select **Insert / Block**, select the Name (▼).
    A list of all the blocks, in this drawing, will appear.

ADDITIONAL DEFINITIONS OF OPTIONS

Block Units
You may define the units of measurement for the block. This option is used with the "Design Center" to drag and drop with Autoscaling. The Design Center is an advanced option and is not discussed in this book.

Scale Uniformly
Specifies whether or not the block is prevented from being scaled non-uniformly during insertion.

Allow Exploding
Specifies whether or not the block can be exploded after insertion.

Description
You may enter a text description of the block.

Hyperlink
Opens the **insert Hyperlink dialog box** which you can use to associate a hyperlink with the block.

HOW LAYERS EFFECT BLOCKS

If a block is created on Layer 0:

1. When the block is inserted, it will take on the properties of the current layer.
2. The inserted block will reside on the layer that was current at the time of insertion.
3. If you Freeze or turn Off the layer the block was inserted onto, the block will disappear.
4. If the Block is **Exploded**, the objects included in the block will revert to their original properties of layer 0.

If a block is created on Specific layers:

1. When the block is inserted, it will retain its own properties. It **will not** take on the properties of the current layer.
2. The inserted block **will reside** on the current layer at the time of insertion.
3. If you **freeze** the layer that was current at the time of insertion the block will disappear.
4. If you turn **off** the layer that was current at the time of insertion the block will not disappear.
5. If you **freeze** or turn **off** the blocks original layers the block will disappear.
6. If the Block is **Exploded**, the objects included in the block will go back to their original layer.

Let's review "How to Insert a Block"

Now we need to review "How to insert the Block".

1. Select the INSERT command using one of the following:

   TYPE = DDINSERT
   PULLDOWN = INSERT / BLOCK
   TOOLBAR =DRAW

The INSERT dialog box will appear.

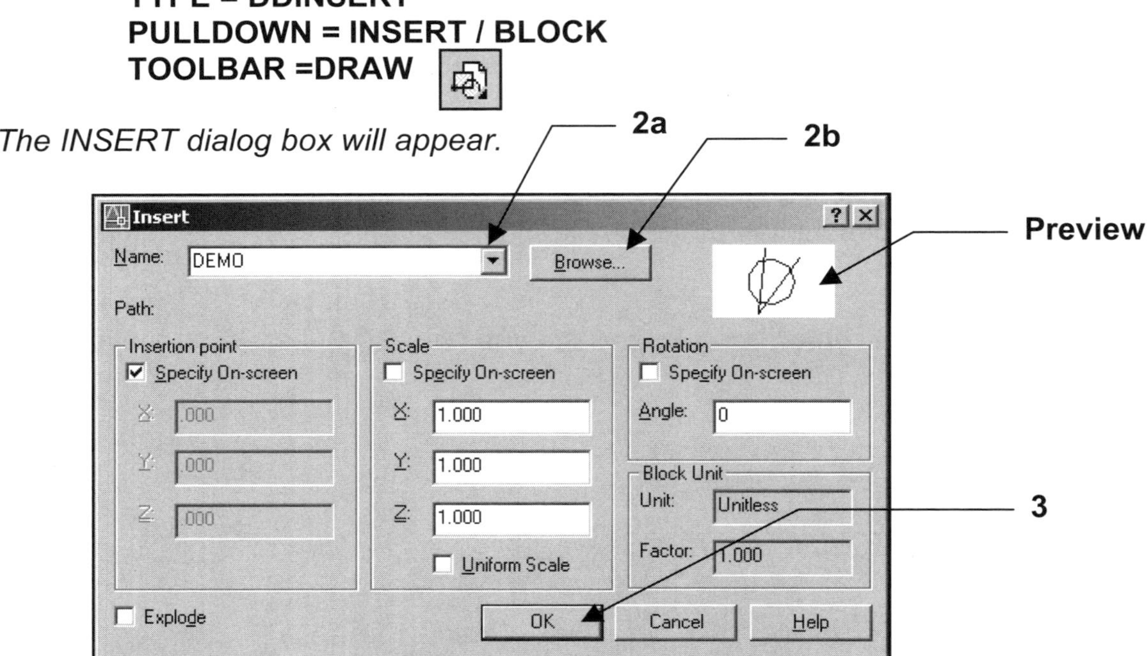

2. Select the **BLOCK** name.
   a. If the block is already in the drawing that is open on the screen, you may select the block from the drop down list shown above
   b. If you want to insert an entire drawing, select the Browse button to locate the drawing file.

3. Select the **OK** button.

   This returns you to the drawing and the selected block should be attached to the cursor.

4. Select the insertion location for the block by pressing the left mouse button or typing coordinates.

   Command: _insert
   Specify insertion point or **[Scale/X/Y/Z/Rotate/PScale/PX/PY/PZ/PRotate]:**

NOTE: If you want to scale or rotate the block before you actually place the block, press the right hand mouse button and you may select an option from the menu or select an option from the command line menu shown above.

Remember you may also "preset" the insertion point, scale or rotation.

PRESETTING THE INSERTION POINT, SCALE or ROTATION

You may preset the **Insertion point, Scale or Rotation** in the **INSERT** box instead of at the command line.

1. Remove the check mark from any of the **"Specify On-screen"** boxes.
2. Fill in the appropriate information describe below:

Insertion point
Type the X and Y coordinates from the Origin. The Z is for 3D only.
The example below indicates the block's insertion location will be 5 inches in the X direction and 3 inches in the Y direction, from the Origin.

Scale
You may scale the block proportionately by typing the scale factor in the X box and then check the Uniform Scale box.
If the block will be scaled non-proportionately, type the different scale factors in both X and Y boxes.
The example below indicates that the block will be scale proportionate at a factor of 2.

Rotation
Type the desired rotation angle relative to its current rotation angle.
The example below indicates the block will be rotated 45 degrees from its originally created angle orientation.

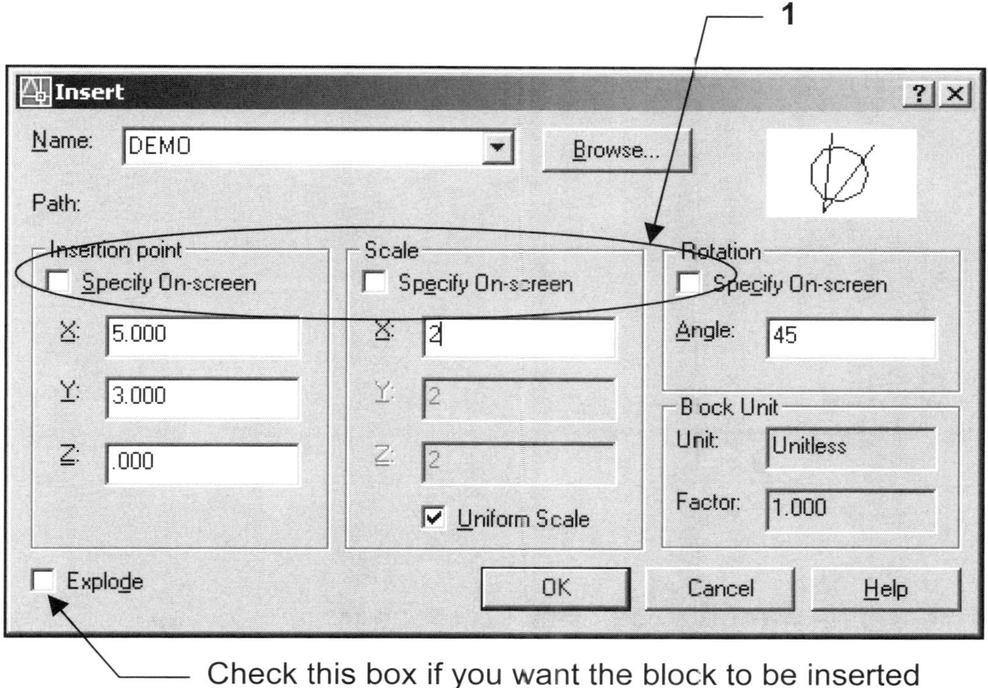

Check this box if you want the block to be inserted already exploded.

ATTRIBUTES

The **ATTRIBUTE** command allows you to add text data to a block. You define the attribute and attach it to a block. Every time you insert the block, the attributes are also inserted.

For example, if you had a block in the shape of a tree, you could assign information (attributes) about this tree, such as name, size, cost, etc. Each time you insert the block, AutoCAD will pause and prompt you for "What kind of tree is this?" You respond by entering the tree name. Then another prompt will appear, "What size is this tree? You respond by entering the size. Then another prompt will appear, "What is the cost"? You respond by entering the cost. The tree symbol will then appear in the drawing with the name, size and cost displayed.

Another example could be a title block. If you assign attributes to the text in the title block, when you insert the title block it will pause and prompt you for "What is the name of the drawing?" or "What is the drawing Number?". You respond by entering the information. The title block will then appear in the drawing with the information already filled in.

You will understand better after completing the following.

CREATING BLOCK ATTRIBUTES

1. Draw a rectangle 2 " X 1"

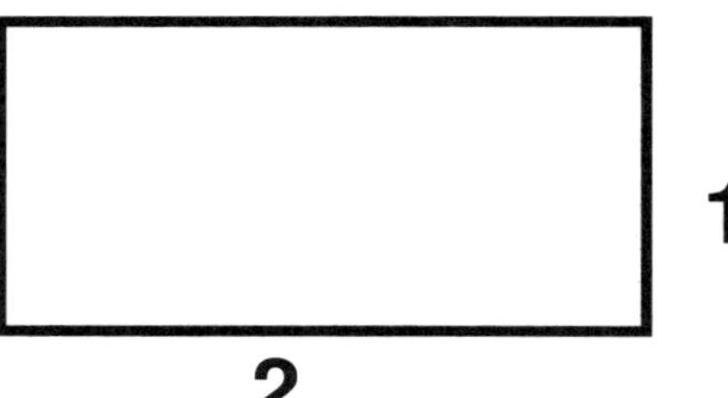

2. Select the **Define Attribute** command using one of the following:

   TYPE = ATTDEF
   PULLDOWN = DRAW / BLOCK / DEFINE ATTRIBUTES
   TOOLBAR = ATTRIBUTE

The following dialog box will appear.

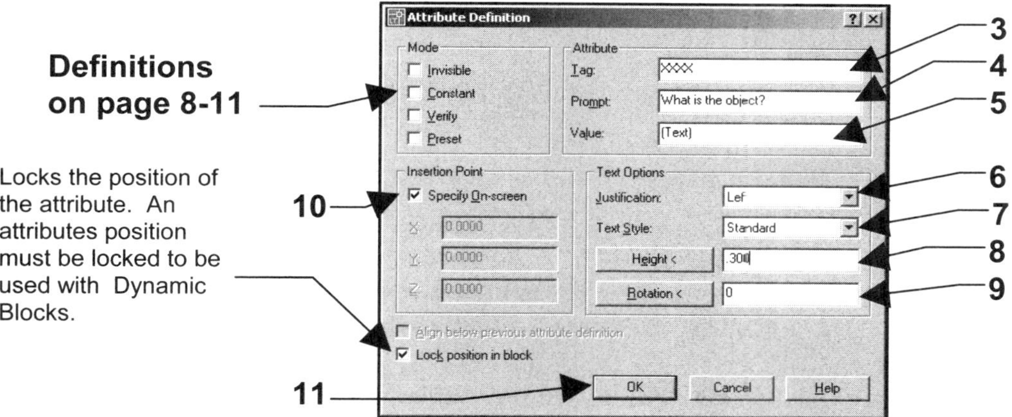

3. Enter the Attribute tag.
   A "Tag" is only a place saver. The "XXX's" will be replaced with your answer to the prompt when the block is inserted.

4. Enter the Attribute prompt.
   This is the prompt that will appear when AutoCAD pauses and prompts you for information. (You create the prompt)

5. Enter a value if necessary.
   The value will appear beside the prompt, in brackets, indicating what format the prompt is requesting. Example: What is the cost? <$0.00> The $0.00 inside the brackets is the "Value".

6. Select the Justification for the text.
   This will be used to place the Attribute text.

7. Select a Text Style.
   The Text Style selected will be used when displaying the Attribute.

8. Enter the text Height.
   This will be the height of the Attribute text.

9. Enter the Rotation.
   This will be the rotation angle of the Attribute text.

10. Insertion Point.
    You may uncheck the "Specify on Screen" box and enter the X, Y and Z coordinates if you know them or select the "Specify on Screen" box to place the insertion point manually.

11. Select the **OK** button.

12. The "**tag**" will be attached to the cursor. It is waiting for you to place it somewhere on the drawing to establish the insertion point.

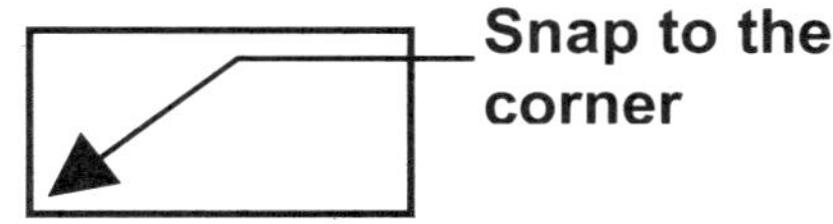

13. Select **DRAW / BLOCK / MAKE**. (Refer to page 8-2)
Make a block of the rectangle and the attribute text.
Name = abc
Pick Point = upper right corner of rectangle
Select Objects = Select rectangle and attribute text with a crossing window.

14. Select **OK** button.

NOTE: Set the "ATTDIA" variable to 0 or 1
Type: Attdia <enter>
Enter 0 or 1 <enter>

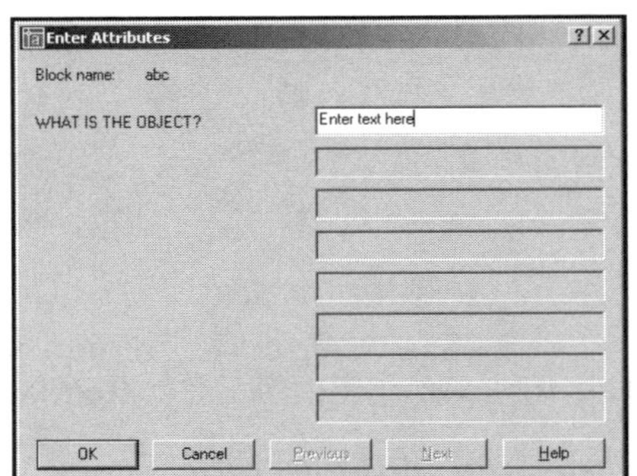

If the variable **ATTDIA** is set to **0**, the prompt will appear on the command line.

If the variable **ATTDIA** is set to **1**, the dialog box shown here will appear.

It is strictly personal preference.
It will not affect the input.
(Personally, I prefer the "dialog box")

15. Select **INSERT / BLOCK** (Refer to page 8-6)
    a. Select block name **abc** and then **OK** button.
    b. Place the block somewhere on the screen and press the left mouse button.
    The "Enter Attributes" dialog box will appear.
    c. Type: **BOX <enter>**

16. Select **INSERT / BLOCK again.**
    a. Select block name **abc** and then **OK** button.
    b. Place the block somewhere on the screen and press the left mouse button.
    c. Type: **SQUARE <enter>**

17. Select **INSERT / BLOCK again.**
    a. Select block name **abc** and then **OK** button.
    b. Place the block somewhere on the screen and press the left mouse button.
    c. Type: **DESK <enter>**

You should now have 3 rectangles that look like the ones below. You inserted the same block but the text is different in each. Think how this could be useful in other applications.

ATTRIBUTE MODES

Invisible

If this box is checked, the attribute will be invisible. You can make it visible later by typing the ATTDISP command and selecting ON.

Constant

The attribute stays constant. It never changes and you will not be prompted for the value when the block is inserted.

Verify

After you are prompted and type the input, you will be prompted to verify that input. This is a way to double check your input before it is entered on the screen. This option does not work when ATTDIA is set to 1 and the dialog box appears.

Preset

You will not be prompted for input. When the block is inserted it will appear with the default value. You can edit it later with **ATTEDIT**. (See Editing Attributes page 9-2.)

EXERCISE 8A

Assigning Attributes to a Block

The following exercise will instruct you to draw a Box, Assign Attributes, create a block of the Box including attributes, then insert the new block and answer the prompts when they appear on the screen.

1. Open **My Decimal Setup** and select the **11 X 17 (1 to 1)** tab.

2. Draw the Box shown on the next page.
   L = 3.75 W = 2.25 H = 3.00

3. Assign Attributes for **Length**
   a. Tag = L
   b. Prompt = What is the Length of the Box?
   c. Value = inches
   d. Justification = Middle
   e. Text Style = Standard
   f. Height = .250
   g. Rotation = 0
   h. Select OK and place as shown on 8-11.

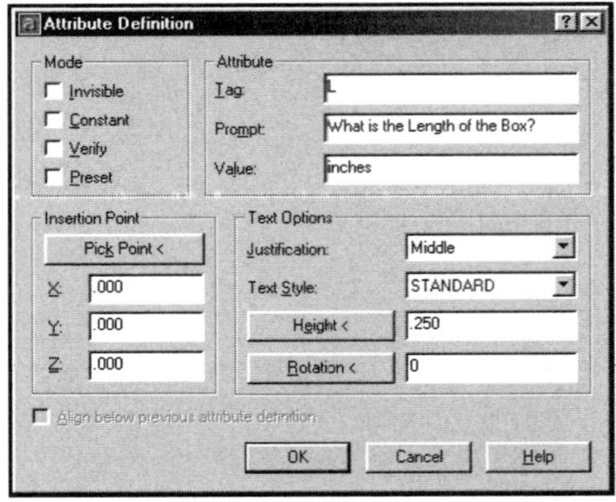

4. Assign Attributes for **Width**
   a. Tag = W
   b. Prompt = What is the Width of the Box?
   c. Value = inches
   d. Justification = Middle
   e. Text Style = Standard
   f. Height = .250
   g. Rotation = 0
   h. Select OK and place as shown on 8-11.

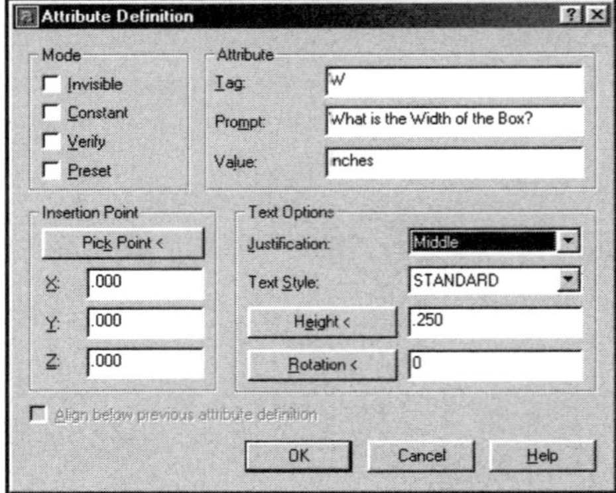

5. Assign Attributes for **Height**.
   a. Tag = H
   b. Prompt = What is Height of the Box?
   c. Value = inches
   d. Justification = Middle
   e. Text Style = Standard
   f. Height = .250
   g. Rotation = 0
   h. Select OK and place as shown on 8-11.

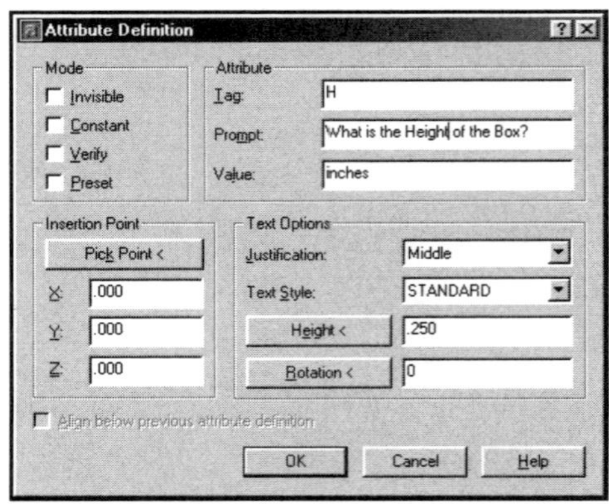

6. Assign Attributes for **Manufacturer**.
   a. Tag = MMMMM
   b. Prompt = Who is the Manufacturer?
   c. Value = (leave blank)
   d. Justification = Left
   e. Text Style = Standard
   f. Height = .250
   g. Rotation = 0
   h. Select OK and place as shown below.

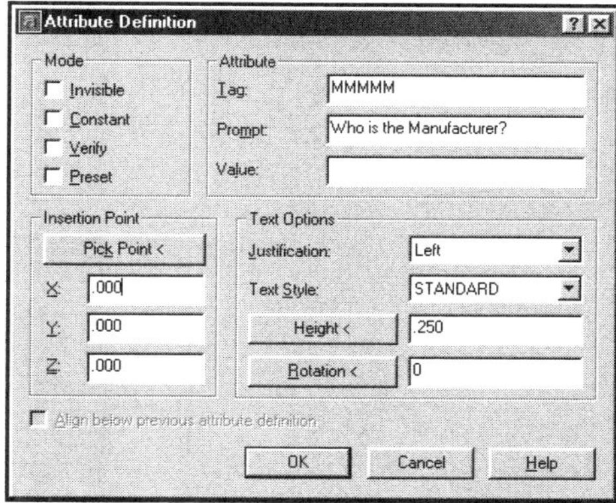

7. Assign Attributes for **Cost**.
   a. Tag = $$$$
   b. Prompt = What is the cost?
   c. Value = $0.00
   d. Select "Align below previous attribute definition" box.
   e. Select OK. But you do not have to place the attribute tag. It automatically "Aligned below previous attribute tag, MMMMM".

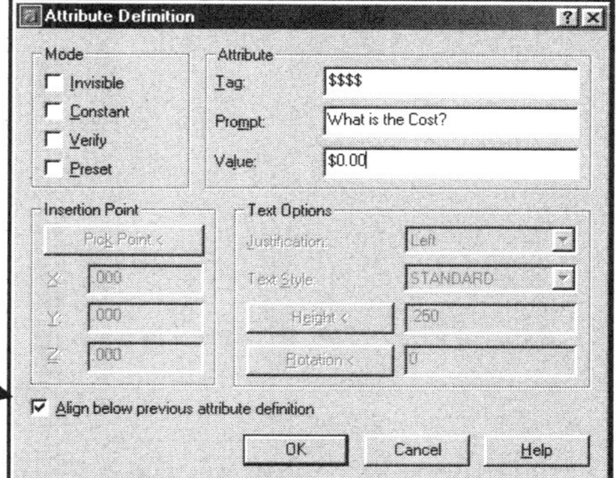

Your drawing should look approximately like the example below.

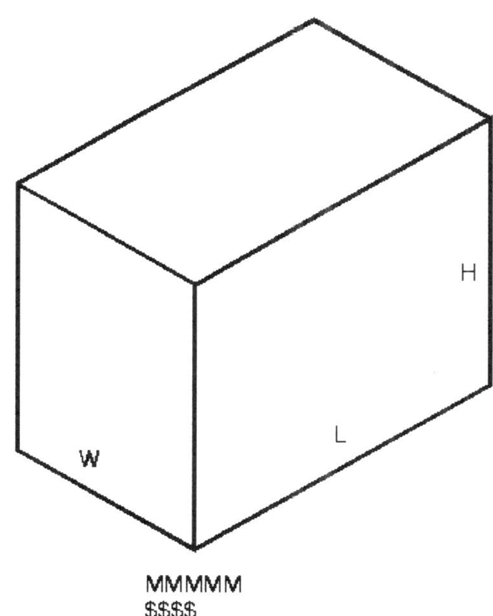

8. When you have finished assigning all of the attributes, create a **BLOCK**. (Select the Box and the Attribute text.) (Refer to page 8-2 for instructions if necessary)

9. Now **Insert** the new block anywhere on the screen using
   a. **INSERT / BLOCK**.
   b. Browse to find the block
   c. Select OK

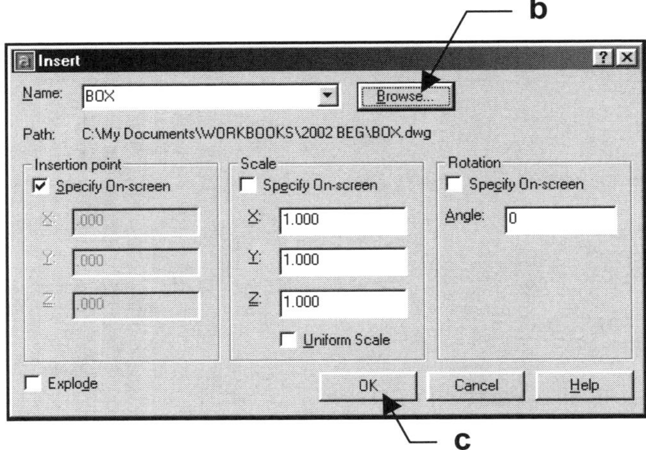

10. Type the answers in the box beside the attribute prompts as shown below.

Note: If the dialog box shown below does not appear, at the command line type ATTDIA <enter> and 1 <enter>.

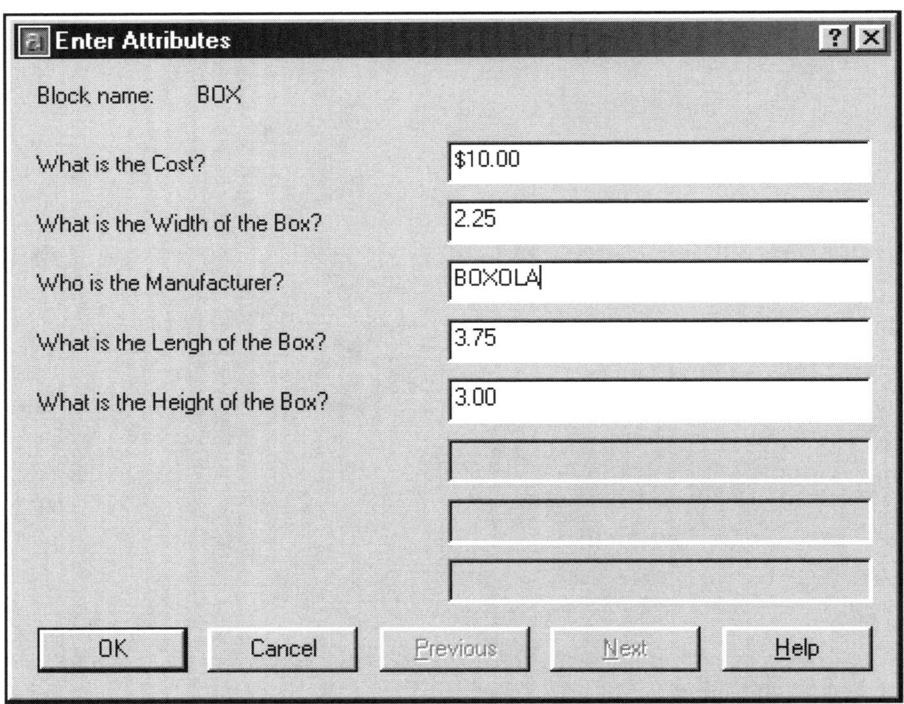

Notice that the prompts are not in any specific order. You can control the initial order of the prompts, when creating the block, by selecting the attribute text one by one with the cursor instead of using a window to select all objects.
In Lesson 9 you will learn how to rearrange the order after the block has been created.

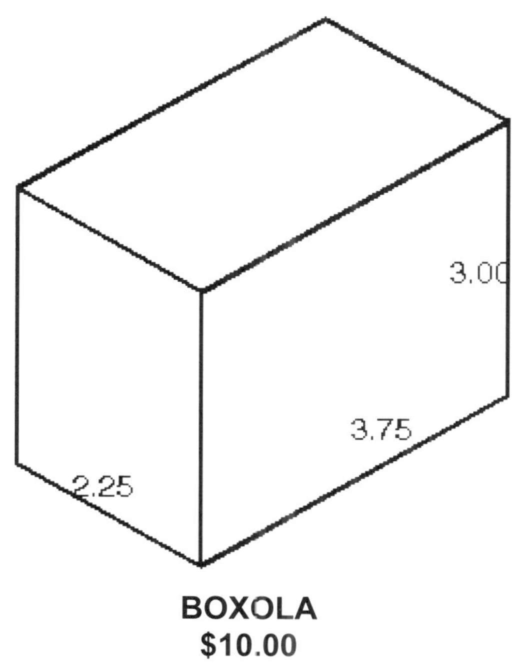

BOXOLA
$10.00

Does your box look like the example shown above?

11. Save as **EX-8A.**
12. **Do not plot.**

EXERCISE 8B

Assigning multiple Attributes to multiple Blocks

The following exercise will help you learn how to assign multiple attributes to 3 different objects. This drawing will be used in lesson 9 to extract the information and place it in a spreadsheet such as Excel.

A. Open **EX-5D (or EX-5E if you have AutoCAD LT)**

B. Assign the attributes to the Sofa in the lobby, the Chair behind the office desk and one File Cabinet in the office.

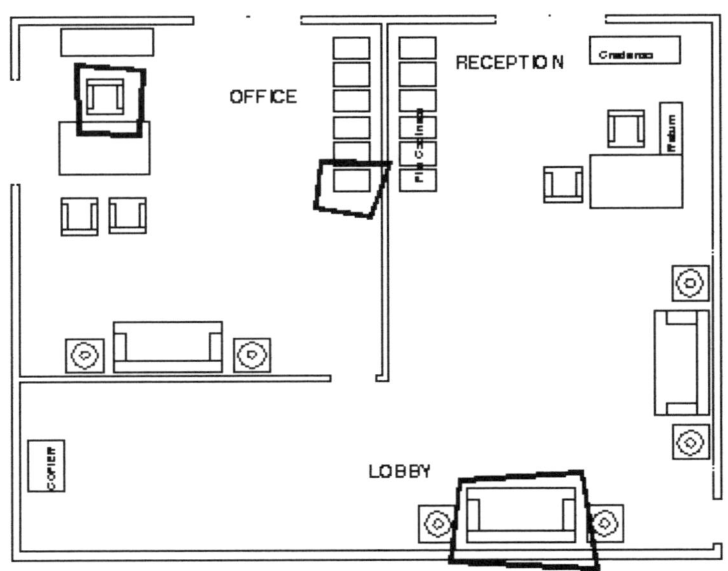

NOTE: The Attribute details for each object are on the next pages.

Before you start, check the ATTDIA setting. (Refer to 8-8)

C. LET'S START WITH THE SOFA

This is what the sofa should look like after you have completed the Attribute definitions shown below.

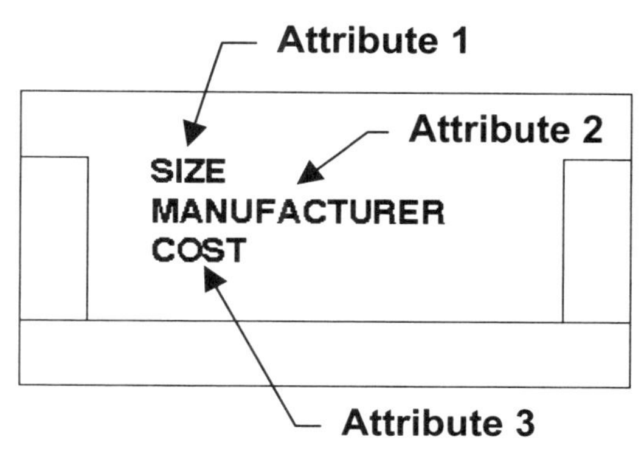

STEP 1

1. Select **Draw / Block / Define Attributes**
2. Fill in the boxes for Attribute 1.
3. Select the **OK** button and place the Tag (Size) approximately as shown.
4. Repeat steps 1, 2 and 3 for Attributes 2 and 3.

ATTRIBUTE 1

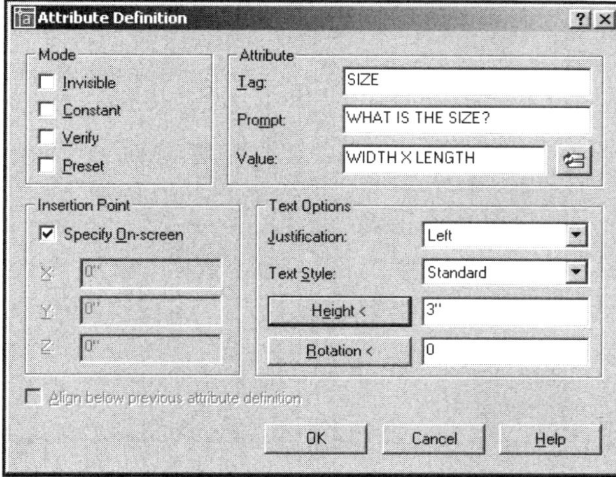

ATTRIBUTE 2

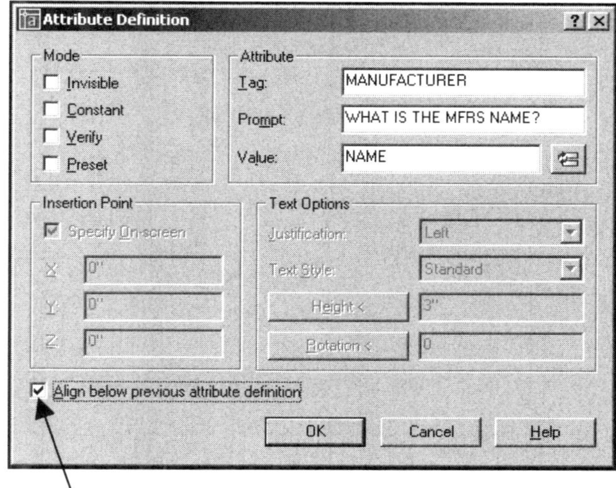

This option will align the new attribute directly below the previous and duplicate the Text options.

ATTRIBUTE 3

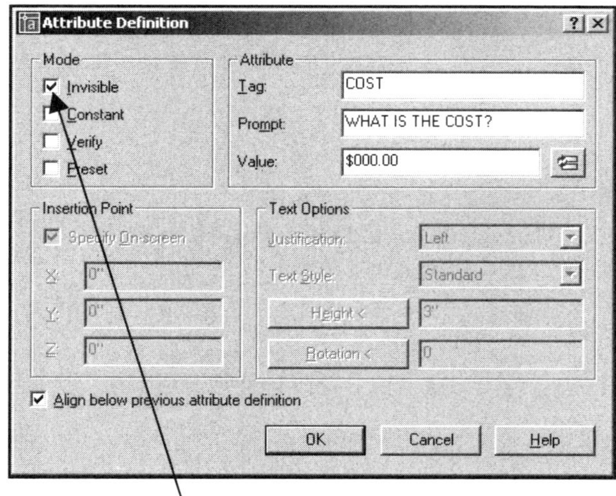

Note: the tag will not be invisible until you insert the block.

STEP 2

5. Now **create a block**.
   a. Select **Draw / Block / Make**
   b. Name = Sofa
   c. Select objects = select the sofa and the attribute text.
   d. Pick Point = Any corner of the sofa.

D. NOW DO THE CHAIR

This is what the chair should look like after you have completed the Attribute definitions shown below.

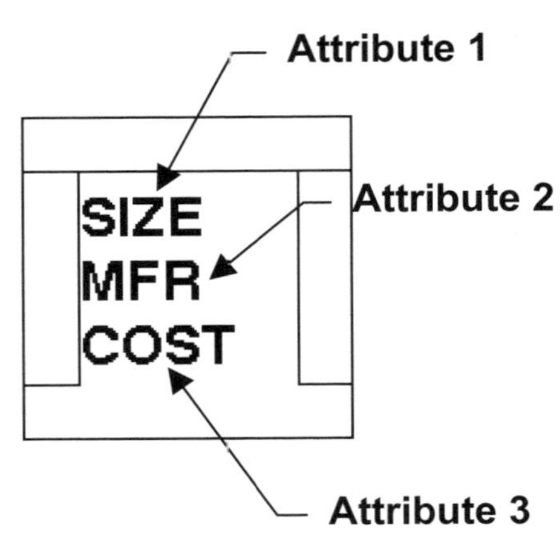

STEP 1

1. Select **Draw / Block / Define Attributes**
2. Fill in the boxes for Attribute 1.
3. Select the **Pick Point** button and place the Tag (Size) approximately as shown.
4. Repeat steps 2 and 3 for Attributes 2 and 3.

ATTRIBUTE 1

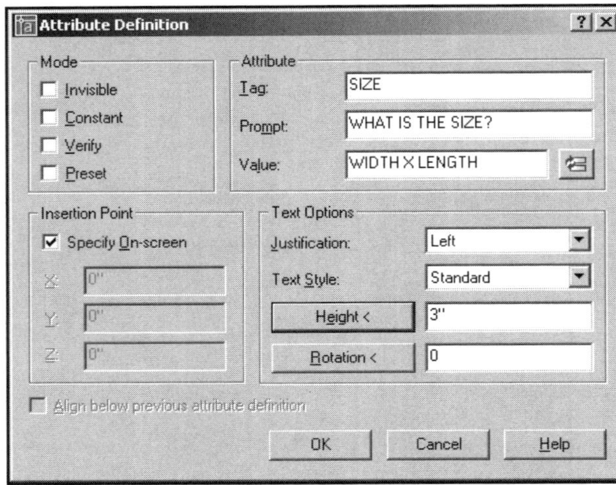

ATTRIBUTE 2

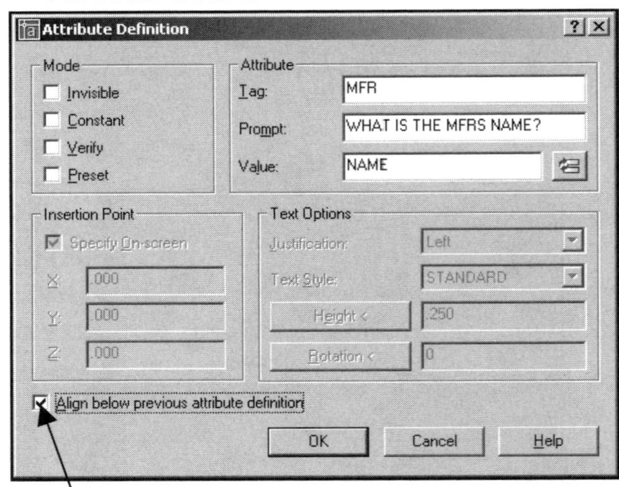

This option will align the new attribute directly below the previous and duplicate the Text options.

ATTRIBUTE 3

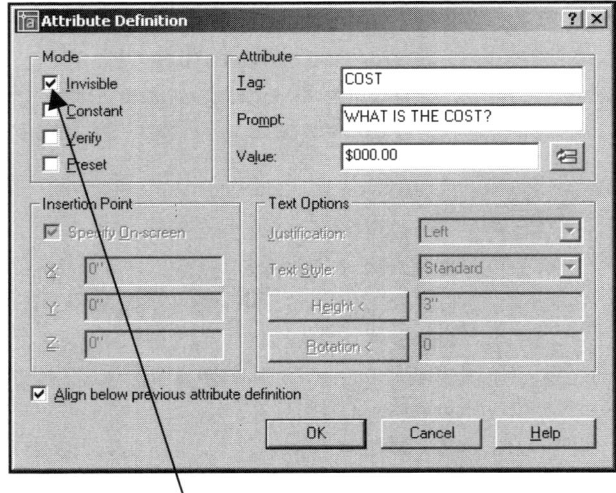

Note: the tag will not be invisible until you insert the block.

STEP 2

5. Now **create a block**.
   a. Select **Draw / Block / Make**
   b. Name = Chair
   c. Select objects = select the Chair and the attribute text.
   d. Pick Point = Any corner of the Chair.

E. NOW DO THE FILE CABINET

This is what the file cabinet should look like after you have completed the Attribute definitions shown below.

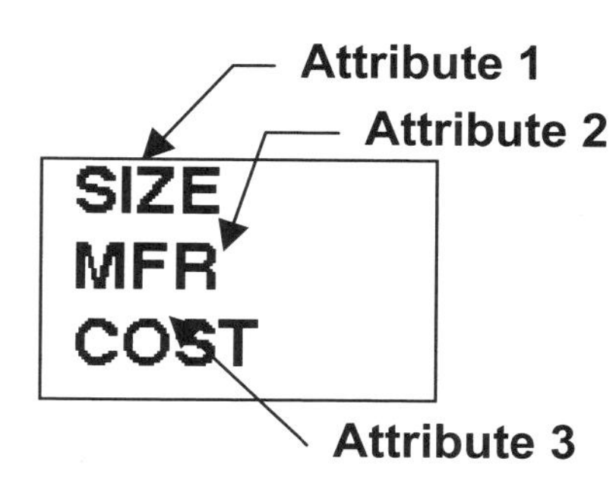

STEP 1

1. Select **Draw / Block / Define Attributes**
2. Fill in the boxes for Attribute 1.
3. Select the **Pick Point** button and place the Tag (Size) approximately as shown.
4. Repeat steps 2 and 3 for Attributes 2 and 3.

ATTRIBUTE 1

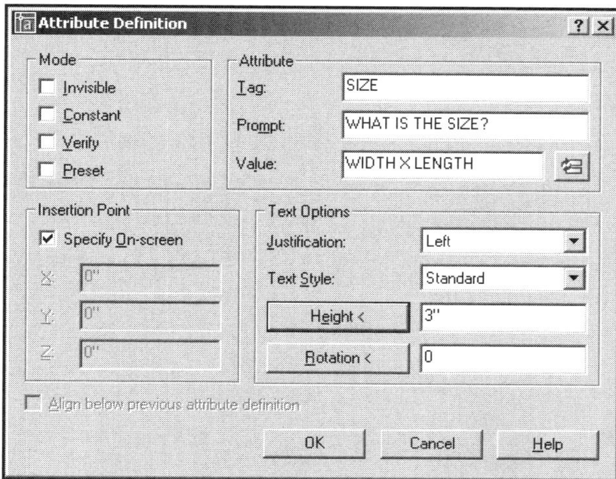

ATTRIBUTE 2

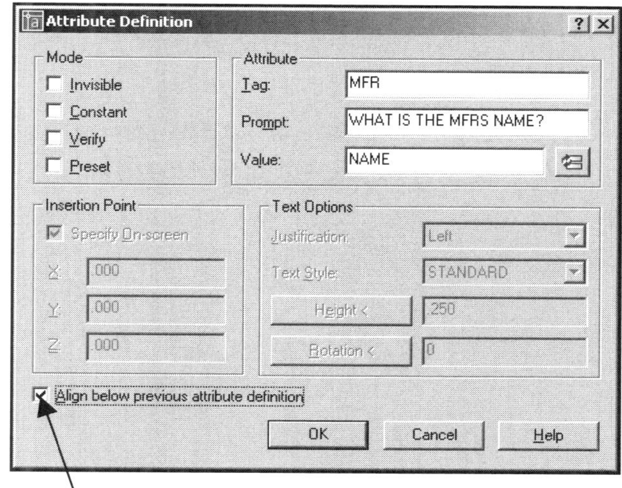

This option will align the new attribute directly below the previous and duplicate the Text options.

ATTRIBUTE 3

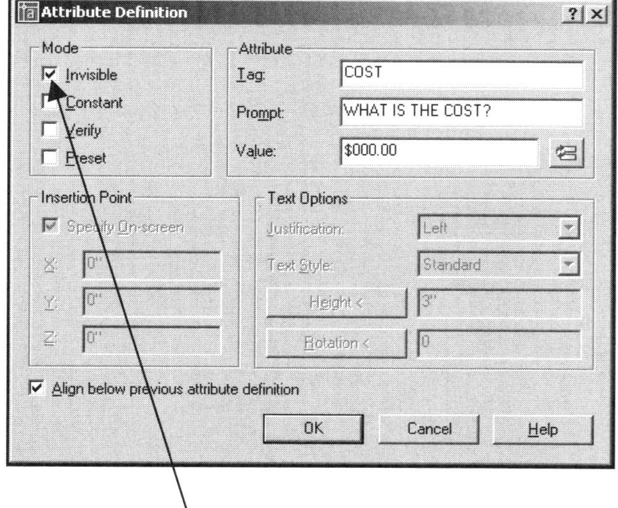

Note: the tag will not be invisible until you insert the block.

STEP 2

5. Now **create a block**.
   a. Select **Draw / Block / Make**
   b. Name = File Cabinet
   c. Select objects = select the File and the attribute text.
   d. Pick Point = Any corner of the File.

<u>Save this drawing at this point to be safe, so you don't lose anything.</u>

F. ERASE all of the Sofas, Chairs and File Cabinets in the drawing.

G. INSERT the **Sofa block** into the same location as ONE of the sofas that you previously erased.

1. Select **INSERT / BLOCK**
2. Select the **SOFA** block from the drop down list.
3. Select **OK**
4. Answer the Attribute prompts:

What is the SIZE?	3' X 6'
Who is the Manufacturer?	Sears
What is the Cost?	$500.00

5. Select **OK**.

Now your sofa should look approximately like the sofa below.
<u>Notice the Cost is invisible</u>.

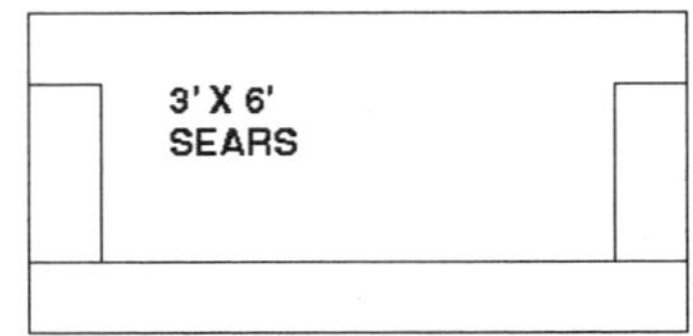

H. INSERT the **Chair block** into the same location as **ONE** of the chairs that you previously erased, as follows.

1. Select **INSERT / BLOCK**
2. Select the **CHAIR** block from the drop down list.
3. Select **OK**
4. Answer the Attribute prompts:

What is the SIZE?	2' X 2'
Who is the Manufacturer?	LAZYBOY
What is the Cost?	$200.00

5. Select OK.

Now your Chair should look approximately like the chair below.
<u>Notice the Cost is invisible.</u>

2' X 2'
LAZYBOY

I. **INSERT** the **File cabinet block** into the same location as **ONE** of the File Cabinets that you previously erased as follows.

1. Select **INSERT / BLOCK**
2. Select the **File Cabinet** block from the drop down list.
3. Select **OK**
4. Answer the Attribute prompts:

   What is the SIZE? 15" X 24"
   Who is the Manufacturer? HON
   What is the Cost? $40.00

5. Select **OK**.

Now your File cabinet should look approximately like the File Cabinet below.
Notice the Cost is invisible.

15" X 24
HON

Your drawing should look approximately like the example on the next page.

J. Save this drawing as: **EX-8B**

Do Not plot this drawing.

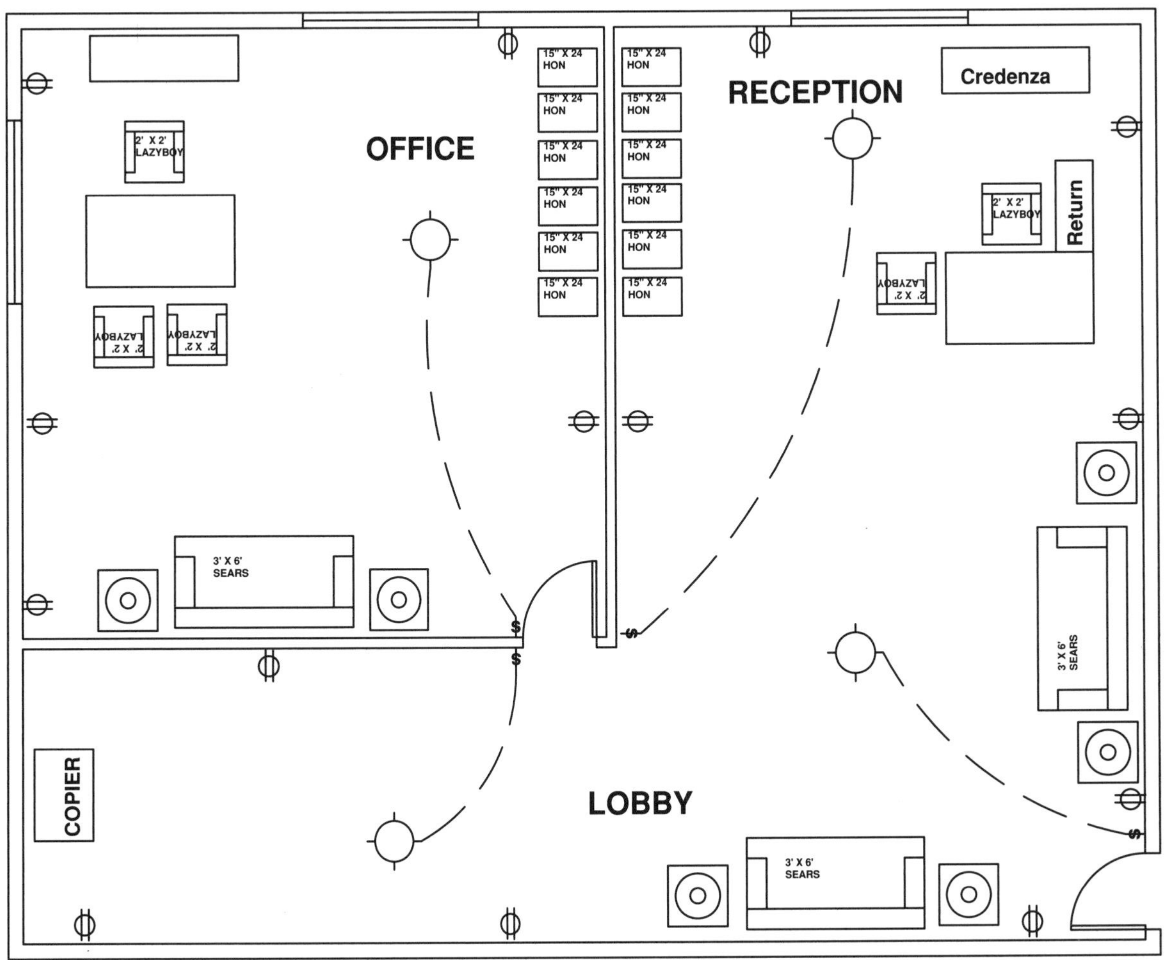

OFFICE
RECEPTION
LOBBY
Credenza
Return
2' X 2' LAZYBOY
15" X 24 HON
3' X 6' SEARS
COPIER
$

EXERCISE 8B

DYNAMIC BLOCKS

What are Dynamic Blocks?
Dynamic Blocks are blocks that can be dynamically modified to represent different variations of the block.

An example could be a window or a screw. You will probably have many different sizes of windows in a floor plan or length of screws in an assembly. If you created a block for every variation you would have a long list of symbols.

Dynamic blocks allow you to create only one block but then define different lengths, size, rotation etc.

What makes a Block Dynamic?
Blocks are made dynamic by adding Parameters and Actions. The Block Editor is used to add these features. Once these features are added, the block can be changed easily after inserting. You will modify the blocks appearance using the custom grips or properties.

Previous to the Block Editor, blocks with many variations, built in, had to be created using AutoCAD's programming code AutoLisp. Those of us that are not programming types found this a bit difficult and would prefer to create all of the variations by drawing them and adding them to the symbol library. But now AutoCAD has made it possible for non-programming types to create Dynamic Blocks. Although I have to admit that it is not a piece of cake. It takes some studying to master this feature.

The following is the Block Authoring Palettes: **Parameters** and **Actions**.

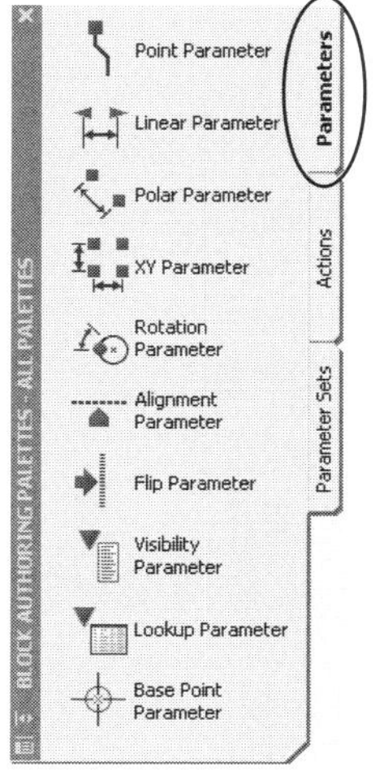

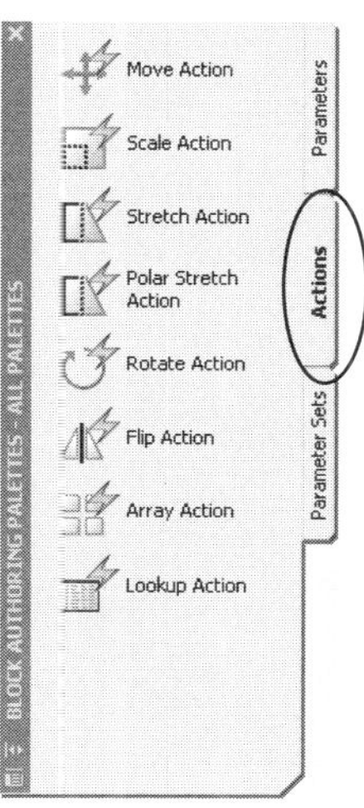

DEFINING A DYNAMIC BLOCK

Let's try creating a very simple Dynamic Block. If you find this too difficult, Dynamic Blocks are not for you. If you find it fascinating and can visualize great advantages by using this new feature you will want to learn more.
The instructions below will show you how to add dynamic stretch to an existing block.

First you need to create a block or insert an existing block.

1. Open **My Feet-inches setup**.

2. Select the **24 X 18 (Qtr-ft)** layout tab.

3. Draw the window shown below. Do not dimension.

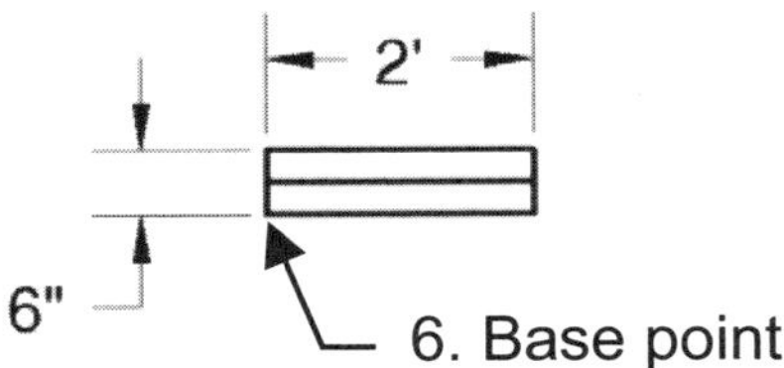

4. Select the **Make Block** command.

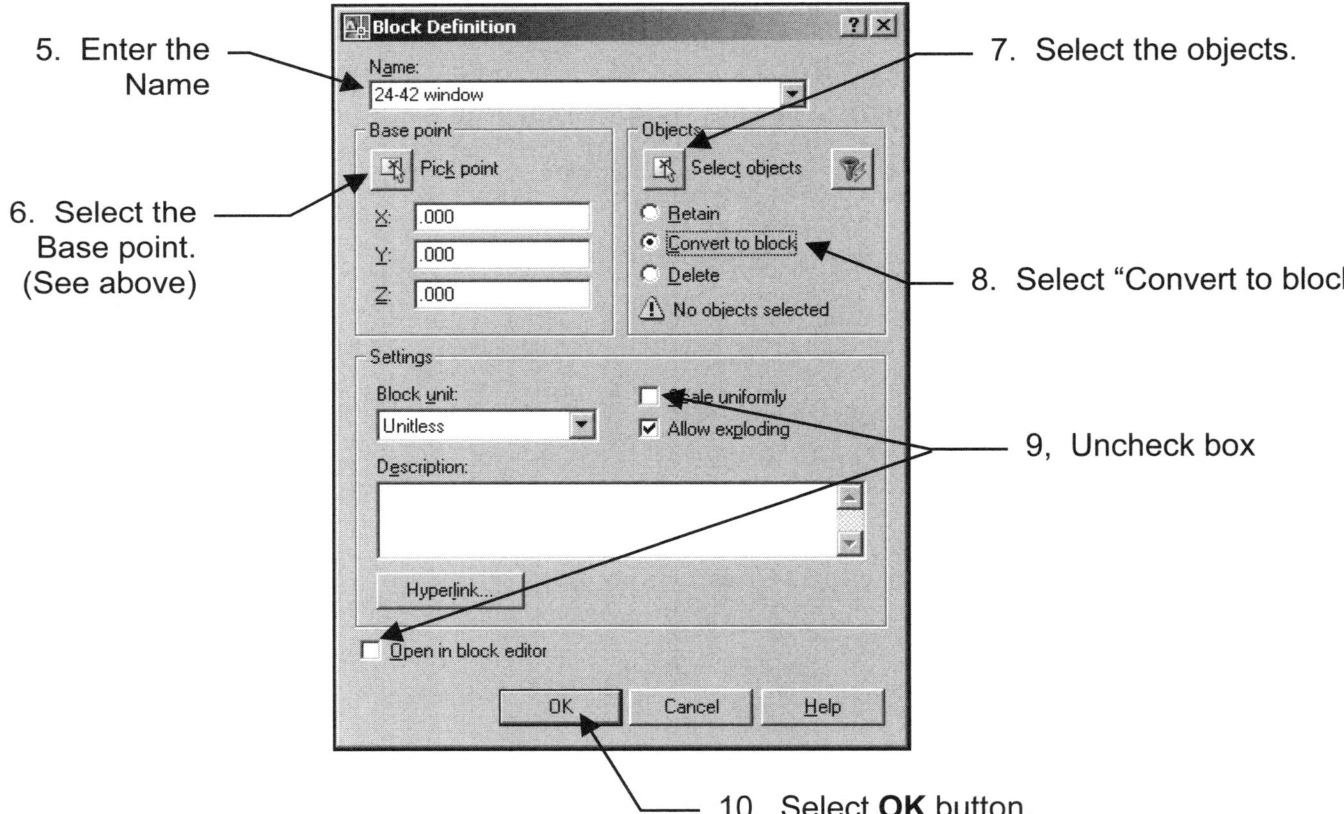

Next you have to define the Parameter.

11. Select **Tools / Block Editor** or type **BEDIT** on the command line.

    The **Edit Block Definition** dialog box should appear.

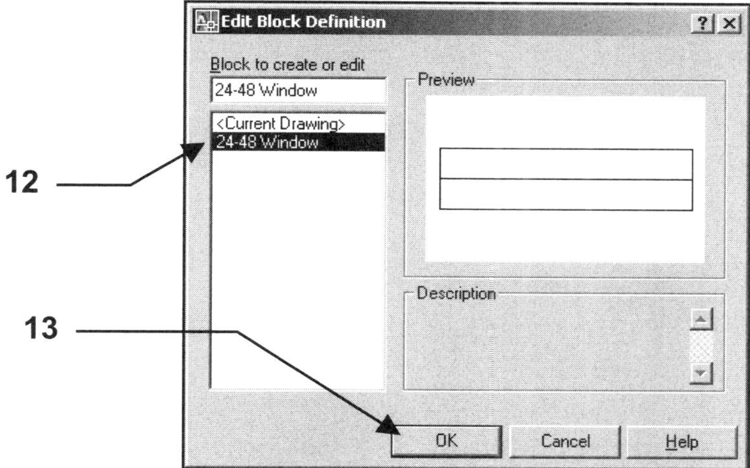

12. Select the block.

13. Select the **OK** button.

> ***Note***: *Click on* ***NO*** *to view AutoCAD's demo on "How to define a dynamic block"*
> *You can always see it later.*

The **Block Authoring Palettes** and the **Block Editor** tool bar will appear.

14. Parameter Sets tab.

15. Linear Stretch tool.

24-48 Window

14. Select the **Parameters** tab.

15. Select the **Linear** parameter tool.

16. Select "**Value Set**" option from the command line by typing **V <enter>.**

17. Select "**Increment**" option from the command line by typing **I <enter>.**

18. Enter the following values:
    Distance Increment : **3**
    Minimum distance: **24**
    Maximum distance: **48**

19. Specify **Start** and **End** points by snapping to the corners shown below.

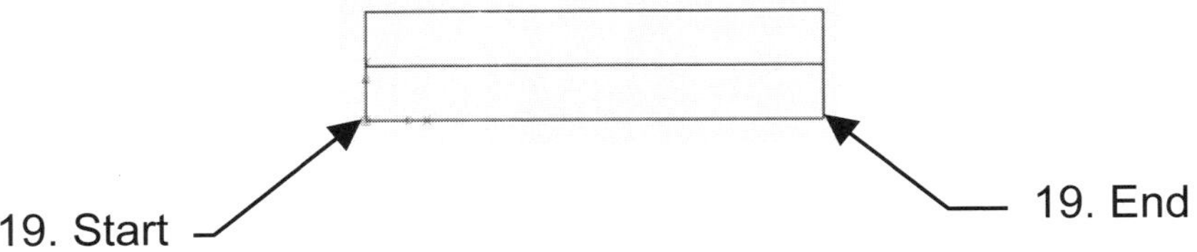

20. Specify Label location.

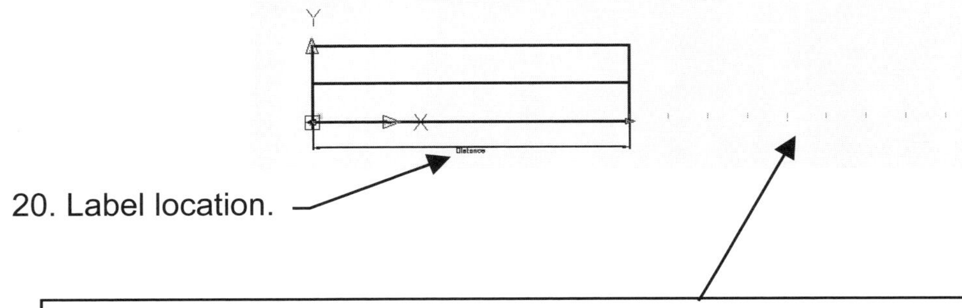

Notice the 8 little bars on the right hand end of the window. Those are the 3" incremental distances you defined in #18 above. When you have completed the following steps, you will be able to stretch the block from 24" to 48".

21. Change the parameter to have only 1 grip.
    a. Click on the **Parameter**. (Distance)
    b. Right click and select **<u>Grip Display</u>** then **1**.

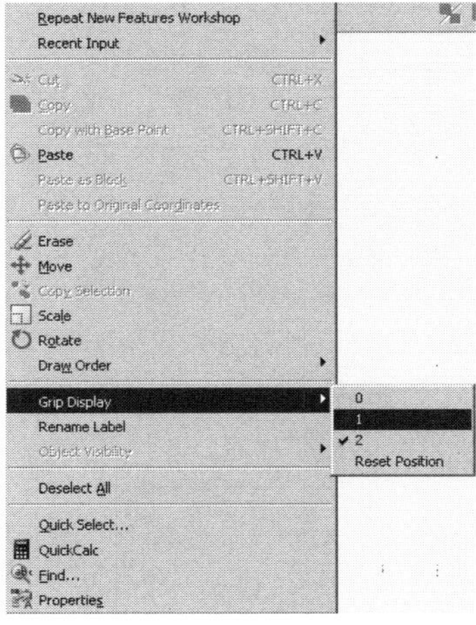

Now you need to assign the Action to the Parameter.

22. Select the **Action tab**.

23. Select the **Stretch Action** tool.

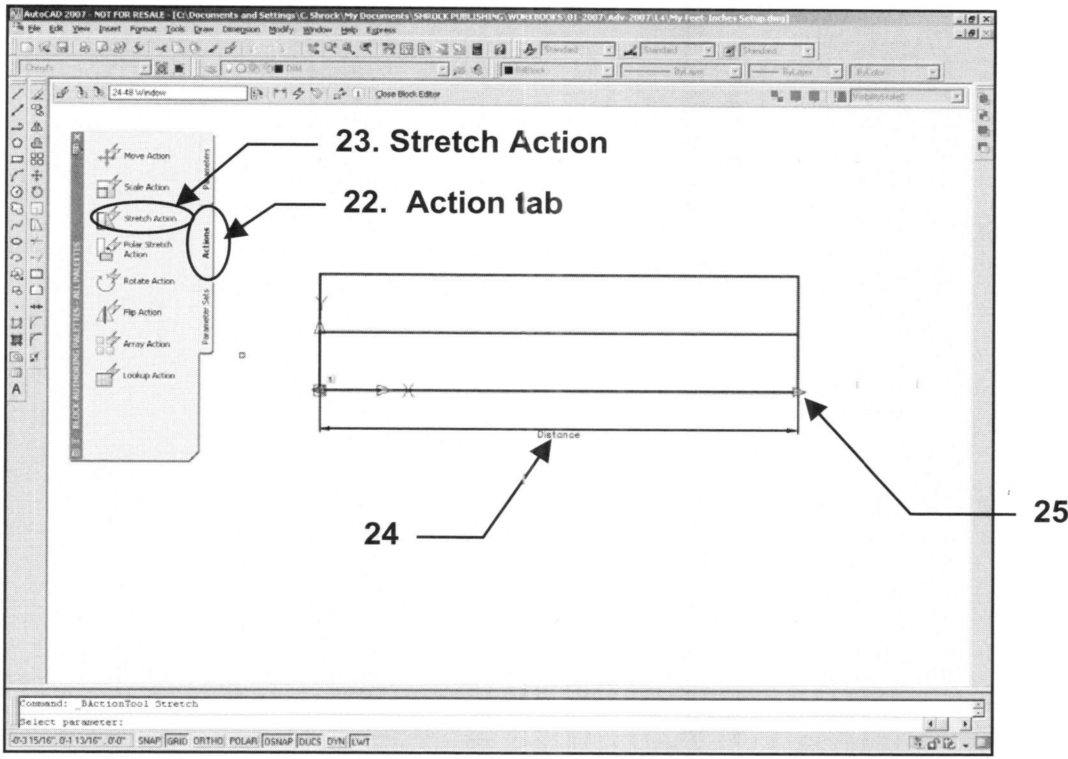

24. Select the **Parameter** to assign the Action. (Click on Distance)

25. Click on the **Grip**.

26. Define the **stretch frame** using a **crossing window**.

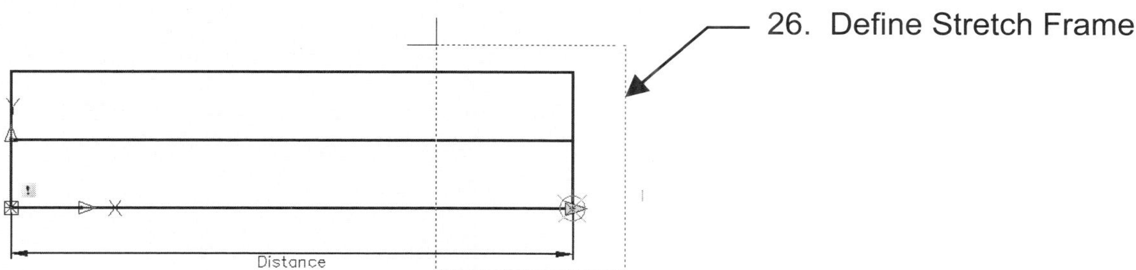

27. **Select the objects** to stretch using a **crossing window** again.
Press <enter> to stop "selecting objects".

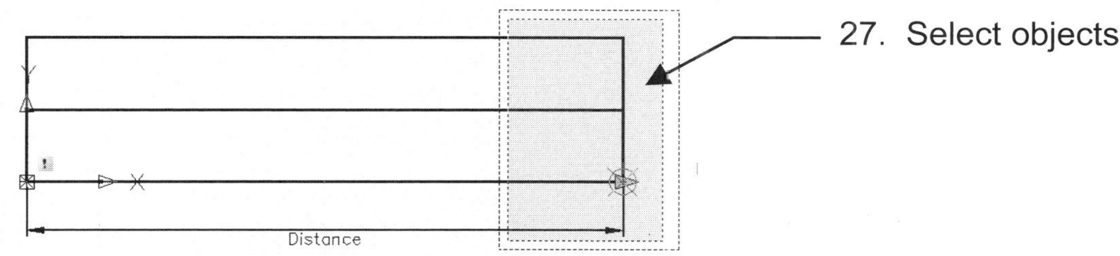

28. Place the **Action label**.

29. Select the "**Save Block Definition**" button.

30. Select the "**Close Block Editor**".

31. Click on the **24-48 Window** block.
 Notice the grips appear.

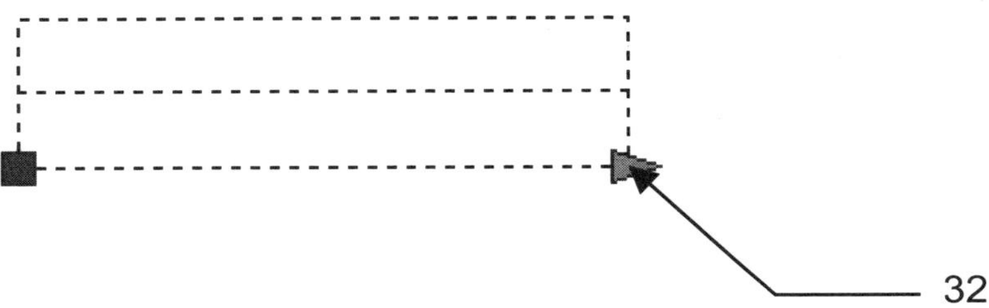

32. Click on the "**Block Stretch** " grip.

 You may slide the cursor to the right in 3" increments to resize the window from 24" to 48".

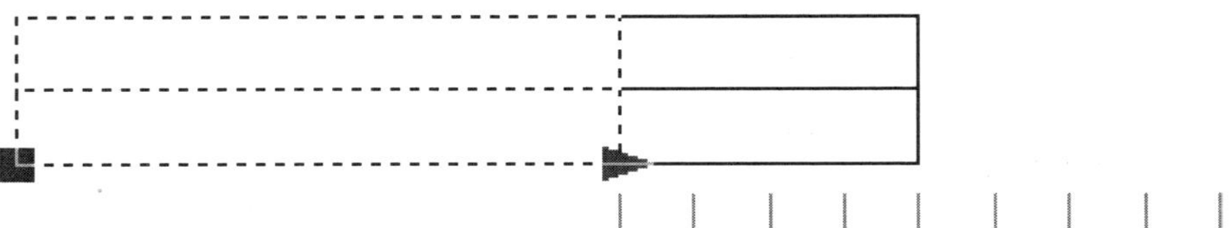

Creating a Lookup Table

Now lets attach a List to the 24-48 Window Block from which you, or who ever use this block, can make a selection. The list will have 3 different window lengths with descriptions. When you make a selection from the list the block will automatically change to the selected length. The List is called a Lookup Table.

1. Insert the **24-48 Window Block** into the current dwg if it is not already on the screen.

2. Select the **Block Editor**.

3. Select the **24-48 Window** from the list and select **OK**.

4. Answer **NO** to AutoCAD's invitation to view the Dynamic Block demo.

5. Select the Parameters tab from the Block Authoring Palette and select the Lookup Table parameter.

6. Place the Lookup Parameter approximately as shown below.

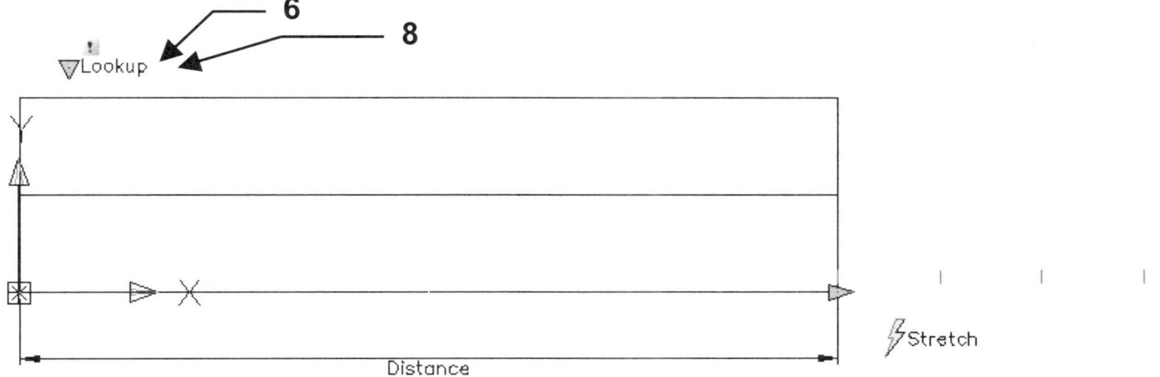

7. Select the Actions tab on the Block Authoring Palette and select the Lookup Action.

8. Click on the Lookup parameter that you previously placed in #6.

9. Place the Lookup Action approximately as shown below.

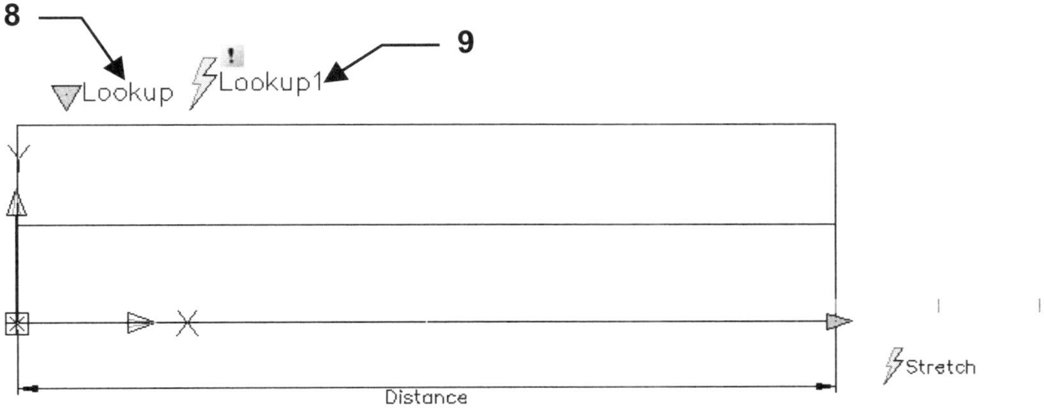

The Property Lookup Table should appear.

10. Select the **Add Properties** button.

11. Select the Parameter from the Parameter Name list. (If you had more parameters applied to this block they would be listed here.)

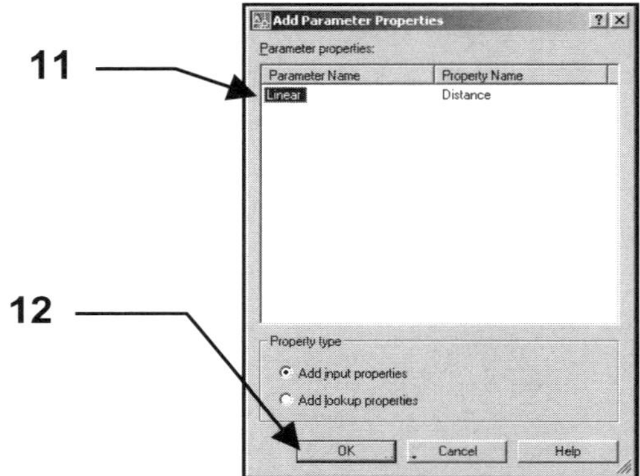

12. Select **OK** button.

The Property Lookup Table dialog box returns but this time it has “Distance” as a header for the Input Properties column.

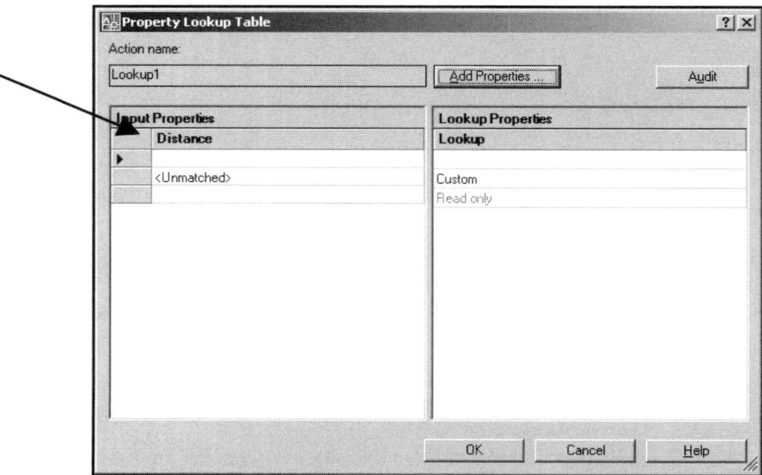

Now you will assign the Distances and the Descriptions.

13. Click on the first Input cell and enter the 2'. (Do not press <enter>)

14. Click on the first Lookup cell and enter 24" Anderson, Dual Pane.
 (Do not press <enter>)

15. Enter the remaining 2 rows using the same method.

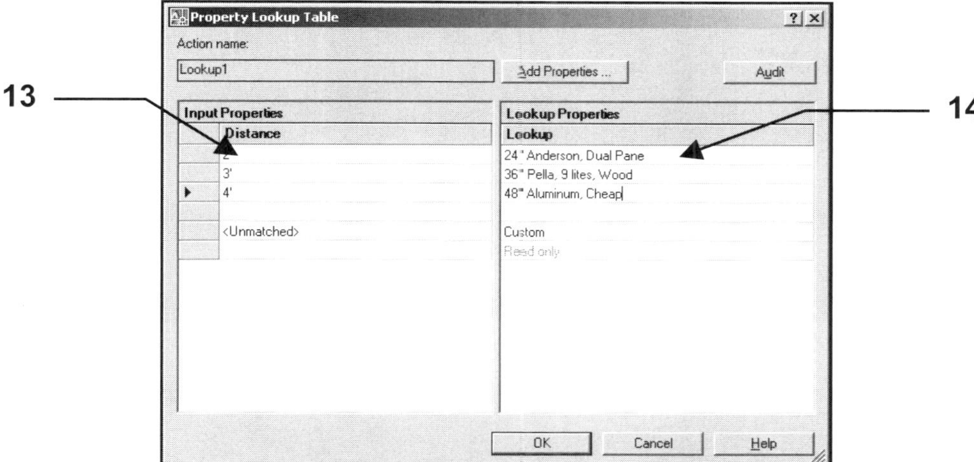

16. Select **Read Only**. It should change to "**Allow reverse lookup**".
 (Note: All the Lookup rows must be unique for this to happen)

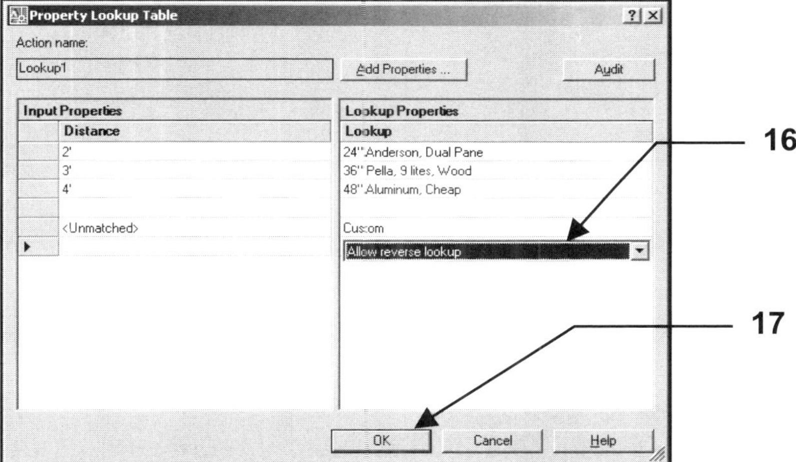

17. Select **OK** button.

18. Select the **Save Block Definitions** button on the **Block Editor** toolbar.
 (Refer to #29, pg. 8-5)

The following warning will appear. <u>Select the **Yes** button</u>.

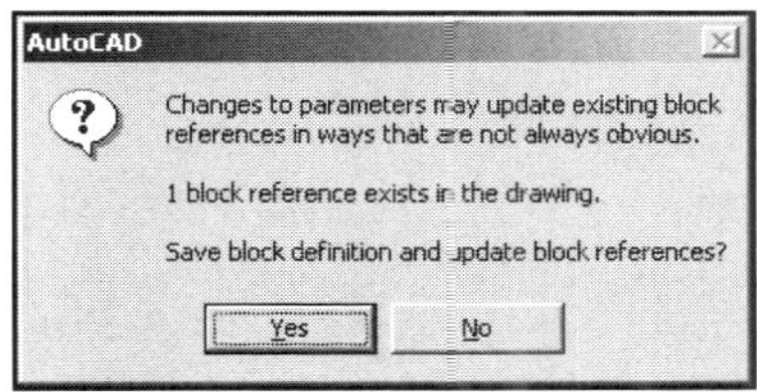

19. Select **Close Block Editor**. (Refer to #29, page 8-5)

20. Click on the 24-48 Window block that should be on your screen right now.

The Stretch grips and the Lookup should appear.

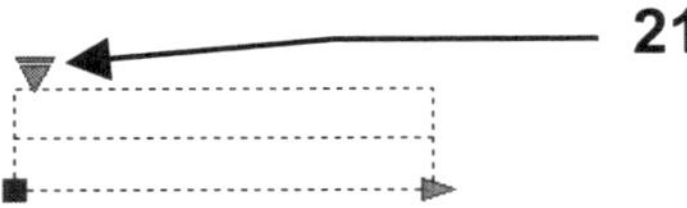

21. Click on the Lookup.

The Lookup drop down list should appear.

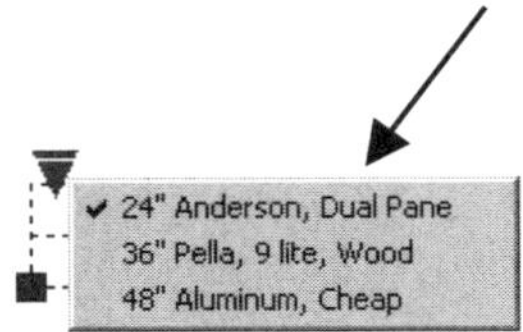

22. Select the 48" Aluminum, Cheap.

The block should have automatically lengthened to the 4' length that you assigned in the Distance column in #15 on page 8-8.

Now you can use the Stretch grip or the Lookup Table to change the length of this block.

As you have probably noticed Dynamic Blocks take a lot of time to master. The above example was merely to introduce you to Dynamic Blocks. There are many more tools that you may find useful. Go through the New Features demo and experiment with some of the other parameters and actions.

Additional notes about Dynamic Blocks.

After defining a Parameter an Exclamation Point appears.
This alerts you that you have not added an Action to the Parameter.

Sometimes the Parameter and Action labels become too large. If this happens, click the update Parameter and Action Text Size button on the Block Editor's toolbar.

LEARNING OBJECTIVES

After completing this lesson, you will be able to:

1. Edit Attributes
2. Extract Attributes
3. Extract data to an External file.

LESSON 9

EDITING ATTRIBUTES

After a block with attributes has been inserted into a drawing you may want to edit it. AutoCAD has many ways to edit these attributes.

Sorry LT users, you can only use the ATTEDIT command shown on page 9-4 to edit the attribute text.

1. Select the **BLOCK ATTRIBUTE MANAGER** command using one of the following:

TYPE = BATTMAN
PULLDOWN = MODIFY / OBJECT / ATTRIBUTE / BLOCK ATTRIBUTE MANAGER
TOOLBAR = MODIFY II

2. Select the block that you want to edit.

The following dialog box will appear.

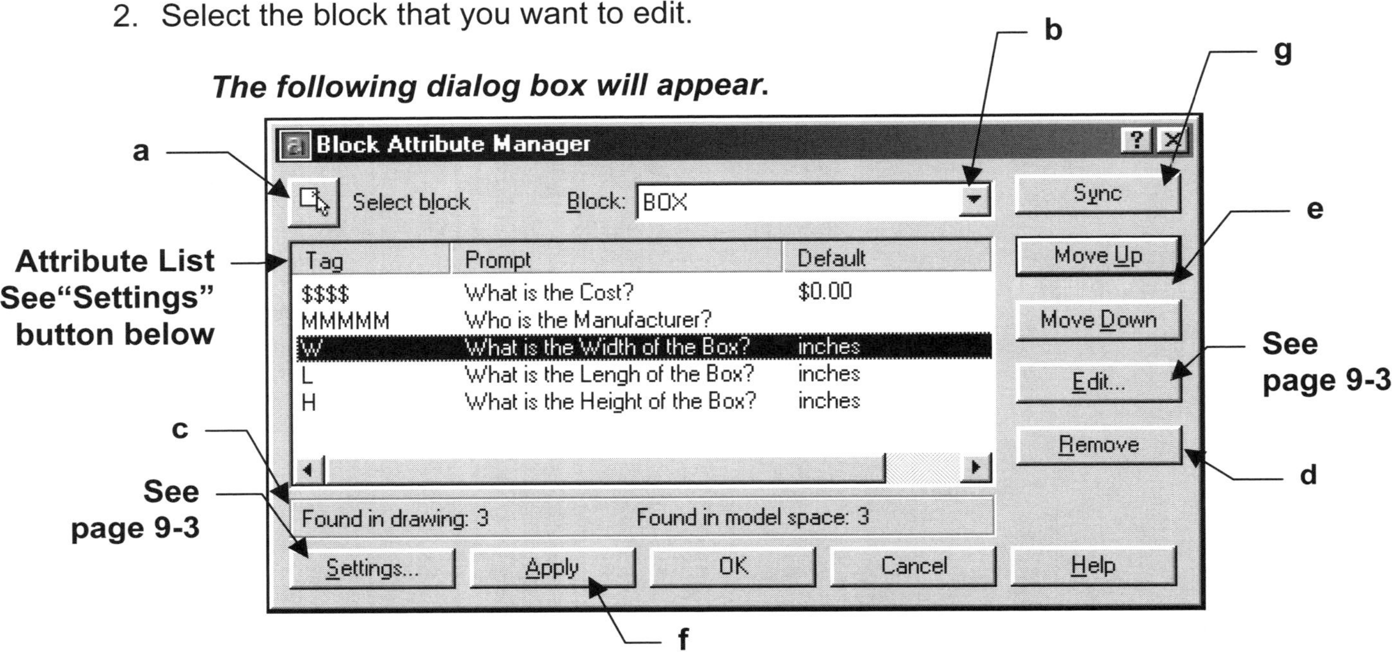

a. **Select Block** – Allows you to select another block – takes you back to the drawing so you can select another block to edit.

b. **Block down arrow** – Allows you to select another block –from a list.

c. **Found** – lists how many of the selected block were found in the drawing and how many are in the space you are currently in. (You must select the model "tab" or Layout "tab". It will not find model space attributes if you are not in model space.)

d. **Remove** - Allows you to remove an Attribute from a block. (See "Apply and Sync" below)

e. **Move Up and Down** - Allows you to put the prompts in the order you prefer.

f. **Apply** – After you have made all the changes, select the Apply button to update the attributes. (The Apply button will be gray if you have not made a change.)

g. **Sync** – If you explode a block, make a change and redefine the block, Sync allows you to update all the previous blocks with the same name.

SETTINGS 

The Settings dialog box controls which attributes are displayed. On the previous page TAG, PROMPT, DEFAULT and MODE are displayed. TAG values are always displayed.

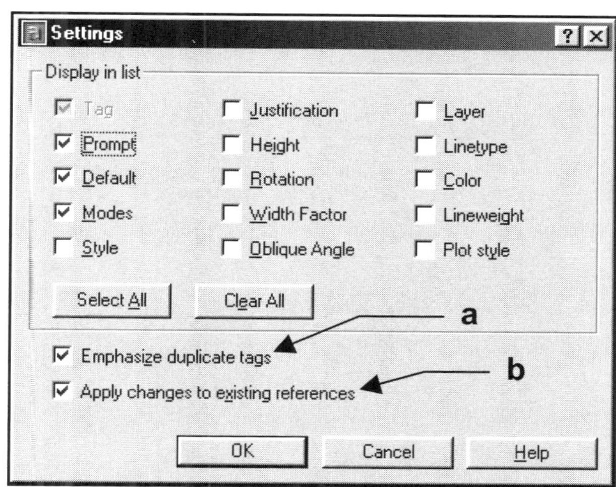

a. Emphasize duplicate tags
If this option is ON, any duplicate tags will display RED.

b. Apply changes to existing refs
If this option is ON, the changes will affect all the blocks that are in the drawing and all the blocks inserted in the future.

Note, very important: If you only want the changes to affect **future** blocks, do not select "Apply changes to existing ref".
If you want all blocks changed,
select the SYNC button. (see page 9-2)

EDIT ATTRIBUTE 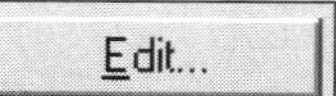

The Edit Attribute dialog box has 3 additional tabs, ATTRIBUTE, TEXT OPTIONS and PROPERTIES.

NOTE: If you select the **Auto preview changes** you can view the changes as you make them.

ATTRIBUTE tab
Allows you to change the MODE, TAG, PROMPT and DEFAULT.

TEXT OPTIONS tab
Allows you to make changes to the Text.

PROPERTIES tab
Allows you to make change to the properties.

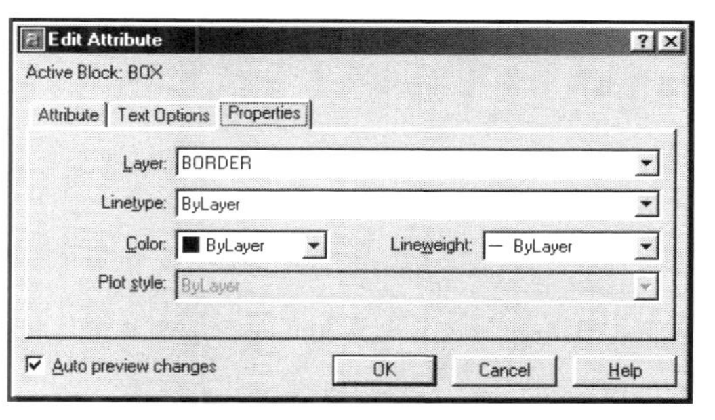

WHAT IF YOU JUST WANT TO EDIT ONE ATTRIBUTE IN ONE BLOCK?

The following two commands allow you to edit the attributes in only one Block at a time and the changes will not affect any other blocks.

Select one of the commands below and then select the block that you wish to change.

ATTENTION LT USERS: You may use the ATTEDIT command.

To change the Attribute value:

TYPE = ATTEDIT
PULLDOWN = NONE
TOOLBAR = NONE

This command allows you to edit the attribute values for only one block.

This command will not affect other blocks.

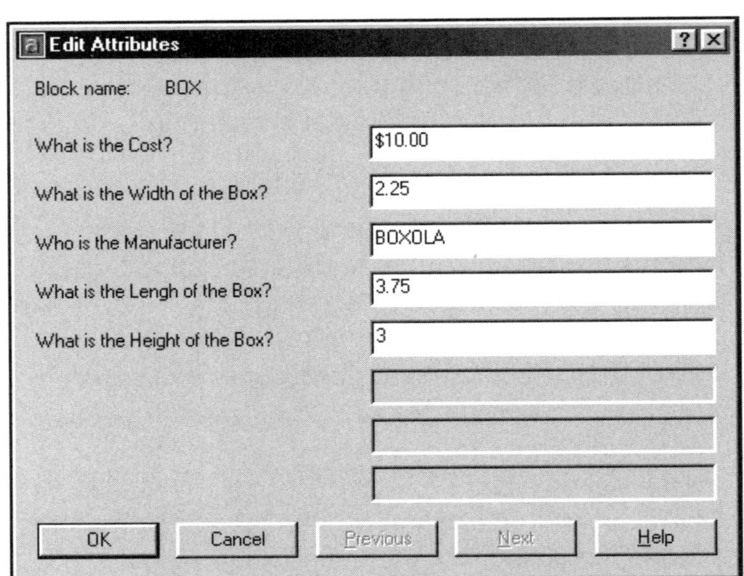

To change the Attribute structure:

TYPE = EATTEDIT
PULLDOWN = MODIFY / OBJECT / ATTRIBUTE / SINGLE
TOOLBAR = MODIFY II

or double click on the block

This command allows you to edit the Attributes, Text Options, Properties and the Values of only one block.

This command will not affect other blocks.

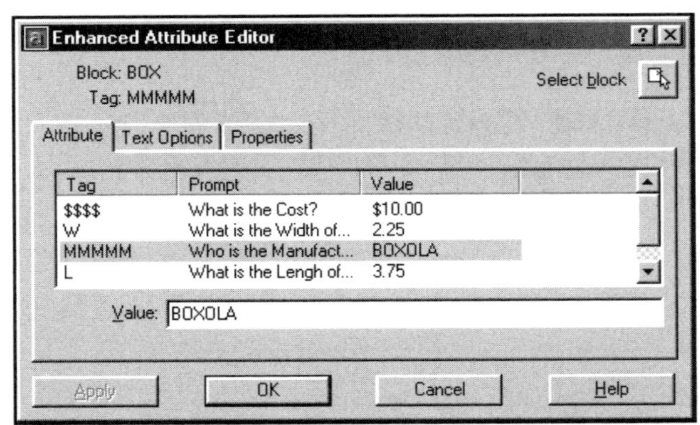

WHAT IF YOU WANT TO EDIT THE OBJECTS IN A BLOCK?

If you would like to add, delete or change objects within an existing block, you may do it easily with the command **Refedit**.
Refedit allows you to make changes to a block, saves the changes and updates all other previously inserted blocks within the drawing automatically.
The changes only affect the current drawing.

1. Select the Refedit command using one of the following:

   TYPE = Refedit
   PULLDOWN = Tools / Xref & Block In-Place Editing / Edit Ref In-Place
   TOOLBAR = MODIFY II

2. Select the Block you wish to edit. (Click on it, can't use a window)

The Reference Edit dialog box should appear.

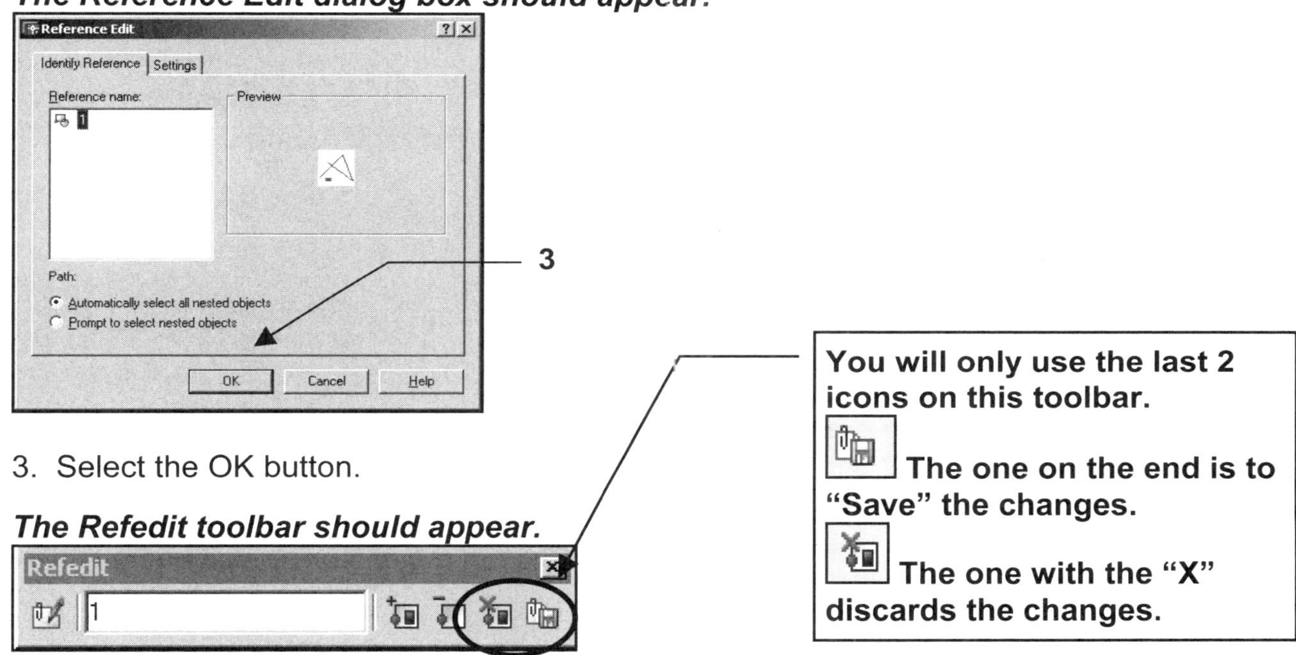

3. Select the OK button.

The Refedit toolbar should appear.

You will only use the last 2 icons on this toolbar.

The one on the end is to "Save" the changes.

The one with the "X" discards the changes.

Note: The selected block will remain bold but all other blocks of the same name will fade to gray. This is to emphasis the selected block.

4. Make the changes to the selected block.

Note: As long as the "Refedit" toolbar is open, anything you add, erase or change will affect the selected block.

5. Select the "**Save**" icon on the far right end of the toolbar.

6. A warning will appear. Select the OK button.

The block has been redefined and all of the existing blocks have been updated to reflect the changes you made. (These changes affect the current drawing only)

EXTRACT DATA FROM BLOCK ATTRIBUTES

Sorry LT Users, you can't use this command.

Your blocks can now contain attribute information (data) such as size, manufacturer, cost or maybe even a bill of materials. The next step is to learn how to **extract data** and save it to a table or to an external file such as a Microsoft Excel spread sheet.

This is a very simple process using the **Attribute Extraction Wizard**.

1. Select the **ATTRIBUTE EXTRACTION WIZARD** using one of the following:

   TYPE = EATTEXT
   PULLDOWN = TOOLS / ATTRIBUTE EXTRACTION
   TOOLBAR = MODIFY II

The Attribute Extraction – Begin (Page 1 of 6) dialog box should appear.

2. Select one of the following:

   Create table or external file from scratch:
   Uses the settings you specify as you proceed through the wizard.

   Use template (schedule, parts list, etc):
   Uses settings previously saved in an attribute extraction template file.

3. Select **Next >**

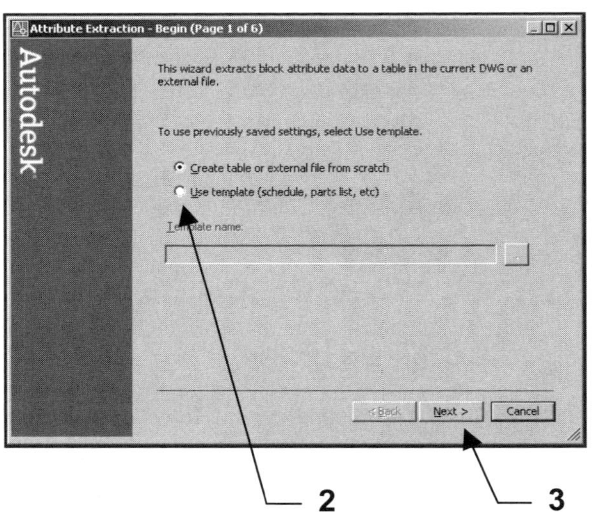

The Attribute Extraction—Select Drawings (Page 2 of 6) dialog box should appear.

4. Select the **Additional Settings..** button.

4

5. Select **Block** and **Count** settings.

6. Select **OK**

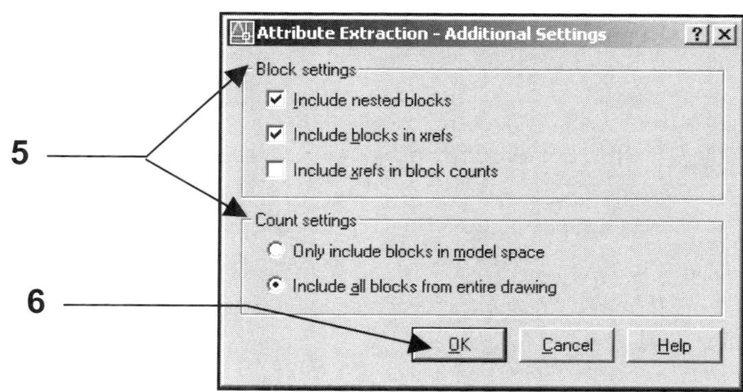

The Attribute Extraction- Select Drawings (Page 2 of 6) dialog box should re-appear.

7. Select one of the following:

   Select Objects: You can select individual blocks within the current drawing.

   Current Drawing: Selects all of the Attributes in the current drawing.

   Select Drawing: You can browse and select multiple drawings.

8. Select **Next >.**

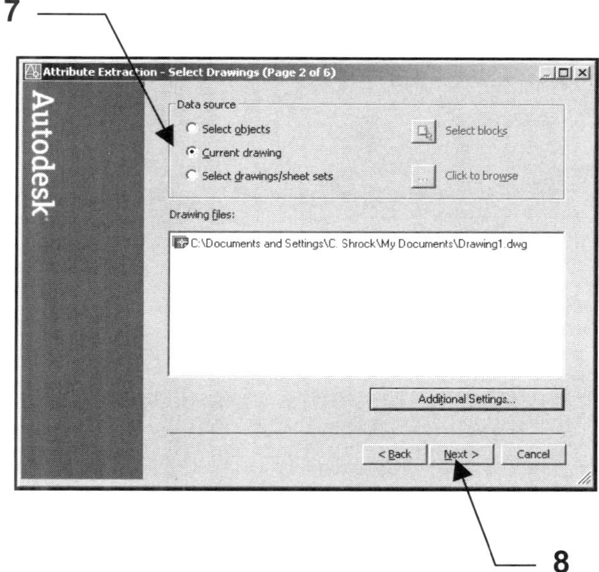

The Attribute Extraction-Select Attributes (Page 3 of 6) dialog box should appear.

9. Check the **Exclude** boxes.
   This allows you to display only the attribute information.

10. Click the **Type** column header.
    This arranges the types alphabetically.

11. Check the boxes you wish to extract.

12. Change **Display Name**.
    Right click on the name and select edit name.

13. Select **Next>.**

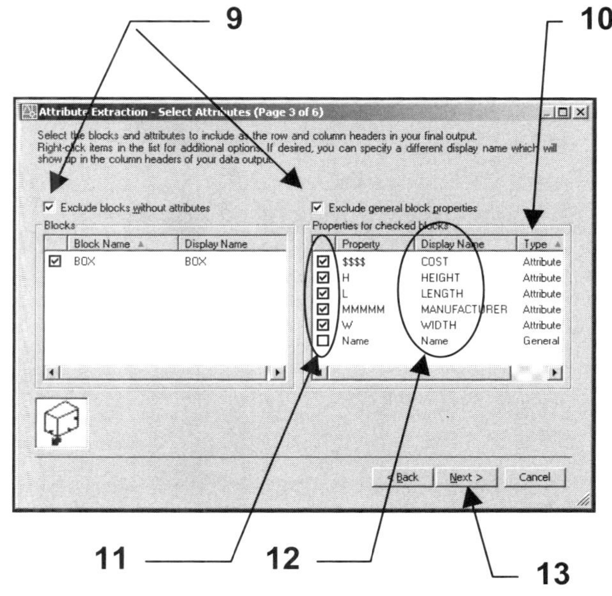

The Attribute Extraction – Finalize Output (Page 4 of 6) dialog box should appear

14. Select if where you want the data extracted:

a. AutoCAD table within the current drawing. **(If you select this option continue onto step 15)**

b. An External file such as Microsoft Excel. **(If you select this option skip to page 9-10)**

Note: Read the options at the top of the dialog box to rearrange information.

15. Select Next >.

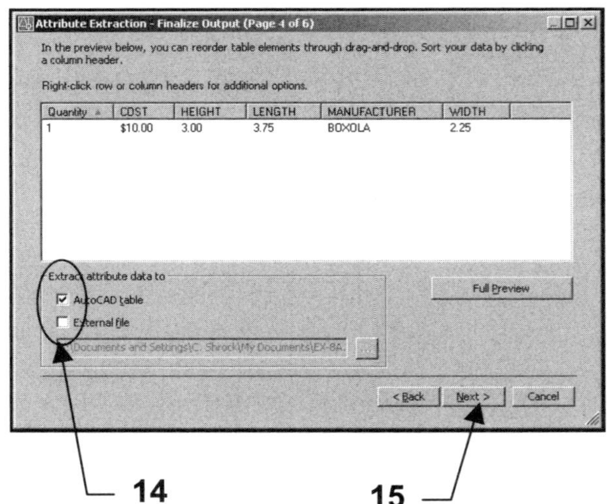

The Attribute Extraction – Table Style (Page 5 of 6) dialog box should appear

16. Enter a Title for the Table.

17. Select a table style from the drop down list or select the [...] button to create a new table style.

18. Select Next >

Note:
TABLES will be discussed in lesson 21.

The Attribute Extraction – Finish (Page 6 of 6) dialog box should appear

Read the information at the top of the dialog box and make your selections depending on where you want to extract the data.
You may wish to experiment with the different options.

19. Select Finish >

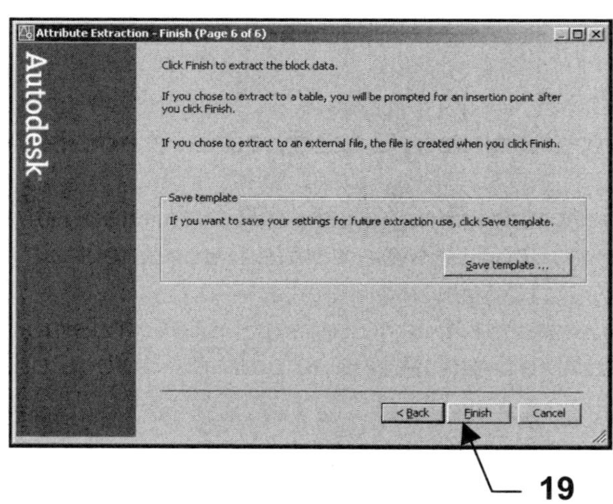

The Extracted Attribute data table is complete and will be inserted into the current drawing on the current layer.

BOX					
Quantity	COST	HEIGHT	LENGTH	MANUFACTURER	WIDTH
1	$10.00	3.00	3.75	BOXOLA	2.25

To extract the attribute data to an External Spread sheet such as Microsoft Excel, refer to the next page.

EXTRACT DATA FROM BLOCK ATTRIBUTES TO AN EXTERNAL FILE.

1. Follow instructions 1 through 13 on the previous pages.
2. Select External File.
3. Select the [...] button.

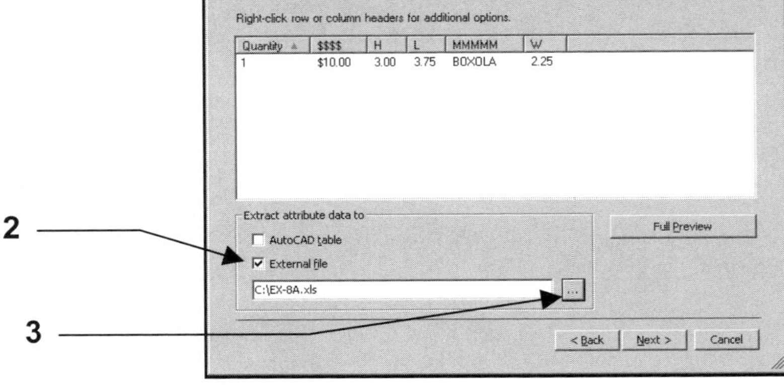

4. Locate where you wish to save the file.
5. Change **File of Type** to:

.xls for Microsoft Excel
.txt for Word Processing Program or Notepad.

6. Enter a name for the file.
7. Select **Save** button.

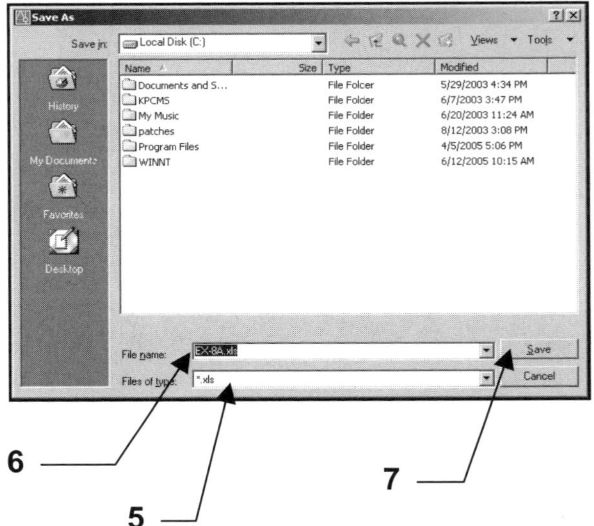

8. Select **Next >**

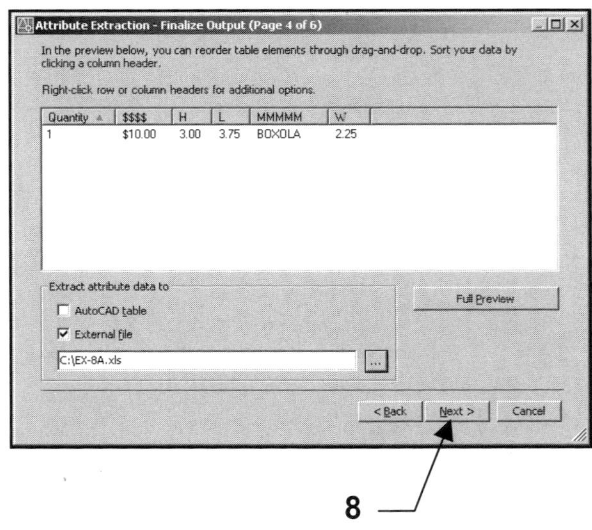

9. Select **Finish**.

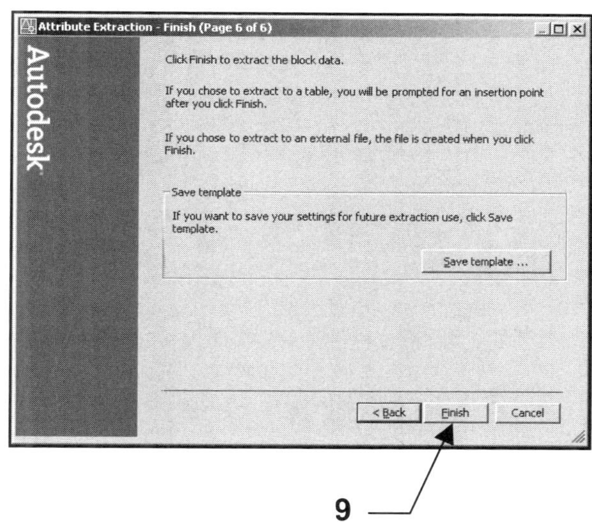

10. Go to the External Program software.

11. Select **File / Open**.

12. Locate the file.

Example of Microsoft Excel

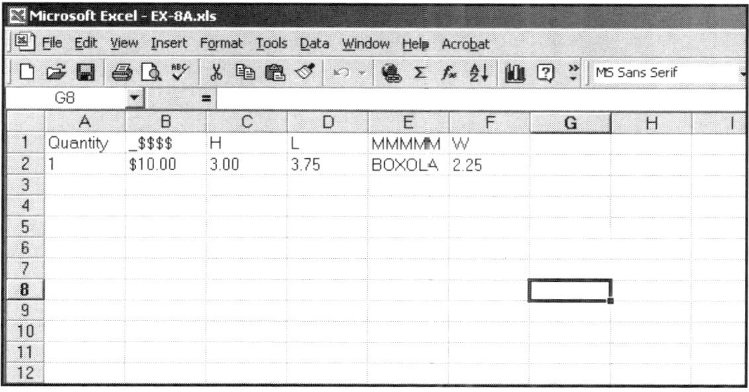

EXERCISE 9A

Extracting Attributes to an AutoCAD table.

The following exercise will take you through extracting attributes. You will open an existing drawing and extract the attribute data to an AutoCAD table.

1. Open **Ex-8B**.
2. Select the **24 X 18 Qtr-Ft** layout tab.
3. Make sure you are in **Paper space**.
4. Select **Tools / Attribute Extraction**
5. Select **Create Table or external file from scratch.**
6. Select **Current Drawing**
7. Make the following selections.

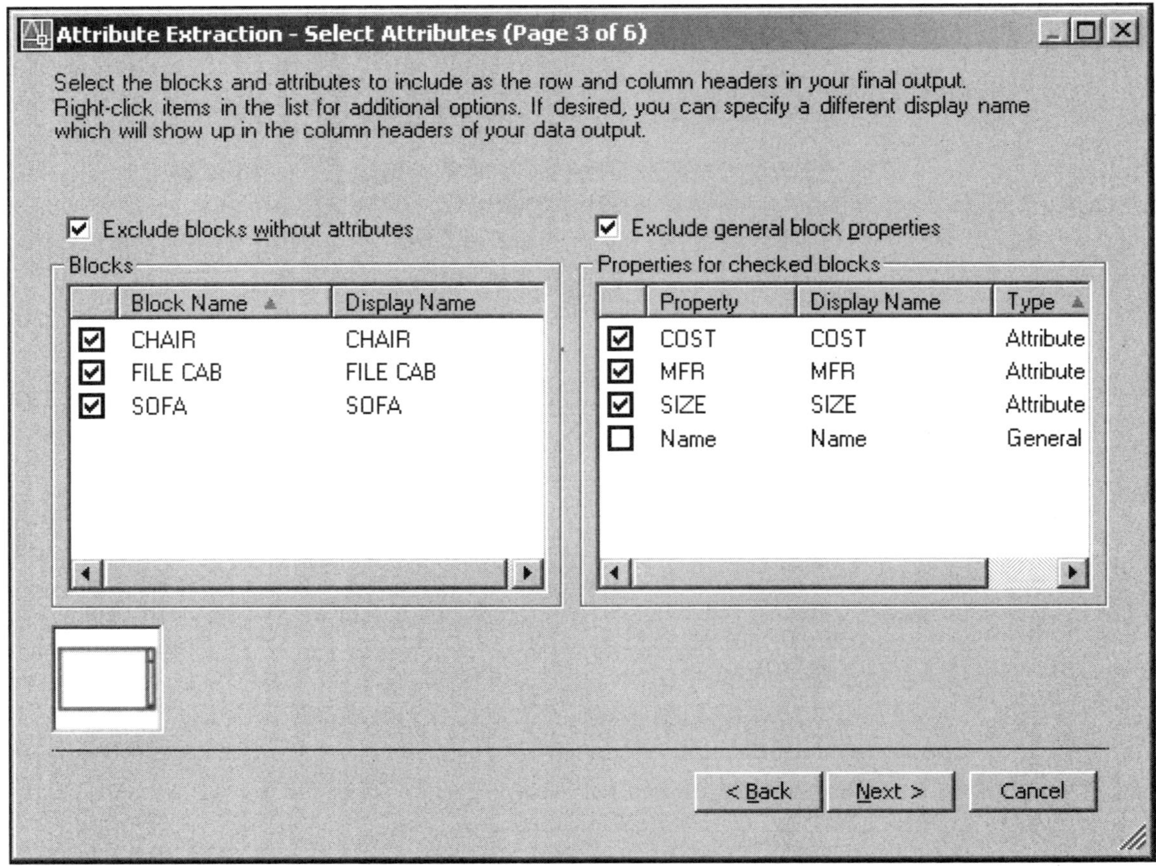

Your Extraction should look approximately like the dialog box shown below.

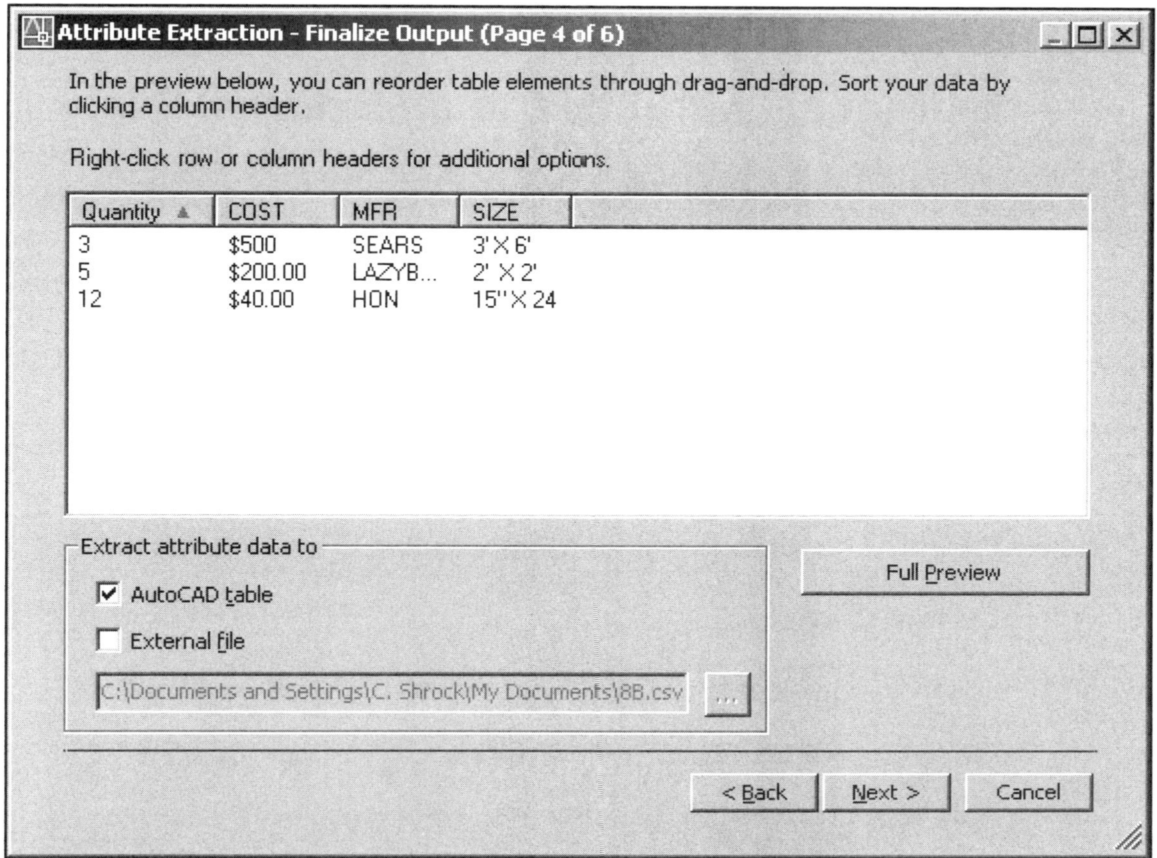

8. Select **AutoCAD table** to Extract attribute data to.

9. Enter the Title: **FURNITURE TABULATION**

Your table should appear as follows:

FURNITURE TABULATION			
Quantity	COST	MFR	SIZE
3	$500	SEARS	3' X 6'
5	$200.00	LAZYBOY	2' X 2'
12	$40.00	HON	15" X 24

EXERCISE 9B

Extracting Attributes to an External File.

The following exercise will take you through extracting attributes. You will open an existing drawing and extract the attribute data to an external program.

1. Open **Ex-8B**.
2. Select the **24 X 18 Qtr-Ft** layout tab.
3. Select **Tools / Attribute Extraction**
4. Select **Create Table or external file from scratch.**
5. Select **Current Drawing**
6. Make the following selections.

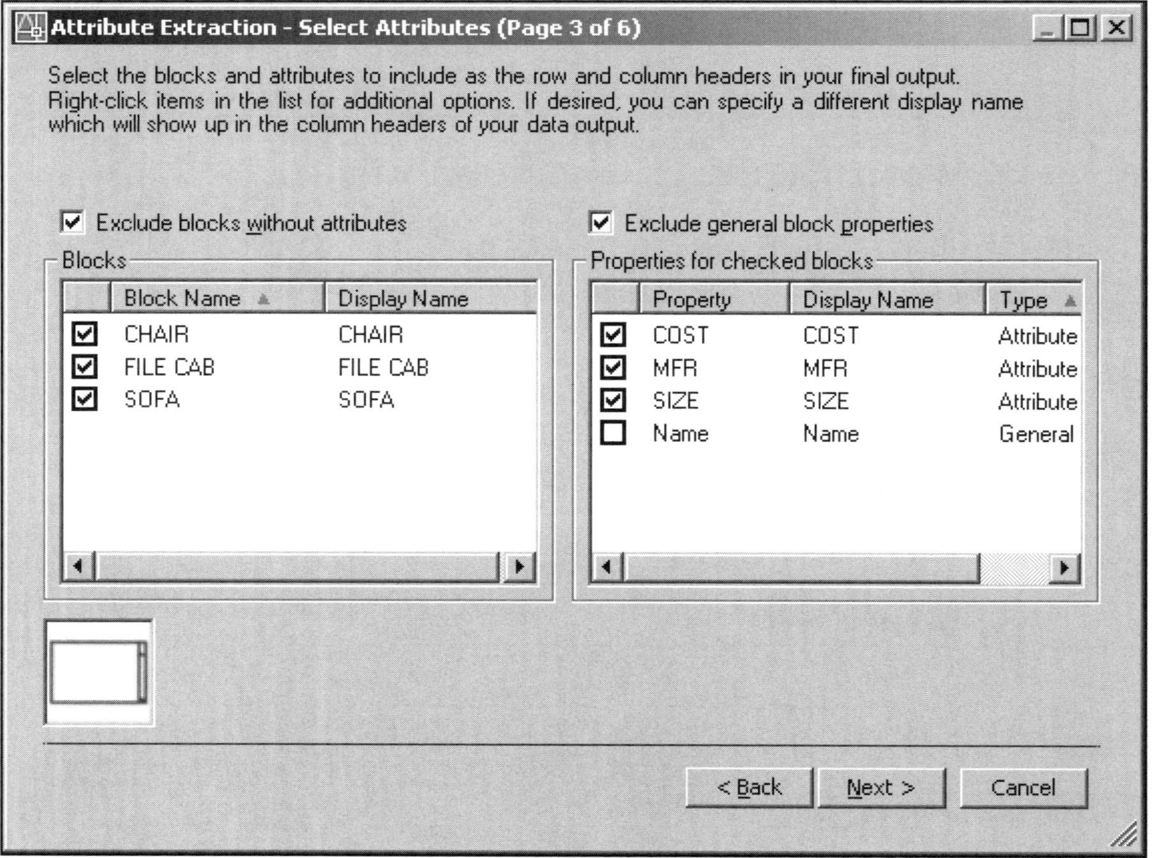

Your Extraction should look approximately like the dialog box shown below.

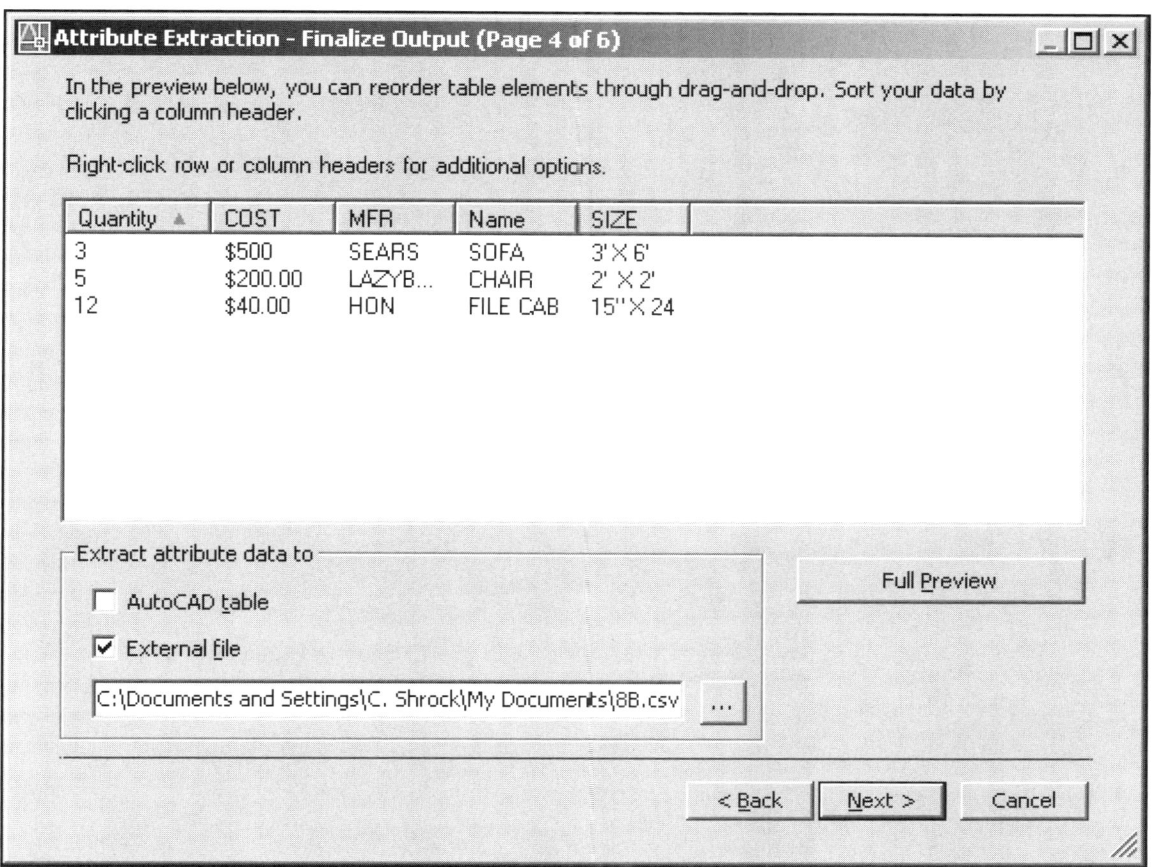

8. "Uncheck" **AutoCAD table** to Extract attribute data to.

9. Select **External file** to Extract attribute data to.

10. Select the **File Type** depending on which external program you will be using.

A text file should appear as follows, with a little adjusting.

Quantity	COST	MFR	SIZE
3	$500	SEARS	3' X 6'
5	$200.00	LAZYBOY	2' X 2'
12	$40.00	HON	15" X 24

An Excel file should appear as follows, with a little adjusting.

Quantity	COST	MFR	SIZE
3	$500	SEARS	3'X6'
5	$200.00	LAZYBOY	2' X2'
12	$40.00	HON	15"X24

Notes:

LEARNING OBJECTIVES

After completing this lesson, you will be able to:

1. Navigate in the DesignCenter Palette
2. Open a drawing from the DesignCenter Palette
3. Insert a block from the DesignCenter Palette
4. Drag and drop hatch patterns.
5. Drag and drop Symbols from the Internet
6. Open the Tool Palette Window
7. Control Tool Palette Properties
8, Create a Tool Palette
9. Export and Import a Tool Palette

LESSON 10

DesignCenter

The AutoCAD **DesignCenter** allows you to find, preview and drag and drop Blocks, Dimstyles, Textstyles, Layers, Layouts and more, from the DesignCenter to an open drawing. You can actually get into a previously saved drawing file and copy any of the items listed above into an open drawing.

Opening the DesignCenter palette

To open the **DesignCenter** palette select one of the following:

TYPE = DC or Ctrl + 2
PULLDOWN = TOOLS / DesignCenter
TOOLBAR = STANDARD

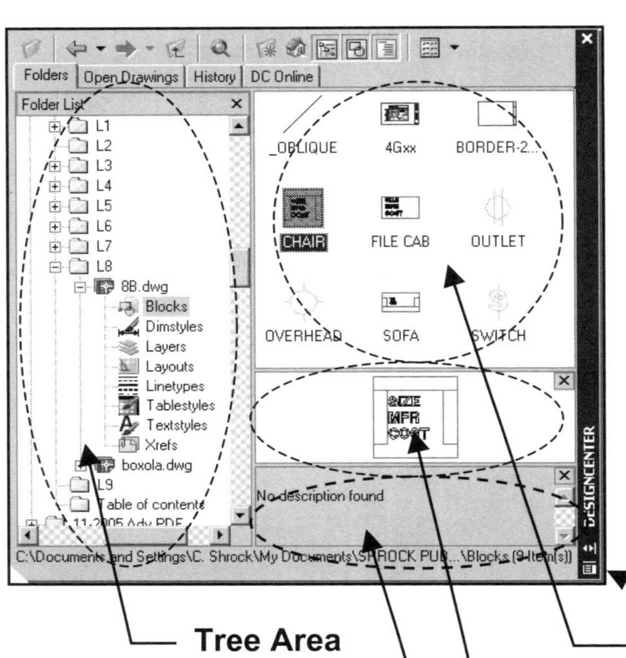

Hide tool

Tree Area

Content Area

Preview Area

Description Area

How to:

Resizing – You can change the width and height. Rest the cursor on an edge until the pointer changes to a double ended arrow. Click and drag to desired size.

Dock – Click the title bar, then drag it to either side of the drawing window.

Hide – You can hide the DesignCenter palette using the Auto-hide option. Click on the "properties" button and select "Auto-hide". When Auto-hide is ON, the palette is hidden, only the title bar is visible. The palette reappears when you place the cursor on the title bar.

<u>**VIEWING TABS**</u>

At the top of the palette there are 4 tabs that allow you to change the view. They are: Folders, Open Drawings, History and DC Online.

Folders tab – Displays the Directories and files similar to Windows Explorer. You can navigate and locate content anywhere on your system.

Open Drawings tab – Displays all open drawings. Allows you to select content from an open drawing and insert it into another open drawing. (Note: the target drawing must be the "active" drawing)

History tab – Displays the last 20 file locations accessed with DesignCenter. Allows you to double click on the path to load it into the "Content" area.

DC Online tab – Connects you to the Internet and gives you access to symbols.

BUTTONS

At the top of the DesignCenter palette, there is a row of buttons. (Descriptions below) To select a button, just click once on it.

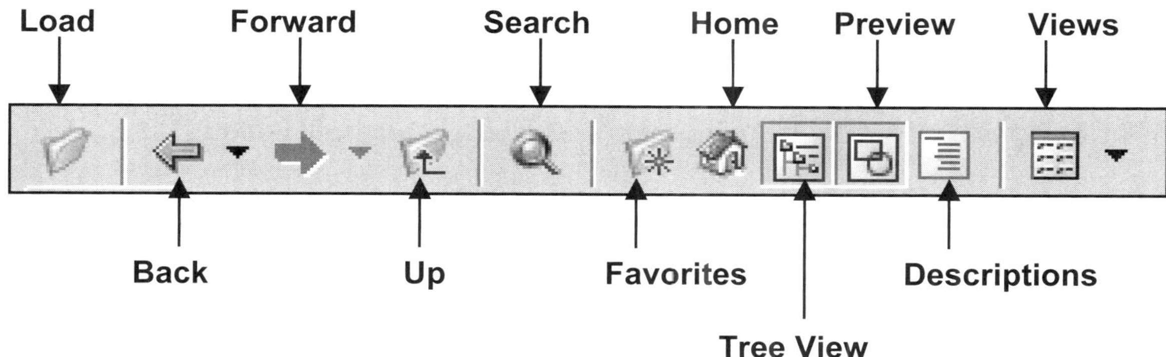

Load – This button displays the "Load" dialog box. (It is identical to the "Select File" dialog box.) Locate the drawing content that you want loaded into the "Content" area. You may also locate the drawing content using the Tree view and Folders tab.

Back and Forward – Allows you to cycle through previously selected file content.

Up – Moves up one folder from the current folder.

Search – Allows you to search for drawings by specifying various criteria.

Favorites – Displays the content of the Favorites folder. Content can be added to this folder. Right click over an item in the Tree View or Content areas then select "Add to Favorites" from the menu that appears. (You may add a drive letter, folder, drawing, layers, blocks etc.)

Home – Takes you to the DesignCenter folder by default. You can change this. Right click on an item in the Tree view area and select "Set as Home" from the menu. (You many select a drive letter, folder, drawing etc.)

Tree View – Toggles the Tree View On and OFF. Only works when the "Folders" or "Open Drawings" tab is current.

Preview – Toggles the "Preview" area On and OFF.

Description – Toggles the "Description" area On and OFF. (A description must have been given at the time the block was created.)

Views – Controls how the "Content" area is displayed. The choices are: Large or small icons, a list view or a detailed view.

How to open a drawing from DesignCenter Palette.

1. Open a "New" drawing, "Start from Scratch".
   (Typically you would want to "open" a drawing into a "new" drawing. But you may also open a drawing into an already open drawing. Note: Only objects in Model Space will come. None of the objects in Paper space.)

2. Locate the drawing you want to open.

3. Make sure it is in the "Content" area. (You cannot open a drawing from the Tree area.)

4. Click and drag the drawing onto the blank drawing area.

5. You will be prompted for the insertion point. Typically you should use 0, 0 .

6. You will then be prompted for the scale for X and Y and then the rotation angle.

Does this sound familiar. Yes, this is just like inserting a block. In fact, the drawing has actually been inserted as a block. If you want to modify the drawing you must Explode it.

How to Insert a Block using DesignCenter.

There are two methods of inserting a block using DesignCenter.

Method 1: Drag and Drop
Locate the Block in the Content area and drag and drop it into the drawing area of an open drawing. You will not be prompted for the insertion point.

Method 2: Specified Coordinates, Scale and Rotation
Double click on the block in the Contents area. The Insert dialog box will appear. Specify coordinates for insertion point, scale factor and rotation angle.

To Close the DesignCenter palette

The DesignCenter palette will remain open until you close it.

(If you have "Autohide" ON, only the title bar will remain visible.)

To Close the DesignCenter palette, select the "X" in the upper right corner of the Palette.

Drag and drop Hatch Patterns

AutoCAD allows you to drag and drop hatch patterns from the DesignCenter directly into a closed area in your drawing.

1. Open the DesignCenter.
2. <u>Find and select</u> AutoCAD's hatch pattern file **acad.pat** as follows:
   a. Select the Search tool

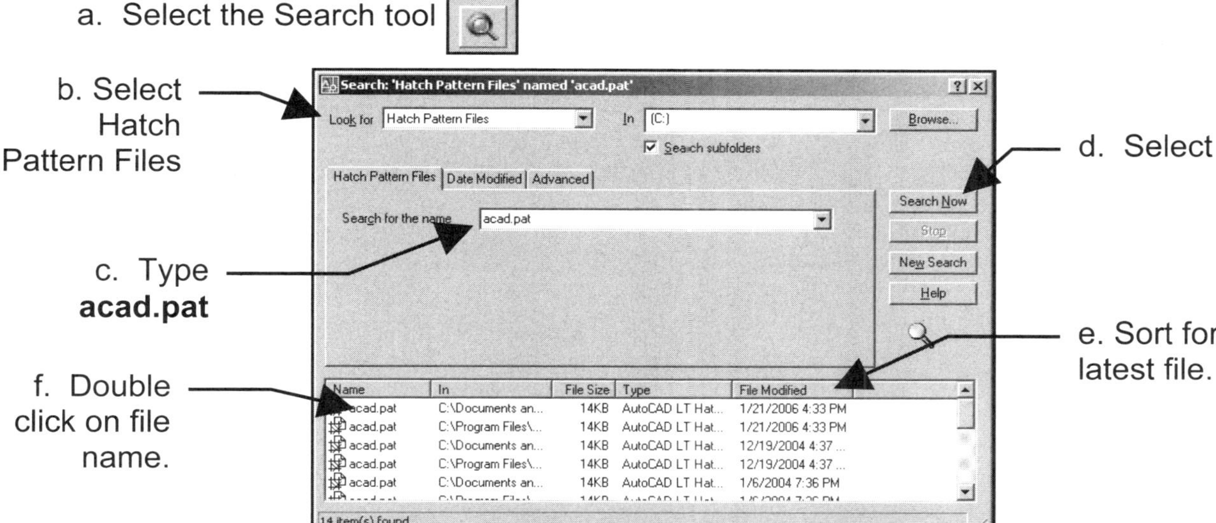

3. Click on the file within Design Center.
4. You should be able to see the hatch pattern squares in the <u>Content area</u>.
5. **Important**: First select the layer that you want the hatch pattern to reside on.
6. Now simply click on a hatch pattern square and drag it on to your drawing and drop it on the area where you want the hatch.

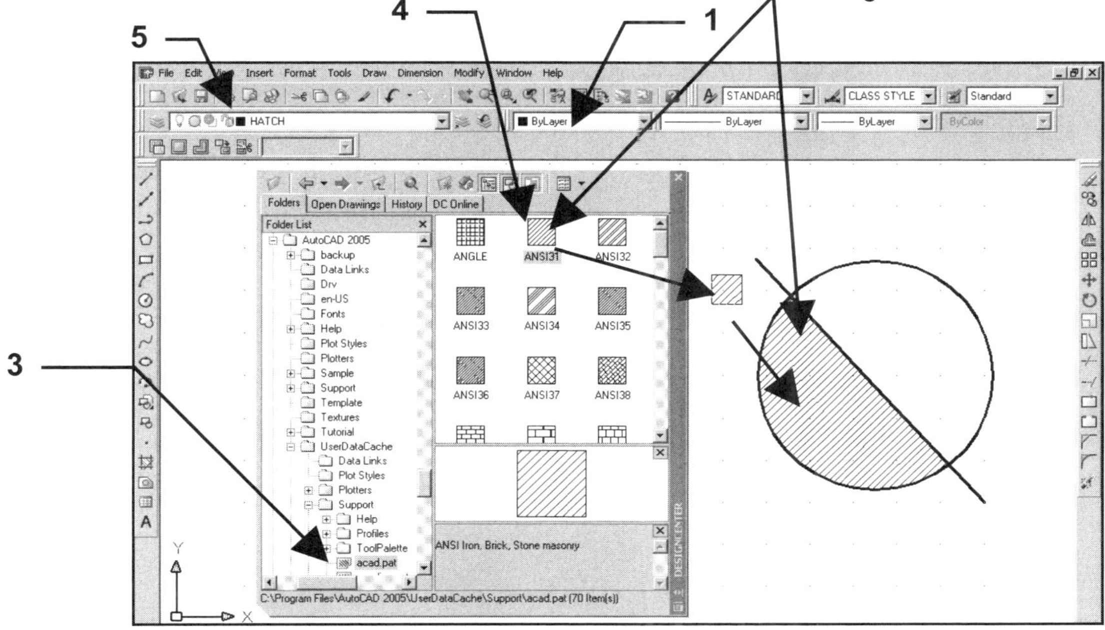

Drag and drop Layouts, Layers, Text Styles etc.

You can drag and drop just about anything from an existing file, listed in the Tree view "Folder List", to an open drawing file. This is a significant time saver. Just locate the source file in the tree area and drag and drop into an open drawing.

Some examples are shown below.

BLOCKS

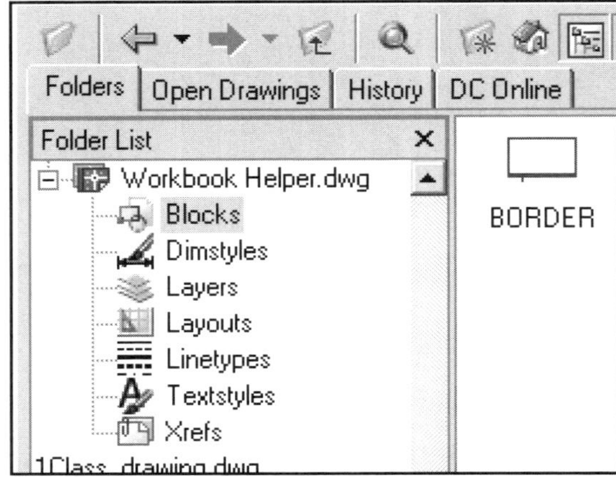

DIMENSION STYLES

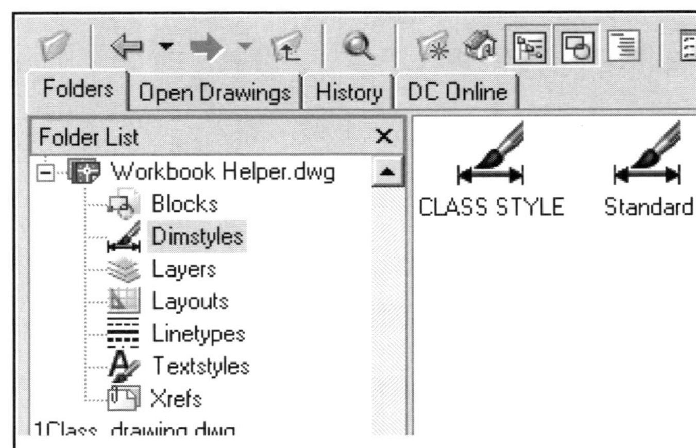

LAYERS

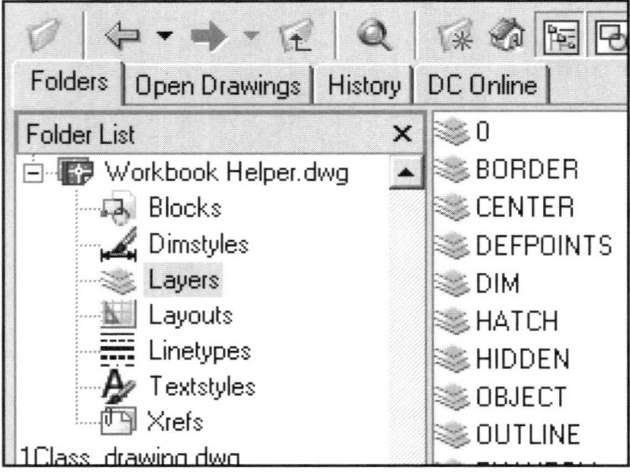

LAYOUTS

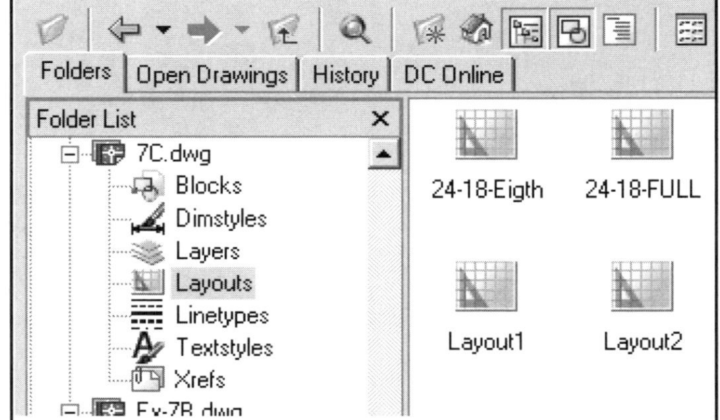

DC Online

DC Online connects you to the Internet and gives you access to thousands of symbols and manufacturer's product information.

Note: You must have an Internet connection to use this feature.

Just select the "DC Online" tab and AutoCAD automatically opens your Internet connection to the correct Internet address.

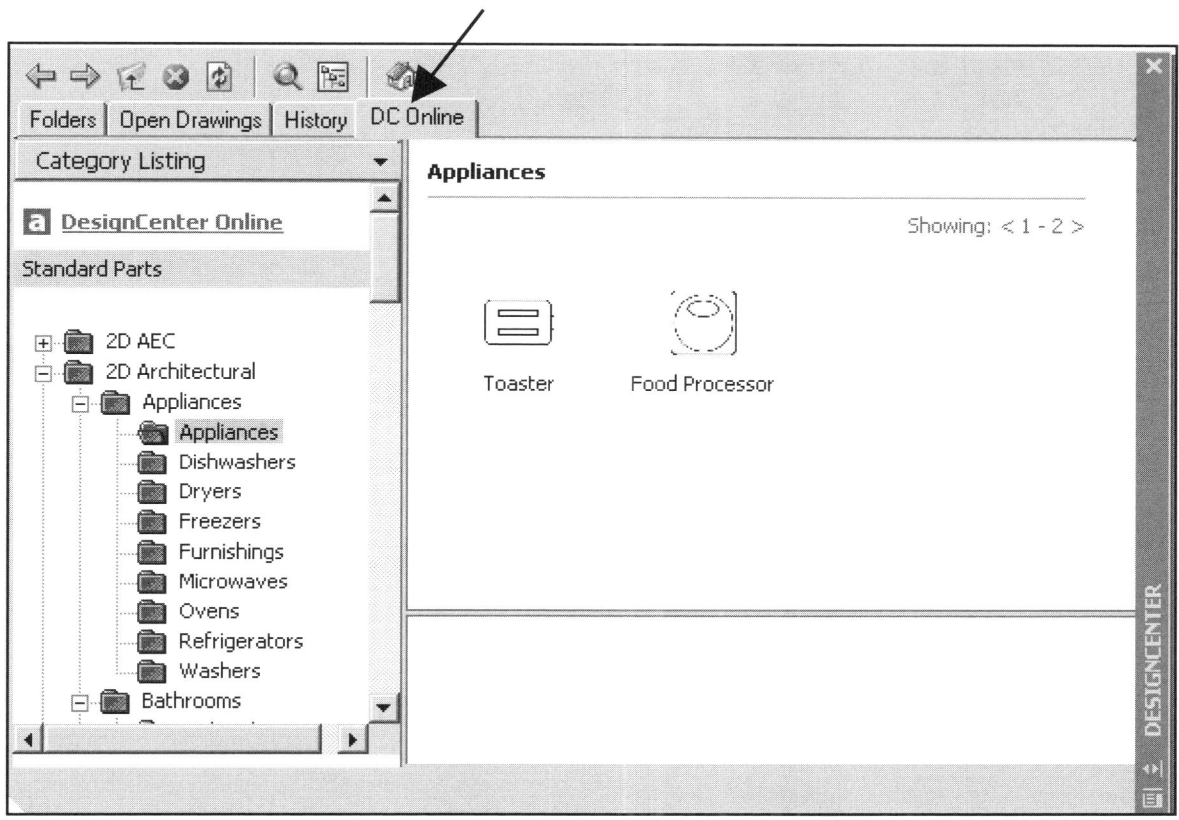

Tool Palettes

AutoCAD has developed another feature to use for storing and inserting frequently used features. This feature is "**Tool Palettes**".
You can easily create new tabs and add your own blocks, commands etc.
Any item on a palette is a Tool and is represented by an icon.

How to Open the Tool Palette

To open the Tool Palette, select one of the following:

> **TYPE = toolpalettes or Ctrl + 3**
> **PULLDOWN MENU = Tools / Palettes / Tool Palettes**
> **TOOLBAR = STANDARD**

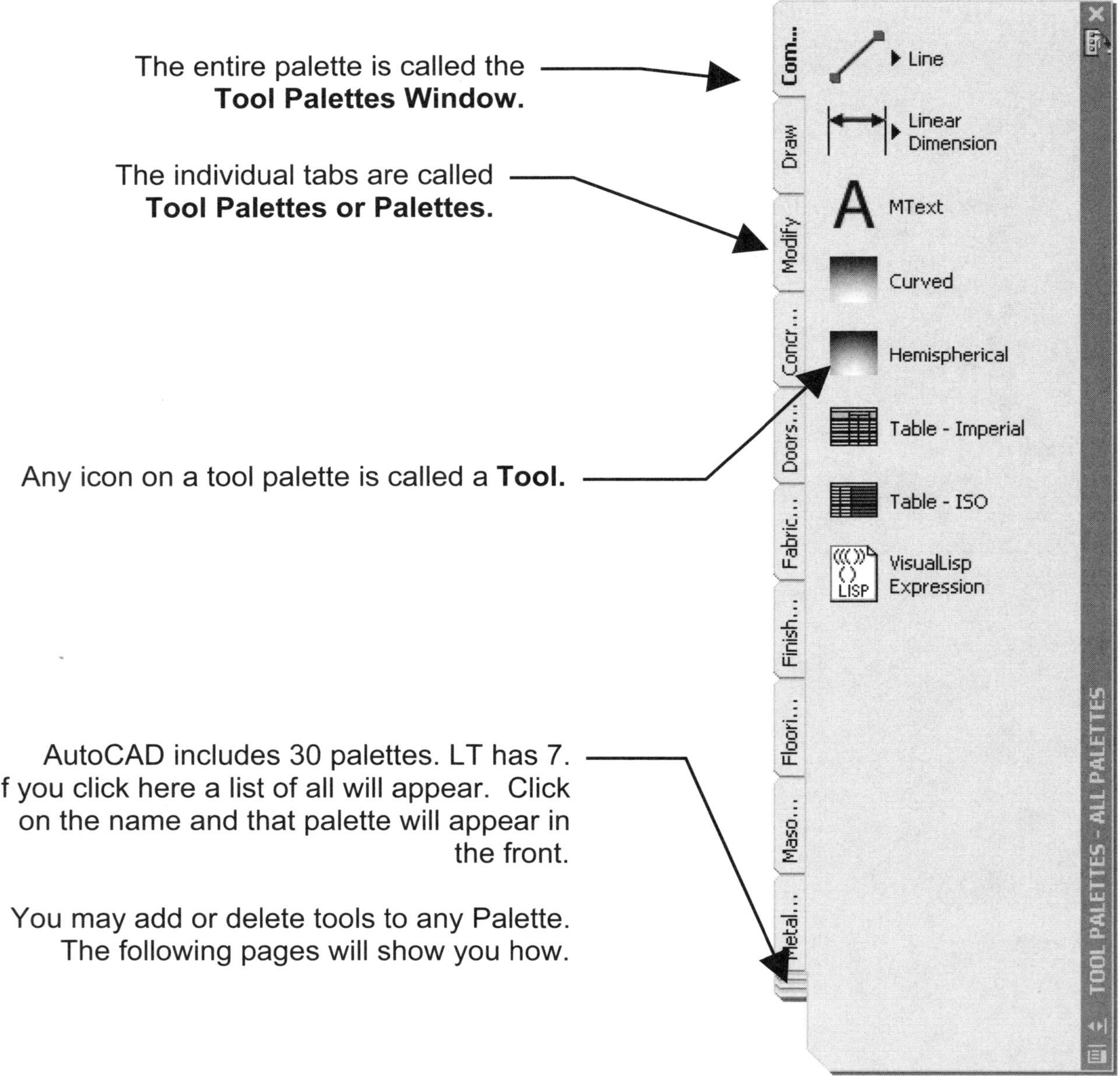

PALETTE PROPERTIES

You can control the appearance and operation of a Tool Palette.

Place you cursor on the tool palette and right click.
The menu shown below should appear:

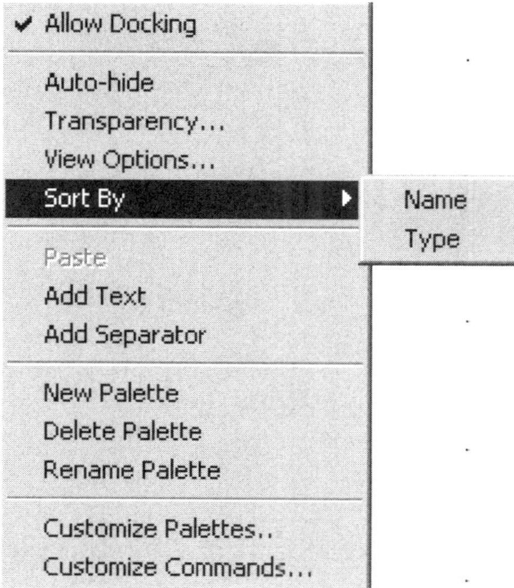

ALLOW DOCKING
This feature allows you to dock the Tool Palettes window on the left or right side of the AutoCAD drawing area. Note: A docked window will never cover your drawing, but it will reduce the drawing area.

AUTO-HIDE
When Auto-hide is ON (with check mark), the Tool Palettes Window is hidden and only the title bar is visible. The Tool Palettes Window reappears when you place the cursor on the title bar. When Auto-hide is OFF (no check mark), the Tool Palette Window remains fully visible.

TRANSPARENCY
Use this option to make the Tool Palette window transparent so you can "see through" the window to the drawing underneath. Slide the pointer to the desired level of transparency.

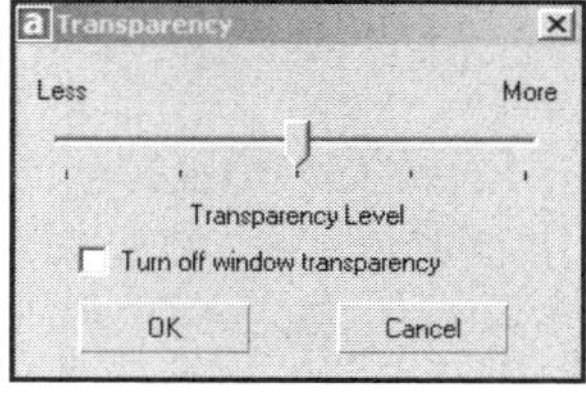

Note:

1. *Transparency feature is available only when hardware acceleration is off. To select Software acceleration, select Tools / Options / System tab. Select the Properties button from the Current 3D Graphics Display section. In the Acceleration section, select Software. Apply and Close.*
2. *Transparency is only available when the palette is not docked.*

Note: Transparency may slow performance. You may prefer to turn it off.

VIEW OPTIONS

This option allows you to specify the *"Image Size"* and "View Style" for the individual tools. Also you need to specify whether to apply the options to the "Current Palette" or "All Palettes".

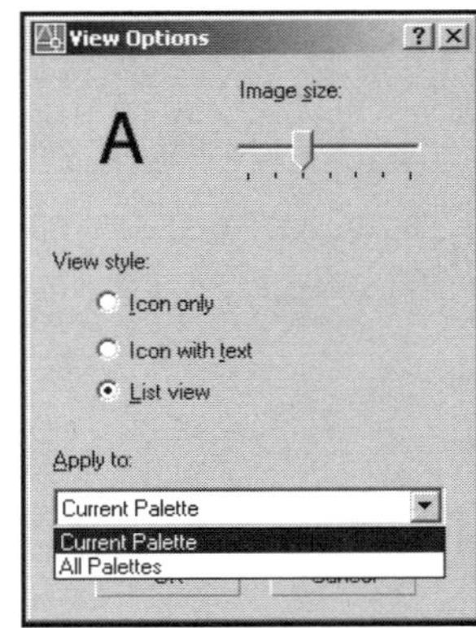

Icon only – Icons with no text

Icon with Text – icons with text in multiple columns.

List View – icons with text always in a vertical column.

ADD TEXT

This option allows you to insert text by a tool.

ADD SEPARATOR

This option allows you to insert a line as a separator.

NEW PALETTE

This option allows you to create a new Tool Palette.
This option is explained on page 10-11.

DELETE PALETTE

This option allows you to delete an existing Tool Palette. First select the Tool Palette to be deleted, right click and select "Delete Tool Palette". A warning will be displayed basically saying "Are you sure?". If you select the OK button, the Tool Palette will be permanently deleted.

RENAME PALETTE

This option allows you to rename an existing Tool Palette. First select the Tool Palette to be renamed, right click and select "Rename Tool Palette". Enter the new name in the box displayed and <enter>. The new name will be shown on the tab.

CUSTOMIZE PALETTES

Palettes

Lists all available palettes. Click and drag a palette to move it up or down in the list. Right-click a palette in the list to rename, delete, or export the palette. (When you export a palette, it's saved to a file with an .xtp extension.) Right-click in the Palettes area to import a palette or to create a new, blank one.

Palette Groups

Displays the organization of your palettes in a tree view. Click and drag a palette to move it into another group. Right-click a palette group, and then click Set Current on the shortcut menu to display that group of palettes.

> The **shortcut menu** also provides options to Create a new group, Delete an existing group, Rename an existing group, Remove a tool palette from a group, Export a group (as an XPG file) and Import a group.

CUSTOMIZE COMMANDS

Allows you to customize commands. Refer to Lesson 2 for Customizing User Interface.

HOW TO CREATE A TOOL PALETTE

Creating a new tool palette means adding a "tab" to the Tool Palettes Window. There will only be one Tool Palette Window but you may have many palette tabs.

How to create an "NEW" tool palette

1. Right click in the Tool Palettes window.
2. Choose "New Tool Palette" from the menu.
3. Enter a name in the box that appears, then press <enter>.

How to add command tools to your new tool palette. (Auto-hide must be off)

1. Select your tool palette tab.
2. Right click on the tool palette window.
3. Select Customize Palettes.

Note: the Customize dialog box will open but you will not be making any changes within it. But is must remain open while customizing any tool palettes.

4. Select a command tool icon, such as "erase." Drag the command icon to the tool palette. A black bar will appear to indicate where the icon will appear. Now drop the command icon on the tool palette.

How to create a tool palette and its content using the DesignCenter

1. Open the DesignCenter
2. Locate one of the following:
    a. Folder – to add all the drawings in the folder
    b. Drawing file – to add all the blocks in the drawing
    c. Block icon – to add the block
    d. Hatch icon - to add all hatches in the .pat file
3. Drag and drop the item to the tool palette.

How to move or copy a tool to another palette

1. Click the palette tab that contains the tool you want to move or copy. (Source)
2. Right click the tool you want to move or copy.
3. Choose "Cut" (to move) or "Copy" (to copy)
4. Click the palette tab on which you want to place the item. (Target)
5. Right click any blank area on the palette and choose "Paste".

Note: Tool Palettes are not saved with the drawing file If you always use the same computer your specific set of Tool Palettes will always appear. But if you move to another computer, and you need your Tool Palettes, it will be necessary for you to Export and Import your Tool Palettes. Refer to page 10-11.

Tool Properties

Before you start using your new tools, you should set their properties. Tool properties specify how a drawing, block or hatch is inserted. For example, you can specify that a hatch be inserted at a certain scale or a block be inserted on a specific layer.

To set Tool properties, right click any tool and choose Properties from the menu. The Tool Properties dialog box will vary slightly depending on whether it is for a block, drawing or hatch pattern.

Updating the Icon for a Tool

If you change a Block you must update its icon in the Tool Palette. The easiest way to accomplish this is "delete" the icon and replace it using the DesignCenter. The more difficult way is to Right Click on the tool and delete the "Source File" field and enter the correct path.

Block Master Drawing (My personal preference)

I like to create and maintain a "Master" drawing for all of my blocks. I named it "Library". I create all of my blocks in the drawing. If I need to edit the block, I do it within the "Library" drawing. If I want to add a block to a Tool Palette, I know exactly where to find the block using DesignCenter. The "path" to the original Tool never changes unless I change the location of my "Library" drawing.

HOW TO EXPORT YOUR CUSTOMIZED TOOL PALETTE

1. Right click on the Tool Palette.
2. Select Customize Palettes…
3. Select the Palette that you wish to Export.
4. Right click and select Export.
5. Find the location where you choose to save the palette file.
6. Select Save button. (The file will have an .XTP extension.

HOW TO IMPORT YOUR CUSTOMIZED TOOL PALETTE

1. Right click on the Tool Palette.
2. Select Customize Palettes…
3. Right click in the Palettes area on the left and select Import.
5. Find the palette file.
6. Select Open button.

EXERCISE 10A

Inserting Blocks from the DesignCenter.

1. Open **EX-5D or 5E** (depending on which one you completed).

2. Make blocks of the chair, desk, sofa, stand with lamp, door, window and file cabinet in the boss's office.

   These blocks must be saved in the current drawing file. Use Retain or Delete when selecting objects. Do not use Convert to block.

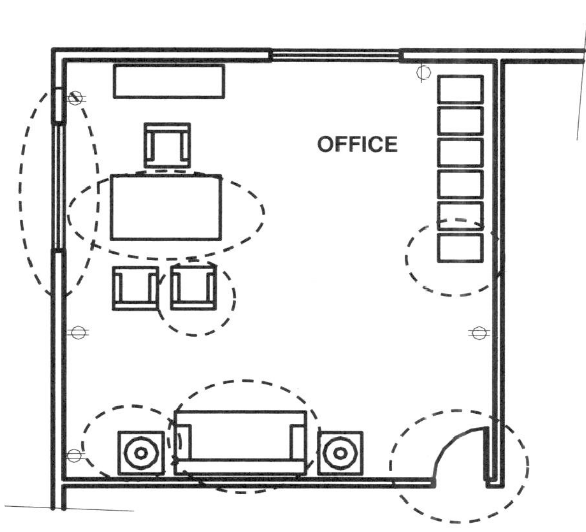

3. **Save** the drawing as **EX-10X (Note: 10X will not appear in the tree until you close and then re-open the directory in the DesignCenter)**

4. **Open** My Feet-Inches Setup and select the **24 X 18 (Qtr-Ft)** tab.
5. Draw the simple floor plan shown below. (Walls = 6" wide)
6. **Save** as **EX-10A** and continue on to page 10-14.

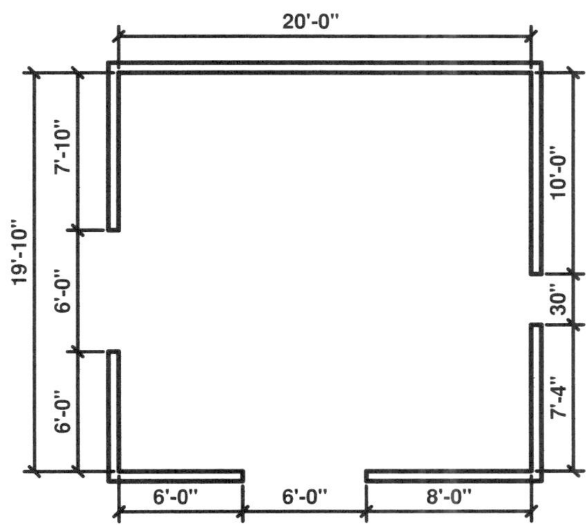

7. **Open** the **DesignCenter** (Refer to page 10-2 if necessary.)
8. **Find** your drawing **EX-10X** in Tree View (left side.)
9. Select the **plus box** beside the file to display the contents of **EX-10X**.
10. Select **"Blocks"** to display the blocks in the Content area (right side).

Note: Experiment with the 4 ways you can display the blocks. Large icon, small icon, List, and details. (Refer to page 10-3, Views.)

12. **Insert** the Blocks from the Content area to create the drawing below.
    (Refer to page 10-4, How to Insert a Block using DesignCenter, if necessary.)
    a. Use the **drag and drop** method for the blocks that are not rotated.
    b. Use **Specify Coordinate** method for the blocks that are rotated.

13. Save as EX-10A.

EXERCISE 10B

Inserting Symbols from the DC Online.

Note: This exercise can only be completed if you have access to the Internet.

1. Open **EX-10A.**
2. Select the **Furniture** layer.
3. Open the DesignCenter.
4. Select the **DC Online** tab.
5. Locate a telephone plan view in the following directory and select it:
   2D Architectural / Telephone / Modems / Phone Modem Plan View
6. Drag and drop the symbol into the open drawing.
7. You will be prompted for an insertion point, first enter "R" for **Rotate.**
8. Enter the Angle: **225 <enter>**.
9. Now place the insertion point on the desk.

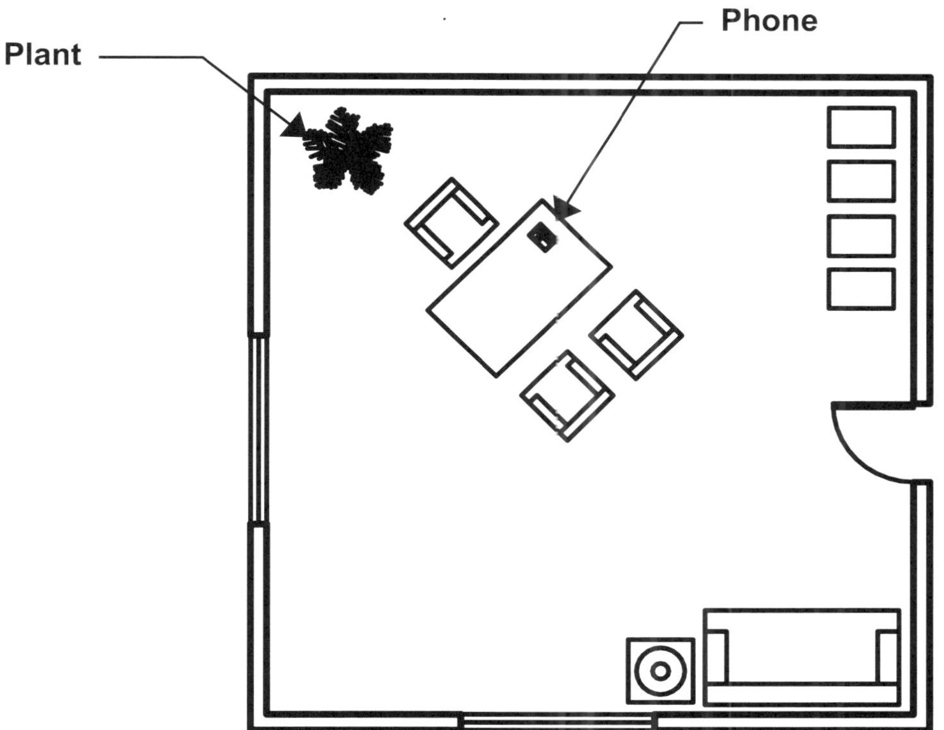

10. Select the **Misc** layer.
11. Locate a Tree Palm in the following directory and select it:
    2D Architectural / Landscaping / Trees / Tree Palm
12. Drag and drop the symbol into the open drawing.
13. You will be prompted for an insertion point, first enter "S" for **Scale.**
14. Enter the Scale: **.3 <enter>**.
15. Now place the insertion point in the corner.
16. Save as **EX-10B**

EXERCISE 10C

Inserting a Hatch pattern from the DesignCenter.

1. Open **My Decimal Setup**

2. Draw the objects shown below.

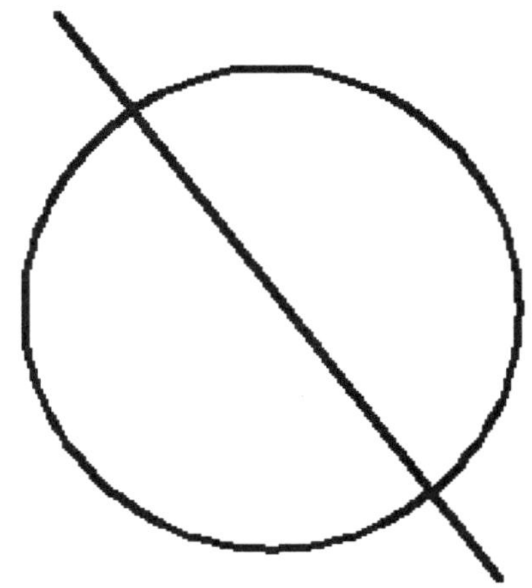

3. Open the DesignCenter

4. Locate AutoCAD's **acad.pat** files (Refer to page 10-5)

5. Apply Ansi 31 pattern on the right side and Ansi 32 on the left side.
   Set the rotation angle of Ansi 32 to appear as the example below.

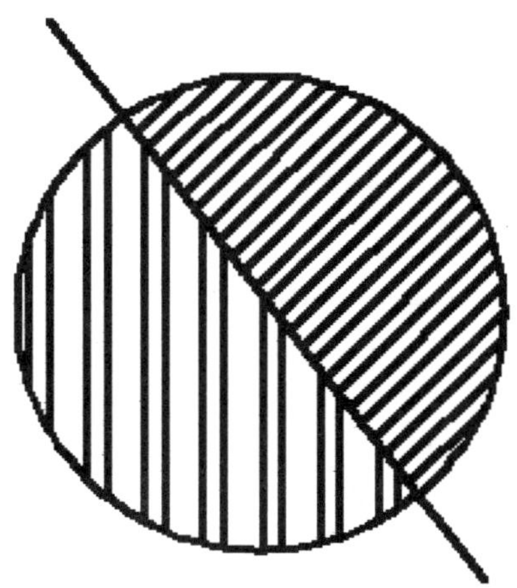

6. Save as **Ex-10C**.

EXERCISE 10D

Create a Tool Palette

1. Open **My Feet-Inches Setup**

2. Create a new empty Tool Palette.
   a. Name **= MINE**

3. Add a Hatch pattern.

4. Add a block using Design Center.

5. Add 4 commands.

6. Move the “MINE” Tool Palette tab name up to the top of the Tool Palettes.

7. Now Export “Mine” Tool Palette. (Refer to page 10-12)

8. Now delete “Mine” Tool Palette from the Tool Palettes. (Refer to page 10-10)

9. Now Import “Mine” Tool Palette. (Refer to page 10-12)

NOTES:

LEARNING OBJECTIVES

After completing this lesson, you will be able to:

1. Understand the use of Externally Referenced drawings
2. Insert an Xref drawing.
3. Use the Xref Manager.
4. Bind an Xref to a drawing.
5. Clip an Xref drawing.
6. Edit an Xref drawing using the XOPEN command.
7. Convert an object to a Viewport.
8. Create multiple Viewports and multiple xrefs.
9. Understand the PSLTCALE command.
10. Create multiple viewports quick and easy.

LESSON 11

External Reference drawings (xref)

The **XREF** command is used to insert an image of another drawing into the current drawing. AutoCAD calls the host drawing the "Parent" and the Xref dwg the "Child". This command is very similar to the INSERT command. But when you externally reference (XREF) the Child dwg into the Parent dwg, the image of the Child dwg appears but only the **path** to where the original Child dwg is stored is loaded into the Parent dwg. The Child dwg does not become a permanent part of the Parent dwg.

Each time you open the Parent dwg, it will look for the Child dwg (via the path) and will load the image of the **current version** of the Child dwg. If the Child dwg has been changed, the new changed version will be displayed. So the Parent dwg will always display the most current version of the Child dwg.

Since only the path information of the Child dwg is stored in the Parent dwg, the amount of data in the Parent dwg does not increase. This is a great advantage.

If you want the Child dwg to actually become part of the Parent dwg, you can **BIND** the Child to the Parent. (Refer to page 11-6)

If you want the image of the Child dwg to disappear temporarily, you can **UNLOAD** it. To make it reappear simply **RELOAD** it. (Refer to page 11-5 & 6)

If you want to delete the image and the path to the Child dwg, you can **DETACH** it. (Refer to page 11-5)

Examples of how the XREF command could be useful?

1. **If you need to draw an elevation**. You could XREF a floor plan and use it basically as a template for the wall, window and door locations. You can get all of the dimensions directly from the floor plan by snapping to the objects. It will not be necessary to refer back and forth between drawings for measurements. If there are any changes to the floor plan at a later date, when you re-open the elevation drawing, the latest floor plan design will appear automatically. To make the floor plan drawing become invisible, simply UNLOAD it. When you need it again, RELOAD it.

2. **If you want to plot multiple drawings on one sheet of paper.** Many projects require "standards" or "detail" drawings that are included in every set of drawings you produce. Currently you probably drew each detail on the original drawing package, or created those transparent "sticky backs". Now, using the XREF command, you would do the following. 1. Make individual drawings of each standard or detail. 2. XREF the standard or detail drawing into the original drawing package. This would not only decrease the size of your file but if you made a change to the standard or detail, the latest revision would automatically be loaded each time you opened the drawing package.

3. **Working with a team of drafters all working on a segment of the project.** (This process works best if your office is networked.) Let's say you are responsible for the furniture layout on an architectural project and the floor plan is not quite finished. You could XREF the current version of the floor plan and start working on the furniture layout. Every couple of hours you can RELOAD the XREF floor plan and your drawing will be automatically updated to the latest work saved on the floor plan.

HOW TO INSERT AN XREF DRAWING

1. Open the Parent drawing. (For example, open My Decimal Setup)

2. Select a Layout tab.

3. Select a Viewport. (double click inside the viewport)
   Note: You should always insert an Xref into Model Space not Paper Space. The viewport can be locked or unlocked, it does not matter.

4. Select the XREF command using one of the following:

TYPE = XATTACH
PULLDOWN = INSERT / DWG Reference...
TOOLBAR = REFERENCE

The following "Select Reference File" dialog box should appear.

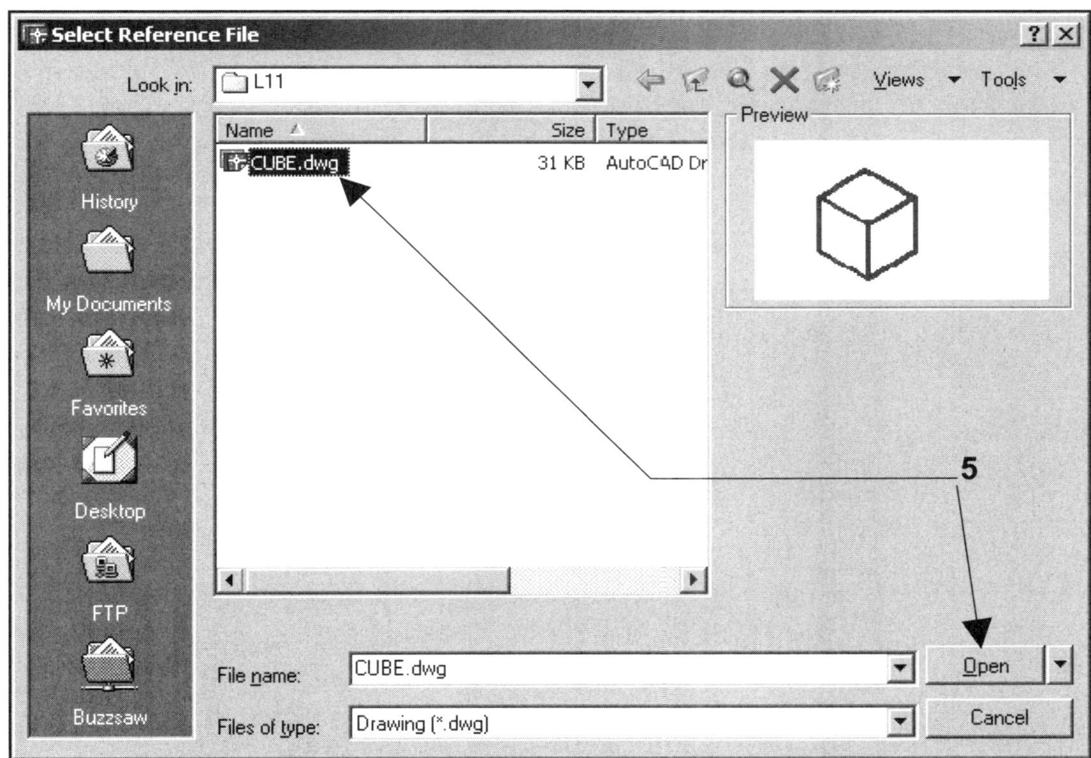

5. Select the file you wish to **XREF** and select **Open.**

continue on the next page....

*The following "**External Reference**" dialog box should appear.*

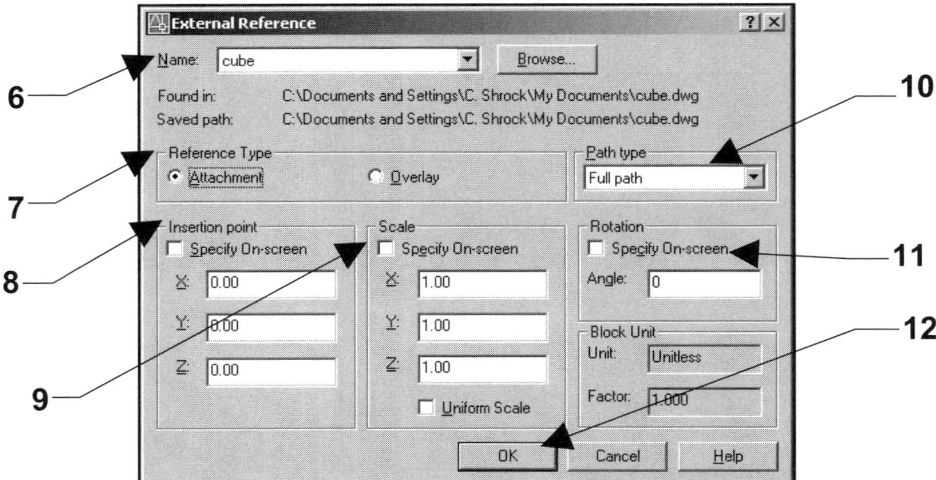

6. The name of the file you selected should appear in the Name box.

7. Select **"Reference Type".**

   Attachment - This option attaches one drawing to another. If you attach the child to the Parent then attach the Parent to a Grandparent, the **Child remains attached**. (Daisy Chain of Child, Parent and Grandparent)

   Overlay - This option is exactly like Attachment except, if you Overlay a Child to a Parent, then attach a Parent to a Grandparent, **only Parent remains attached**. The Child is not. (Parent and Grandparent only)

8. **Insertion Point -** Specify the X, Y and Z insertion coordinates in advance or check the box "Specify on-Screen" to locate the insertion point with the cursor.

9. **Scale -** Specify the X, Y and Z scale factors in advance or check the box "Specify on-screen" to type the scale factors on the command line. You may select the **Uniform Scale** box if the scale is the same for X, Y and Z.

10. **Path -** Select the Path type to save with the xref.

11. **Rotation -** Specify the Angle or check the box "Specify on-screen" to type the scale factors on the command line.

12. Select the **OK** button.

13. Place the insertion point with your cursor if you have not preset the insertion point.

Note: If you xref a drawing and it does not appear, try the following:
1. Verify that you xref-ed the drawing into Model space, not paper space.
2. Unlock the viewport, use View / Zoom / Extents to find the drawing in a viewport, adjust the scale and re-lock the viewport.
3. Go back to the original drawing to verify that it was drawn in model space. Objects drawn in paper space will not xref.
4. Go back to the original drawing and verify that the layers are not frozen.

XREF MANAGER

The **XREF MANAGER** allows you to view and manage the xref drawing information in the current drawing.

Select the XREF Manager using one of the following:

TYPE = XREF
PULLDOWN = INSERT / External References...
TOOLBAR =

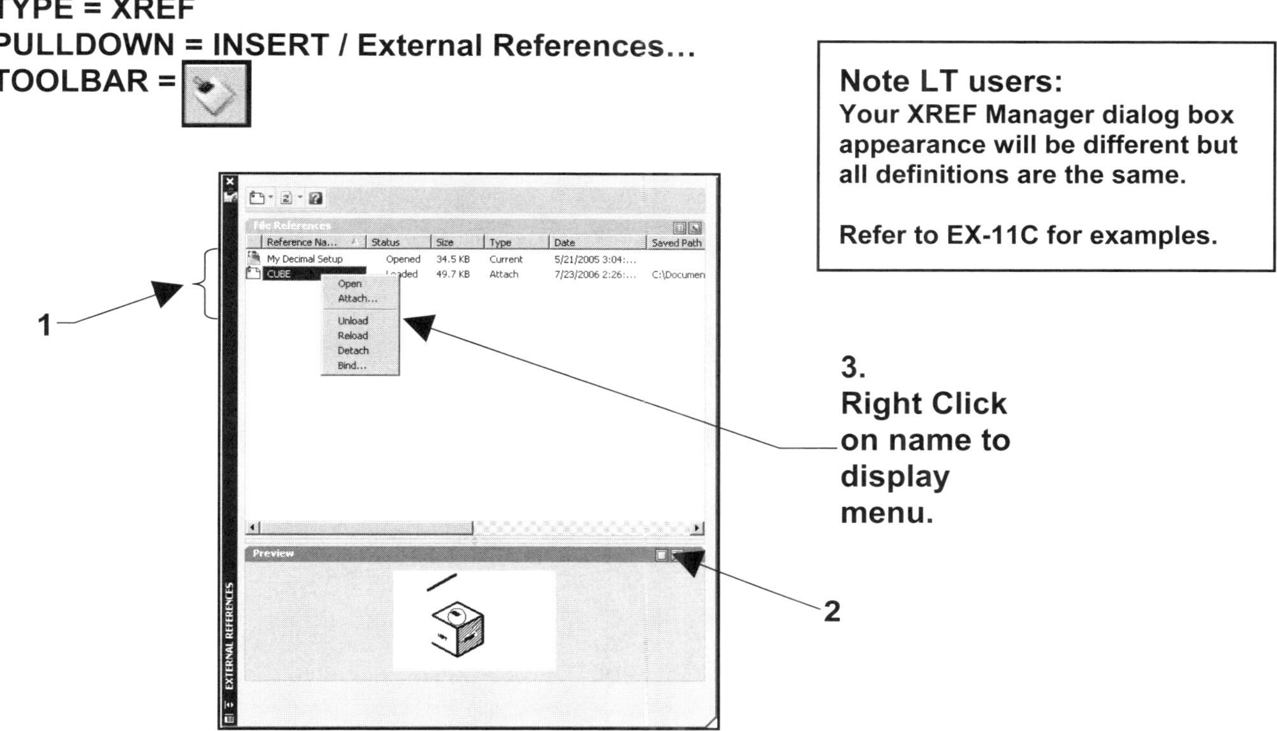

Note LT users:
Your XREF Manager dialog box appearance will be different but all definitions are the same.

Refer to EX-11C for examples.

1. **Title Area** - This describes each drawing in the current drawing.
   Reference Name = The original drawing file name of the drawing
   Status = This lists whether the drawing was opened, loaded, not found or unloaded.
   Size = Indicates the size of the drawing.
   Type = Lists whether the xref drawing is an Attachment or Overlay.
   Date = Lists when the xref drawing was originally referenced.
   Saved Path = This is the path the computer will follow to find the xref drawing listed and load it each time you open the current drawing.

2. **Preview button:** - Displays a preview of the drawing selected.

3. **MENU**

   Open - Opens the selected xref for editing in a new window. The new window is displayed after the Xref Manager is closed.

   Attach - Takes you back to the External Reference dialog box so you can select another drawing to xref.

   Unload - Unload is not the same as Detach. An unloaded xref drawing is not visible but the information about it remains and it can be reloaded at any time.

Reload - This option will reload an unloaded xref drawing or update it. If another team member is working on the drawing and you would like the latest version, select the Reload option and the latest version will load into the current drawing.

Detach - This will remove all information about the selected xref drawing from the current file. The xref drawing will disappear immediately. Not the same as Unload.

Bind – When you bind an xref drawing, it becomes a permanent part of the current base drawing. All information in the Xref manager concerning the xref drawing selected will disappear. When you open the base drawing it will not search for the latest version of the previously inserted xref drawing.

When a drawing is XREFed, its Blocks, Layers, Linetypes, Dimension Styles and Text Styles are kept separate from the current drawing. The name of the xref drawing and a pipe (I) symbol is automatically inserted as a prefix to the newly inserted xref layer. This assures that there will be no duplicate layer names.

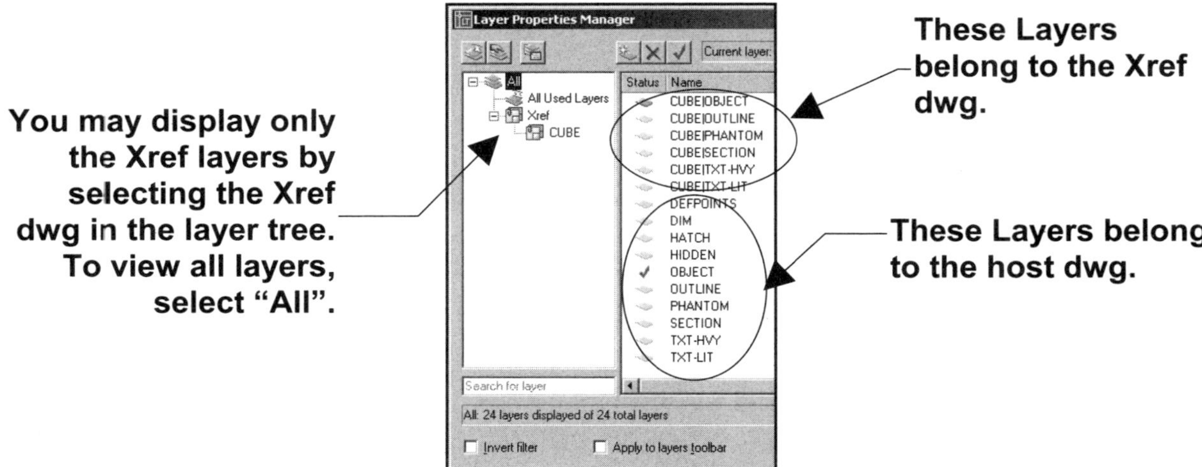

XBIND

It is important to understand that these new names are listed but they cannot be used. Not unless you bind the entire xref drawing or use the Xbind command to bind individual objects.

The following is an example of the Xbind command with an individual layer. You may also use this command for Blocks, Linetypes, Dimension Styles and Text Styles.

1. Type **XBIND** <enter> at the command line.

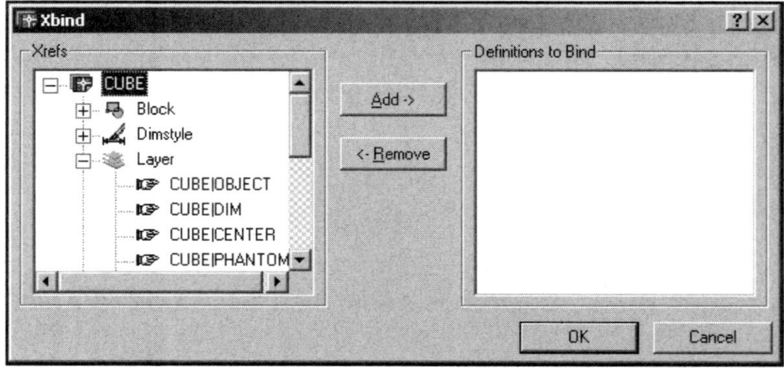

2. Expand the Xref file information list by clicking on the **+** sign beside the Xref drawing file name; then click on the **+** sign beside "Layer". The + sign will then change to a - sign, as shown below.

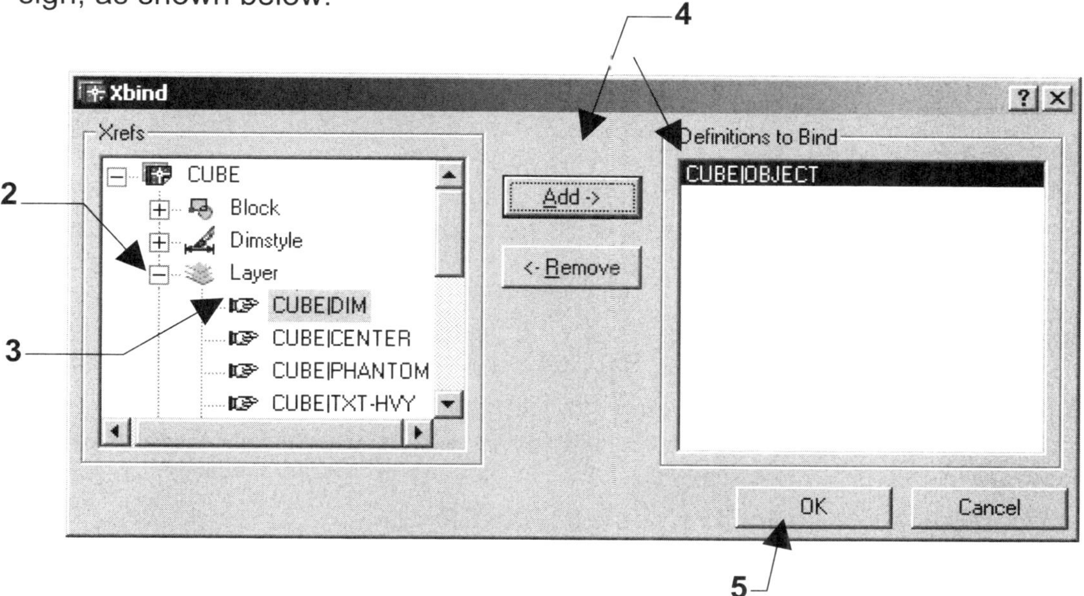

3. Select the layer name to bind.

4. Select the **ADD** button. The xref layer will now appear in the "Definitions to Bind" window.

5. Select the **OK** button.

6. Now select Format / Layers and look at the xref layer. The Pipe (|) symbol between the xref name and the layer name has now changed to 0. This layer is now usable.

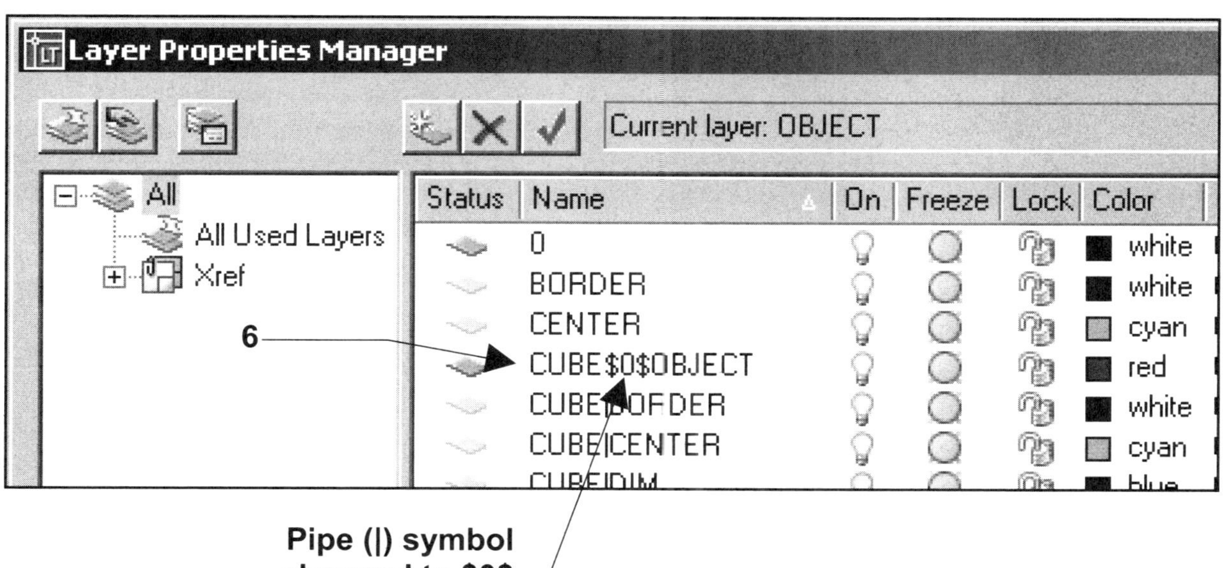

Pipe (|) symbol changed to 0

Clipping an External Reference (Not available in LT)

When you Xref a drawing, sometimes you do not want the entire drawing visible. You may only want a portion of the drawing visible. For example, you may Xref an entire floor plan but actually only want the kitchen area visible within the viewport.

This can be easily accomplished with the Xclip command. After you have inserted an External Referenced drawing, select the Xclip command. You will be prompted to specify the area to clip by placing a window around the area. **All objects outside of the window will disappear.**

How to use the **Xclip** command

1. Select the **Xclip** command using one of the following:

   TYPE = XC
   PULLDOWN MENU = MODIFY / CLIP / XREF
   TOOLBAR = MODIFY

2. Select objects. ***select the xref you want to clip <enter>***

   Enter clipping option
   [ON/OFF/Clipdepth/Delete/generate Polyline/New boundary] <New>: ***N <enter>***

   Specify clipping boundary:
   [Select polyline/Polygonal/Rectangular] <Rectangular>: ***R <enter>***

   Specify first corner: ***select the location for the first corner of the "Rectangular" clipping boundary.***

   Specify opposite corner: ***select the location for the opposite corner of the "Rectangular" clipping boundary.***

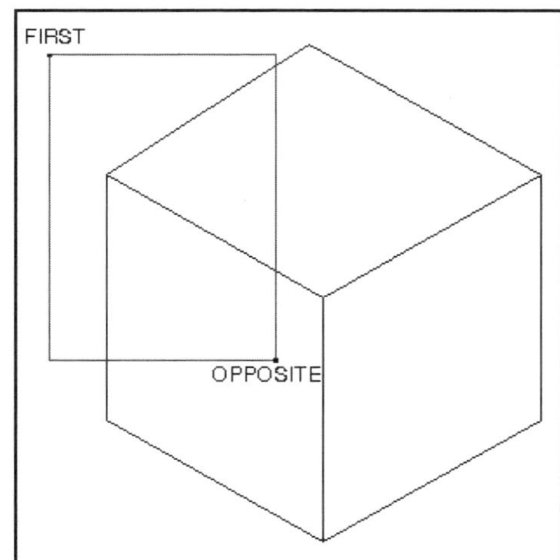

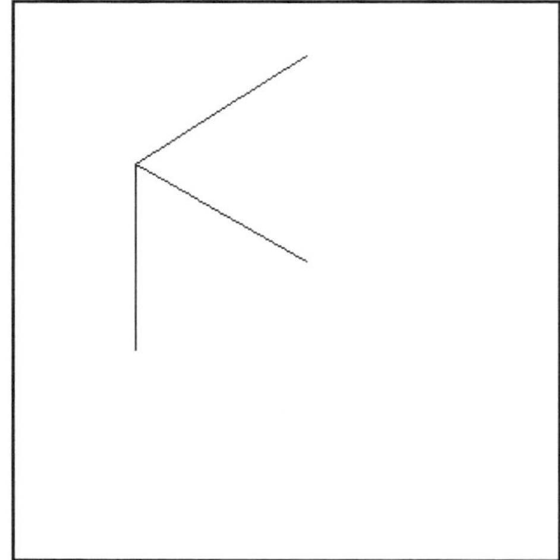

CLIPPING OPTIONS

1. Select the XCLIP command.
2. Click on the clipped image.

The following is a description for each option.

Enter clipping option
[ON/OFF/Clipdepth/Delete/generate Polyline/New boundary] <New>:

The description for the options are shown below:

ON / OFF – If you have "clipped" a drawing, you can make the clipped area visible again using the "OFF" option. To make the clipped area invisible again, select the "ON" option.

CLIPDEPTH – This option allows you to select a front and back clipping plane to be defined on a 3D model.

DELETE – To remove a clipping boundary completely.

GENERATE POLYLINE – This option creates a polyline object to represent the clipping border of the selected Xref. (The boundary is invisible by default)

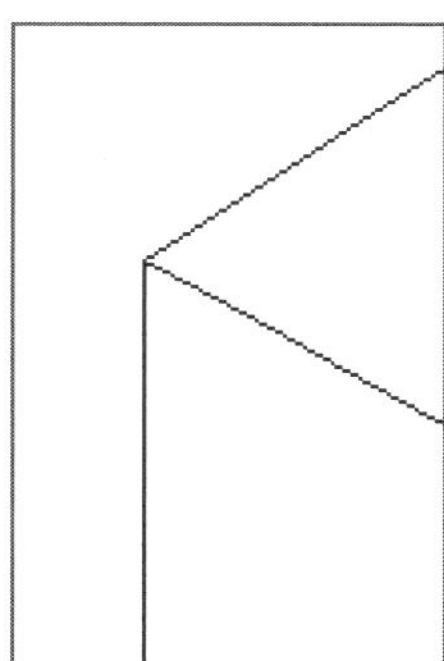

NEW BOUNDARY – Allows you to select a new boundary.

After the New Boundary option is selected, the following prompt appears:

Specify clipping boundary:
[Select polyline/Polygonal/Rectangular] <Rectangular>:

The description for the options are shown below:

SELECT POLYLINE – Allows you to select an existing polyline as a boundary.

POLYGONAL –Allows you to draw and irregular polygon, with unlimited corners, as a boundary.

RECTANGULAR – Allows a rectangular shape, with only 2 corners, for the boundary.

EDIT AN EXTERNAL REFERENCED DRAWING

An Xref drawing can be edited very easily using the Xopen command. The Xopen command allows you to open the Xref drawing (the child) in a separate window, make the changes and save those changes to the original of the Xref (Child) drawing. Select "Reload" to view the changes in the Parent drawing.

Note: This command is not available in AutoCAD LT. To edit an Xref drawing within LT you must: 1. Save the combined dwg (Parent and Child). 2. Open the original file for the Xref dwg (Child). 3. Make the change and save. 4. Re-open the Parent dwg. The changes to the Child should appear.

How to use the Xopen command.

1. Open the drawing that contains the External Referenced drawing.

2. Select the **Xopen** command using:

   TYPE = XOPEN

3. Select the Xref (Child) drawing, within the Parent drawing, to be changed.

   Important: The Xref (Child) original drawing will open in a separate window.

   (Note: the "Single-drawing compatibility mode" must be "off", unchecked. Ref. page 1-2) Verify by selecting <u>Window / Tile vertically</u>.

4. Make the necessary changes.

5. Save the drawing. ***(Note: you must use the same name.)***

6. Close the drawing, using **File / Close**.

   The Parent drawing will reappear with a "balloon message" notifying you that the Xref drawing has been changed and it needs to be "Reloaded".

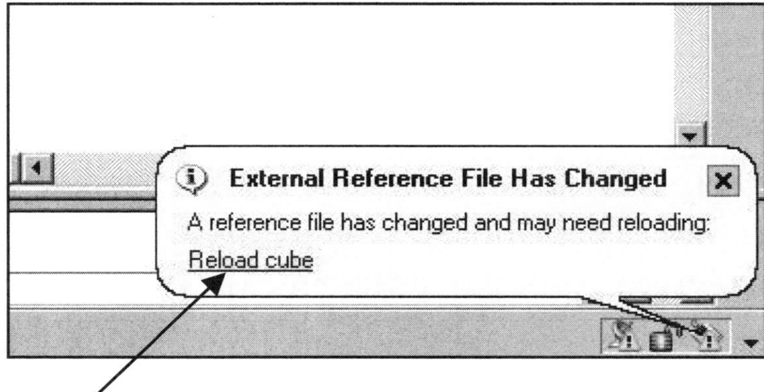

7. Click on the drawing name (underlined and blue)

Note: If the Change Balloon does not appear select <u>Insert / External Reference</u>...

If it was necessary to select Insert / External Reference... the following dialog box should appear.

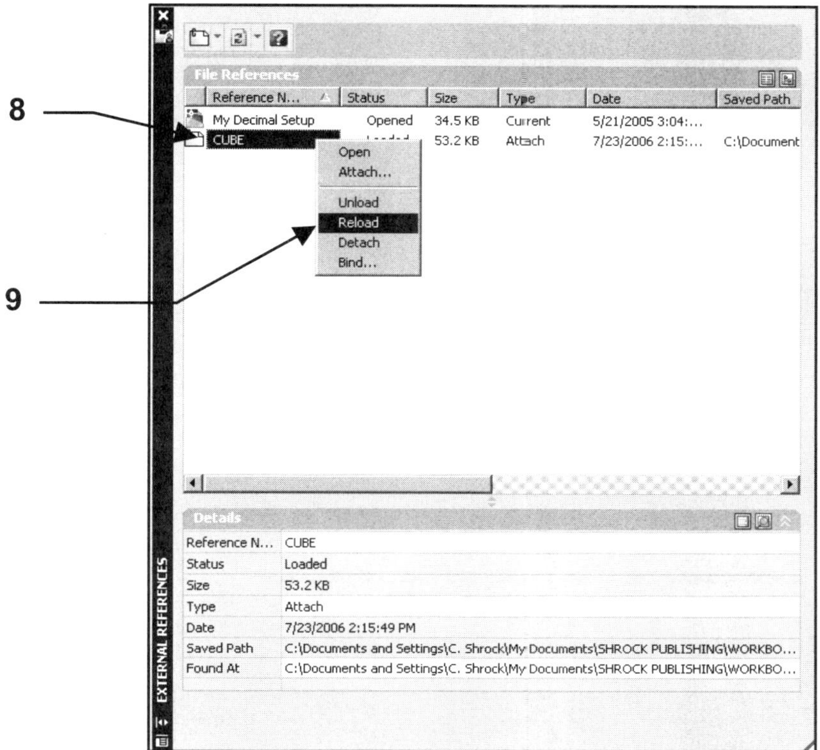

8. Select the drawing to reload.

9. Right click on the name and select the ***Reload*** from the drop down menu.

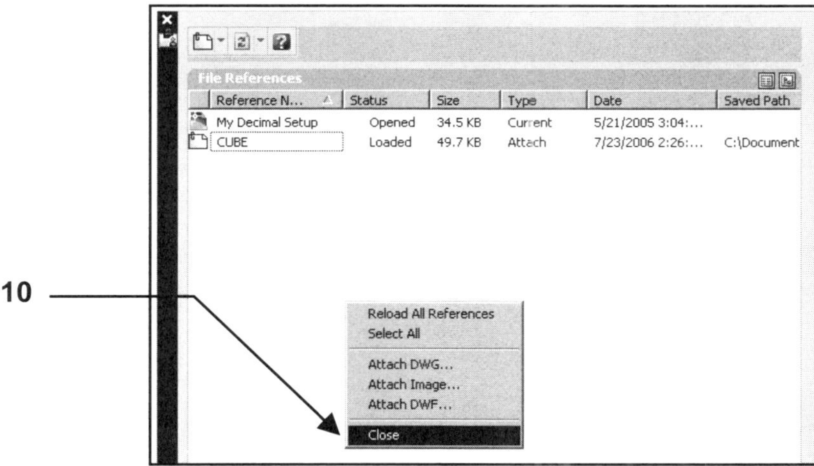

10. Right click again and select ***Close.***

Convert an Object to a Viewport (Not available in LT)

You may wish to have a viewport with a shape other than rectangular. AutoCAD allows you to convert Circles, Rectangles, Polygons, Ellipse and Closed Polylines to a Viewport.

1. You must be in Paperspace.
2. Draw the shape of the viewport using one of the objects listed above.

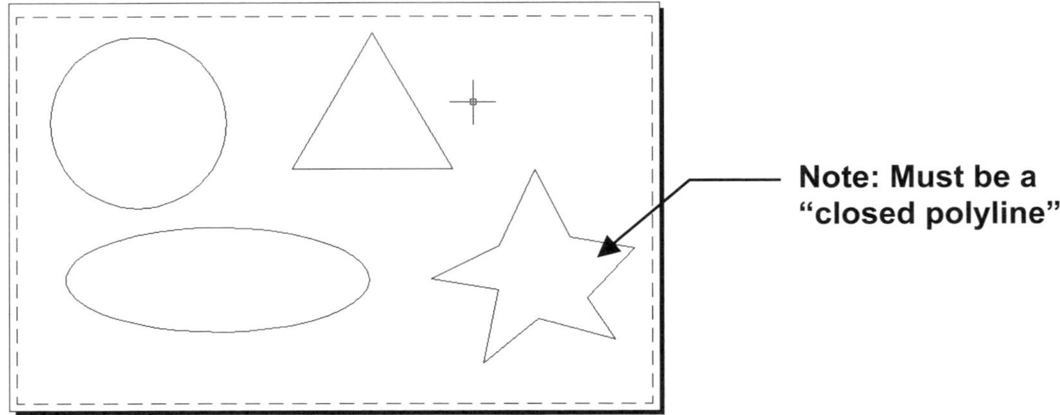

3. Select the Convert Object to Viewport icon on the Viewports toolbar **or**

   View / Viewports / Object.

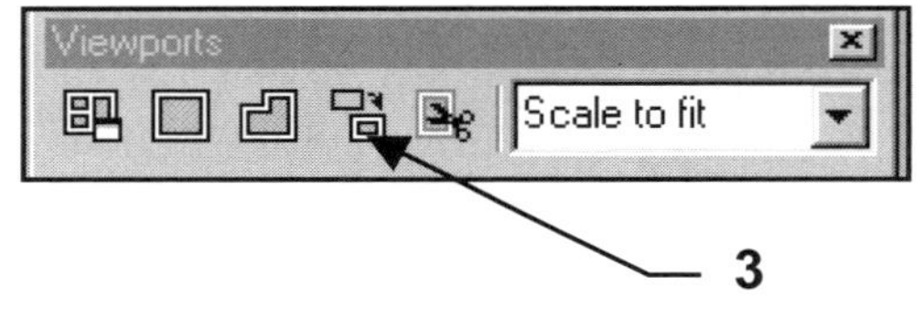

4. Select the object to convert.

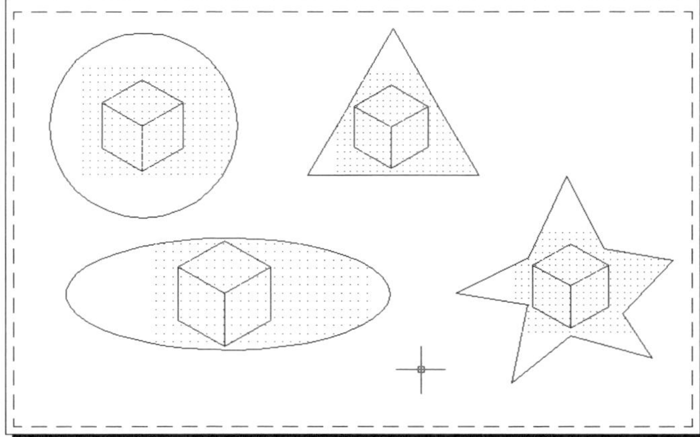

Now you can see through to model space. This is like looking out all the windows in your living room and seeing the same tree growing in your front yard.

Creating Multiple Viewport and Multiple Xrefs

When using Xrefs you will usually want multiple drawings Xref-ed. And you will also want multiple viewports. It will not be difficult to understand how to create multiple viewports but you will have to think a little about multiple xrefs.

The following is a example of how to control the viewing of multiple xrefs. The actual exercise, with step-by-step instructions, is EX-11A

1. Open My Decimal Setup
2. Select the 11 X 17 (1 to 1) layout tab.
3. Delete any viewports that already exist. (Click on each frame then select Erase)
4. Select the viewport layer.
5. Create 2 viewports approximately as shown below.
   First one, than the other.
   Simple so far, huh?

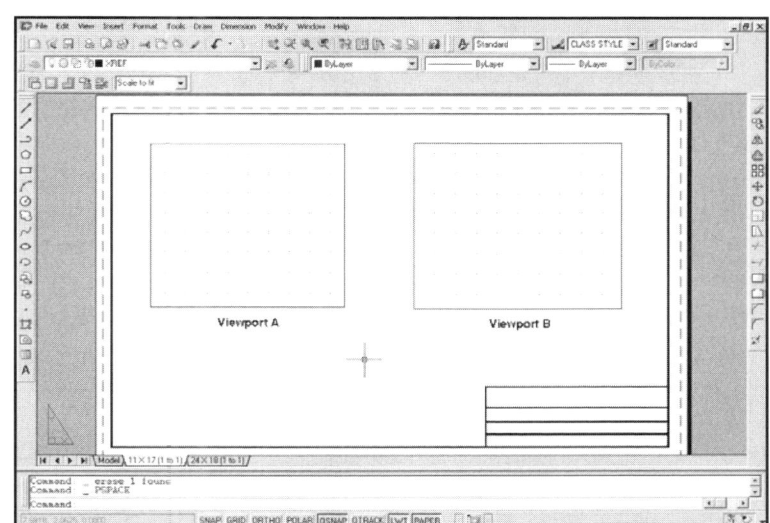

6. Double click inside the Viewport A, on the left.
7. Select the XREF layer.
8. Xref a drawing into this viewport. (7D would be a good one to use)
9. Select Zoom / Extents (You should see the entire drawing inside Viewport A)
10. Now activate Viewport B (double click inside Viewport B)
11. Select Zoom / Extents (You should see the same Xref drawing inside Viewport B also)

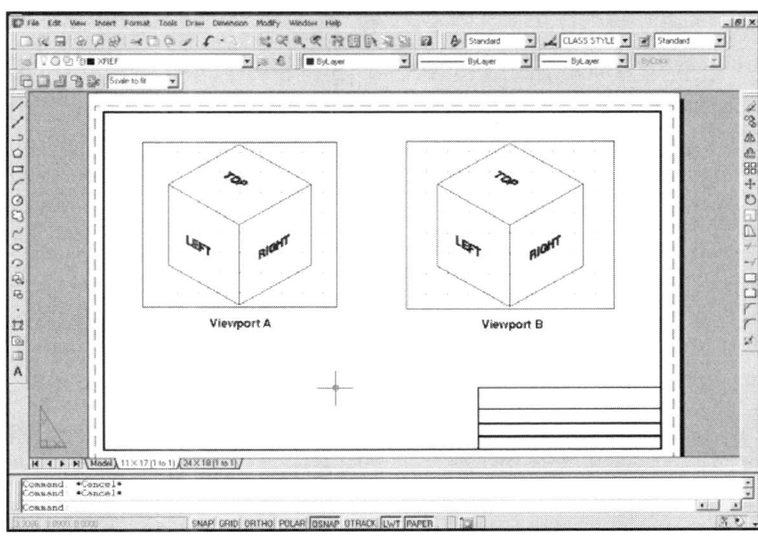

Now take a minute to think about this. This is an important concept to understand. *Pretend that the Viewport frames are 2 windows in your living room and the xref drawing is a tree in your front yard. If you stand in front of one of the windows you can see the tree. Then if you walk to the other window you see the same tree. It is the same tree, right?*

This is basically what is happening with AutoCAD. You have one xref drawing in Model Space (your front yard) and you are in Paper Space (your living room) looking through the viewports (your windows) to Model Space.

Now let's make the xref drawing in Viewport B disappear.

If you do not want to see the xref drawing in Viewport B you must "**Freeze in Current Viewport**" the layers of the xrefed drawing inside Viewport B.
This is easier than is sounds.

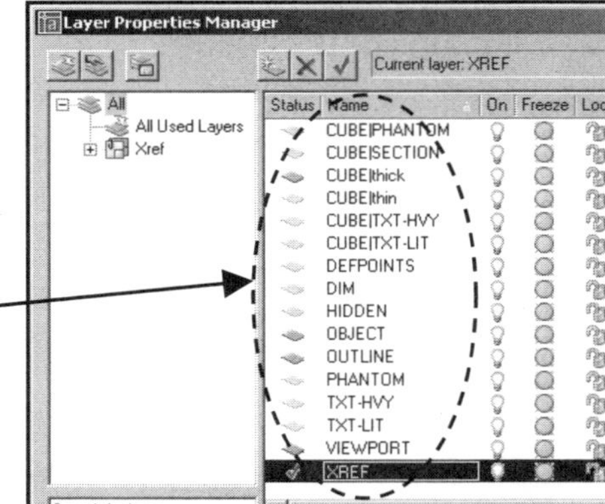

12. Activate Viewport B. (click in it)

13. Select Format / Layer…

Scroll up and down and look at the layer names. Notice some layers have been added. (Refer to page 11-6)

When the drawing was Xrefed, the layers came with that drawing.

14. To view only the layers that belong to the Xref drawing select the xref drawing name in the **Filter Tree** area.

15. Select all of the layers that belong to the XREF drawing. (highlight them)

16. Select the "**Freeze in Current Viewport**" box. Now this is very important: **Do not select "Freeze"**. Slide all the way over to the right hand side of the list of layers and select the "**Freeze in Current Viewport**" column. (Displayed as: **Cur**)

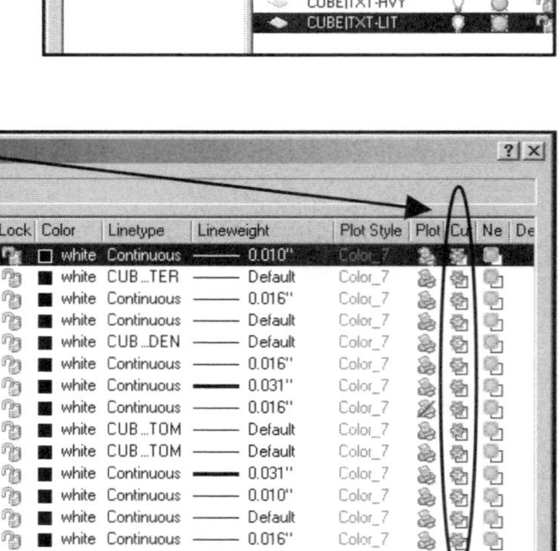

17. Select OK

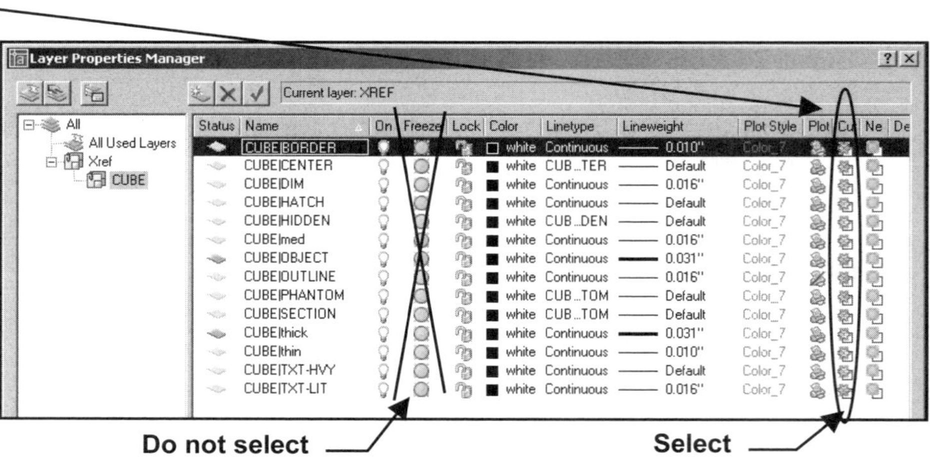

18. The Xrefed drawing should have disappeared in Viewport B. If it didn't do items 12 through 16 again.

19. Now you should adjust the scale in Viewport A and lock it.

Now think about this. *If you were to Xref another different drawing into Viewport B, it would also be visible in Viewport A (just another tree in your front yard). So you would have to activate Viewport A, select the layers that belong to the newly Xrefed drawing and select "Freeze layers in Current Viewport". Then only one drawing would be visible in each viewport.*

CREATING MULTIPLE VIEWPORTS – A QUICK METHOD

1. Select **VIEW / VIEWPORTS / NEW VIEWPORTS**

2. Select **Four: Equal** from the column on the left.
   This will divide the paperspace into 4 equal viewports

3. Enter **.50 Viewport Spacing**
   This the spacing between each of the new viewports.

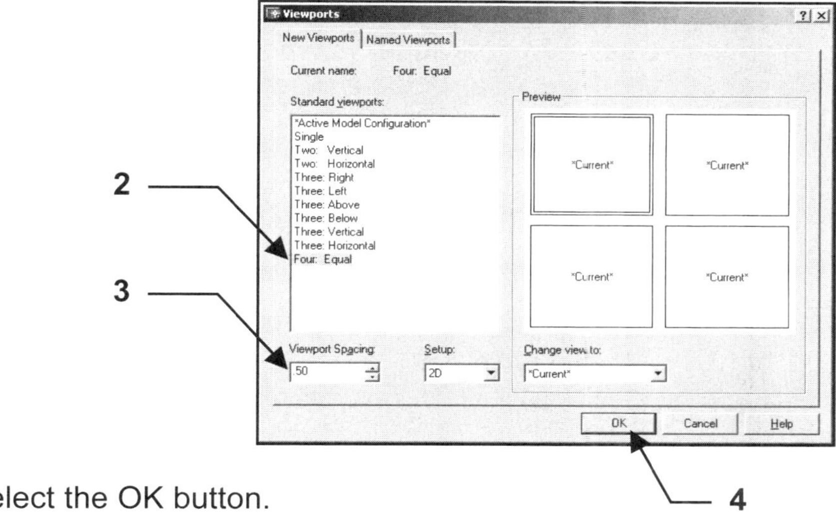

4. Select the OK button.

5. The following prompt will appear:

   Specify first corner or [Fit] <Fit>: ***Enter the location of the first corner of the area to divide and then the opposite corner.***

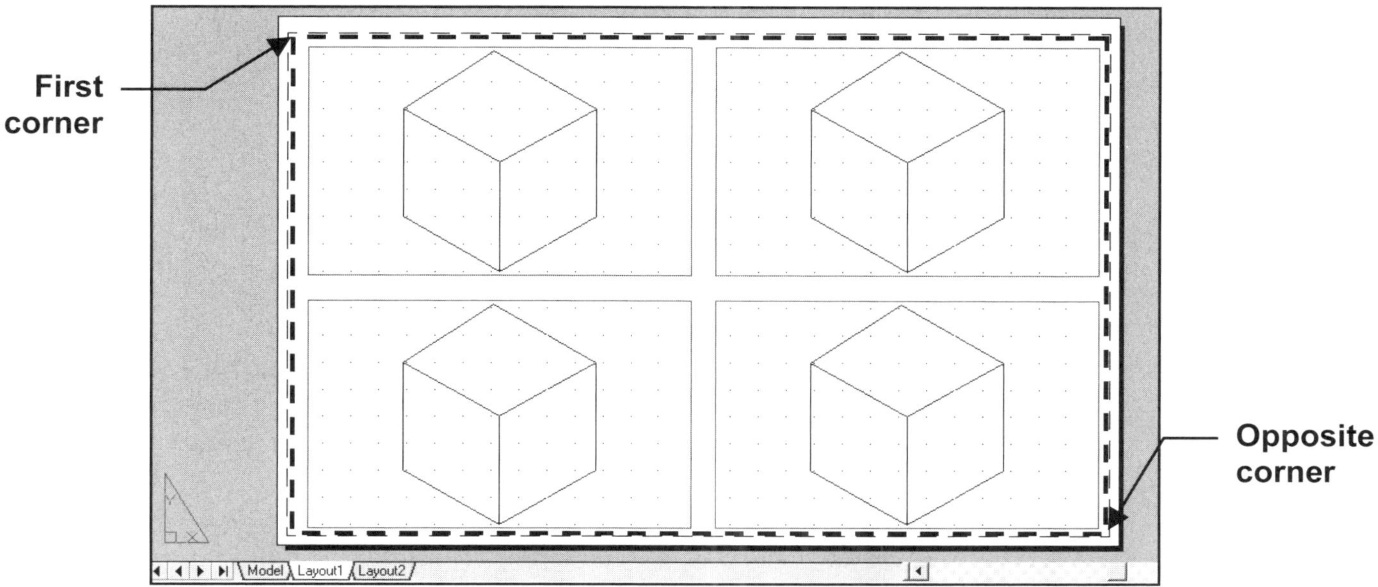

Note: If you have objects in Modelspace, they will appear in each viewport.

Non-continuous Linetype scales within Viewports

When you adjust the scale within a viewport, the image inside the viewport becomes larger or smaller. This you already know. But what you probably didn't notice is the non-continuous linetypes also change in scale. AutoCAD has a command that will control the linetype scale, within each viewport and paper space, based on the viewport scale or the paper space scale. This command is PSLTSCALE. (Paper space linetype scale) You may control the appearance of these linetypes by setting the PSLTSCALE to 1 or 0. (on or off)

PSLTSCALE set to 0 (off)
The linetype scale is based on the scale of the individual viewports and paperspace. All non-continuous linetype (Center, dashed, etc.) can appear different if the scales in the individual viewports and paper space are different.

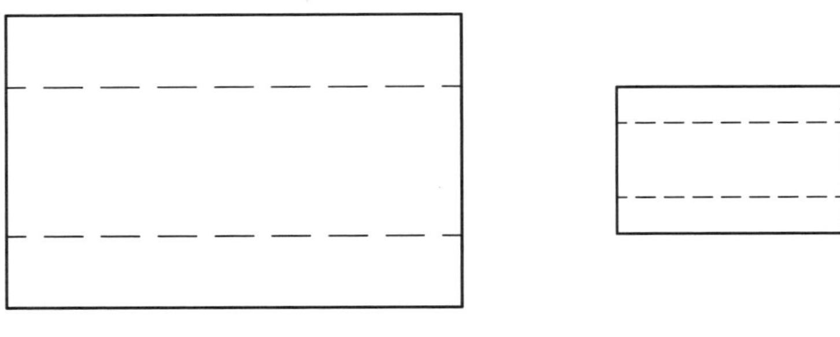

Scale = 1 :1 Scale 1 : 2

PSLTSCALE set to 1 (on; default setting)
The linetype scales for both model space and paper space are scaled to paper space units. **All non-continuous linetypes will appear equal**.

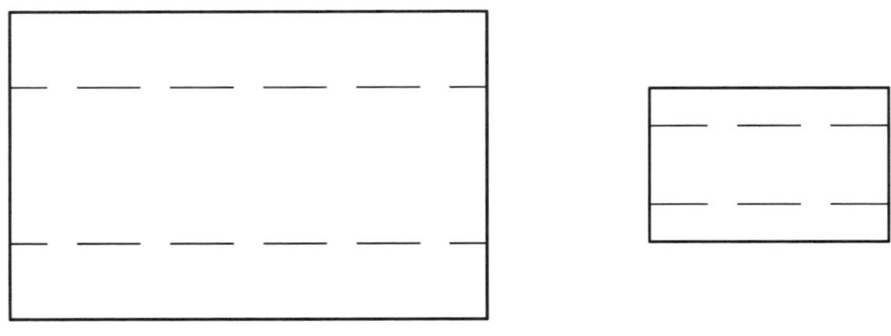

Scale = 1 : 1 Scale 1 : 2

Note, VERY IMPORTANT: After you change the PSLTSCALE you must select View / REGENALL to display the effects of the new setting.

To change PSLTSCALE:

Command: PSLTSCALE <enter>
Enter new value for PSLTSCALE: ***type 1 or 0 <enter>***
Command: REGENALL <enter>

EXERCISE 11A

XREF MULTIPLE DRAWINGS

1. Open **My Decimal Set Up**.

2. **Draw** a Rectangle (5" L X 3" W) in models space and **save as "RECT"**

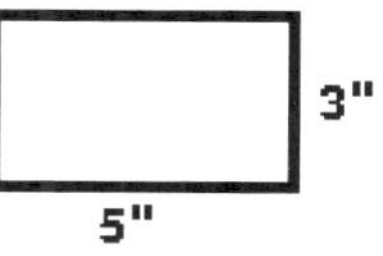

3. **Close** the drawing.

4. **Open** My Decimal Set Up **again**.

5. **Draw** a Circle (2" Radius) in model space and **save** as **"CIRCLE"**.

6. **Close** the drawing.

7. **Open** My Decimal Setup **again.**

8. Select the **11 X 17 (1 to 1)** tab.

9. Erase the Viewport frame. (Click on it, select erase and <enter>)

10. Select the Viewport layer.

Note: Lt users cannot convert objects to viewports. But you can still do this exercise. Instead of drawing circles, draw two single viewports approximately the size of the circles shown below. Then skip to instruction 12.

11. Create 2 new Viewports as shown below as follows:
    a. Draw 2 circles, **R3.50**, as shown below.
    b. Select the "**Convert Object to Viewport**" icon from the Viewport toolbar.
    c. Select each new Circle. (They are now viewports.)

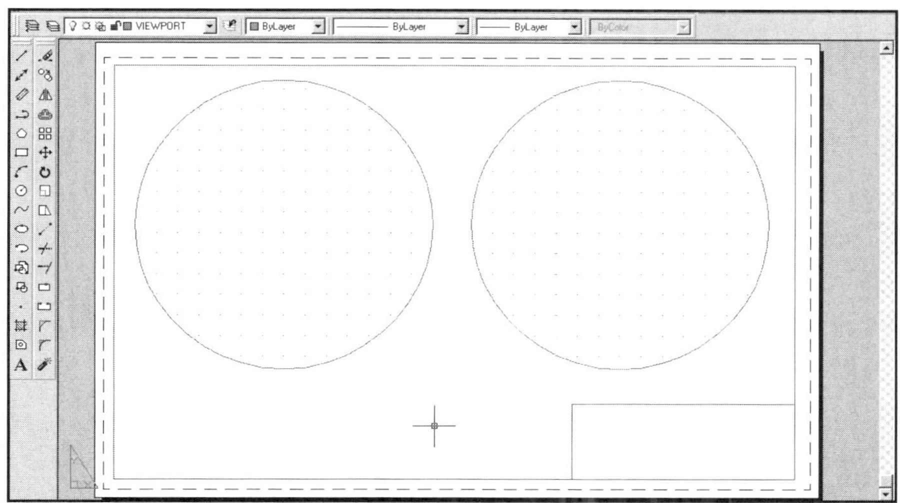

12. **Add Viewport Titles** as shown below. **(<u>In paper space</u>)**
    a. Use "Single Line Text"
    b. Style = Class Text
    c. Height = .35
    d. Layer = Text Heavy

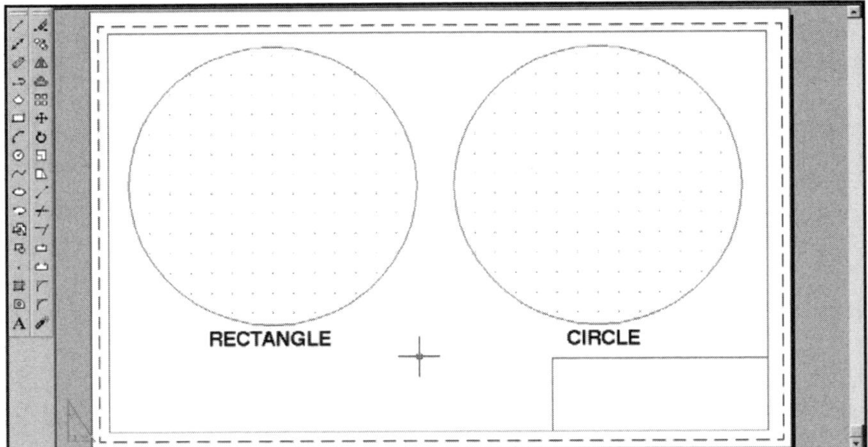

13. Double click inside of the Rectangle Viewport (on the left) to **activate Model space**.

14. Change to the XREF layer.

15. **Xref** the **Rect** drawing as follows:
    a. Select **INSERT / DWG. Reference…**
    b. Find and Select **Rect** drawing.
    c. Select the **OPEN** button.
    d. Select the **OK** button.
    e. Click inside the "Rectangle" Viewport.

16. Use **Zoom / All** inside each viewport so you can see the rectangle.

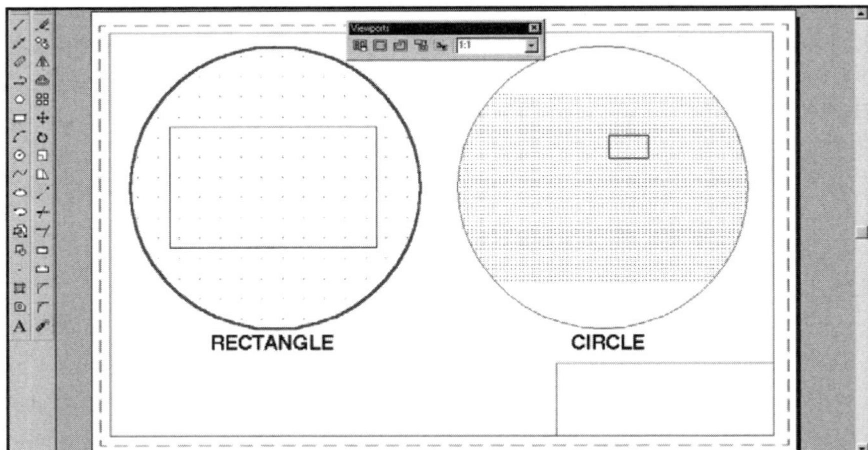

17. **Adjust the scale** in the **Rectangle** Viewport to **1 : 1.**

18. **"Pan"** the **Rect** drawing inside of the Rectangle viewport to display it as shown.
    <u>(Do not use the Zoom commands or you will have to re-adjust the scale)</u>

Now we are going to make the Rect drawing disappear in the Circle viewport.

19. Freeze all the layers that belong to **Rect** drawing in the **Circle Viewport** as follows:
    a. **Activate** the Circle Viewport
    b. Select **FORMAT / LAYER**
    c. Select **Rect** in the Filter tree.
    d. Select all the layers that begin with **Rect.**

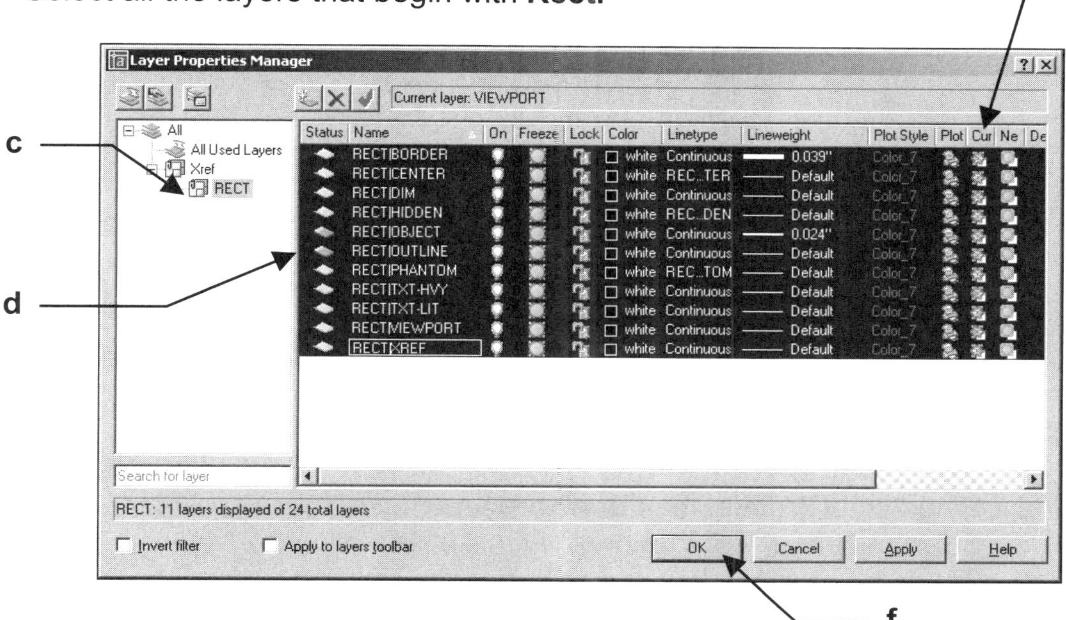

    e. Select the **"Freeze in current Viewport"** column.
    f. Select **OK** button.

The Rect drawing, inside the Circle Viewport, should have disappeared.

The next step is to XREF a second drawing into Model Space.
Then you will make the newly xref-ed drawing disappear in the Rectangle viewport.
When you have completed step 24 you should see only the Rectangle in the Rectangle viewport and only the Circle in the Circle viewport.

20. **Xref** drawing **Circle** into the **Circle Viewport** as follows:
    a. Select the **XREF** layer.
    b. Select **INSERT / DWG Reference…**
    c. Find and Select the **Circle** drawing.
    d. Select the **OPEN** button.
    e. Select the **OK** button.
    f. Click inside the Circle Viewport.

21. Use **Zoom / All** inside the Circle viewport if you can't see the Circle drawing.

22. **Adjust the scale** in the **Circle** Viewport to **1 : 1.**

23. **"Pan"** the **Circle** drawing inside of the Circle viewport to display it as shown.
    <u>(Do not use the Zoom commands or you will have to re-adjust the scale)</u>

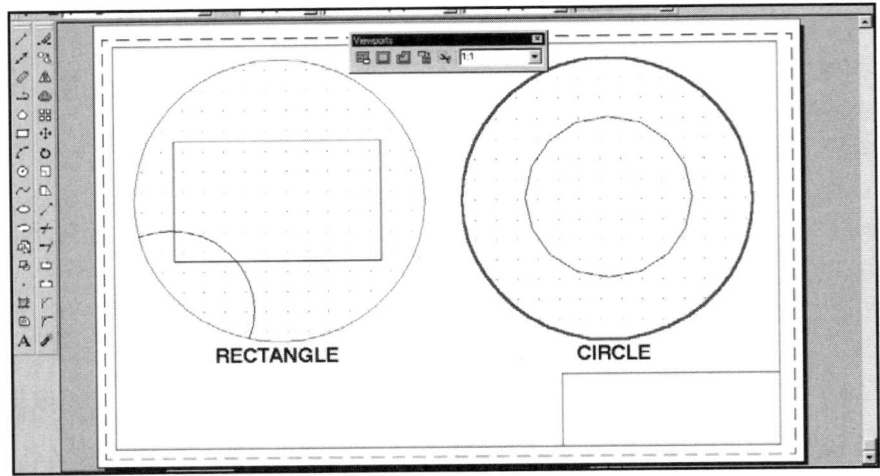

Now we will make the Circle drawing disappear in the Rectangle viewport.

24. "Freeze in Current Viewport" all the layers that belong to the Circle drawing, in the Rectangle Viewport as follows:
    a. **Activate** the Rectangle Viewport.
    b. Select **FORMAT / LAYER.**
    c. Select the **Circle** in the **Filter tree** area.
    d. Select all the layers that begin with **Circle**.
    e. Select the "**Freeze in current Viewport**" column.
    f. Select **OK** button.

The Xref drawing, Circle, should have disappeared in the Rectangle Viewport.

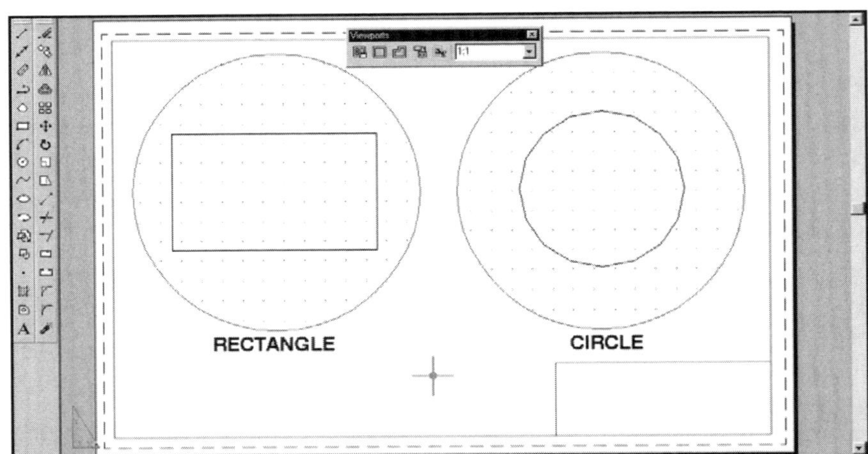

25. **Change to Paper space and change the Title Block**
    a. Title = Multiple Xref drawings
    b. Scale = 1 - 1

26. Save as **EX-11A.**

27. **Plot** using Page Set up **11 X 17 (MONO).**

EXERCISE 11B

Creating Multi-scaled Views

1. Open **My Decimal Setup.**
2. Select the **11 X 17 (1 to 1)** tab.
3. Important: Erase the existing Viewport frame. (Click on it, select erase and <enter>)
4. Select the Viewport layer.
5. Create 4 new Viewports as shown below.

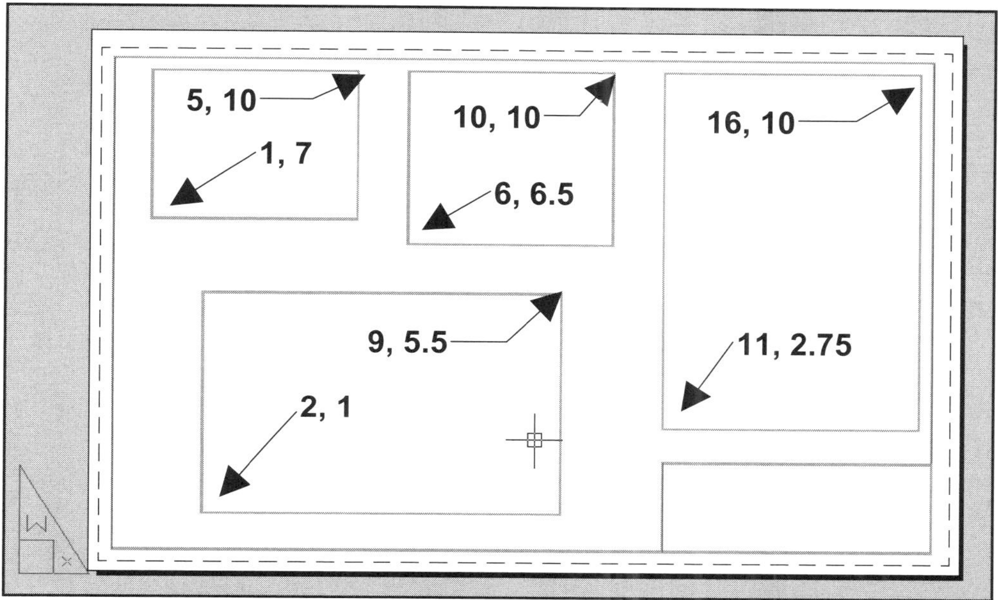

6. Add Viewport Titles as shown below. (In paperspace)
   a. Use "Single Line Text"
   b. Style = Class Text
   c. Height = .35
   d. Layer = Text Heavy

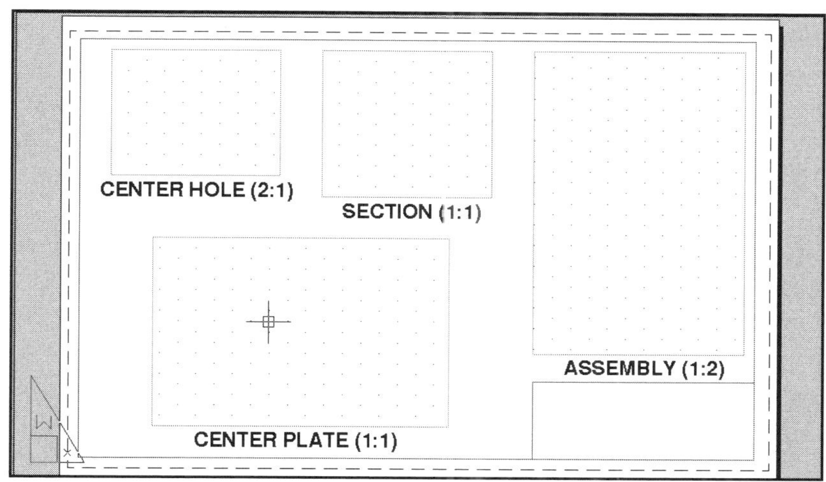

7. Double click inside one of the Viewports to **activate Model space**.

8. **Xref** drawing 6A as follows:
   a. Select **INSERT / DWG Reference...**
   b. Find and Select 6A
   c. Select the OPEN button.
   d. Remove the check mark from all of the "Specify On-Screen" boxes.
   e. Select the OK button.

9. **Adjust the scale** in each Viewport to the scale listed under each viewport as shown below.

10. **"Pan"** the drawing inside each viewport to display the correct area. (Do not use the Zoom commands or you will have to re-adjust the scale)

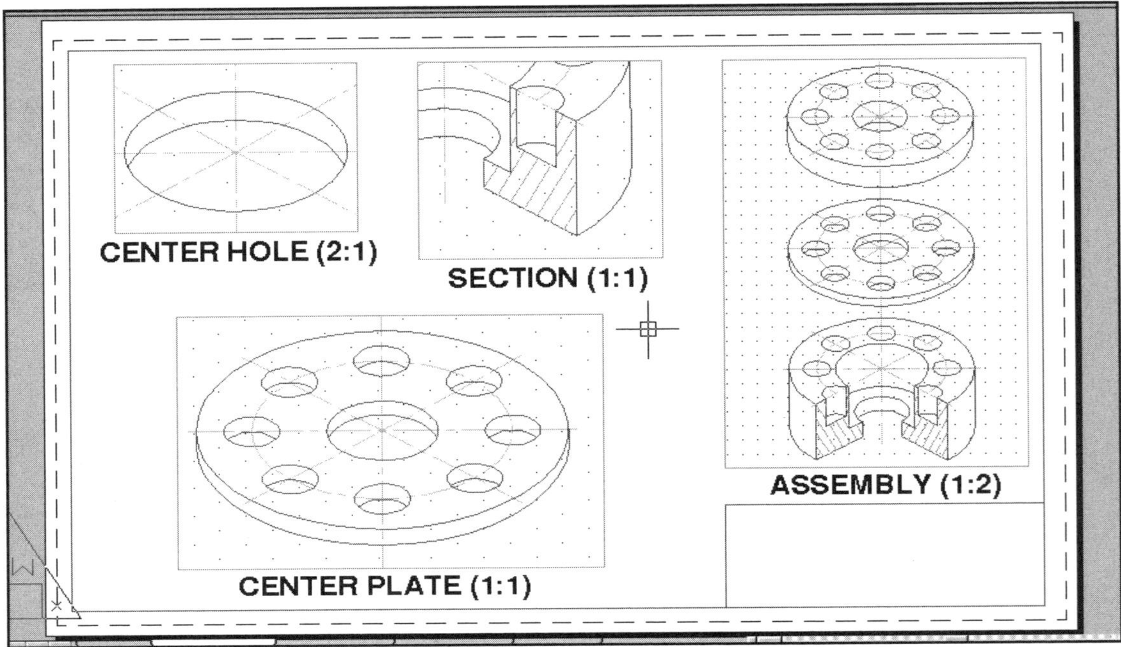

11. Set **"PSLTSCALE"** command to **"0"** and **Regenall.** (Refer to page 11-15)

12. **Return to Paper space and change the Title Block**
    a. Title = Multi-scaled Viewports
    b. Scale = Noted

13. Save as **EX-11B**

14. **Plot** using Page Set up **11 X 17 (MONO).**

EXERCISE 11C

Unload, Reload and Detach an Xref drawing.
<u>FOR LT USERS ONLY</u>

1. Open **EX-11B.**
2. Select the **11 X 17 (1 to 1)** tab.
3. Select **INSERT / EXTERNAL Reference…**

The following dialog box should appear.

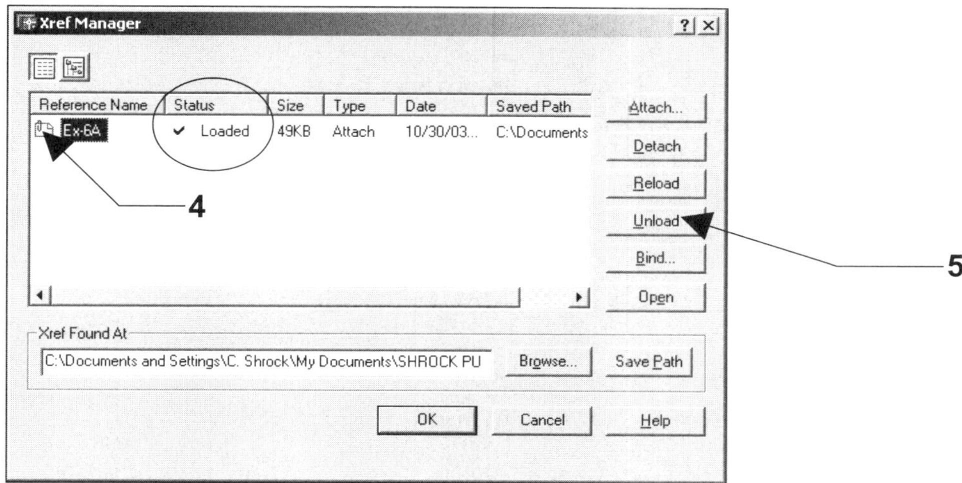

4. Select the xref drawing **EX-6A** from the Reference Name list.
5. Select the **Unload** button and then the **OK** button.

   Did the drawing disappear?

6. Select **INSERT / EXTERNAL Reference…**again.

   Notice EX-6A is still listed.

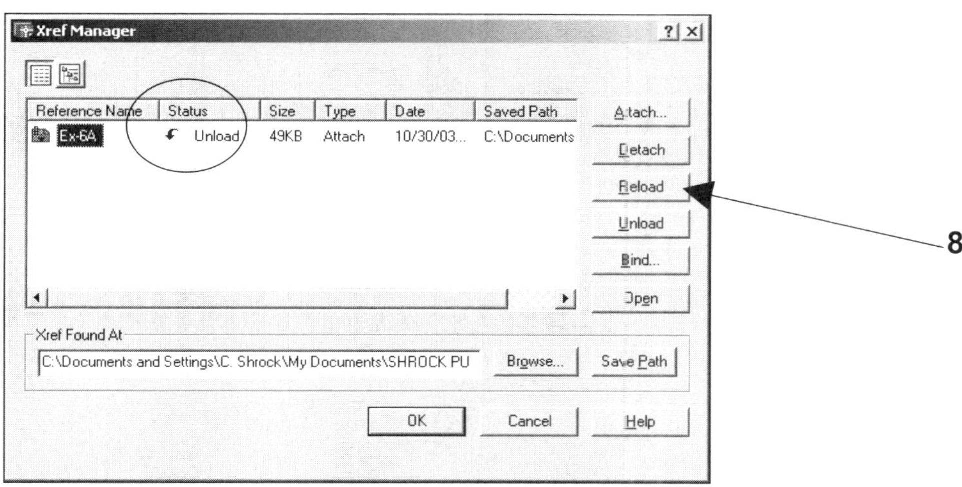

7. Select the **EX-6A** drawing from the Reference Name list again.

8. Select the **Reload** button and then the **OK** button.

   Did the drawing re-appear?

9. Select **INSERT / EXTERNAL Reference…**again.

10. Select the EX-6A drawing from the Reference Name List again.

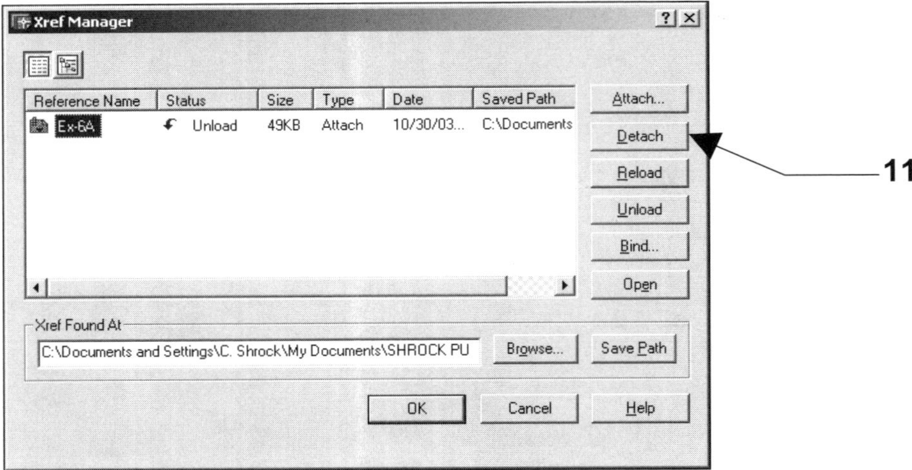

11. Select the **DETACH** button and then the **OK** button.

    Did the drawing disappear?

12. Select **INSERT / EXTERNAL Reference…**<u>one more time</u>.

    This time, EX-6A should no longer appear in the "Reference Name" list.

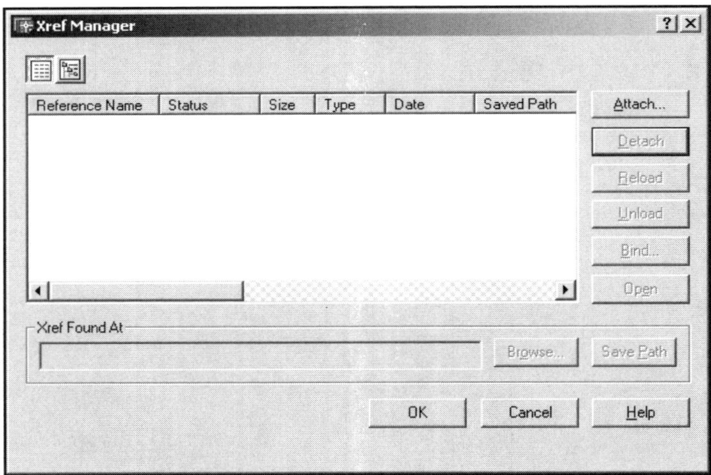

13. **Save as EX-11C. <u>Do not Plot</u>.**

EXERCISE 11D

CLIPPING AN EXTERNAL REFERENCE

1. Open **EX-11B.**
2. Select the **Model** tab.
3. Select **View / Zoom / Extents**

4. Select **XCLIP** command. (Ref to page 11-8)
5. Clip the lower section to appear as **shown below**:
   a. Select objects.
   b. Select New Boundary and Rectangular

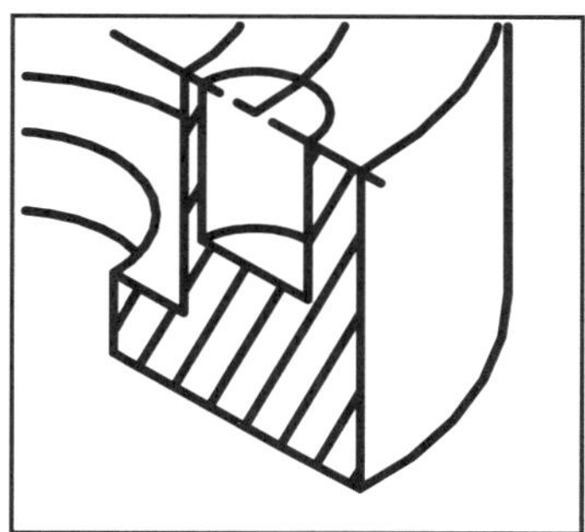

6. **Save as EX-11D1**

7. Select the **XCLIP** command again.
   a. Select Objects.
   b. Select **OFF**
   The entire drawing should reappear.

8. Select the **XCLIP** command again.
9. Clip the lower section to appear approximately as shown **on the right**:
   a. Use New Boundary .
   b. Select "Yes" when prompted to "Delete old boundary"
   c. Use Polygonal.
10. **Save as EX-11D2**

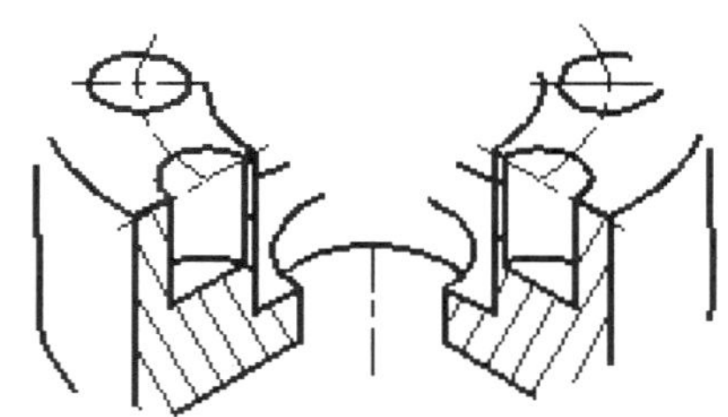

11. Select the **XCLIP** command again.
12. Generate a visible Polyline boundary as shown **below right**.
    a. Use generate Polyline.

13. **Save as EX-11D3, do not plot.**

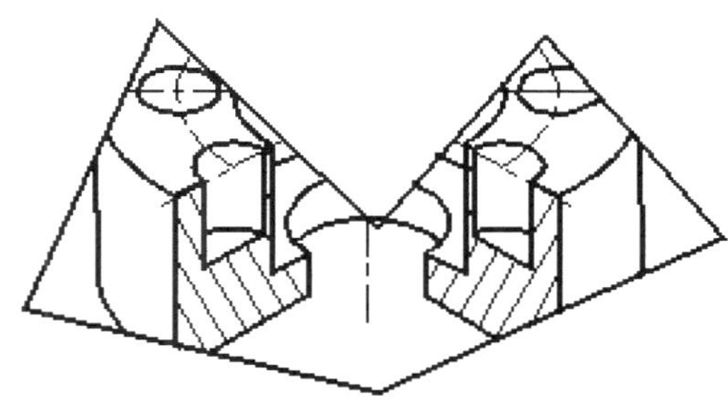

NOTES:

LEARNING OBJECTIVES

After completing this lesson, you will be able to:

1. Dimension using Datum dimensioning.
2. Use alternate dimensioning for inches and M illimeters.
3. Assign Tolerances to a part.
4. Use Geometric tolerances.
5. Typing Geometric Symbols.

LESSON 12

ORDINATE Dimensioning

Ordinate dimensioning is primarily used by the sheet metal industry. But many others are realizing the speed and tidiness this dimensioning process allows.

Ordinate dimensioning is used when the X and the Y coordinates, from one location, are the only dimensions necessary. Usually the part has a uniform thickness, such as a flat plate with holes drilled into it. The dimensions to each feature, such as a hole, originate from one "datum" location. This is similar to "baseline" dimensioning. Ordinate dimensions have only one datum. The datum location is usually the lower left corner of the object.

Ordinate dimensions appearance is also different. Each **dimension** has only one **leader line** and a **numerical value**. Ordinate dimensions do not have extension lines or arrows.

Example of Ordinate dimensioning:

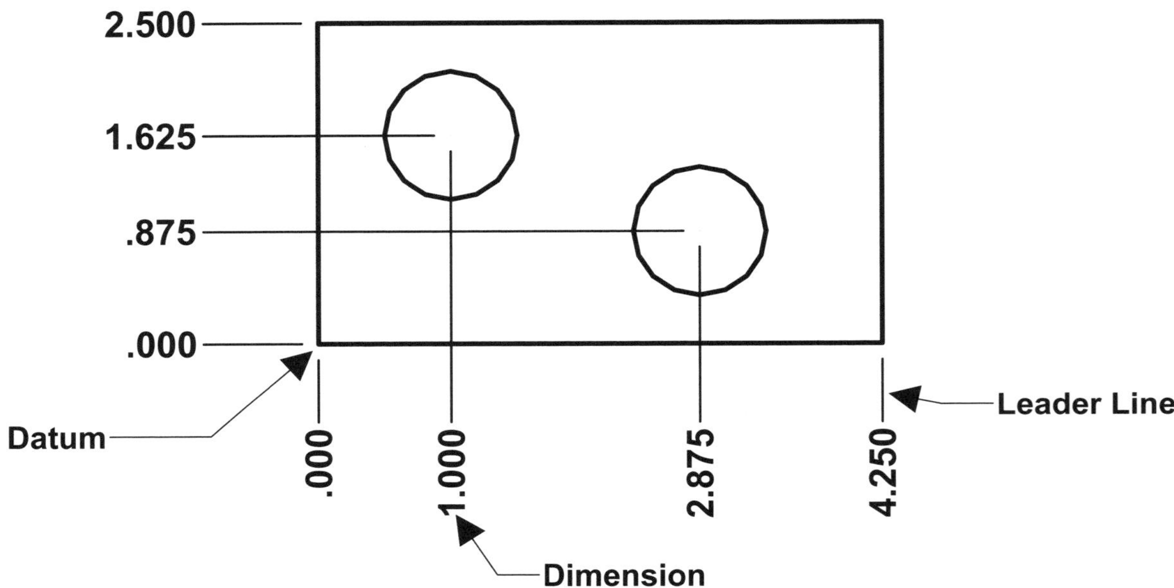

Note:
Ordinate dimensions can be Associative and are Trans-spatial. Which means that you can dimension in paperspace and the ordinate dimensions will remain associated to the object they dimension. (Except for Qdim ordinate)

Refer to the next page for step by step instructions to create Ordinate dimensions.

Creating Ordinate dimensions

1. Move the "Origin" to the desired "datum" location as follows:
   Note: This must be done in Model Space.
   a. Select **TOOLS / NEW UCS / ORIGIN**
   b. Snap to the desired location.

2. Select the Ordinate command using one of the following:
   Note: Ordinate dimensioning may be used in model or paper space.

   TYPE = DIMORDINATE
   PULLDOWN = DIMENSION / ORDINATE
   TOOLBAR = DIMENSION

3. Select the first feature, using object snap.

4. Drag the leader line horizontally or vertically away from the feature.

5. Select the location of the "leader endpoint.
   (The dimension text will align with the leader line)

Use "Ortho" to keep the leader lines straight.

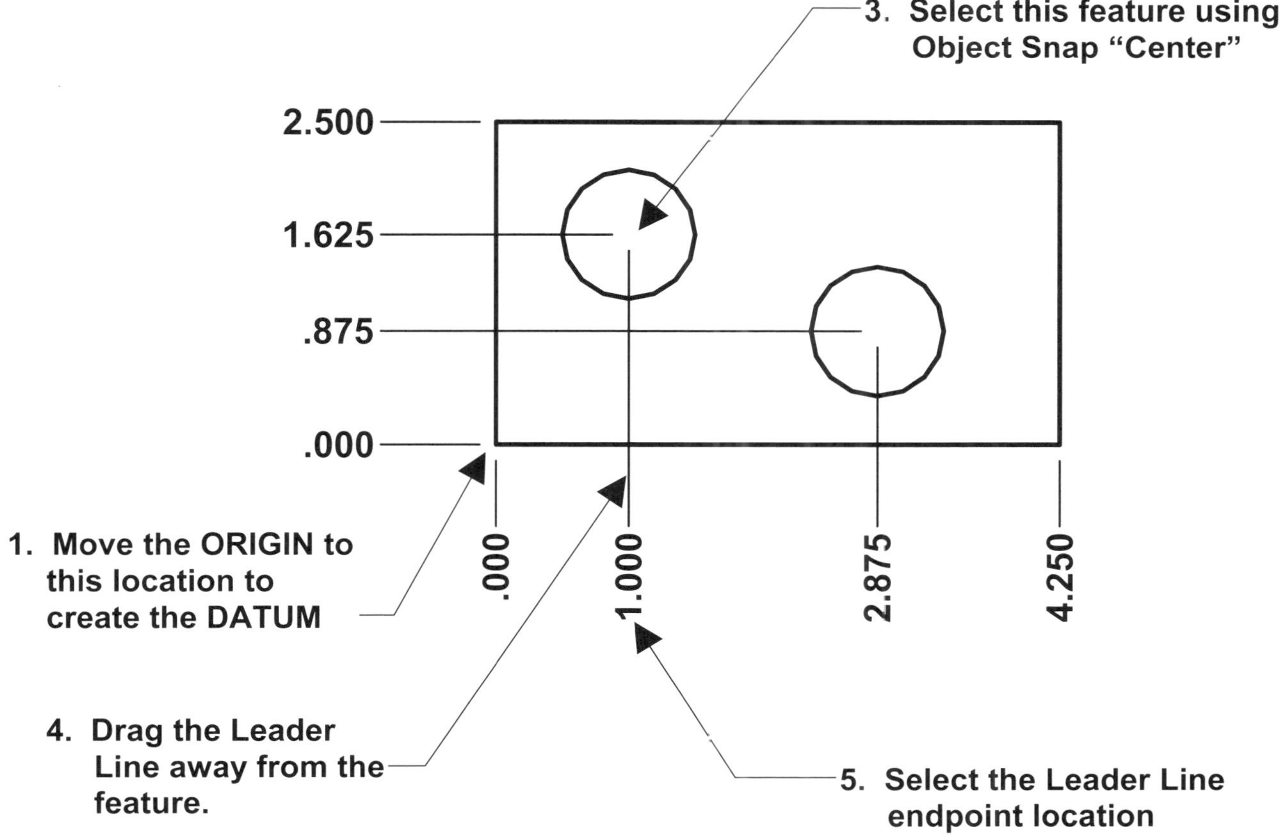

JOG an Ordinate dimension

If there is insufficient room for a dimension you may want to jog the dimension.
To **"jog"** the dimension, as shown below, turn "**Ortho**" **off** before placing the Leader Line endpoint location. The leader line will automatically jog. With Ortho off, you can only indicate the feature location and the leader line endpoint location, the leader line will jog the way it wants to.

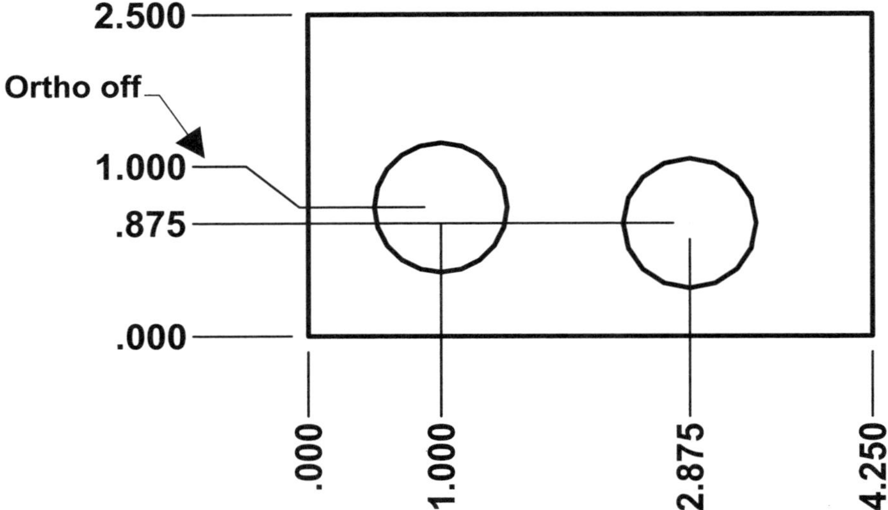

Quick Dimension with Ordinate dimensioning

(Not available in version LT)

1. Select **DIMENSION / Quick Dimension** (Refer to Ex Workbook for Beg Acad, lesson 20)
2. Select the geometry to dimension <enter>
3. Type **"O"** <enter> to select Ordinate
4. Type **"P"** <enter> to select the **datumPoint** option.
5. Select the datum location on the object. (use Object snap)
6. Drag the dimensions to the desired distance away from the object.

> **Note: Qdim can be associative but is not trans-spatial.**
> **If the object is in Model Space, you must dimension in Model Space.**

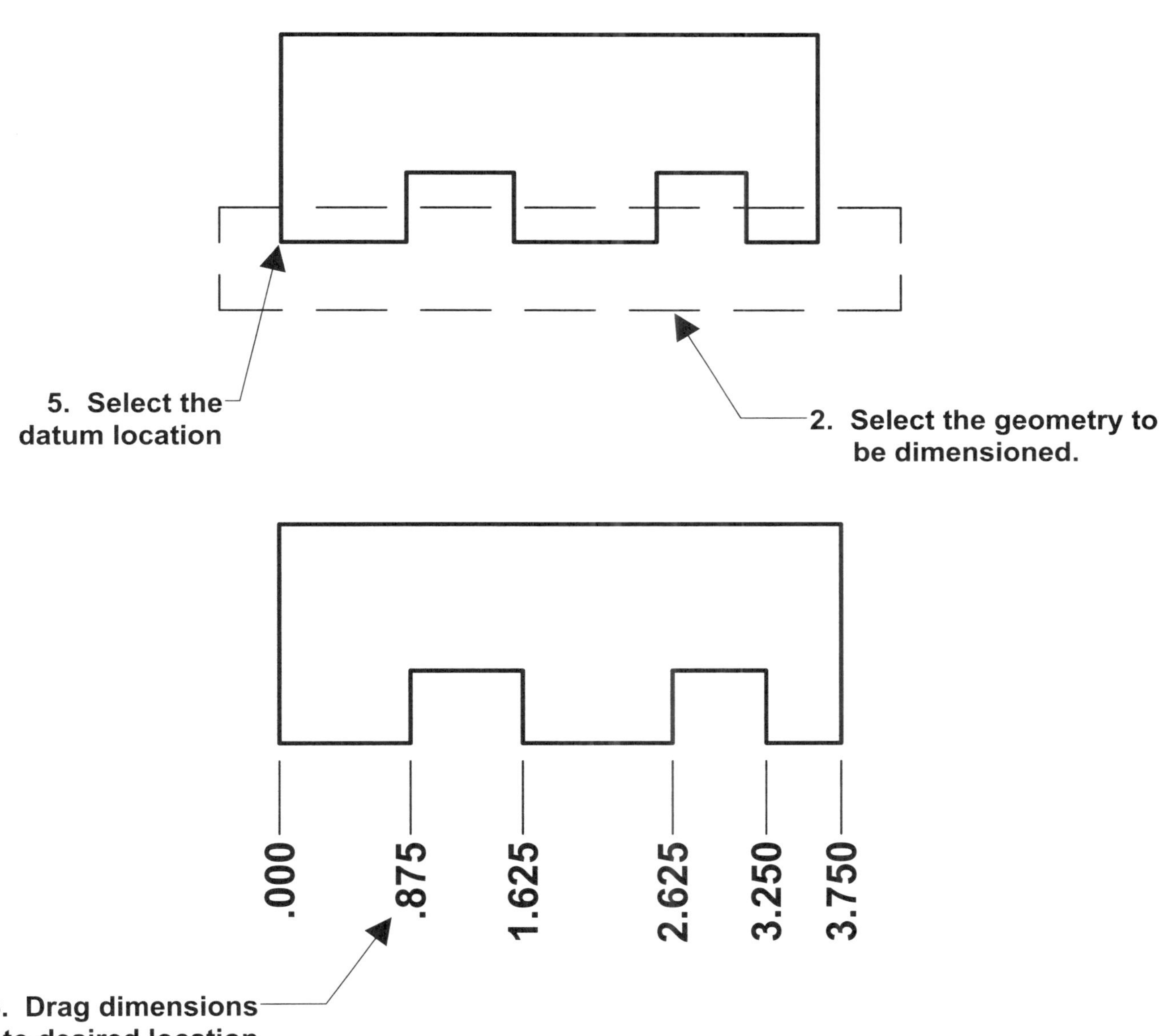

ALTERNATE UNITS

The options in this tab allow you to display inches as the primary units and the millimeter equivalent as alternate units. The millimeter value will be displayed inside brackets immediately following the inch dimension. Example: 1.00 [25.40]

1. Select **DIMENSION / STYLE / MODIFY**
2. Select the **ALTERNATE UNITS** tab.

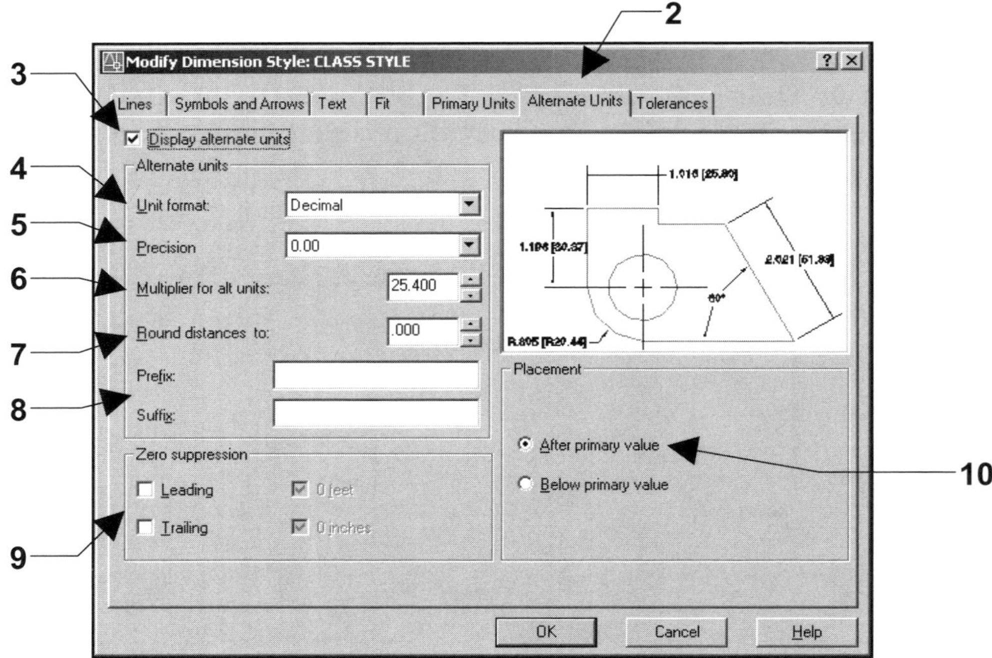

3. **Display alternate units.** Check this box to turn ON alternate units
4. **Unit format**. Select the Units for the alternate units.
5. **Precision** Select the Precision of the alternate units. This is independent of the Primary Units.
6. **Multiplier for all units** The primary units will be multiplied by this number to display the alternate unit value.
7. **Round distance to** Enter the desired increment to round off the alternate units value.
8. **Prefix / Suffix** This allows you to include a Prefix or Suffix to the alternate units. Such as: type **mm** to the Suffix box to display **mm** (for millimeters) after the alternate units.
9. **Zero Suppression** If you check one or both of these boxes, it means that the zero will not be drawn. It will be suppressed.
10. **Placement** Select the desired placement of the alternate units. Do you want them to follow immediately after the Primary units or do you want the Alternate units to be below the primary units?

Note: If alternate units are turned off you may display them by editing the dimension text in properties.
Use **<>** to represent the associative dimension and **[ ]** to display the alternative dimension.
Example: If the dimension is **15.00** you would enter **< > [ ]** in **Text Override** box.
The display would change to **15.00 [381]**.
Must be used with the generated dimension text. Cannot be displayed alone.

TOLERANCES

When you design and dimension a widget, it would be nice if when that widget was made, all of the dimensions were exactly as you had asked. But in reality this is very difficult and or expensive. So you have to decide what actual dimensions you could live with. Could the widget be just a little bit bigger or smaller and still work? This is why tolerances are used.

A **Tolerance** is a way to communicate, to the person making the widget, how much larger or smaller this widget can be and still be acceptable. In other words each dimension can be given a maximum and minimum size. But the widget must stay within that **"tolerance"** to be correct. For example: a hole that is dimensioned 1.00 +.06 -.00 means the hole is nominally 1.00 but it can be as large as 1.06 but can not be smaller than 1.00.

1. Select **DIMENSION / STYLE / MODIFY**
2. Select the **TOLERANCES UNITS** tab.

Note: if the dimensions in the display look strange, make sure "Alternate Units" are turned OFF.

3. **Method**

The options allows you to select how you would like the tolerances displayed. There are 5 methods: None, Symmetrical, Deviation, Limits. (Basic is used in geometric tolerancing and will not be discussed at this time)

Refer to the next page for descriptions of methods.

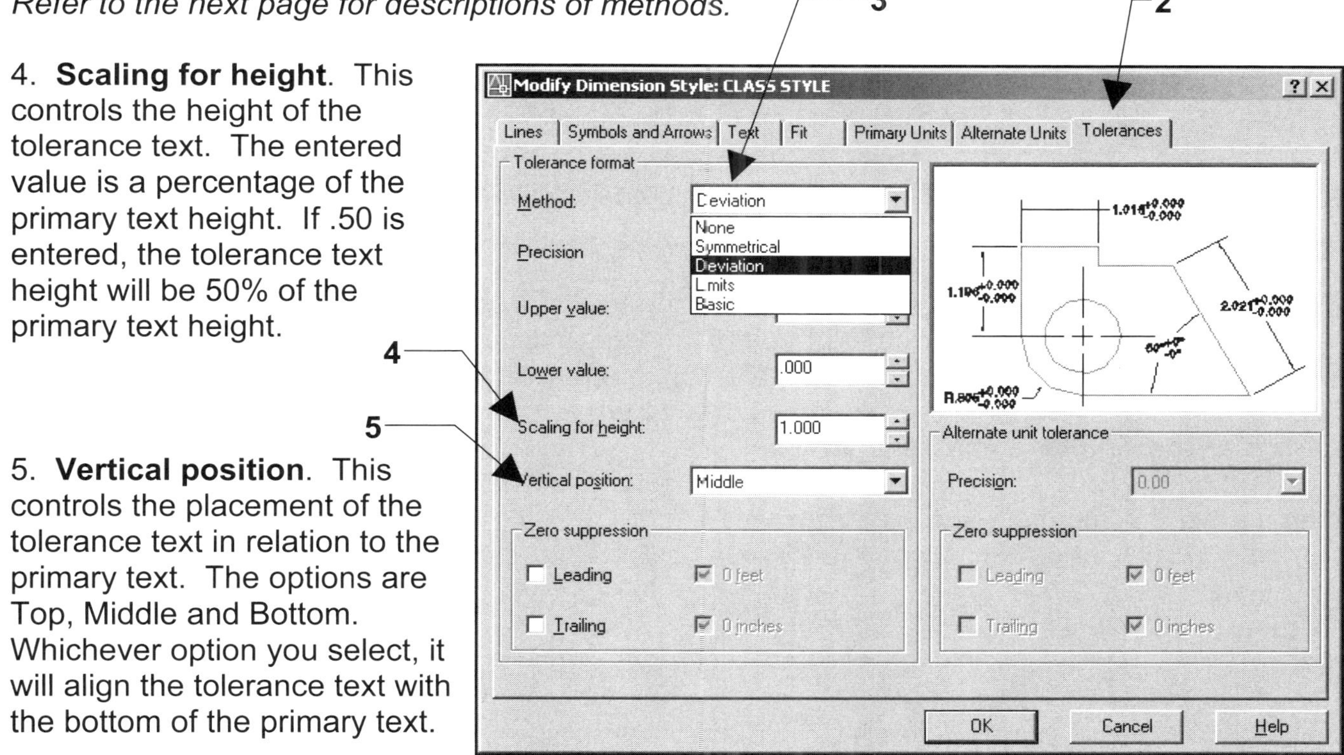

4. **Scaling for height**. This controls the height of the tolerance text. The entered value is a percentage of the primary text height. If .50 is entered, the tolerance text height will be 50% of the primary text height.

5. **Vertical position**. This controls the placement of the tolerance text in relation to the primary text. The options are Top, Middle and Bottom. Whichever option you select, it will align the tolerance text with the bottom of the primary text.

SYMMETRICAL is an equal bilateral tolerance. It can vary as much in the plus as in the negative. Because it is equal in the plus and minus direction, only the "Upper value" box is used. The "Lower value" box is grayed out.

Example of a Symmetrical tolerance
:

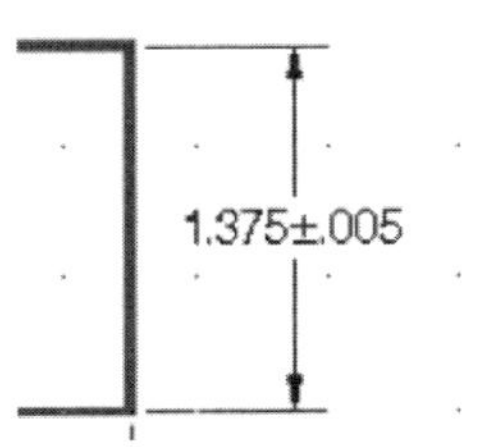

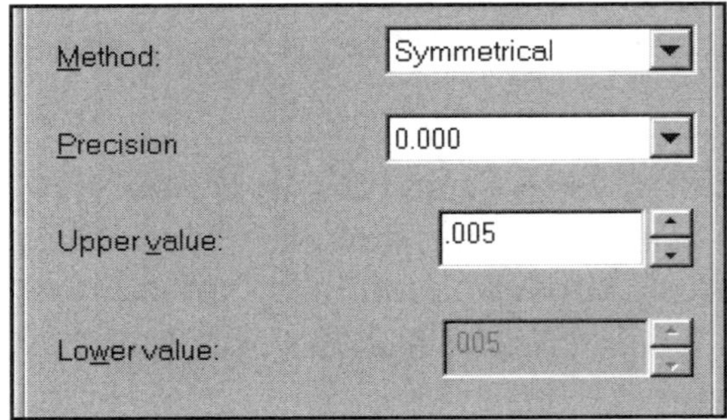

DEVIATION is an unequal bilateral tolerance. The variation in size can be different in both the plus and minus directions. Because it is different in the plus and the minus the "Upper" and "Lower" value boxes can be used.

Example of a Deviation tolerance:

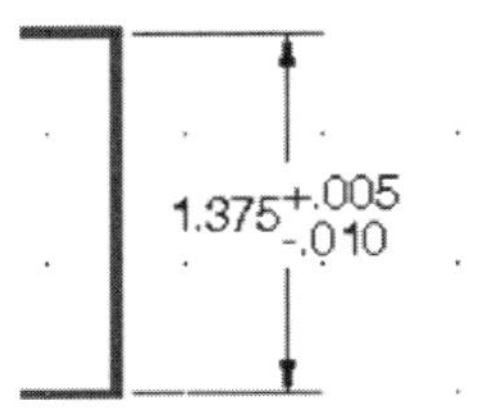

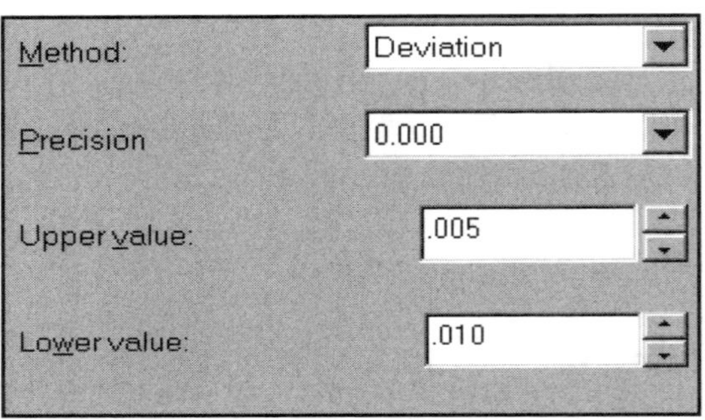

Note: If you set the upper and lower values the same, the tolerance will be displayed as symmetrical.

LIMITS is the same as deviation except in how the tolerance is displayed. Limits calculates the plus and minus by adding and subtracting the tolerances from the nominal dimension and displays the results. Some companies prefer this method because no math is necessary when making the widget. Both "Upper and Lower" value boxes can be used.
Note: The "Scaling for height" should be set to "1".

Example of a Limits tolerance:

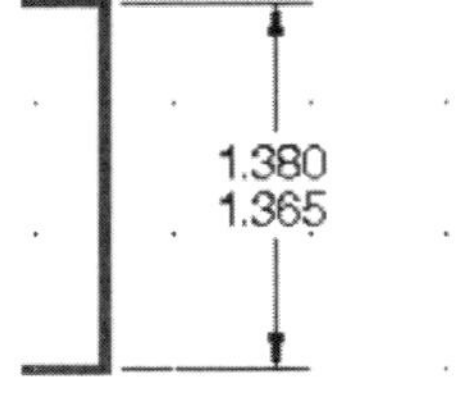

GEOMETRIC TOLERANCING

Geometric tolerancing is a general term that refers to tolerances used to control the form, profile, orientation, runout, and location of features on an object. Geometric tolerancing is primarily used for mechanical design and manufacturing. The instructions below will cover the Tolerance command for creating geometric tolerancing symbols and feature control frames.

If you are not familiar with geometric tolerancing, you may choose to skip this lesson.

1. Select the **TOLERANCE** command using one of the following:

   TYPE = TOL
   PULLDOWN = DIMENSION / TOLERANCE
   TOOLBAR = DIMENSION

   The Geometric Tolerance dialog box, shown below, should appear.

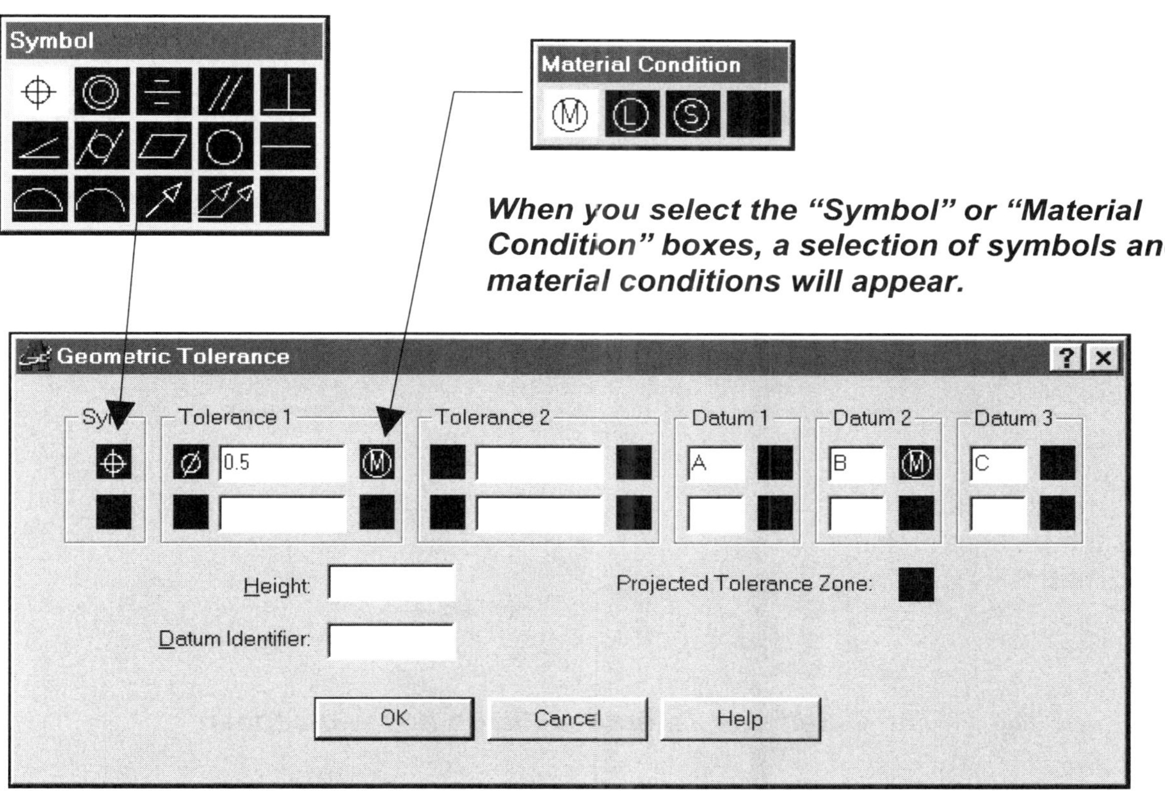

2. Make your selections and fill in the tolerance and datum boxes.
3. Select the **OK** box.
4. The tolerance should appear attached to your cursor. Move the cursor to the desired location and press the left mouse button.

Note: the size of the Feature Control Frame above, is determined by the height of the dimension text.

GEOMETRIC TOLERANCES and QLEADER

The **Qleader** command allows you to draw leader lines and access the dialog boxes used to create feature control frames in one operation.

1. Select **Dimension / Leader**
2. Select the **"Settings"** option. (Right click, and select Settings from the short cut menu)

The Leader Settings dialog box should appear.

3. Select the Annotation tab.
4. Select Tolerance then OK.

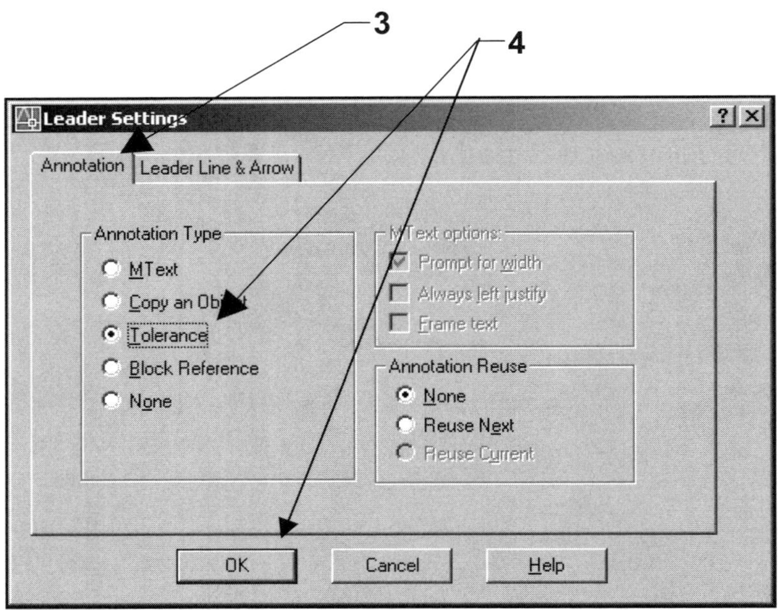

5. Place the first leader point. **P1**
6. Place the next point **P2**
7. Press <enter>

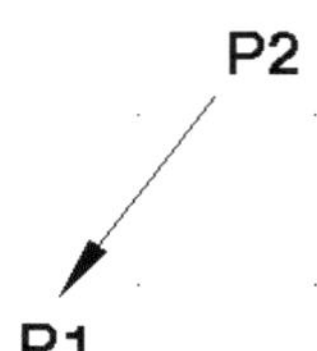

The Geometric Tolerance dialog box will appear.

8. Make your selections and fill in the tolerance and datum boxes.

9. Select the **OK** button.

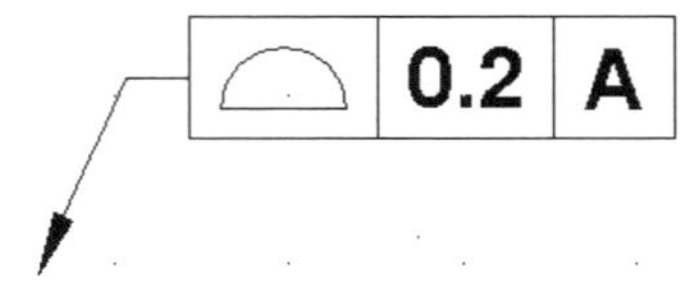

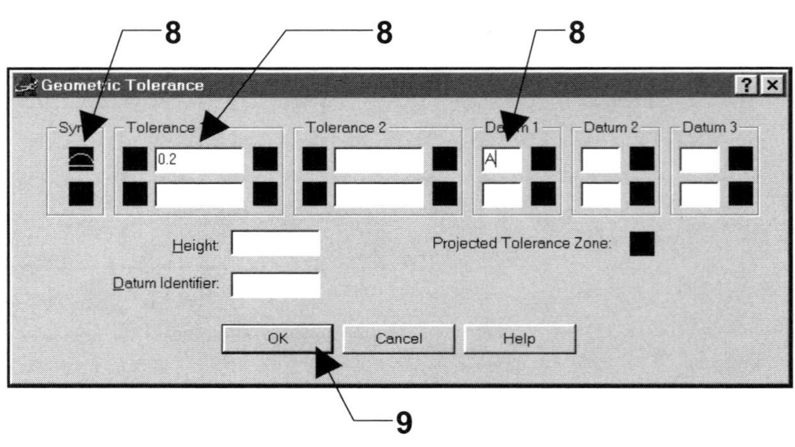

DATUM FEATURE SYMBOL

A datum in a drawing is identified by a **"datum feature symbol".**

To create a ***datum feature symbol***: 

1. Select Dimension / Tolerance
2. Type the "datum reference letter" in the "Datum Identifier" box.
3. Select the OK button.

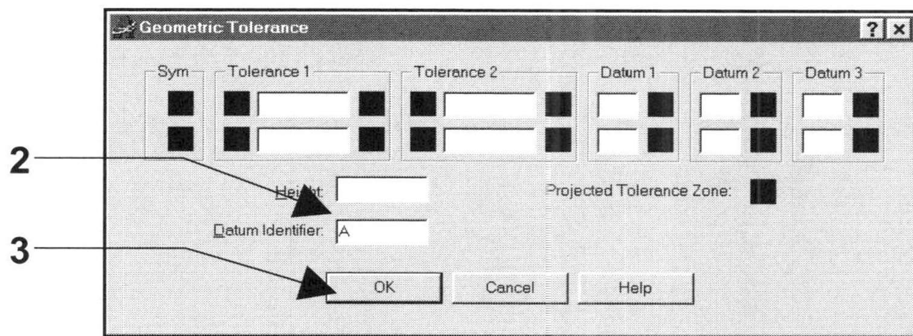

To create a ***datum feature symbol*** combined with a ***feature control frame***:

1. Select **Dimension / Tolerance**
2. Make your selections and fill in the tolerance.
3. Type the "datum reference letter" in the "Datum Identifier" box.
4. Select the **OK** button.

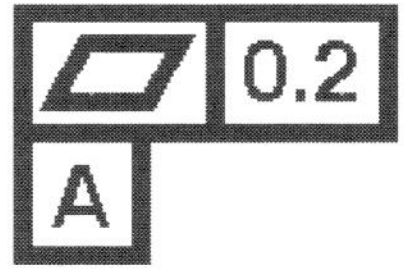

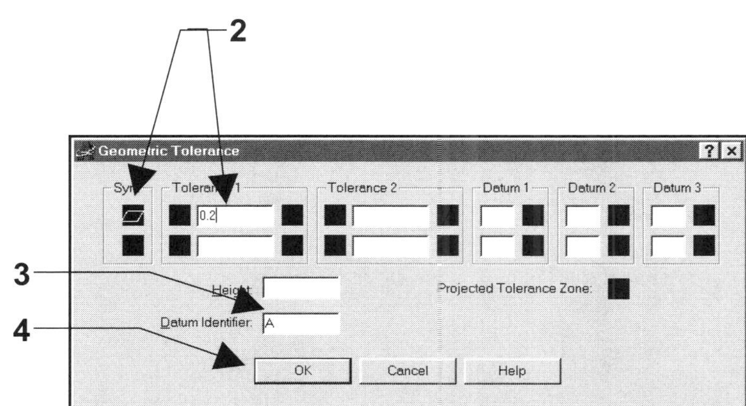

DATUM TRIANGLE

A datum feature symbol, in accordance with ASME Y14.5M-1994, includes a leader line and a datum triangle filled. You can create a wblock or you can use the two step method below using Dimension / Tolerance and Qleader.

1. Select **Dimension / Leader**.
2. Select the **"Settings"** option. (Right click, and select Settings from the short cut menu)

The Leader Settings dialog box should appear.

3. Select the Annotation tab.
4. Select the Tolerance button.
5. Select the Leader Line & Arrow tab.
6. Select Datum Triangle Filled
7. Select the OK button.

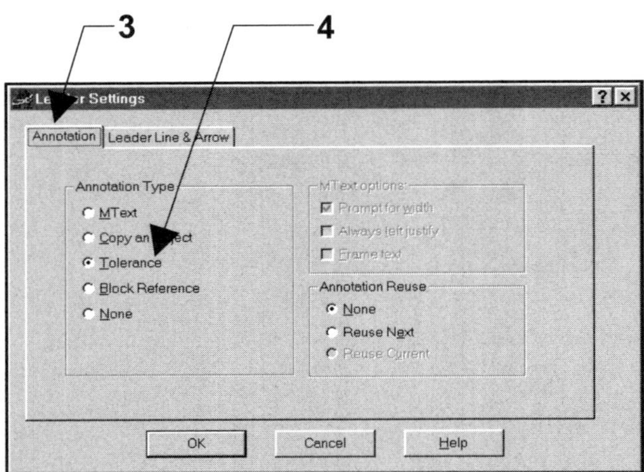

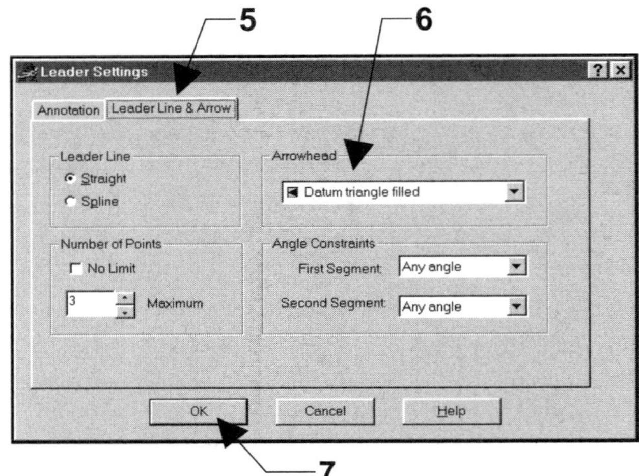

8. Specify the first leader point. (The triangle endpoint)
9. Specify next point then press <enter>
10. When the Geometric Tolerance dialog box appears, select the OK button.

If you were successful,
a <u>datum triangle filled</u> with
a <u>leader line</u> should appear.
(As shown below)

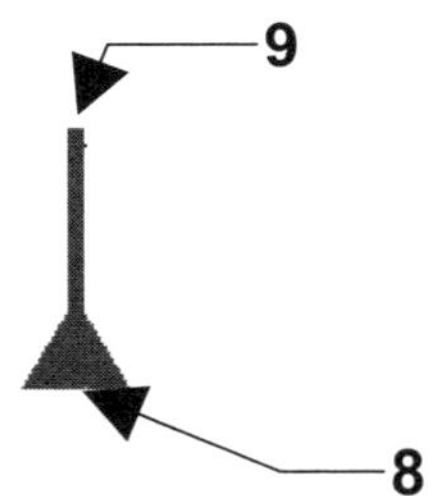

Note: When you change the leader settings AutoCAD automatically creates an "Override" in dimensions and "sets it current". To make changes to the Leader settings you must change the "override".

11. Next create a datum feature symbol. (Follow the instructions on the previous page.)

12. Now move the datum feature symbol to the endpoint of the leader line to create the symbol below left.

You are probably wondering why we didn't just type "A" in the identifier box. That method will work if your leader line is horizontal. But if the leader line is vertical, as shown on the left, it will not work. (The example on the right illustrates how it would appear)

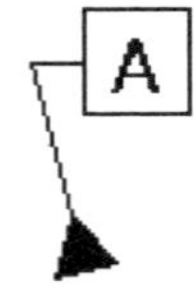

TYPING GEOMETRIC SYMBOLS

If you want geometric symbols in the notes that you place on the drawing, you can easily accomplish this using a font named **GDT.SHX**. This font will allow you to type normal letters and geometric symbols, in the same sentence, by merely toggling the SHIFT key up and down with CAPS LOCK on.

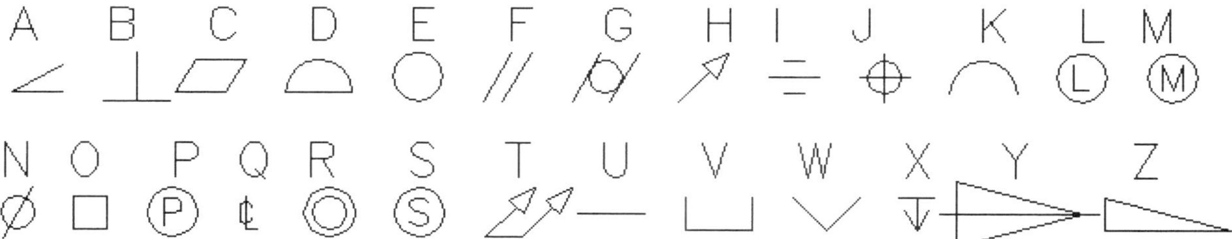

1. First you must create a new text style using the **GDT.SHX** font.

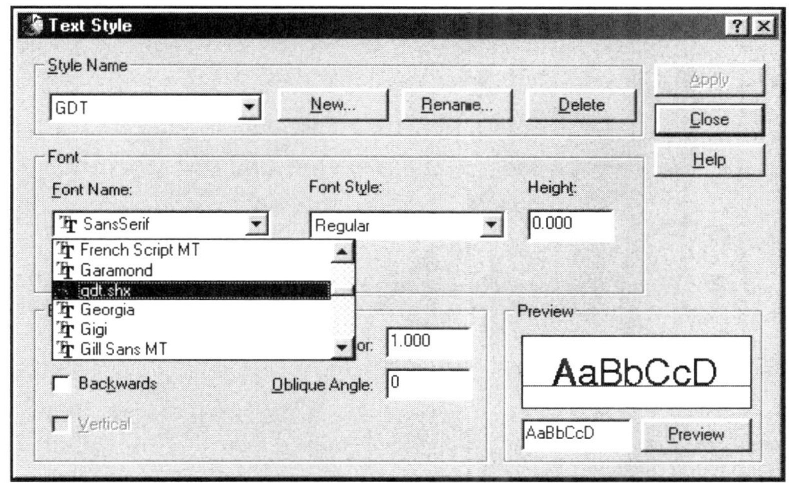

2. **CAPS LOCK** must be **ON**.

3. Select **DRAW / TEXT / Single Line or Multiline**

4. Now type the the sentence shown below. When you want to type a symbol, press the **SHIFT** key and type the letter that corresponds to the symbol. For example: If you want the diameter symbol, press the shift key and the "N" key. (Refer to the alphabet of letters and symbols shown above.)

3X ⌀.44 ⌴⌀1.06 ↧.06

Can you decipher what it says?
(Drill (3) .44 diameter holes with a 1.06 counterbore diameter .06 deep)

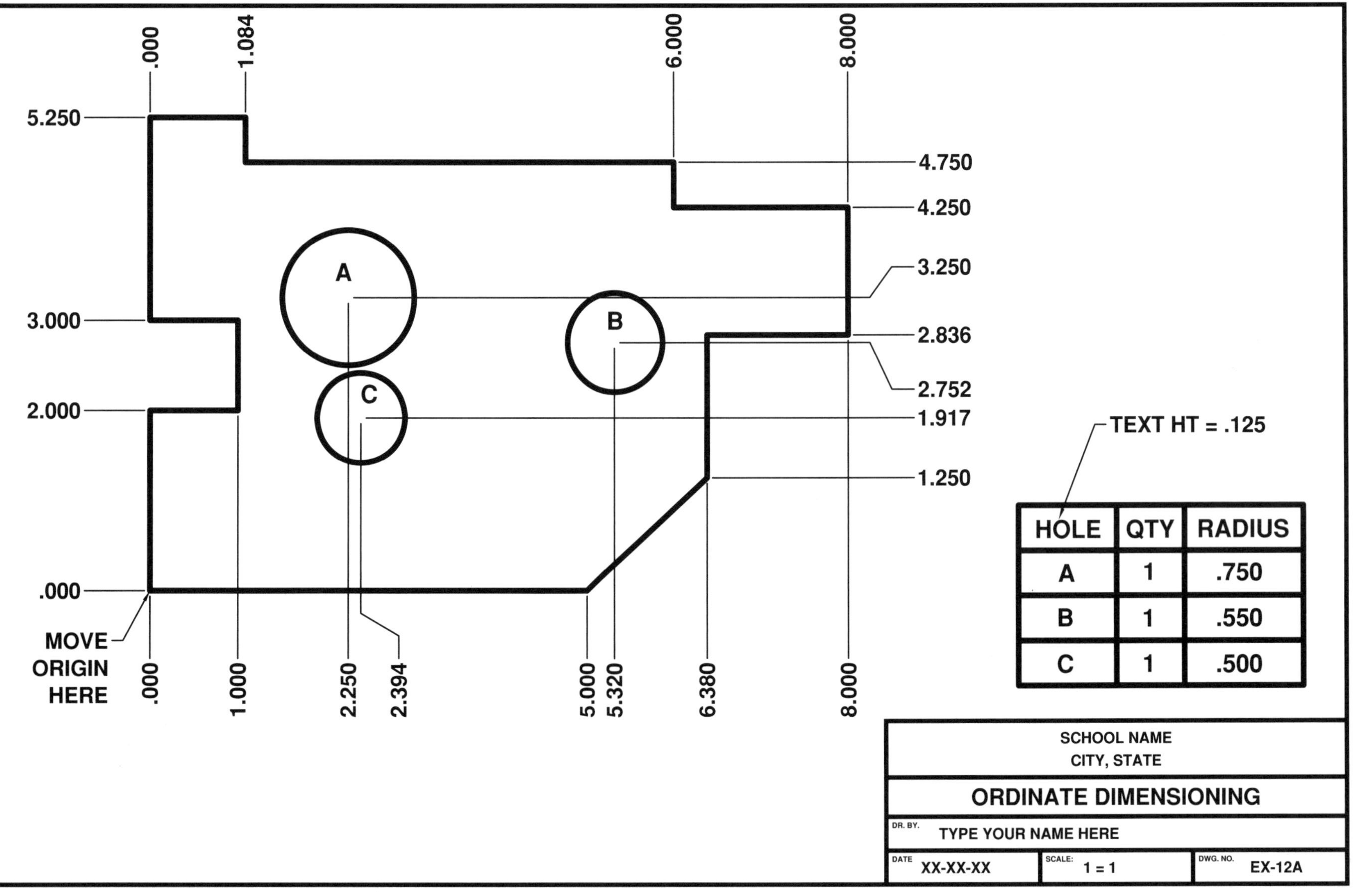

HOLE	QTY	RADIUS
A	1	.750
B	1	.550
C	1	.500

INSTRUCTIONS:

1. Open MY DECIMAL SETUP and select the 11 X 17 (1 to 1) tab.
2. Draw the drawing above inside the viewport, in model space.
3. Dimension using ORDINATE dimensioning (Refer to pages 12-2 thru 5)
4. Make sure the model space scale is 1:1.
5. Save as: EX-12A
6. Plot using "11 X 17 (MONO) " Page Setup.

EXERCISE 12A

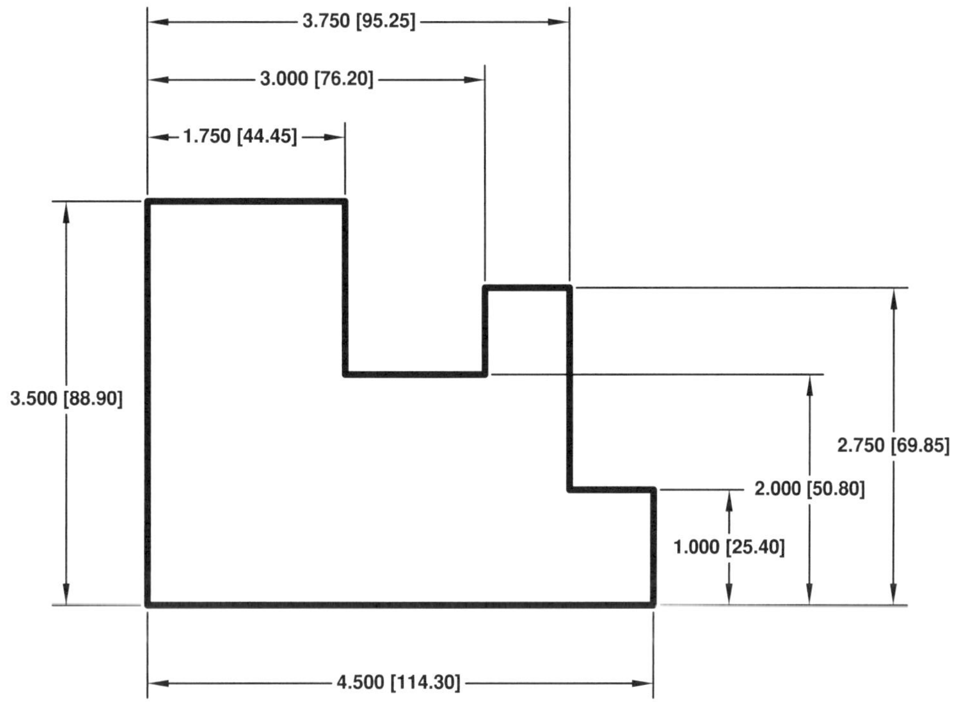

3.750 [95.25]
3.000 [76.20]
1.750 [44.45]
3.500 [88.90]
2.750 [69.85]
2.000 [50.80]
1.000 [25.40]
4.500 [114.30]

SCHOOL NAME CITY, STATE		
DUAL DIMENSIONING		
DR. BY. TYPE YOUR NAME HERE		
DATE XX-XX-XX	SCALE: 1 = 1	DWG. NO. EX-12B

INSTRUCTIONS:

1. Open MY DECIMAL SETUP and select the 11 X 17 (1 to 1) tab.
2. Draw the drawing above inside the viewport, in model space.
3. Dimension using ALTERNATE UNITS (Refer to Page 12-6)
4. Make sure the model space scale is 1:1.
5. Save as: EX-12B
6. Plot using "11 X 17 (MONO)" Page Setup.

EXERCISE 12B

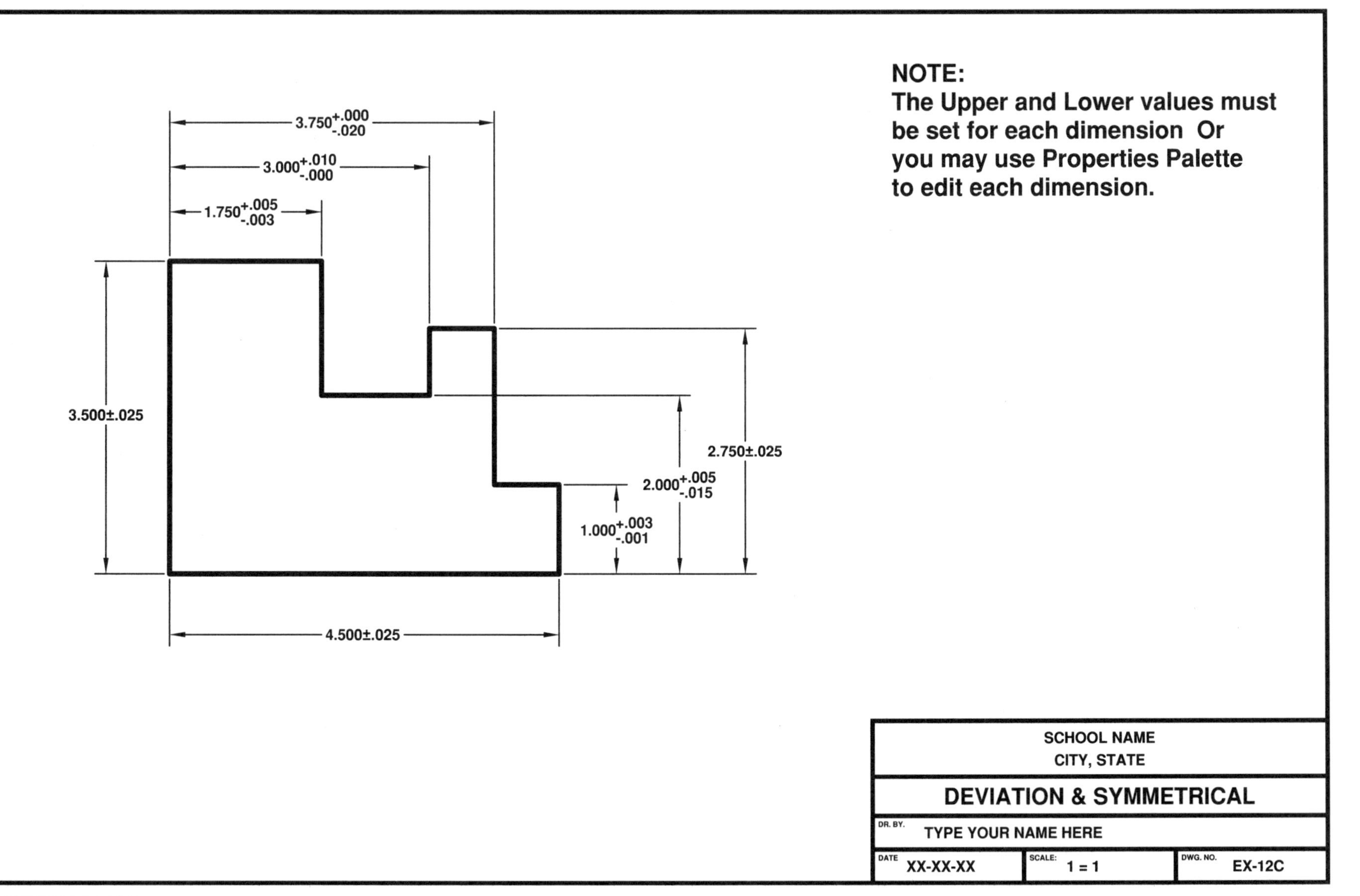

EXERCISE 12C

INSTRUCTIONS:
1. Open EX-12B and save immediately as EX-12C.
2. Change the Dimensions to DEVIATION & SYMMETRICAL (Refer to Page 12-7 & 8)
3. You may erase the dimension or use Modify/Properties.
4. Save as: EX-12C
5. Plot using "11 X 17 (MONO)" Page Setup.

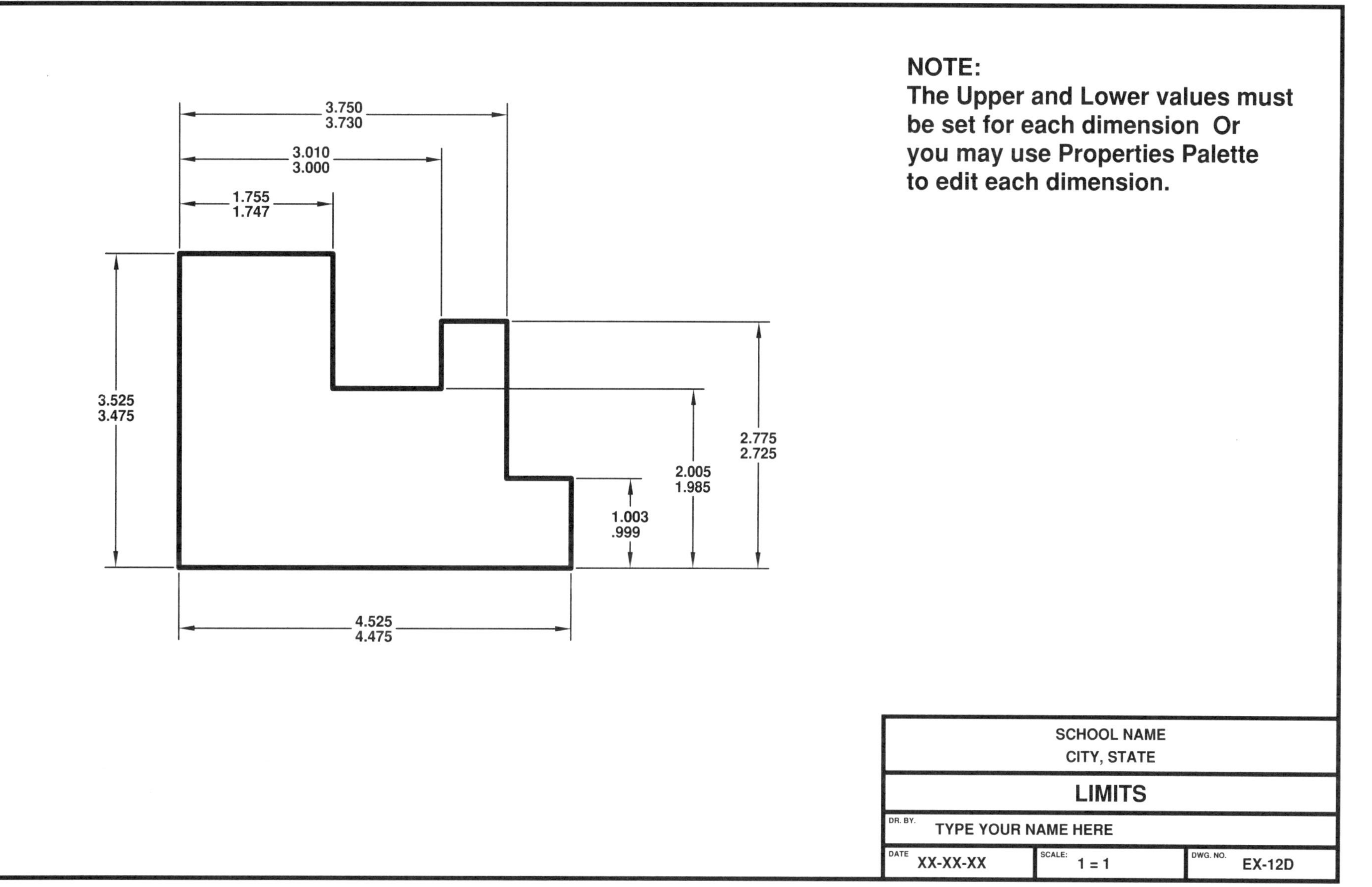

INSTRUCTIONS:

1. Open EX-12B or EX-12C and save immediately as EX-12D.
2. Change the Dimensions to LIMITS (Refer to Page 12-7 & 8)
3. You may erase the dimension or use Modify/Properties.
4. Save as: EX-12D
5. Plot using "11 X 17 (MONO)" Page Setup.

EXERCISE 12D

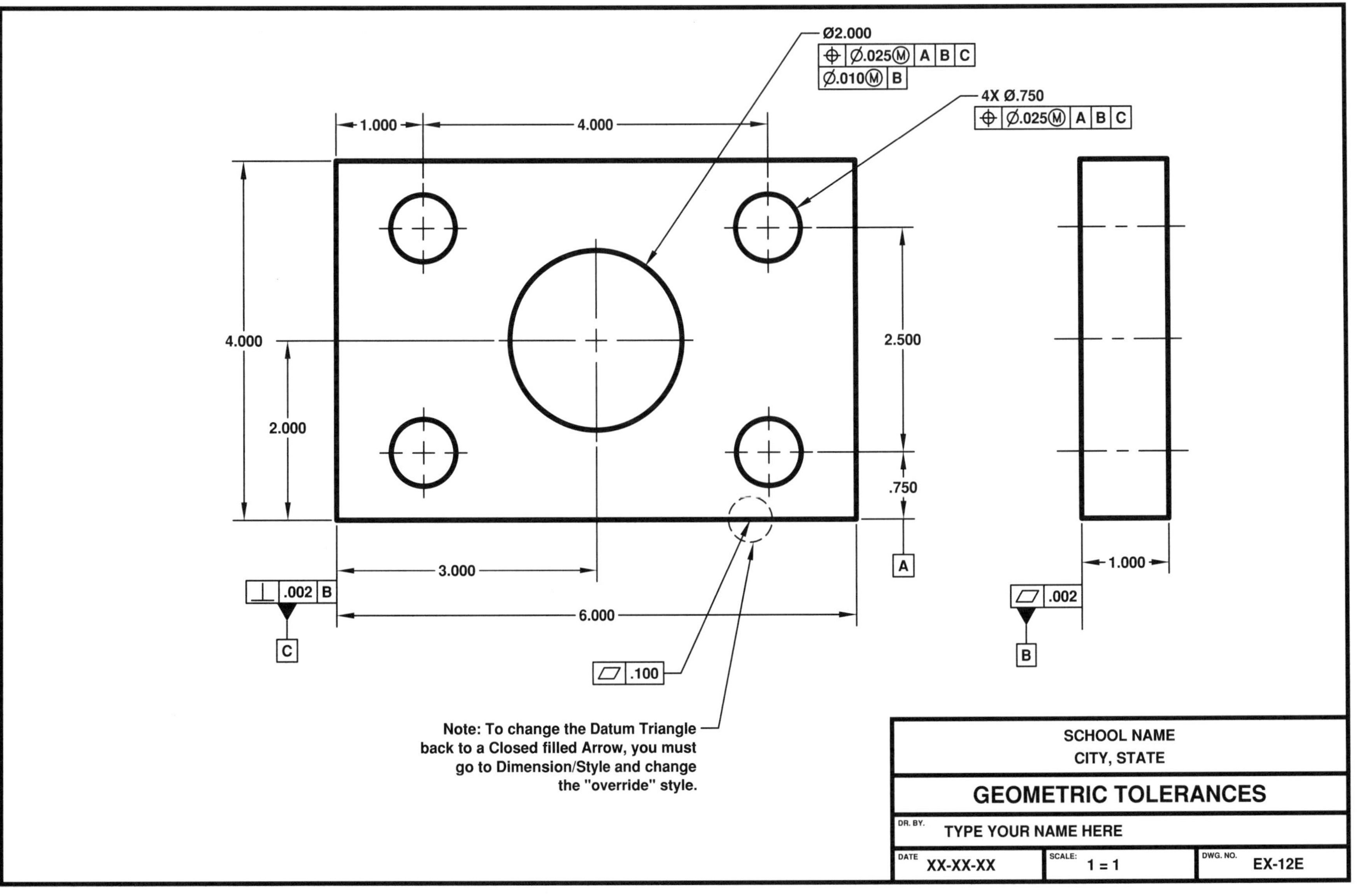

INSTRUCTIONS:

1. Open MY DECIMAL SETUP and select the 11 X 17 (1 to 1) tab.
2. Draw the drawing above inside the viewport, in model space.
3. Dimension using GEOMETRIC TOLERANCES (Refer to Page 12-9 thru 12-12)
4. Make sure the model space scale is 1:1.
5. Save as: EX-12E
6. Plot using "11 X 17 (MONO)" Page Setup.

EXERCISE 12E

LEARNING OBJECTIVES

After completing this lesson, you will be able to:

1. Understand the concept of 3D
2. Change the display as a hidden view
3. Understand the difference between:
   Wireframe, Surface and Solid Modeling

The following lessons are an introduction to 3D. These lessons will give you a good understanding of the basics of AutoCAD's 3D program.

Unfortunately, LT users will not be able to do Lessons 14 through 20. The LT version does Not have solid modeling commands.

Note: Increased RAM requirement
The 3D visualization and rendering abilities of AutoCAD requires much more memory than ever before. It will run with 512MB but Autodesk recommends 2GB of RAM and a video card with at least 128MB of its own memory. Check Autodesk's website (www.autodesk.com) for the list of supported video cards.

LESSON 13

INTRODUCTION TO 3D

We live in a three-dimensional world, yet most of our drawings represent only two dimensions. In Lesson 6 you made an isometric drawing that appeared three-dimensional. In reality, it was merely two-dimensional lines drawn on angles to give the appearance of depth.

In the following lessons you will be presented with a basic introduction to AutoCAD's 3D techniques for constructing and manipulating objects. These lessons are designed to give you a good start into the environment of 3D and encourage you to continue your education in the world of CAD.

DIFFERENCES BETWEEN 2D AND 3D

Axes

In 2D drawings you see only one plane. This plane has two axes, X for horizontal and Y for vertical. In 3D drawings an additional plane is added. This third plane is defined by an additional axis called Z. The direction of the positive Z-axis basically comes out of the screen toward you and gives the object height. To draw a 3D object you must input all three, X, Y and Z axes coordinates.

A simple way to visualize these axes is to consider the X and Y axes as the ground and the Z axes as a Tree growing up (positive coordinates) from the ground or the roots growing down (negative coordinates) into the ground.

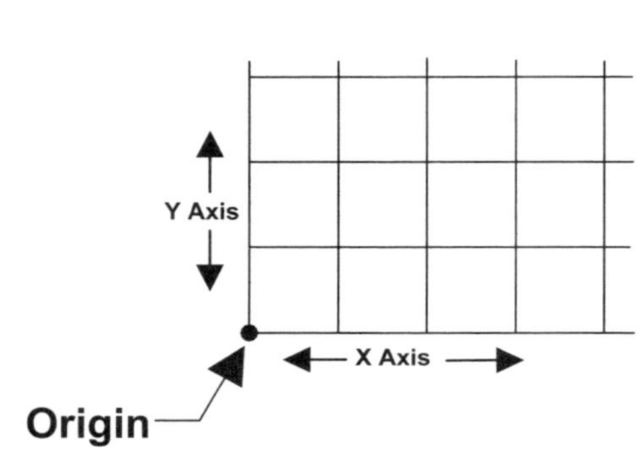

2D Coordinate System

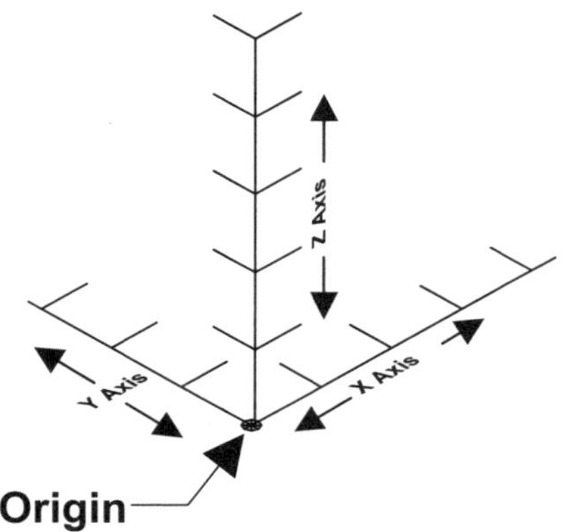

3D Axes Coordinate System

BASIC TYPES OF 3D MODELS

There are <u>3 basic types</u> of 3D models.

1. Wireframe models
2. Surface models
3. Solid models

<u>A description of each starts on page 13-8.</u>

***Note: Increased RAM requirement**
The 3D visualization and rendering abilities of AutoCAD requires much more memory than ever before. It will run with 512MB but Autodesk recommends 2GB of RAM and a video card with at least 128MB of its own memory. Check Autodesk's website (www.autodesk.com) for the list of supported video cards.

ENTER THE AutoCAD 3D WORKSPACE

Note: AutoCAD LT has limited 3D capabilities and does not have the new 3D Workspace. But you will still be able to do the exercises in this lesson without the 3D Modeling Workspace. Skip to page 13-5.

1. Select **Tools / Workspaces / 3D Modeling. (LT users do not change workspaces)**

2. Your workspace may not appear exactly like the example shown below because you need to do one more step. Select **File / New**

3. Select one of the following templates:
   acad3D.dwt (for imperial units)
   acadiso3D.dwt (for metric)

Now your screen should appear like the example shown below. (Not LT users)

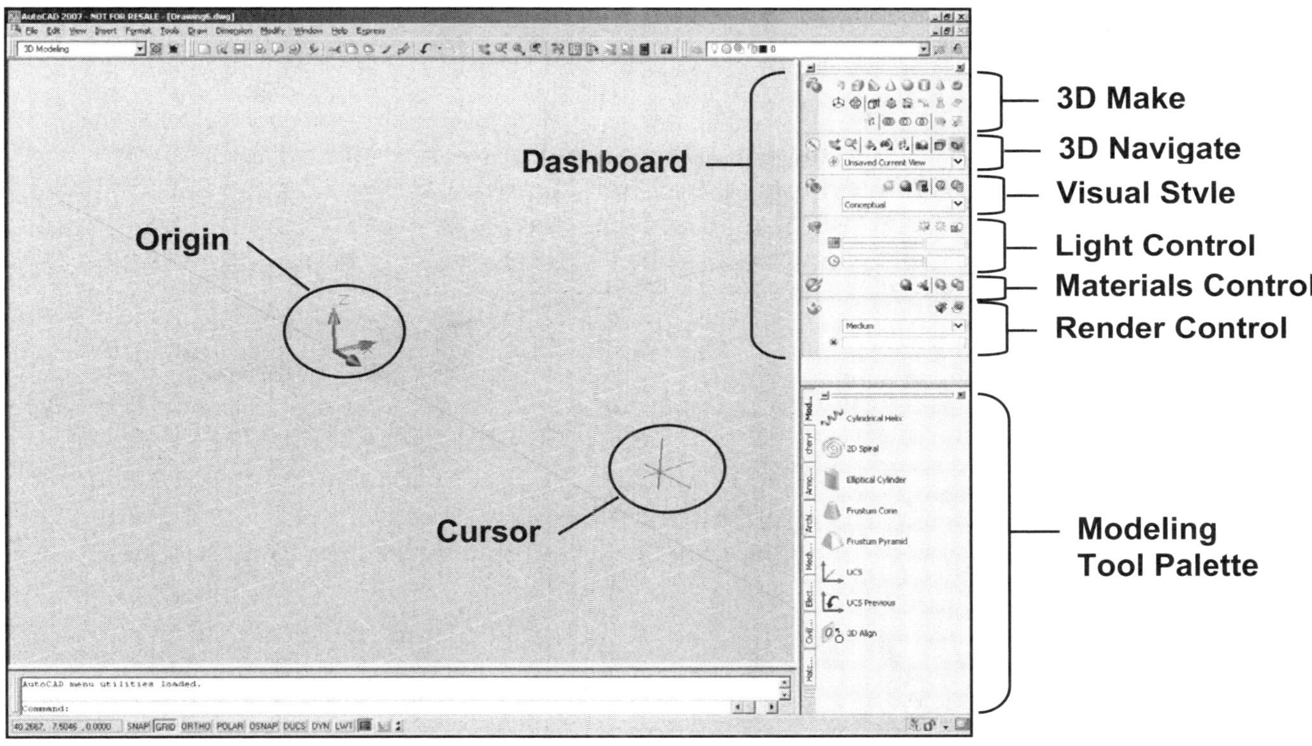

The large palette area on the right is called the **Dashboard**. It contains several panels of 3D tool buttons and controls. (A description of each panel follows on the next page.)

Each panel can be expanded to expose additional buttons and controls. Move the cursor over the dark gray area on the left of the panels and click on the downward-pointing double arrows. When expanded, the dark gray area changes to orange. To collapse the panel, move the cursor over the orange area and click on the upward-pointing arrows.

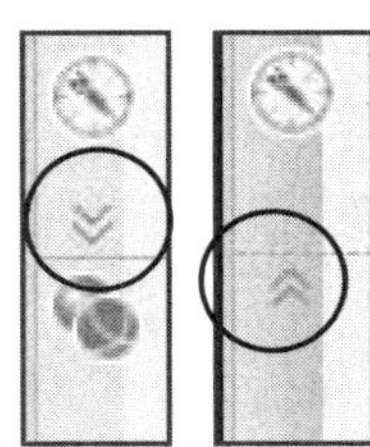

Notice the appearance of the **Origin** and the **Cursor** has changed.
They now display the **"Z" axis**. Note: the arrows are always pointing in the positive direction.

Dashboard Panel descriptions

3D Make Control Panel

Contains tool buttons for creating, manipulating and modifying 3-dimensional objects.

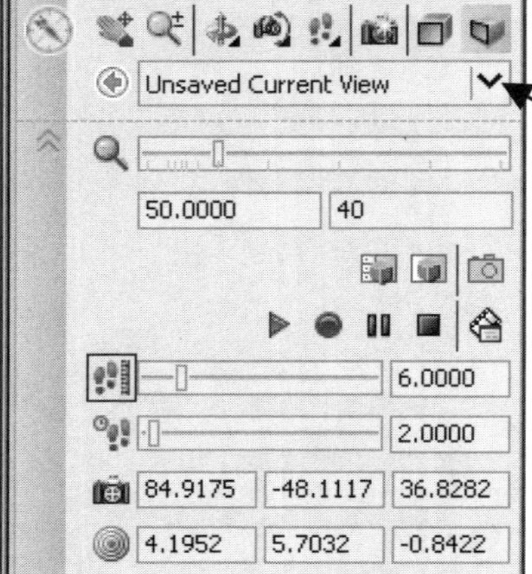

3D Navigate Control Panel

Contains tool buttons for Pan and Zoom, camera, animation walkthrough and perspective.

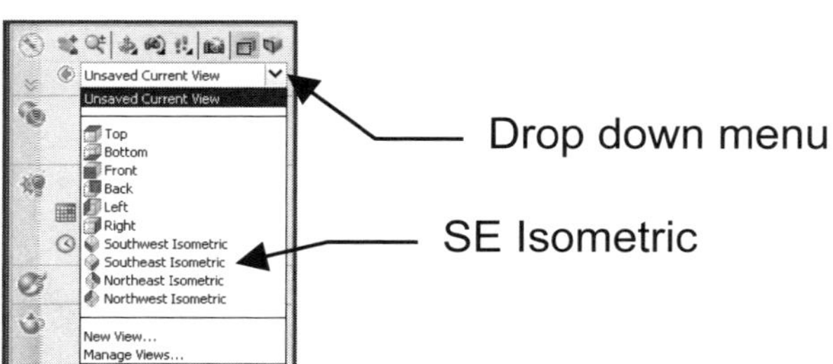

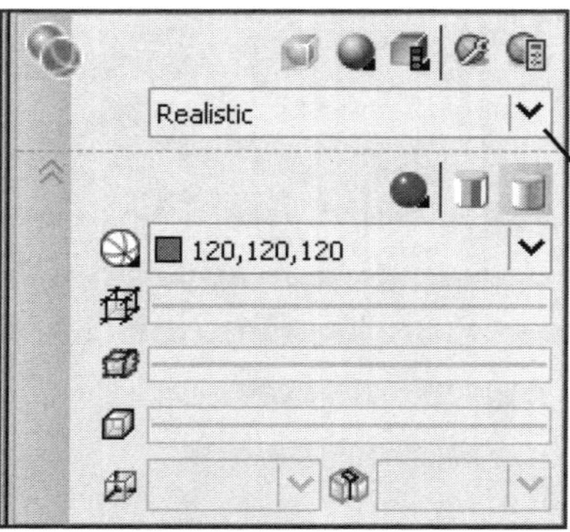

Visual Style Control Panel

Contains tools to control how 3-dimensional objects and displayed.

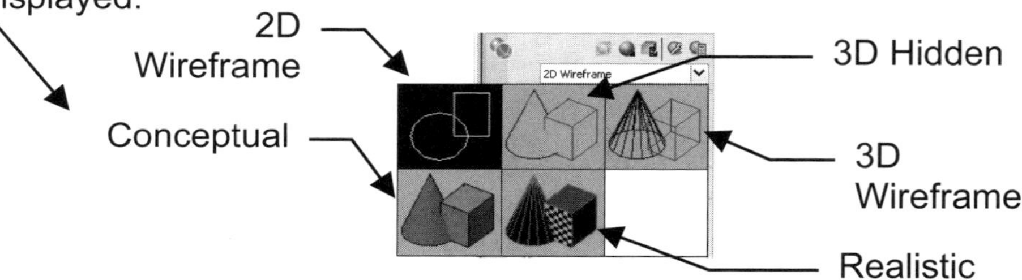

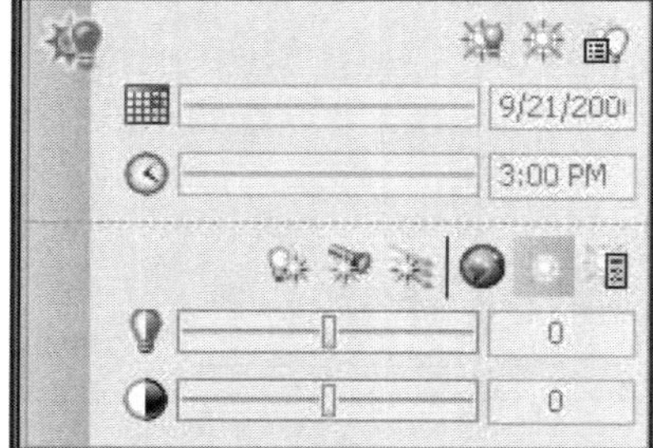

Light Control Panel

Contains tools to control ambient light and sunlight.

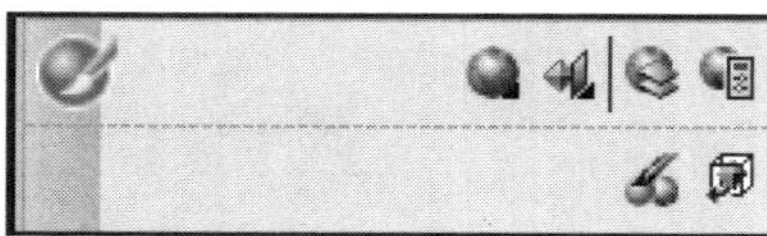

Materials Control Panel

Contains materials to apply to surfaces.

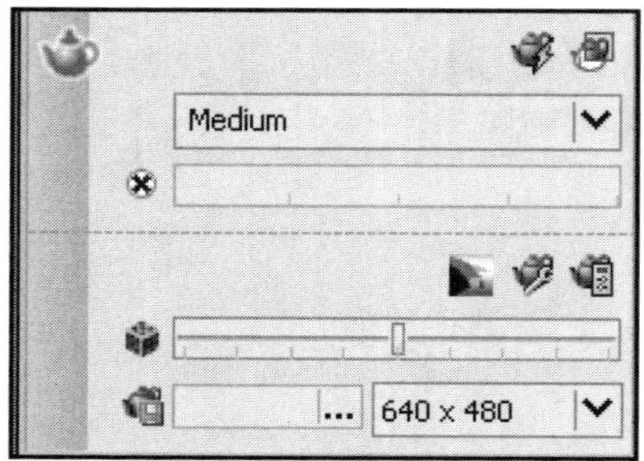

Render Control Panel

Contains tools to Render a 3-dimension object.

A Peek into the 3D World

Just to get you a little excited about 3D, let's experiment a little with the View and Shade options. This will give you a basic understanding of the difference between a 2D and 3D viewing.

Step 1. Open the drawing "**2007-3D demo.dwg**" or "**2007LT-3D demo.dwg**".
(This drawing should be on the CD in the Beginning workbook, or you may download it from the website www."Industrialpress.com" or www.shrockpublishing.com.)

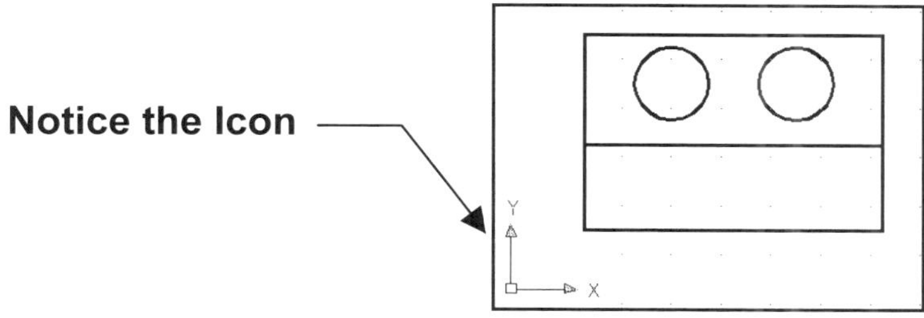

This is a view of the X and Y plane (the ground). This should appear very familiar to you.

Step 2. Select **View / 3D Views / SE Isometric**
(or use the drop down menu shown on pg 13-4)

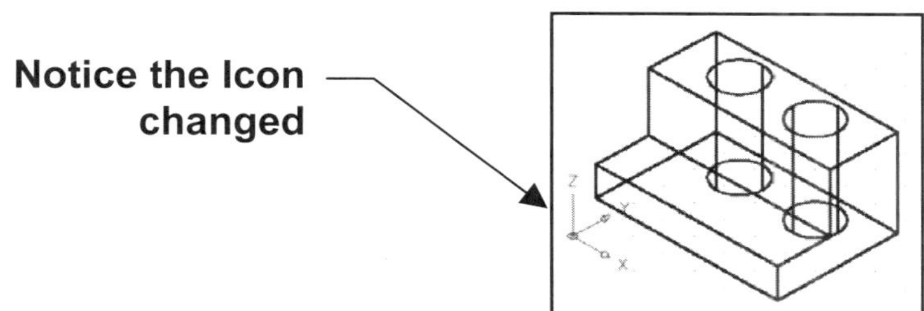

The object is now displayed as a **"3D Model",** not a **"2D Object"**.
You are viewing the X, Y and Z axes (height). The object is shown as a "2D Wireframe".
(You are viewing the **SE (South East) Isometric** view of the **"Model"**.
More about viewing on the next page.

Step 3. Select **View / Visual Styles / Realistic** (or use the drop down menu on pg. 13-4)

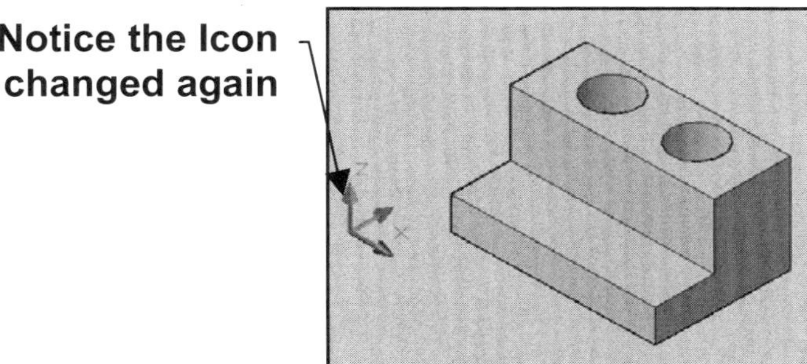

Note: LT users select:
View / Shade / 16 color filled

Now is when the fun starts!

We will cover all of this and more in the Lessons that follow.

Viewing a 3D Model

Viewpoint

It is very important for you to be able to control how you view the 3D Model. The process of changing the view is called **selecting the Viewpoint**. (Note: View**point;** not view**port.**)
The **Viewpoint** is the location where "**YOU**" are standing. ***This is a very important concept to understand***. The Model does not actually move; you move. For example, if you want to look at the South East corner of your house, you need to walk to the South East corner of your lot and look at the corner of your house. Your house did not move, but you are seeing the South East corner of your house. If you want to see the Top or Plan View of your house, you would have to climb up on the roof and look down on the house. The house did not move, you did. So in other words, the Viewpoint is **your "Point of View"**.

Remember when you select a Viewpoint you are selecting where you are standing. The view that appears on the screen is what you would see when you move to the viewpoint and looked at the model.

Select a Viewpoint from **View / 3D Views or**

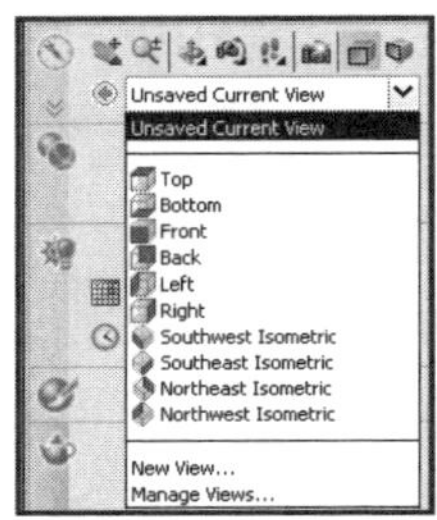

3D Orbit (Sorry, not available in version LT)

3D Orbit allows you to dynamically rotate the viewpoint of the 3D Model. (walk around it)

TYPE = 3do
PULLDOWN = VIEW / ORBIT / Free Orbit or Constrained
CONTROL PANEL = 3D NAVIGATE

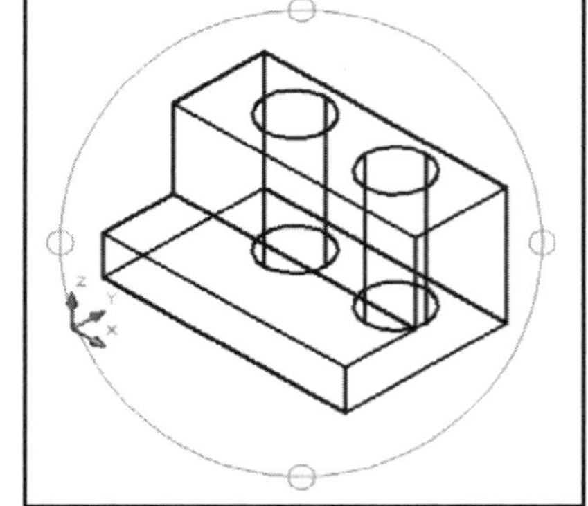

After selecting the 3D Orbit command, a green circle appears if you selected Free Orbit. (No green circle with Constrained.) This is called the Trackball. Place the cursor inside or outside of the trackball, click and drag the cursor. The view is dynamically changed as you drag the cursor.
When done, right click and select Exit or press the ESC key.

To return to an isometric view, select one of the 3D Views (SE isometric) mentioned above.

Note: This command may be used in the middle of any drawing command.

Hiding Lines

You may have noticed when you displayed the model, it displayed all of the lines including the lines that represent the internal features. This is called "**Wireframe Display**". This is not the same as "Wireframe Model" as described on page 13-8. This model is a solid object. But it is displayed with all of the edges visible to make it easier to select an edge.

Sometimes this gets somewhat confusing and you would like the view to be displayed without the internal lines. To create a "**Hidden Display**" in the current viewport you must use the **Hide** command.

1. First, if you are in a layout, you would select the **Viewport** where the model is displayed. (If you are in the "model tab" just continue on to 2.)

2. Select the Hide Command using one of the following:

   TYPE = HI
   PULLDOWN = VIEW / HIDE
   VISUAL STYLE CONTROL PANEL = 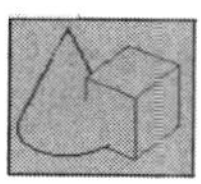 3D HIDDEN

(Note: If you have multiple viewports, you must use the hide command in each viewport)

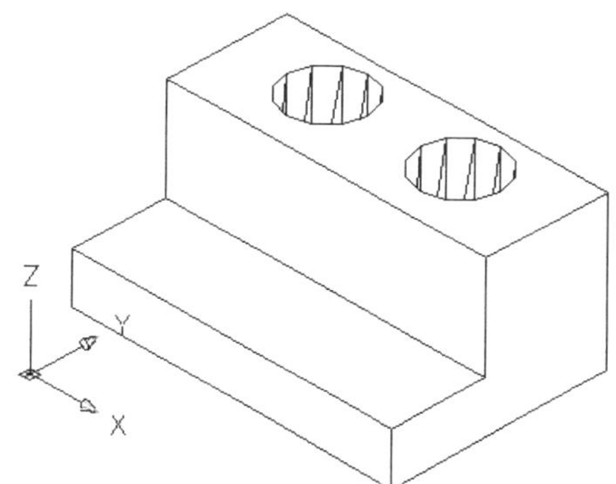

Dispsilh variable = 0

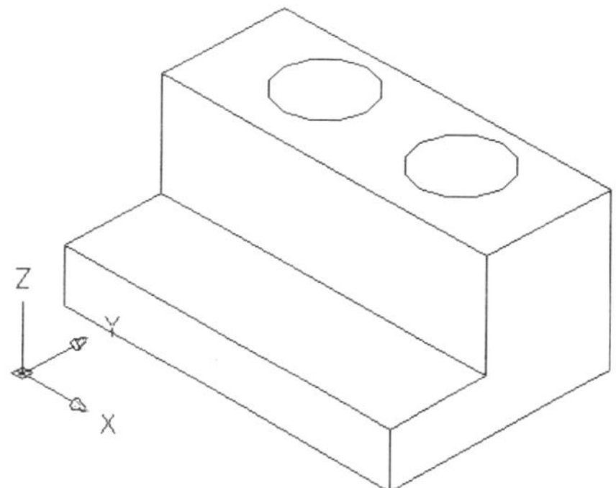

Dispsilh variable = 1

3. To return to a wireframe display, select the 2D or 3D icon.

DISPSILH variable

Notice that the Hide command uses zigzag lines, called mesh, to indicate internal or external curved surfaces. The visibility of the mesh is controlled by the **dispsilh** variable.
(dispsilh = Display Silhouette)
If you do not want the facets visible, change the **dispsilh** variable to 1. This will display 3D solid objects with silhouette **edges** only.

1. Command: ***type: dispsilh <enter>***
2. Enter new value for DISPSILH <0>: ***type 1 <enter>***
3. Select **View / Regenall**
4. Command: ***type: hi <enter>***

WIREFRAME MODEL

A wireframe model of a box is basically 12 pieces of wire (lines). Each wire represents an **edge** of the object. You can see through the object because there are no surfaces to obscure your view. This type of model does not aid in the visualization of the 3D object. Wireframe models have no volume or mass. The hidden line removal command cannot be used on a wireframe model.

How to draw a Wireframe Box

1. Open **2007-3D SETUP**
2. Draw a **Rectangle** just like you would in 2D. **First corner: 0,0,0 Second corner: 6,4**
3. Zoom in closer if the rectangle seems too small.

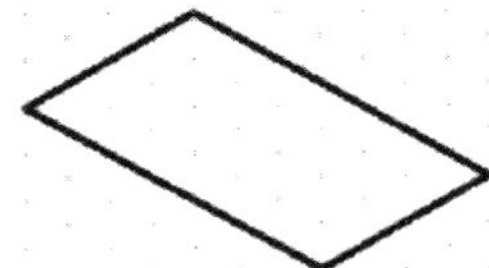

4. Copy the Rectangle 5" **above** the original rectangle as follows:
   a. Select the **Modify / Copy** command.
   b. Click on the Rectangle then <enter>.
   c. Select one of the corners as the **basepoint**
   d. Type the **X, Y, Z** coordinates for the new location: @ **0, 0, 5**

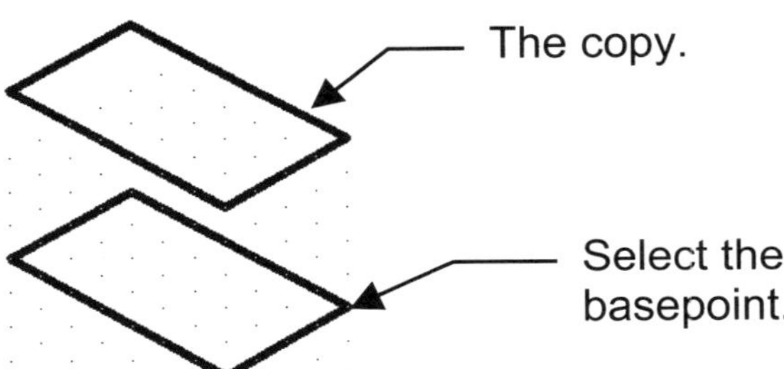

*Now think about the coordinates entered. The "@" sign of course is very important because you want the new rectangle location to be "relative" to the existing rectangle. The coordinates **0, 0, 5** mean that you **do not** want the new rectangle location to move in the **X** or **Y** axis. But you **do** want the new rectangle location to move **5"** in the **positive "Z"** axis.*

5. Now using Draw / Line, draw lines from the corners of the one rectangle to the other rectangle.

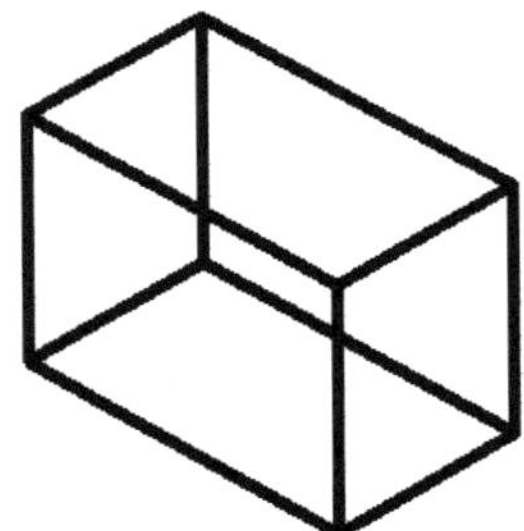

That is all there is to it. You have now completed a Wireframe Box.
This Wireframe can now be used as the structure for a Surface Model described on the next page.

SURFACE MODELS

A **Surface Model** is like an empty cardboard box. All surfaces and edges are defined but there is nothing inside. The model appears to be solid but it is actually an empty shell. The hidden line removal command can be used because the front surfaces obscure the back surfaces from view. A surface model makes a good visual representation of a 3D object.

You may attach a 3D Surface to a Wireframe structure or you may use one of AutoCAD's pre-defined 3D Objects.

(Note: LT users can't use the remainder of this lesson. Please skip to Lesson 21)

How to add a 3D Surface to a Wireframe structure

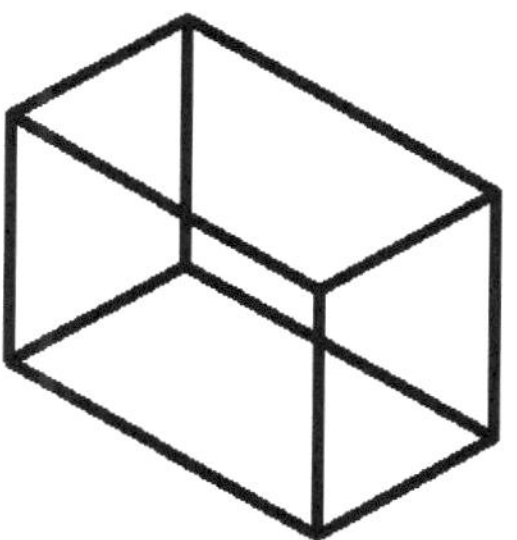

1. Create a Wireframe Model as shown on the previous page.
2. Select **SE Isometric**
3. Select the 3D face command:

    TYPE = 3dface
    PULLDOWN = DRAW / MODELING / MESHES / 3D FACE

4. Draw a 3D Surface by snapping to the 4 corners as shown below and **<enter>.**
   (The Shape will **Close** automatically.)
   Note: Each surface must be created as a separate object.

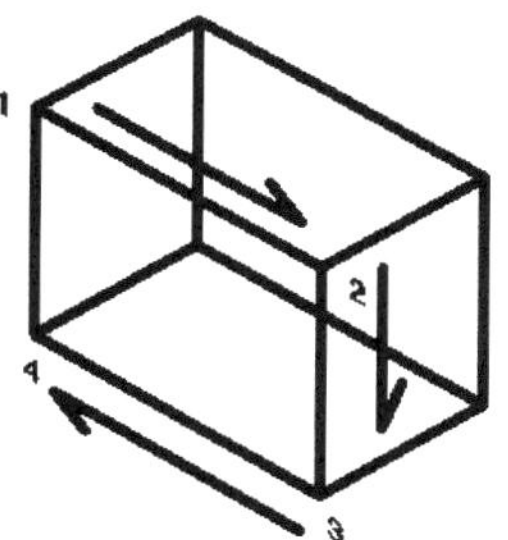

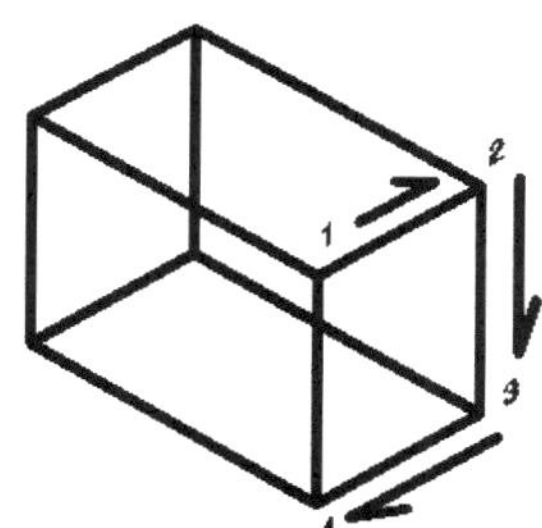

I know that the surfaces do not seem to be there but they are. The next step will make it appear.

5. Type: **hi**

Your 3D Wireframe model now has 2 surfaces attached.
Now you may even add a "Visual Style" to the surface.

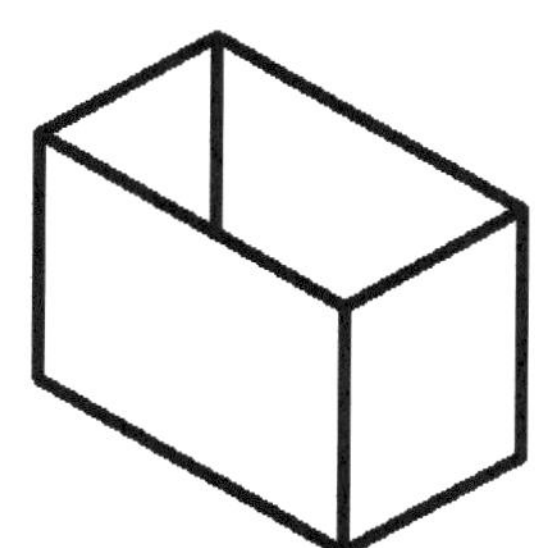

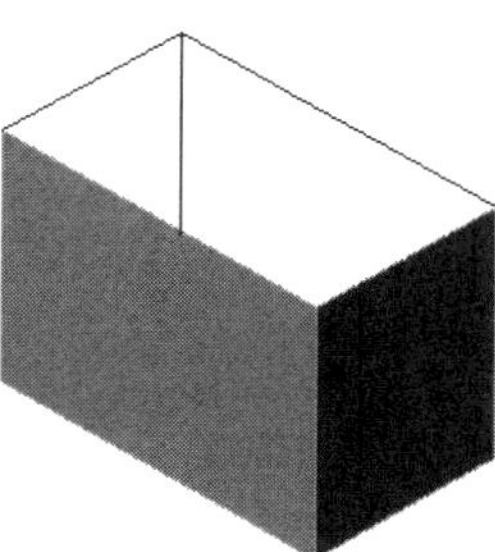

SOLID MODELS *(Refer to Lesson 14 for detailed instructions)*

*It is important that you understand both **Wireframe** and **Surface** Modeling **but** you will not be using either very often. **Solid Modeling** is the most useful and the most fun.*

Solid models have edges, surfaces and **mass**.
You can display the solid model as a wireframe display or hidden display.

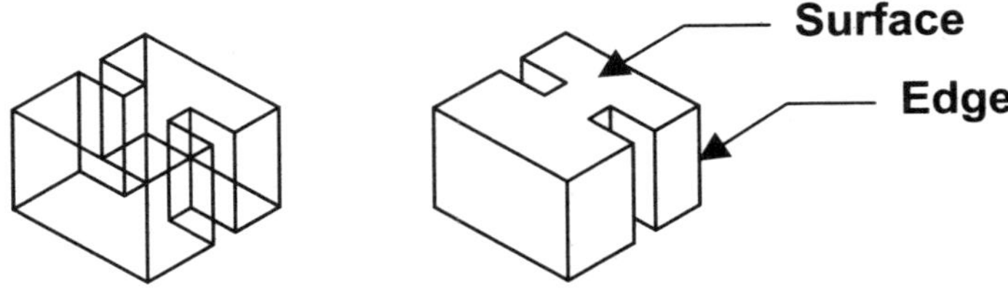

Solid Models can be modified using many **Solid Editing** features such as the ones described below.

Use Boolean operations such as **Union, Subtract and Intersect** to create a solid form.
These commands can be selected with:
Modify / Solid Editing or 3d Make Control Panel or type the command.

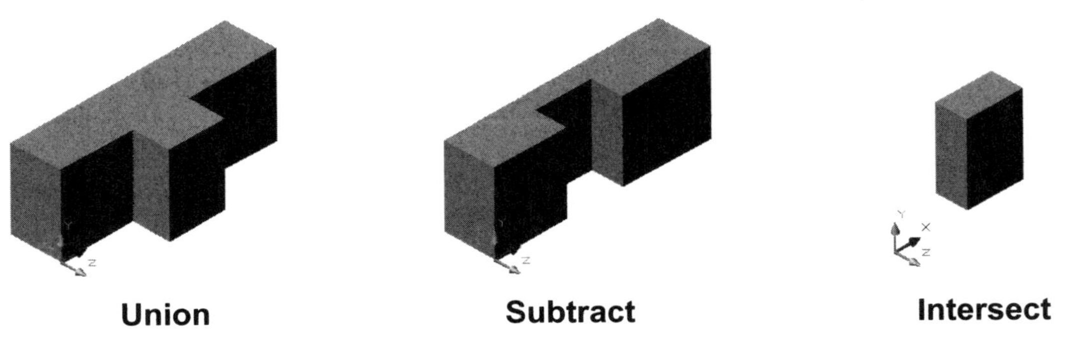

Union **Subtract** **Intersect**

Shapes can be: **Revolved** **Shelled**

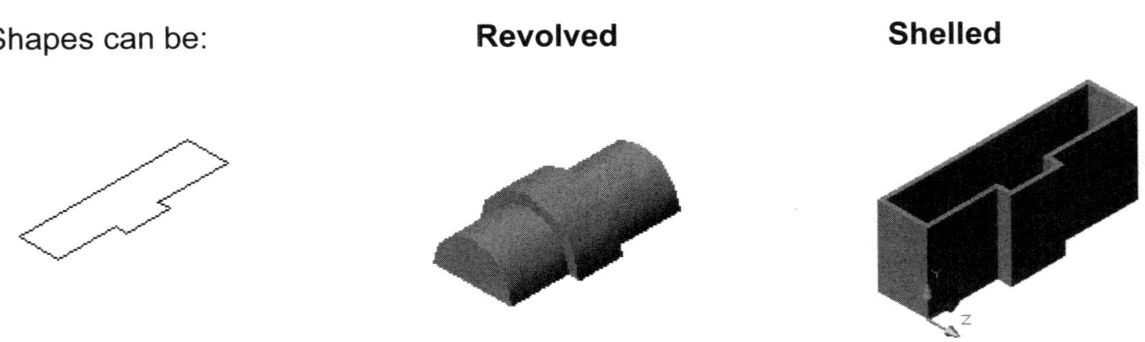

You will learn all of these features and more in the following Lessons. But first do the following Exercises to get some practice with Wireframe and Surface Modeling.

EXERCISE 13A

Create a Wireframe Model

1. **Select File / Open and select <u>2007-3D Setup.dwg</u>**

2. **Create the "Wireframe" Model shown below.**
   (Refer to page 13-8 for instructions if necessary)

3. Select **View / Hide**

   *Notice **none of the lines are hidden**. Why? Because this is a "wireframe" model. There are **no surfaces** to "hide" anything.*

   *Note: If your lineweights changed, select **View / Regen** and they will return.*

4. Save as **EX-13A**

5. **Do not Dimension**

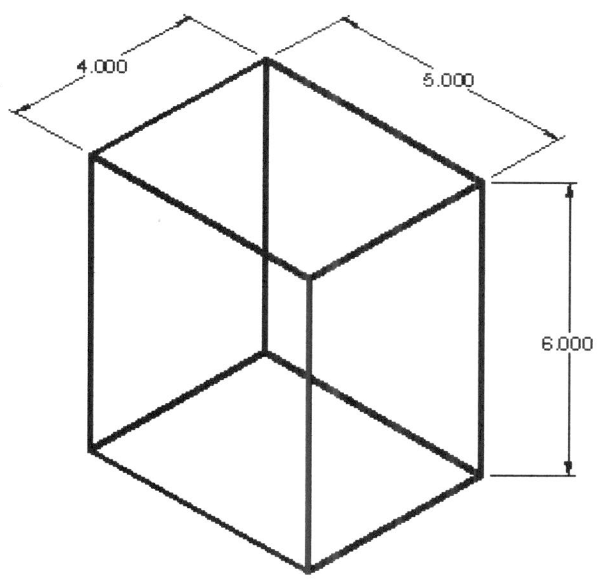

EXERCISE 13B

Create a Surface Model

1. Open **13A**

2. Add Surfaces to all 6 sides. (Refer to page 13-9)
   Use 3d Orbit to rotate the model to access all sides. (Refer to page 13-62

3. Select **View / Hide**
   Now the Hide command works because you now have surfaces.

4. Save as **EX-13B**

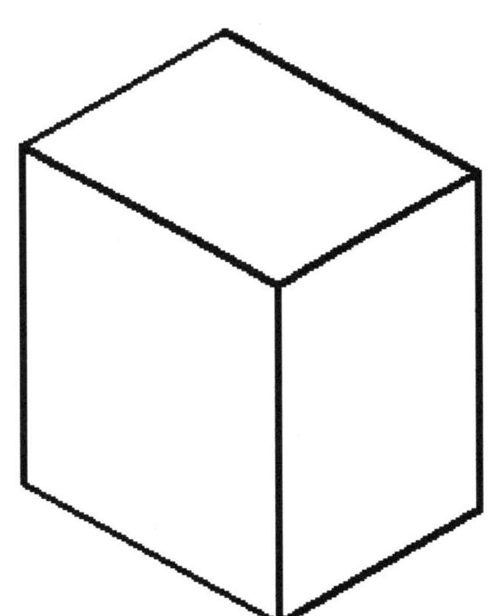

LEARNING OBJECTIVES

After completing this lesson, you will be able to:

1. Construct 7 Solid model Primitives:
   Box, Sphere, Cylinder, Cone, Wedge, Torus and Pyramid

Sorry LT users, you do not have this option.

When you have completed this lesson you will be able to create the 3D Model shown below

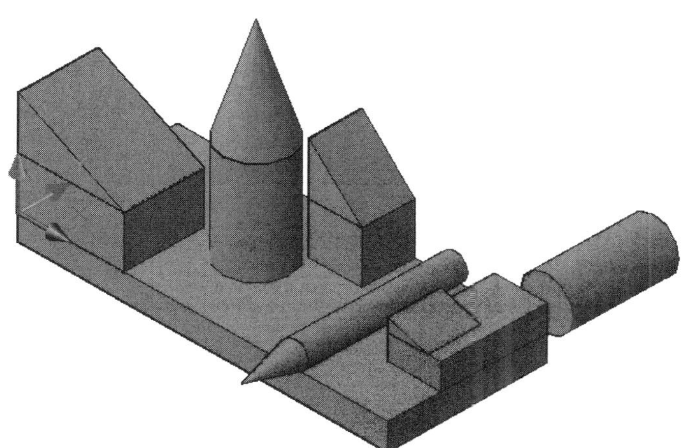

LESSON 14

DRAWING BASIC GEOMETRIC SHAPES

AutoCAD has 7 Solid shapes.
The 6 are, Box, Sphere, Cylinder, Cone, Wedge, Pyramid and Torus.

How to select a solid primitive.

TYPE = TYPE THE NAME
PULLDOWN = DRAW / MODELING
3D MAKE CONTROL PANEL =

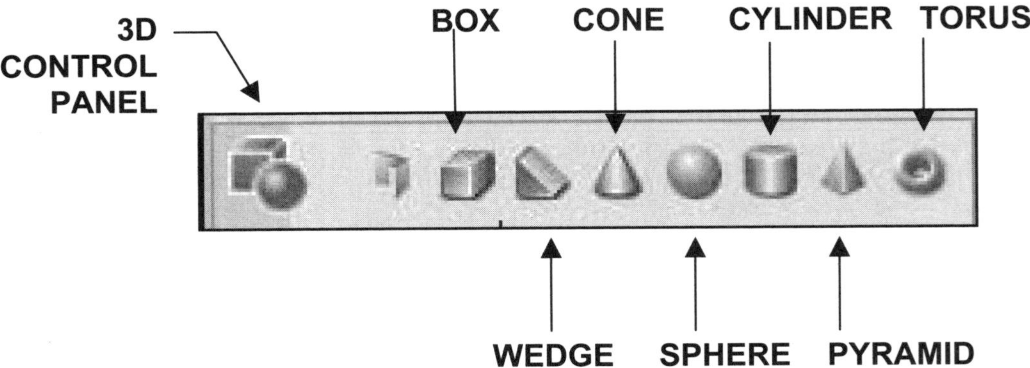

In this lesson you will learn the required steps to construct each of AutoCAD's basic geometric shapes. Each one requires different input information and some have multiple methods of construction.

3D input direction
When drawing in 3D and prompted for Length, Width or Height, each input corresponds to an axis direction as follows:

Length = X Axis
Width = Y Axis
Height = Z Axis

For example, if you are prompted for the Length the dimension that you enter will be drawn in the X axis. So keep your eye on the UCS icon in order to draw the objects in the correct orientation. Although it should be easy to visualize because you will actually see the shape constructed as you draw.

Consider starting the primitives on the Origin. It is useful to know where the primitive is located so it can be moved or rotated easily.

In Lesson 15 you will learn more about moving the UCS around to fit your construction needs. But relax and let's take it one step at a time.

BOX

There are 4 methods to draw a **Solid Box**. Which one you will use will depend on what information you know. For example, if you know where the corners of the base are located and the height, then you could use method 1 or 2.

Start a new file and select Acad3d.dwt

Method 1 (Enter the location for: base corner, diagonal corner and height)

a. Select the **SE Isometric** view.
b. Select the **Box** command. (See page 14-2)
c. Specify first corner [Center] <0,0,0>: ***type coordinates or pick location with cursor (P1).***
d. Specify other corner or [Cube/Length]: ***type coordinates for the diagonal corner or pick location with the cursor (P2).***
e. Specify height or [2Point]: ***type the height.***

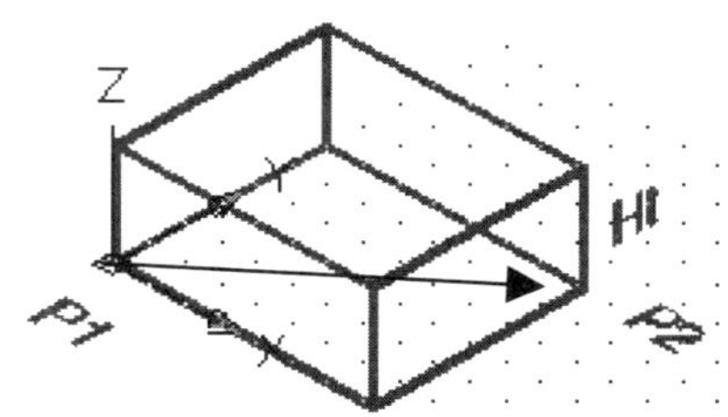

Method 2 (Enter the dimension for L, W, and Ht)

a. Select the **SE Isometric** view.
b. Select the **Box** command.
c. Specify first corner or [Center] <0,0,0>: ***type coordinates or pick location with cursor. (P1)***
d. Specify other corner or [Cube/Length]: ***type "L" <enter>.***
e. Specify length: ***enter the Length (X axis).***
f. Specify width: ***enter the Width (Y axis).***
g. Specify height: ***enter the Height (Z axis).***

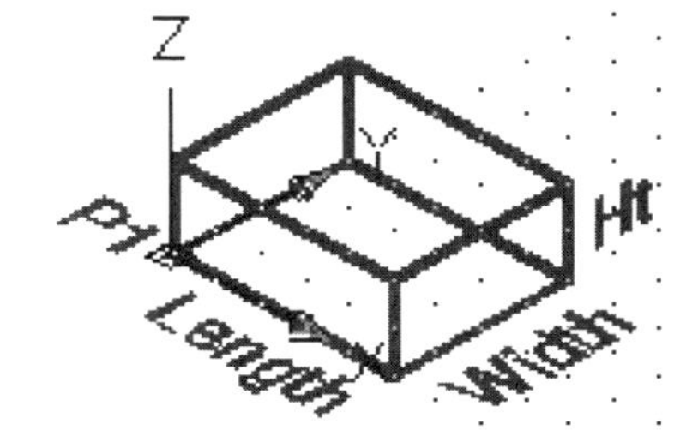

(Note: when entering L, W or H use DDE with Ortho "ON".)

Method 3 (Length, Width & Height have the same dimension)

a. Select the **SE Isometric** view.
b. Select the **Box** command.
c. Specify first corner or [Center] <0,0,0>: ***type coordinates or pick location with cursor. (P1)***
d. Specify corner or [Cube/Length]: ***type "C" <enter>.***
e. Specify length: ***enter the dimension.***

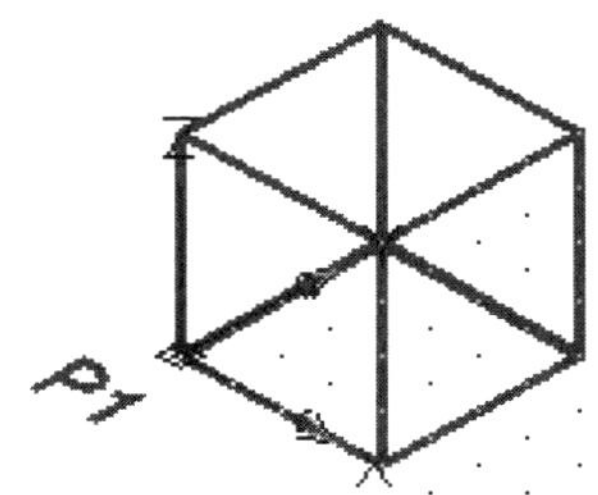

Method 4 (Enter the location for the center, a corner and the height)

a. Select the **SE Isometric** view.
b. Select the **Box** command.
c. Specify corner of box or [Center] <0,0,0>: ***type "C"<enter>.***
d. Specify center of box <0,0,0>: ***type coordinates or pick location with cursor. (P1)***
e. Specify corner or [Cube/Length]: ***type coordinates for a corner or pick location with the cursor. (P2)***
f. Specify height: ***type the height.***

(Note: The total Length, Width and Height straddles the centerpoint.)

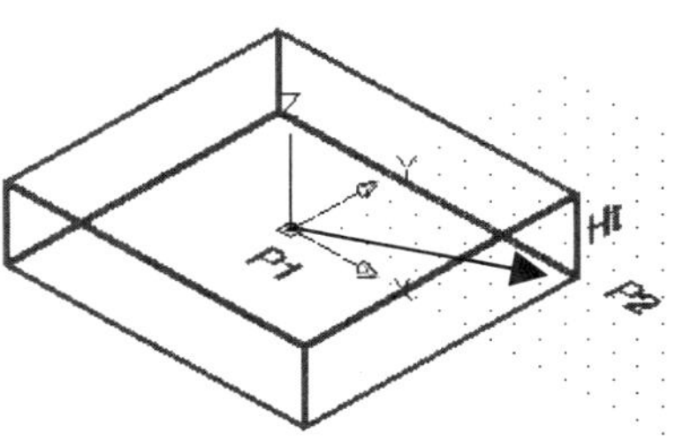

SPHERE

Sphere creates a spherical solid. You define the center point and then define the size by entering either the radius or the diameter.

a. **Start a new file and select <u>Acad3d.dwt</u>**.
b. Select the **Sphere** command. (See page 14-2)
c. Specify center point or [3P/2P/Ttr]: ***type coordinates or pick location with cursor.***
d. Specify radius of sphere or [Diameter]: ***enter radius or D.***

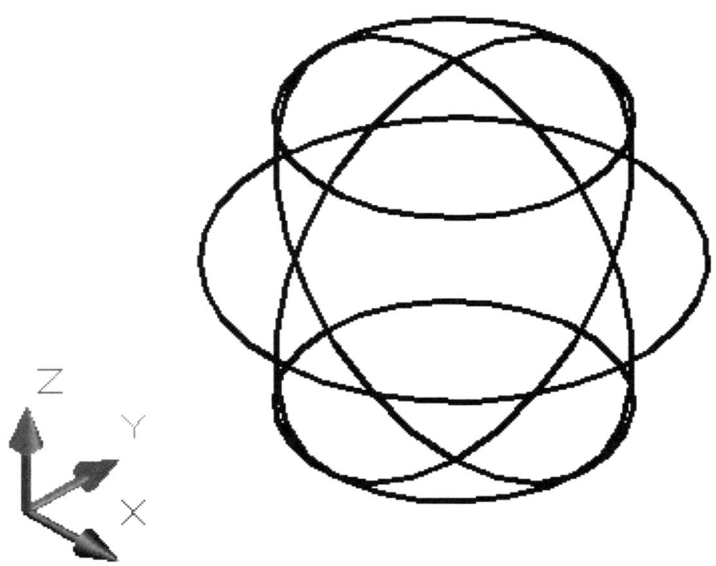

The Sphere above is displayed as “3D Wireframe”.

Try some of the other Visual Styles.

CYLINDER

Cylinder creates a cylindrical solid. You will define the center location for the base and then define the radius or diameter and then the height or location for the other end.
Start a new file and select <u>Acad3d.dwt</u>

The two methods below are the most commonly used.

Method 1
The default orientation of the cylinder has the base on the X,Y plane and the height is in the Z direction. When you enter the height, the cylinder grows in the Z axis direction. You may enter a positive or negative number. It depends upon which direction you want the cylinder to grow.
So remember, keep an eye on the UCS icon.

a. Select the **SE Isometric** view.
b. Select the **Cylinder** command.
c. Specify center point for base or [3P/2P/Ttr/Elliptical] <0,0,0>: ***type coordinates or pick location with cursor (P1)***
d. Specify radius for base of cylinder or [Diameter]: ***enter radius or D.***
e. Specify height of cylinder or [Center of other end]: ***enter the height (Ht).***

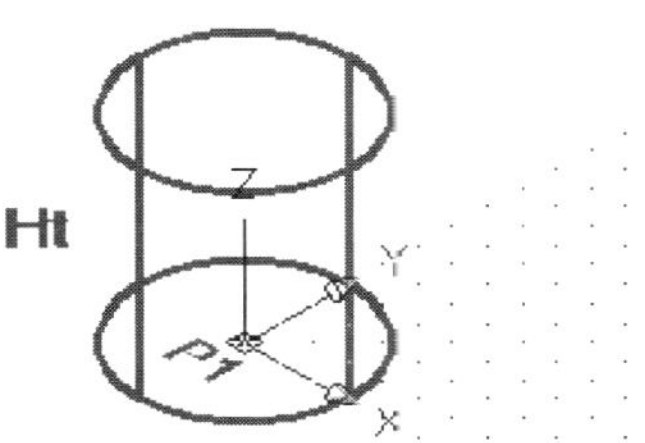

Method 2
The orientation of the Cylinder base depends on the placement of the <u>Center of the other End</u>. This method allows you to tilt the cylinder. Define the center of the base and radius then select the "Axis Endpoint" option. Define the "Axis Endpoint" entering relative coordinates or snapping to an object.

a. Select the **SE Isometric** view.
b. Select the **Cylinder** command.
c. Specify center point of base or [3P/2P/Ttr/Elliptical] <0,0,0>: ***type coordinates or pick location with cursor (P1).***
d. Specify base radius or [Diameter]: ***enter radius or D.***
e. Specify height or [2 point/Axis endpoint]: ***A <enter>.***
f. Specify axis endpoint: ***type coordinates or snap to an object. (P2)***
(See the 2 examples below)

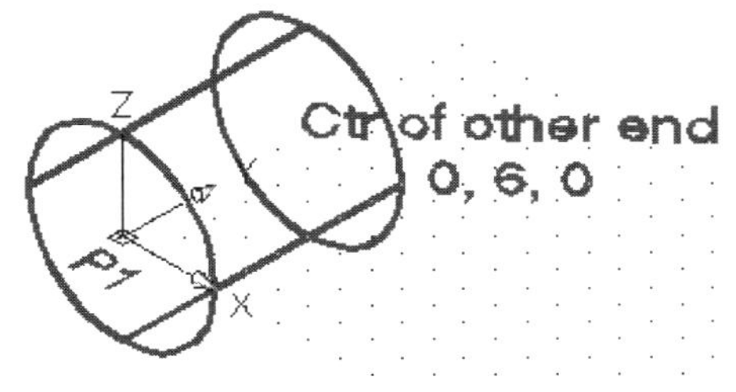

<u>**Enter coordinates**</u>

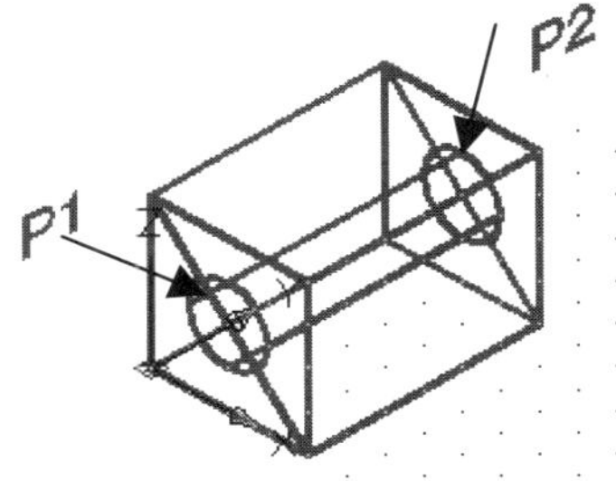

<u>**Snap to an object**</u>

CONE

Cone creates a Conical solid. There are 2 methods to create a Cone.
You will define the center location and radius or Diameter for the base and then define the height or location for the apex. **Start a new file and select Acad3d.dwt**

Method 1
The default orientation for the base is on the X and Y plane and the height is perpendicular in the Z direction.

a. Select the **SE Isometric** view.
b. Select the **Cone** command. (See page 14-2)
c. Specify center point of base [3P/2P/Ttr/Elliptical] <0,0,0>: ***type coordinates or pick location with cursor (P1)***
d. Specify base radius or [Diameter]: ***enter radius or D.***
e. Specify height or [2Point/Axis endpoint/Top Radius]: ***enter the height (can be positive or negative).***

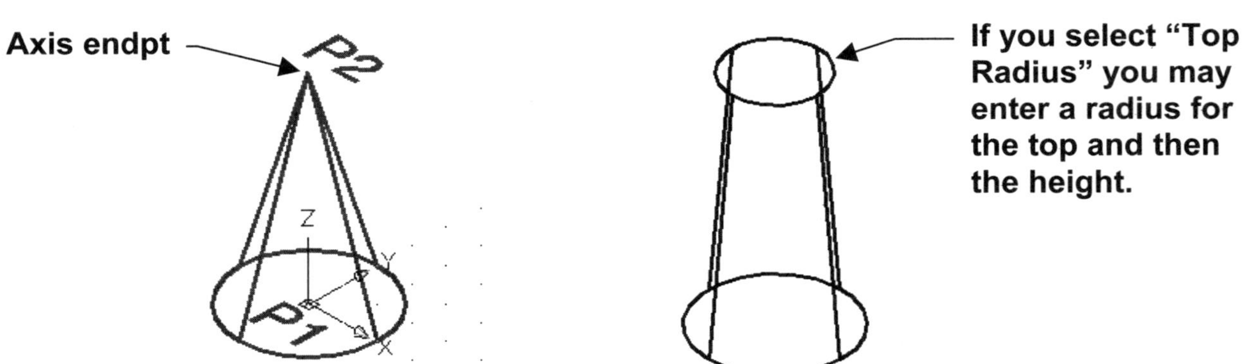

Method 2
The orientation of the Cone depends on the placement of the Apex. Define the center of the base and radius then select the "Apex" option. Define the "Apex" location using coordinates or snapping to an object.

a. Select the **SE Isometric** view.
b. Select the **Cone** command.
c. Specify center point of base or [Elliptical] <0,0,0>: ***type coordinates or pick location with cursor (P1)***
d. Specify base radius or [Diameter]: ***enter radius or D.***
e. Specify height or [2Point/Axis endpoint/Top radius]: ***type "A" <enter>.***
f. Specify axis endpoint: ***type coordinates or snap to an object. (P2)***

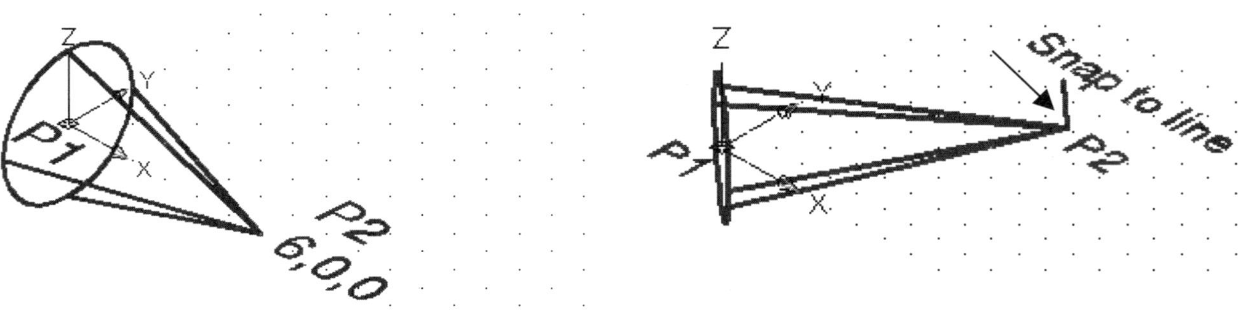

Enter coordinates **Snap to an object**

WEDGE

Wedge creates a wedge solid. There are 4 methods to create a Wedge.
The base is always parallel with the current UCS, XY plane, and the height is always along the Z axis. The slope is always from the Z axis along the X axis.
Start a new file and select <u>Acad3d.dwt</u>

<u>Method 1</u> (Define the location for 2 corners of the Base and then the Height)

a. Select the **SE Isometric** view.
b. Select the **Wedge** command. (See page 14-2)
c. Specify first corner or [Center] <0,0,0>: ***type coordinates or pick location with cursor. (P1)***
d. Specify other corner or [Cube/Length]: ***type coordinates for the diagonal corner or pick location with the cursor. (P2)***
e. Specify height: ***type the height***

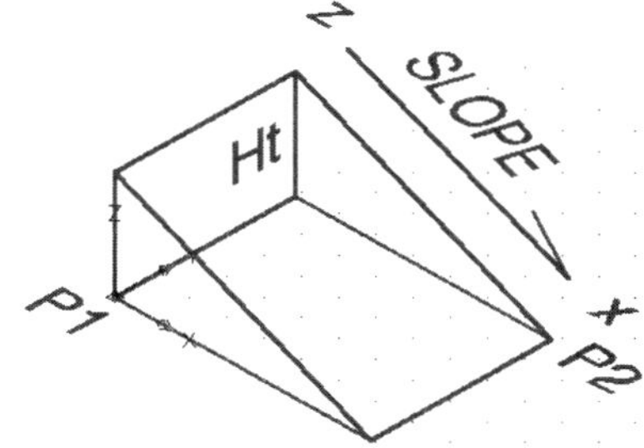

<u>Method 2</u> (Define the location for each: L, W and Ht)

a. Select the **SE Isometric** view.
b. Select the **Wedge** command.
c. Specify first corner or [Center] <0,0,0>: ***type coordinates or pick location with cursor. (P1)***
d. Specify other corner or [Cube/**Length**]: ***type "L" <enter>.***
e. Specify length: ***enter the Length (X axis).***
f. Specify width: ***enter the Width (Y axis).***
g. Specify height: ***enter the Height (Z axis).***

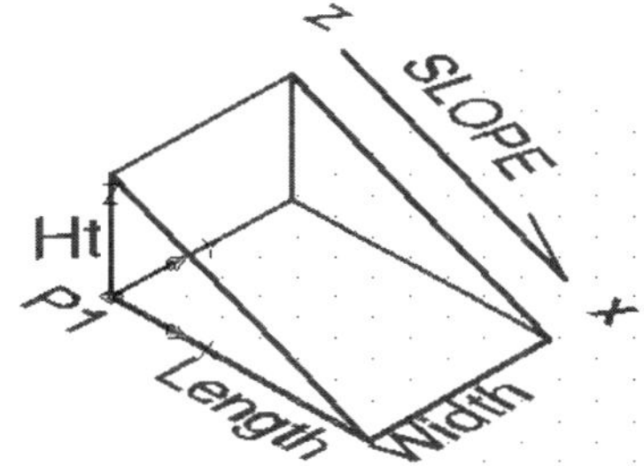

<u>Method 3</u> (Define the same dimension for Length, Width & Height)

a. Select the **SE Isometric** view.
b. Select the **Wedge** command.
c. Specify first corner or [Center] <0,0,0>: ***type coordinates or pick location with cursor. (P1)***
d. Specify other corner or [**Cube**/Length]: ***type "C" <enter.>***
e. Specify length: ***enter the dimension.***

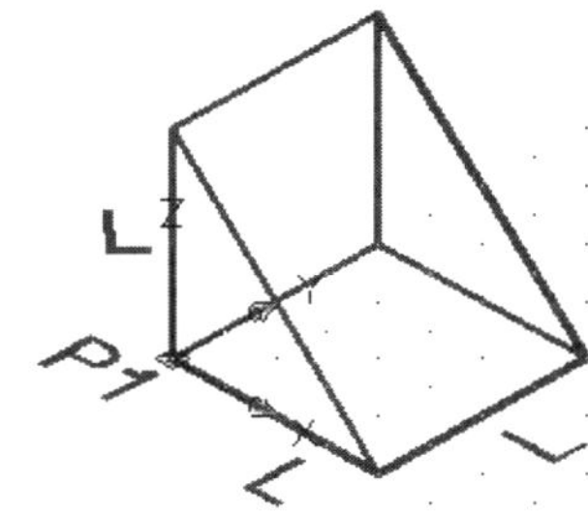

<u>Method 4</u> (Define the location for the Center and the Height)

a. Select the **SE Isometric** view.
b. Select the **Wedge** command.
c. Specify first corner or [Center] <0,0,0>: ***type "C"<enter>.***
d. Specify center <0,0,0>: ***type coordinates or pick location with cursor. (P1)***
e. Specify corner or [Cube/Length]: ***type coordinates for a corner or pick location with the cursor. (P2)***
f. Specify height: ***type the height.***

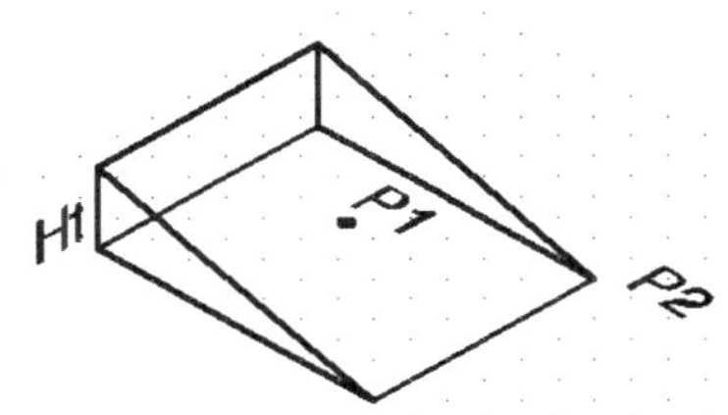

TORUS

Torus can be used to create 3 different solid shapes.

Start a new file and select <u>Acad3d.dwt</u>

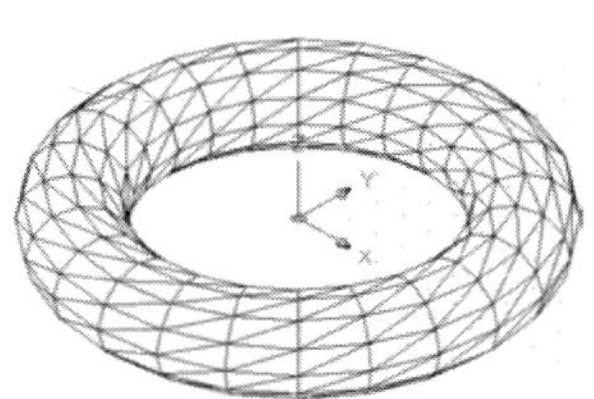

Torus = 3 Tube = 1
Donut shaped

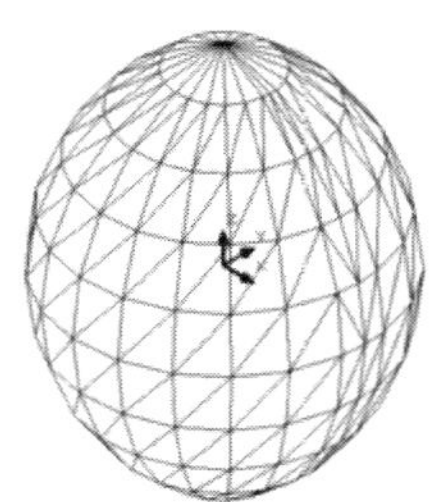

Torus = -2 Tube = 3
Football shaped

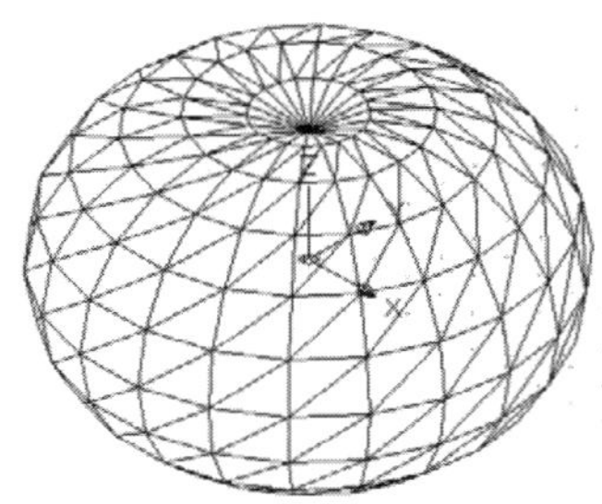

Torus = 1 Tube = 3
Self-Intersecting

The 2 dimensions that are required are the radius or diameter of the **Torus** and the **Tube**.
(Pay close attention to what the Torus and Tube are defining. Example on the right)

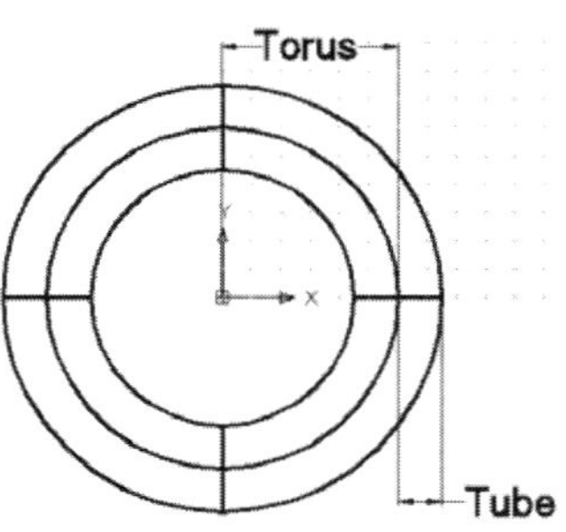

Donut shaped
Note: the <u>Torus radius</u> must be <u>greater than</u> the <u>Tube radius</u>.

a. Select the **SE Isometric** view.
b. Select the **Torus** command. (See page 14-2)
c. Specify center point or [3P/2P/Ttr]: ***type coordinates or pick location with cursor.***
d. Specify radius or [Diameter]: ***(this dim. must be <u>greater than</u> the <u>Tube</u> radius).***
e. Specify tube radius or [2pt/Diameter]: ***(this dim. must be <u>less than</u> the <u>Torus</u> radius).***

Football shaped
Note: the <u>Torus radius</u> must be <u>negative</u> and the <u>Tube radius positive</u> and <u>greater than</u> the <u>Torus radius</u>.

a. Select the **SE Isometric** view.
b. Select the **Torus** command.
c. Specify center point or [3P/2P/Ttr]: ***type coordinates or pick location with cursor.***
d. Specify radius or [Diameter]: ***(this dim. must be <u>negative</u>).***
e. Specify tube radius or [2pt/Diameter]: ***(this dim. must be <u>positive</u> and <u>greater than the Torus radius</u>).***

Self-Intersecting
Note: the <u>Torus radius</u> must be <u>less than</u> the <u>Tube radius</u>.

a. Select the **SE Isometric** view.
b. Select the **Torus** command.
c. Specify center point or [3P/2P/Ttr]: ***type coordinates or pick location with cursor.***
d. Specify radius or [Diameter]: ***(this dim. must be <u>less than</u> the <u>Tube</u> radius).***
e. Specify tube radius or [2pt/Diameter]: ***(this dim. must be <u>greater than</u> the <u>Torus</u> radius).***

PYRAMID

The method for drawing a Pyramid is very similar to drawing a polygon, for the base, and a cone. You will define how many sides the base will need. It may have from 3 to 32 sides. Decide if the base radius is Inscribed or Circumscribed. Then define the height. If you select the Top option you can truncate it. **Start a new file and select <u>Acad3d.dwt</u>**

a. Select **SE Isometric** view.
b. Select the **Pyramid** command.
c. Specify center point of base or [Edge/Sides]: ***type coordinates or pick location with cursor (P1)***
d. Specify base radius or [Inscribed] <1.000>: ***enter radius or D***
e. Specify height or [2Point/Axis endpoint/Top radius] <1.000>: ***enter the height (can be positive or negative).***

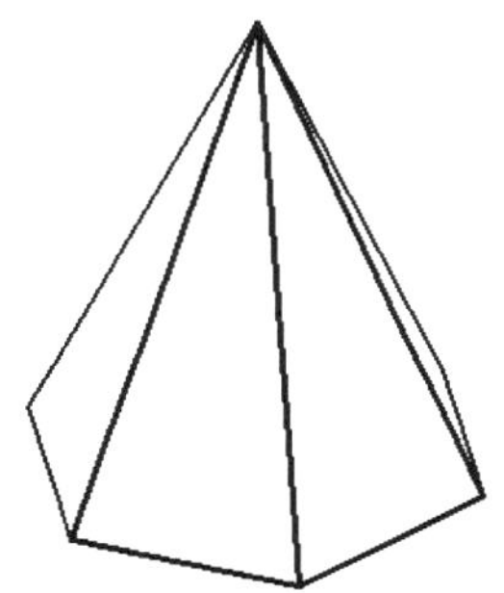

Pyramid with 6 sided base

Truncated Pyramid
Use the "Top radius" option.
Top radius is smaller than base radius.

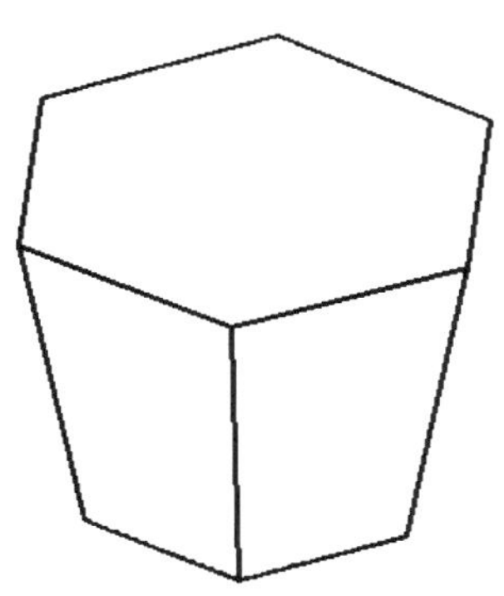

Truncated Pyramid
Use the "Top radius" option.
Top radius is larger than base radius.

EXERCISE 14A

Create 4 Solid Boxes

1. **Select File / New and select <u>Acad3d.dwt</u>**
2. Select the **SE Isometric view**
3. **Create 4 solid boxes as shown below. *<u>You decide which method to use.</u>***
   (Refer to page 14-3 for instructions if necessary)
4. Save as **EX-14A**
5. **Do not Dimension**

BOX A
L = 14 W = 6 HT = -1

BOX B
L = 4 W = 3 HT = 2

BOX C
ALL SIDES 2"

BOX D
L = 2 W = 4 HT = 1 (Think, positive or negative)

Note: Do not add the letters B, C or D. They are for reference only.

REMEMBER L = X AXIS W = Y AXIS HT = Z AXIS

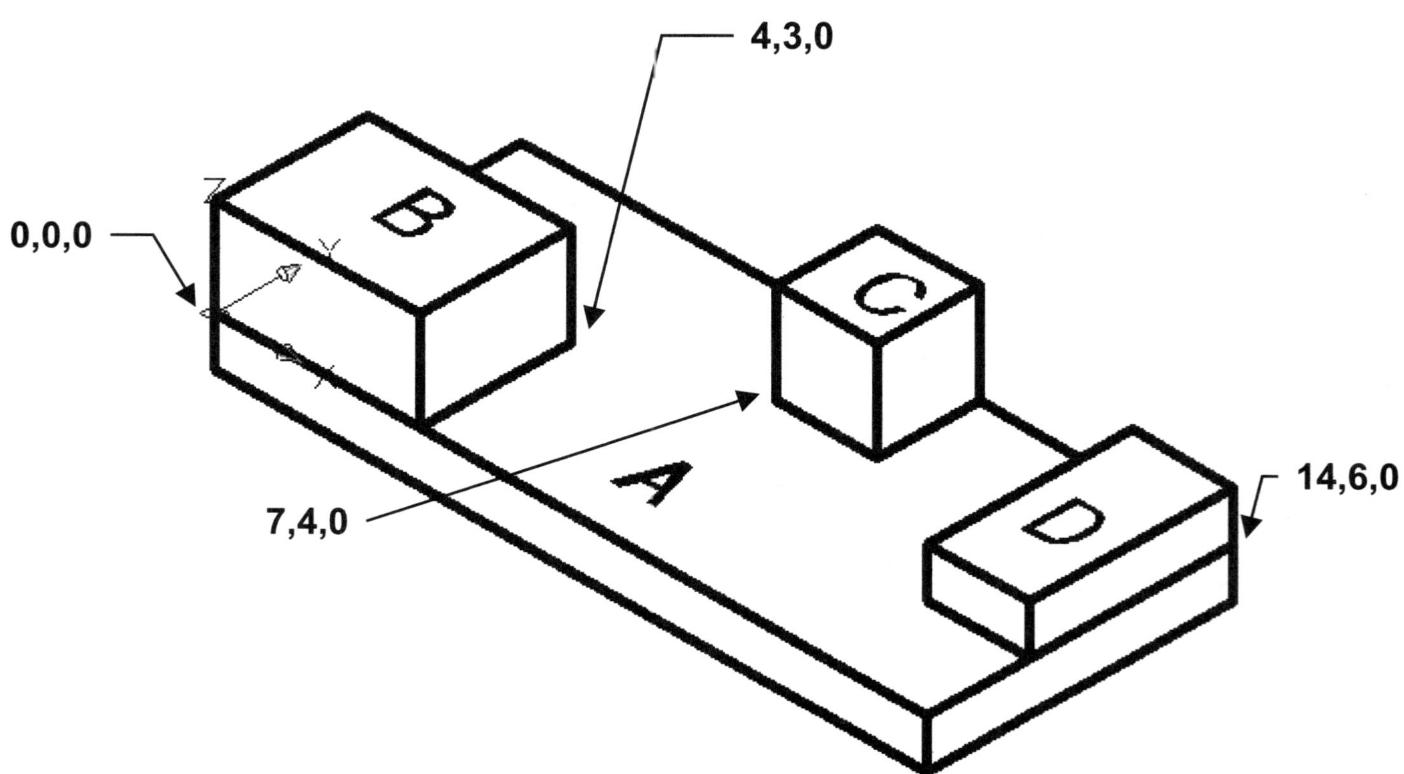

EXERCISE 14B

Create a solid Sphere

1. **Select File / New and select Acad3d.dwt**
2. Select the **SE Isometric view**
3. **Create the solid Sphere shown below.**
   (Refer to page 14-4 for instructions if necessary)
4. Experiment with Visual Styles. (Realistic shown below.)
5. Save as **EX-14B**

   Do not Dimension

Center = 0, 0, 0
Radius = 4

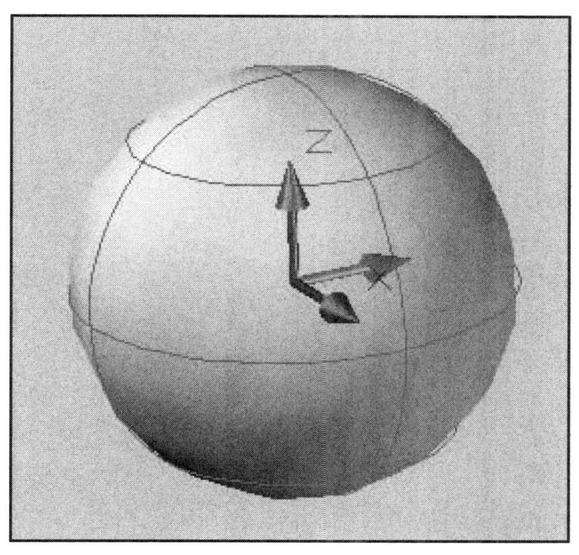

EXERCISE 14C

Create 3 solid Cylinders

1. Open **EX-14A**

2. Add the 3 Cylinders as shown below.
   Use Method 1 or 2. The method will depend on the information you are given below.
   (Refer to page 14-5 for instructions if necessary)

   Read the command line to make sure you are entering the information that AutoCAD is asking for.

3. Use 3D Orbit to make sure you have placed the objects in the correct location.

4. Save as **EX-14C**

   Do not Dimension

CYLINDER E	**CYLINDER G**	**CYLINDER H**
Radius = 1.25	6" from end to end	Radius = 1
Ht = 4"	Dia = 1 (notice "diameter")	Length = 4
	Hint: Use Axis endpoint @0,6,0	

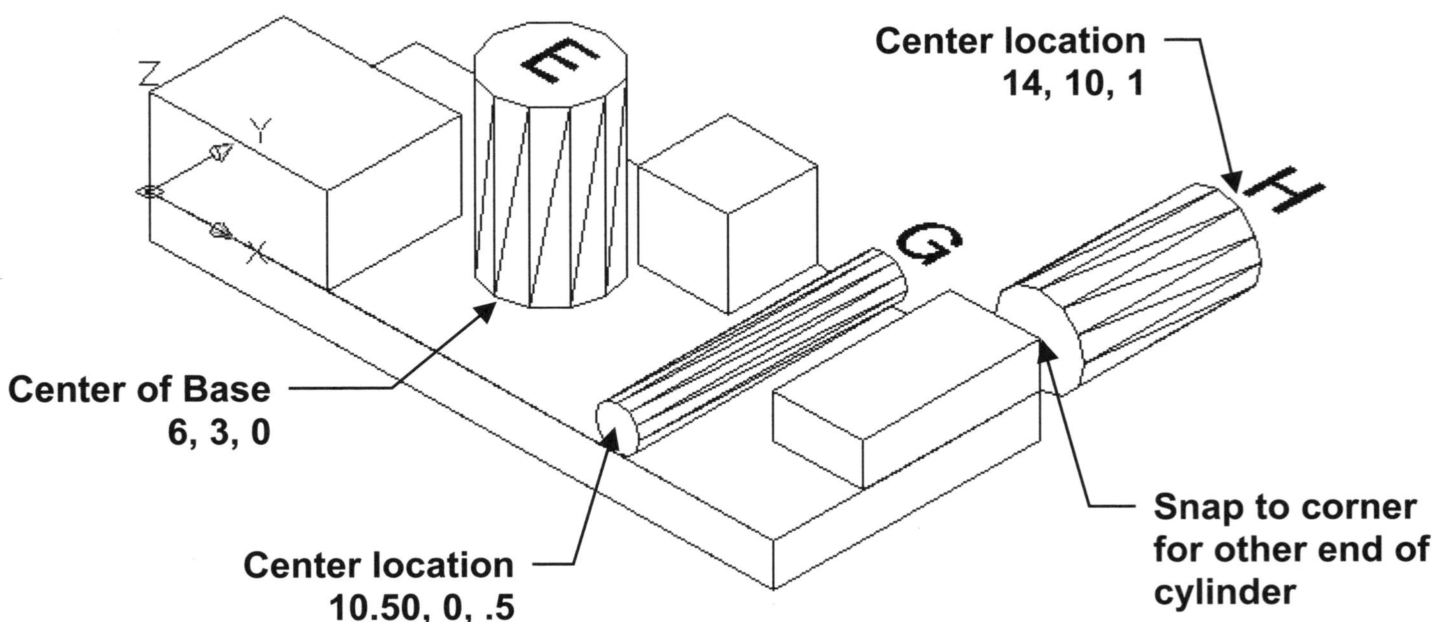

EXERCISE 14D

Create 2 solid Cones

1. Open **EX-14C**
2. Add the 2 Cones as shown below.
   (Refer to page 14-6 for instructions if necessary)
3. Save as **EX-14D**

   Do not Dimension

CONE J
Ht = 4"

CONE K
Ht = 2

Note:
Locate the Center location by snapping to the Center of the Cylinder.
Locate the Radius by snapping to the Quadrant of the Cylinder.

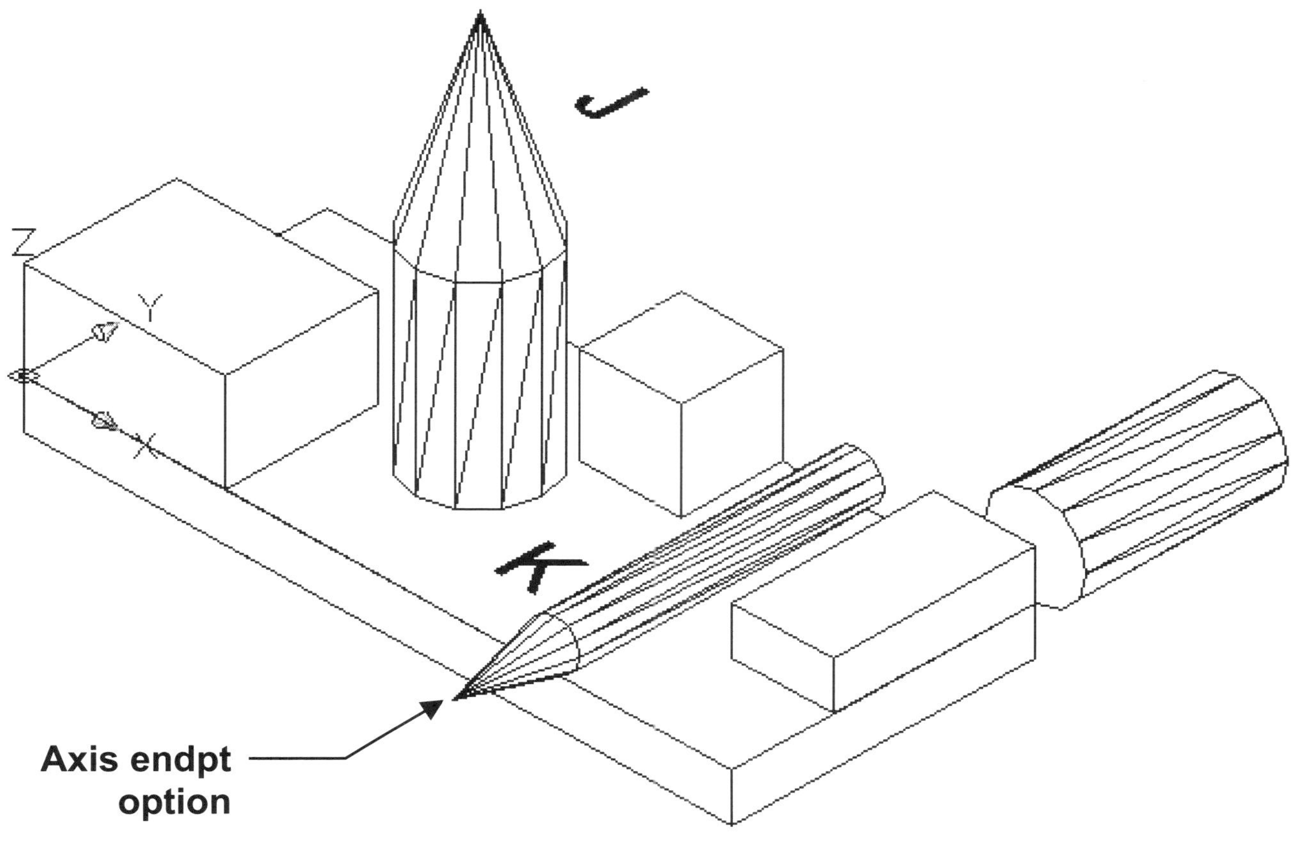

EXERCISE 14E

Create 3 solid Wedges

1. Open **EX-14D**

2. Add the 3 Wedges as shown below.
   (Refer to page 14-7 for instructions if necessary)

3. Save as **EX-14E**

   Do not Dimension

WEDGE L	**WEDGE H**	**WEDGE K**
Ht = 2"	L, W & H = 2	L = 1.5 W = 2 Ht = .5

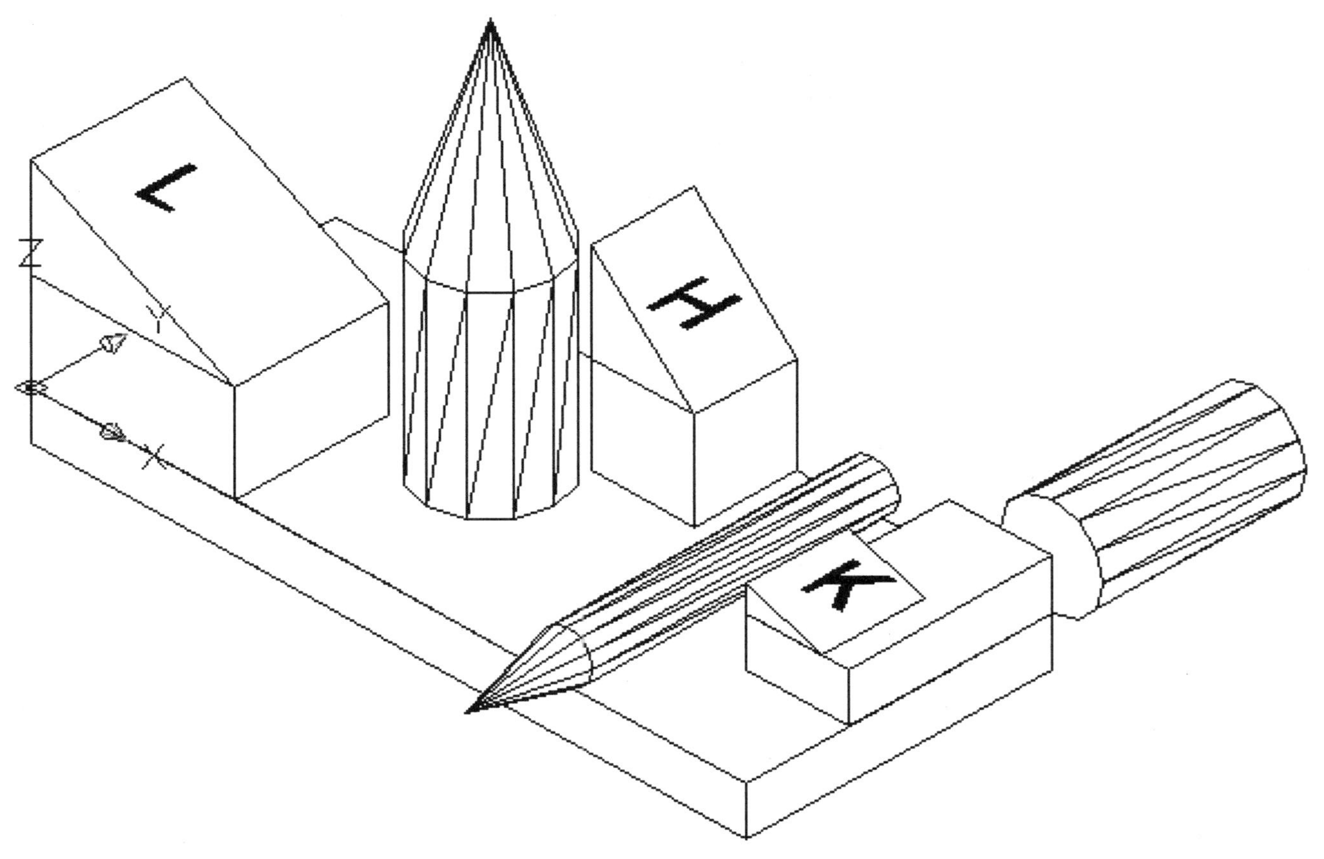

EXERCISE 14F

Create 3 solid Torus'

1. **Select File / New and select Acad3d.dwt**
2. Select the SE Isometric View.
3. Add the 3 Torus' as shown below.
   (Refer to page 14-8 for instructions if necessary)
4. Save as **EX-14F** **Do not Dimension**

DONUT	**FOOTBALL**	**SELF-INTERSECTING**
Center = 4, 3, 0	Center = 8.5, 7, 0	Center = 10.5, 3, 0
Torus Rad = 3	Torus Rad = -3	Torus Rad = 1
Tube Rad = 1	Tube Rad = 5	Tube Rad = 1.50

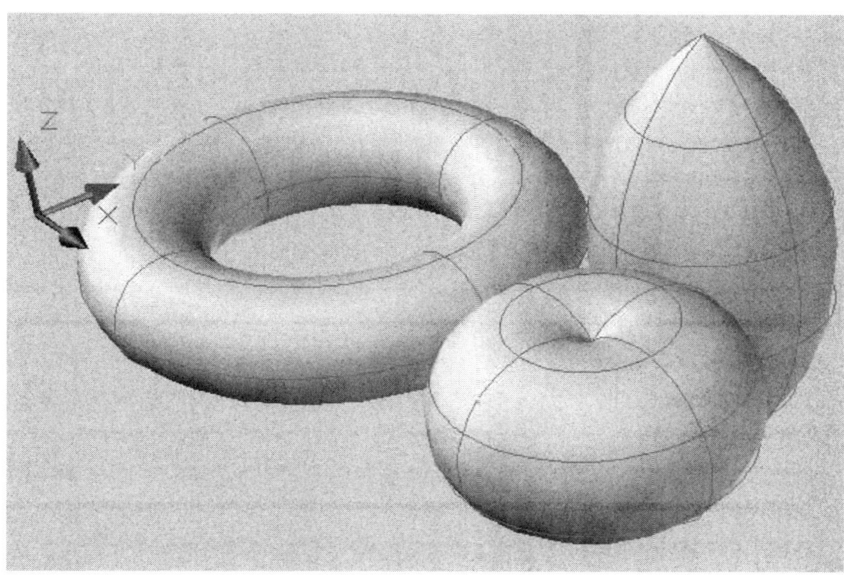

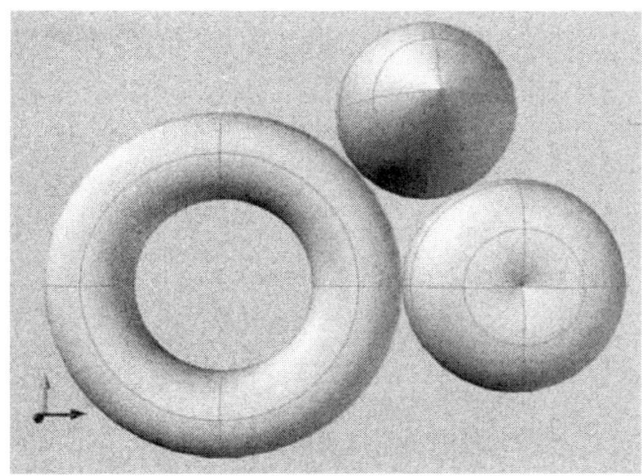

Take a look at the Top View to see if you have them positioned correctly.
Select View / 3D Views / Top.

EXERCISE 14G

Create 2 Pyramids

1. **Select File / New and select Acad3d.dwt**

2. Draw the 2 Pyramids shown below.
   (Refer to page 14-9 for instructions if necessary)

3. Save as **EX-14G**

   Do not Dimension

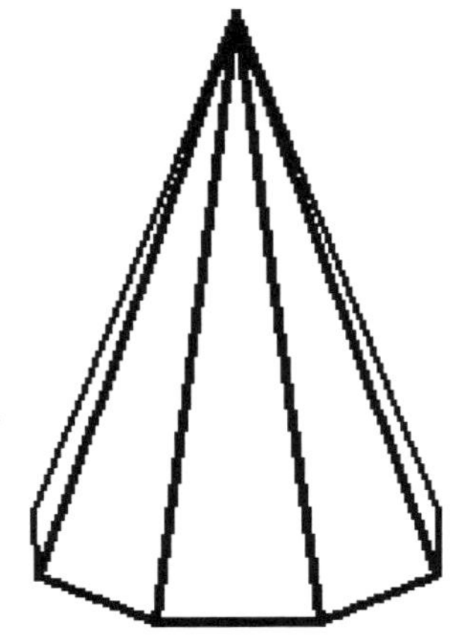

Pyramid 1
Sides = 8
Base Radius = 2
Ht = 6

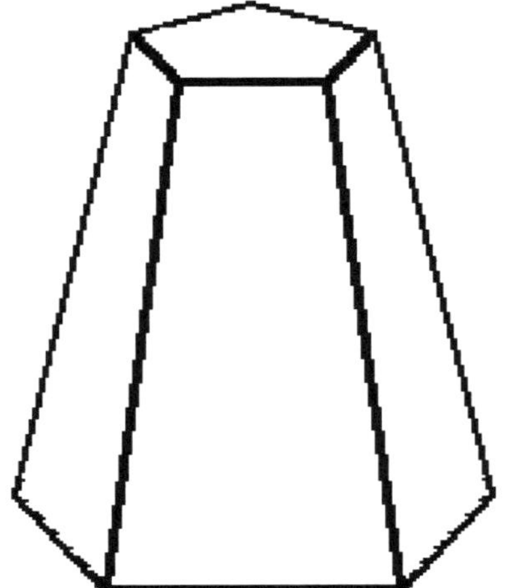

Pyramid 2
Sides = 5
Base Radius = 2
Top Radius = 1
Ht = 6

LEARNING OBJECTIVES

After completing this lesson, you will be able to:

1. Move the UCS to aid in the construction of the model.
2. Split the screen into 2 viewports.
3. Construct model easier using Plan view.
4. Understand and use Boolean operations:
   Union, Subtract and Intersect.

Sorry LT users, you do not have this option.

LESSON 15

UNDERSTANDING THE UCS

In this Lesson you are going to learn how to manipulate the UCS Origin to make constructing 3D models easy and accurate.

World Coordinate System vs. User Coordinate System

All objects in a drawing are defined with XYZ coordinates measured from the 0,0,0 Origin. This coordinate system is **fixed** and is referred to as the **World Coordinate System (WCS)**. When you first launch AutoCAD, the WCS icon is in the lower left corner of the screen. (An easy way to return to WCS is: Type UCS <enter> <enter>.)

The **User Coordinate Sytem (UCS)** allows you to move a temporary Origin to any location. (This is a procedure is referred to as "moving the Origin".)

Why move the UCS Origin?

Objects that are drawn will always be parallel to the XY plane. (Unless you type a Z coordinate) So it is necessary to define which plane is the XY plane. You define the plane by moving the UCS origin to a surface.

Let me give you a few examples and maybe it will become clearer.

Example 1:
This wedge was drawn using L, W, H. When you entered the L, W and H how did AutoCAD know to draw its base parallel to the XY plane? Because all objects are drawn parallel to the XY plane. Length is always in the X axis, Width is always in the Y axis and Height is in the Z axis. Notice the UCS icon displays the plane orientation.

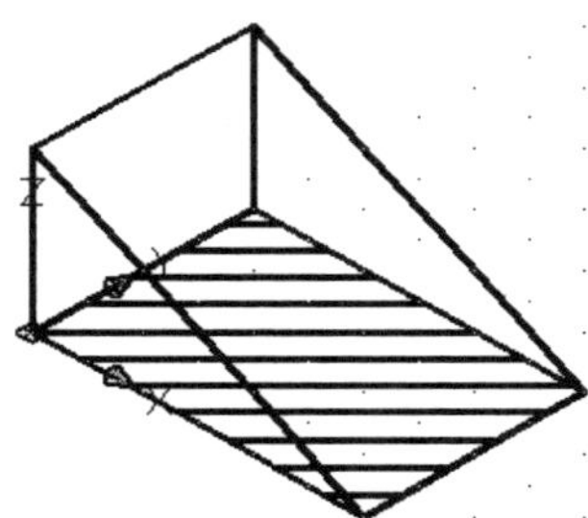

Example 2:
Below I have drawn the base of the cylinder parallel to the sloped surface. How did I do that? Notice the UCS icon. I moved it to the surface I wanted to draw on. (Remember: all objects are drawn parallel to the XY plane.) The base of the cylinder is automatically placed on the XY surface I defined and the height is always drawn in the Z axis.

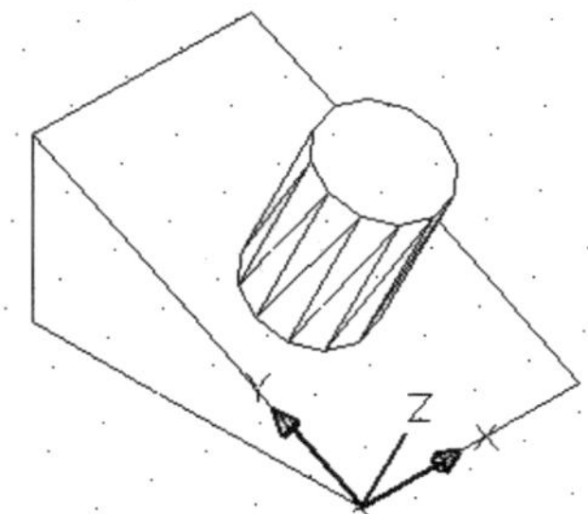

How did I move the UCS icon to the orientation shown in the example above? I will discuss that next.

MOVING the UCS ICON

There are many options available to manipulate the UCS icon. In this Lesson I will only be discussing two. When you get more experienced with AutoCAD's 3D world you may wish to investigate other options.

MOVING THE UCS ICON

Method 1 - 3 Point
This method allows you to select in which direction you want the X and Y axes to point. This is primarily used to attach the UCS to an existing surface. (Such as the slope in Example 2 on the previous page.)

1. Select the **3 point** command using one of the following:

    TYPE = UCS
    PULLDOWN = Tools / New UCS / 3 point
    TOOLBAR = UCS

2. Specify new origin point <0,0,0>: ***Locate where you want the new 0,0,0 (type coordinates or place with the cursor)***

3. Specify point on positive portion of X-axis <1.000,0.000,0.000>: ***define which direction is <u>positive</u> X axis by typing coordinates or with the cursor.***

4. Specify point on positive-Y portion of the UCS XY plane <0.000,1.000,0.000>: ***define which direction is <u>positive</u> Y axis by typing coordinates or with the cursor.***

<u>EXAMPLES</u>

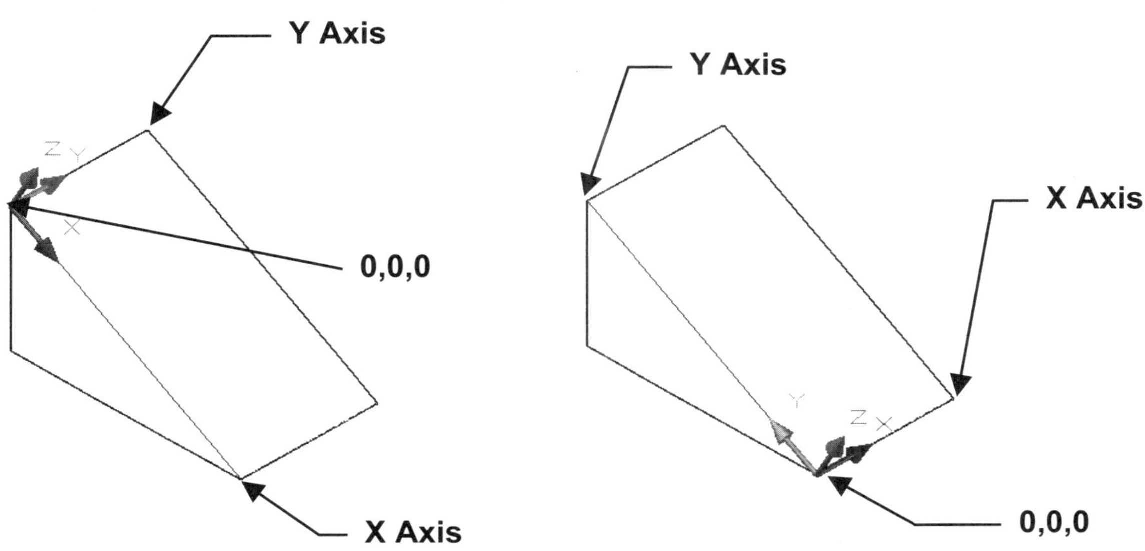

MOVING the UCS ICON continued....

Method 2 – Dynamic UCS

This method is a new feature in 2007. It allows you to dynamically change the XY Plane and then return the XY plane automatically to the previous location.

Note: The Dynamic UCS feature works only on solids.

1. You may turn this feature **ON** or **OFF** with the **DUCS** button on the status bar or **F6** key.

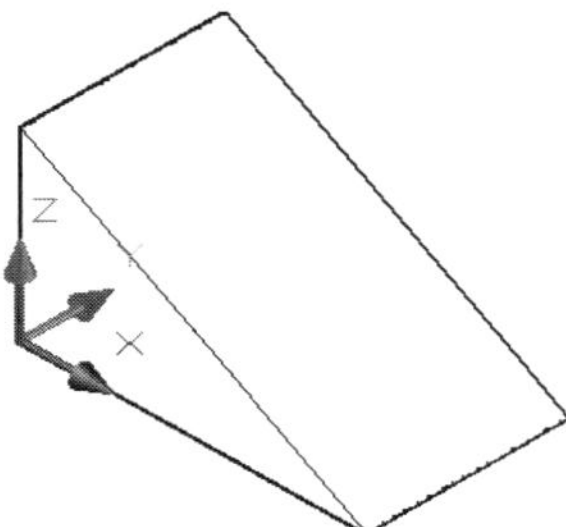

2. Start a command such as Cylinder.

3. Pass the cursor over a face of a solid. (Such as the slope of a wedge) A dashed border appears on the face.

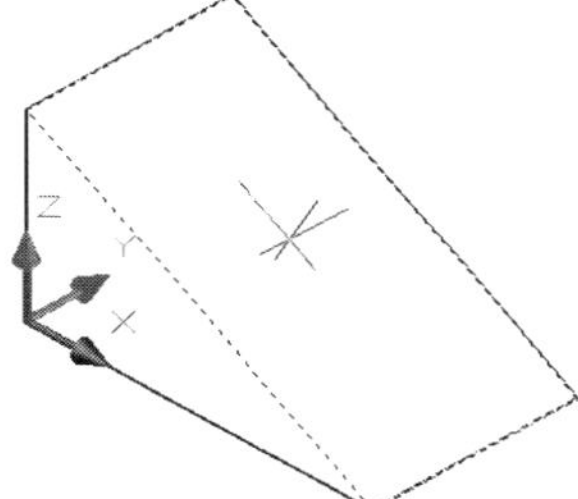

4. Continue with the command. (Such as click to place the center of the cylinder) The UCS icon automatically moves to that location and temporarily orients the XY plane.

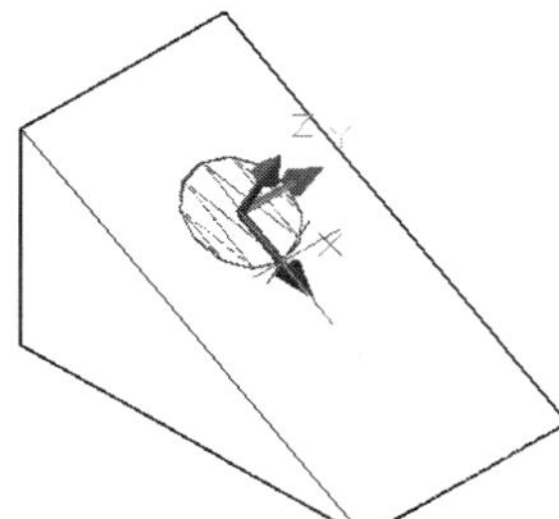

5. Complete the command. The UCS icon automatically returns to its previous location.

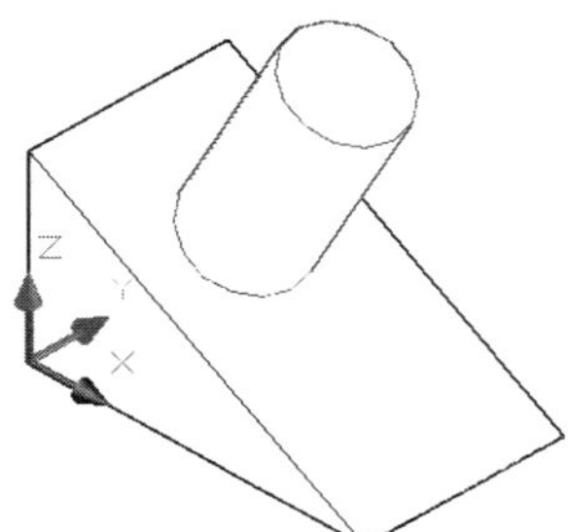

ROTATING the UCS ICON

Rotating the UCS is another option to manipulate the UCS icon. You may rotate around any one of the 3 axes. You may use the cursor to define the rotation angle or type the rotation angle.

1. Select the UCS Rotate command using one of the following:

    TYPE = UCS / New UCS / select X, Y or Z
    PULLDOWN = Tools / select X, Y or Z
    TOOLBAR = UCS

2. Select which Axis to rotate "**Around**".
   *(For Example: If you select **X**, you are rotating the Y and Z-axes around the X axis.)*

3. Specify rotation angle about X axis <90>: ***type the rotation angle or use cursor.***

Understanding the rotation angle
The easiest way to determine the rotation angle is to think of yourself standing in front of the point of the axis arrow, then rotate the other 2 axes.
For example, if you are rotating around the X axis, put the point of the X axis arrow against your chest and then grab the Z axis with your left hand and the Y axis with your right hand. Now rotate your hands Clockwise or Counterclockwise like you are holding a steering wheel vertically. The X axis arrow, corkscrewing into your chest, is the Axis you are rotating around. CW is negative and CCW is positive just like the Polar clock you learned about in the Beginning workbook.

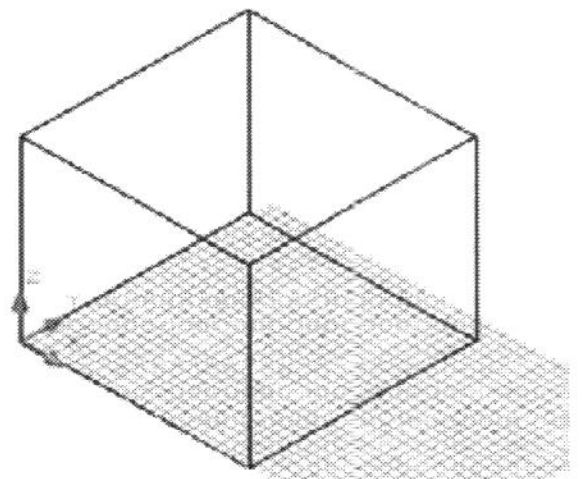

Original UCS Position

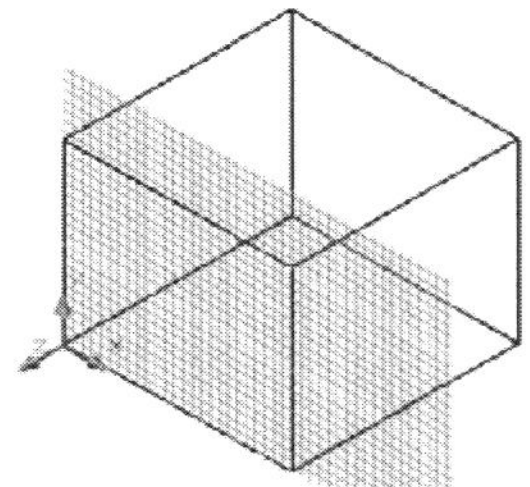

CCW Rotation 90

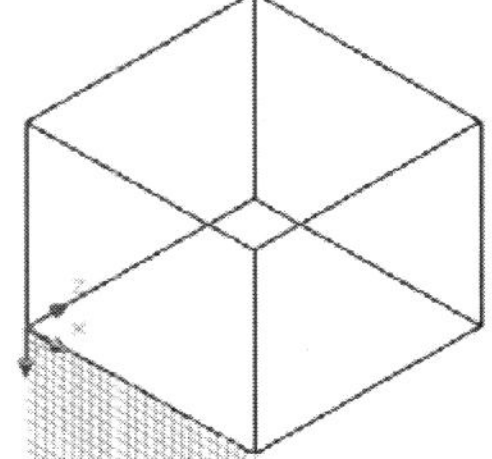

CW Rotation -90

Notice the grid indicates the drawing plane.

Note: To return the UCS to its default location and rotation: type UCS <enter> <enter>.
If its stubborn select the SE Isometric View and then type UCS <enter> <enter>

NEW DIRECTION FOR Z AXIS

Remember that "**Height**" is always the **positive** Z axis. Sometimes the positive Z axis is not oriented to suit your height direction need. The "**Zaxis Vector**" option allows you to change the positive direction of the Z axis. The X and Y axes will change also automatically.

1. Select the **Zaxis** command using one of the following:

    TYPE = UCS / ZAxis
    PULLDOWN = TOOLS / NEW UCS / Z axis Vector
    TOOLBAR = UCS

2. Specify new origin point or object <0,0,0>: ***type the coord or place the cursor***

3. Specify point on positive portion of Z-axis <0.000,0.000,1.000>: ***type the coordinates or place the cursor to select the positive Z direction.***

EXAMPLES
Notice all 3 examples below use the same information to draw the Cylinder. The only difference is the direction of the Z-axis.

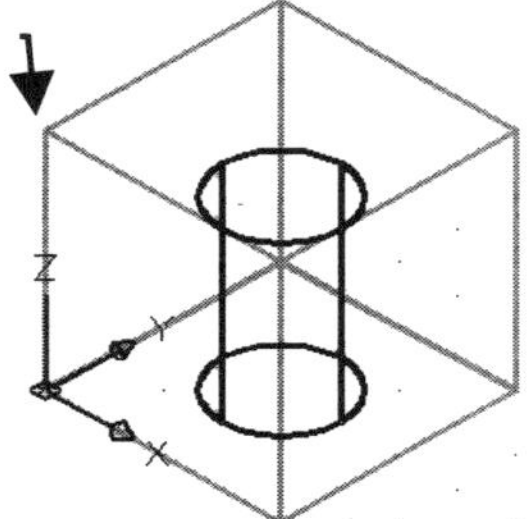

The Original
The XY drawing plane is on the inside bottom.
A Cylinder is drawn using the following:
Base center location: **2, 2, 0**
Radius : **1**
Ht = **3**

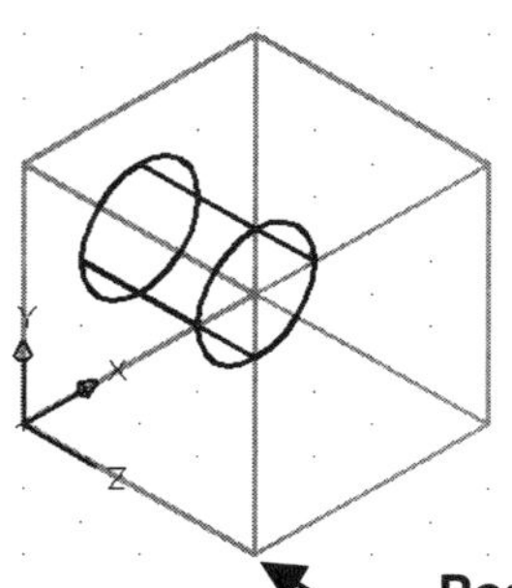

Positive Zaxis direction has been changed
The XY drawing plane is on the inside of the left side.
A Cylinder is drawn using the following:
Base center location: **2, 2, 0 (same as above)**
Radius : **1 (same as above)**
Ht = **3 (same as above)**

Positive Z

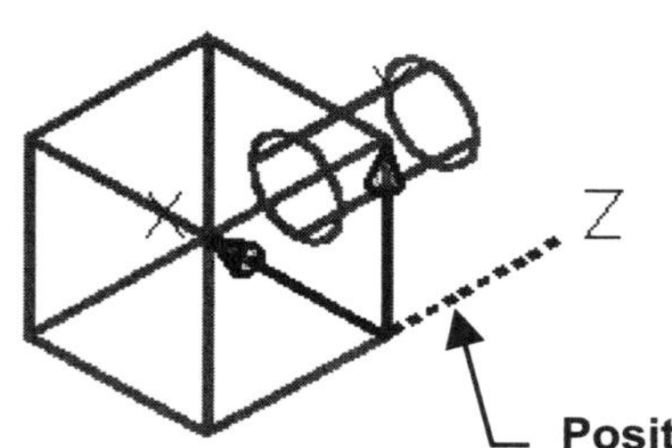

Positive Zaxis direction has been changed again
The XY drawing plane is on the outside of the Back.
A Cylinder is drawn using the following:
Base center location: **2, 2, 0 (same as above**
Radius : **1 (same as above)**
Ht = **3 (same as above)**

Positive Z

DRAWING WITH TWO VIEWPORTS

Sometimes it is easier to work on the model if you split Model space into two viewports. An isometric view can be displayed in one and the plan view in the other. This makes it easy to move the UCS around in the isometric view and draw on the XY plane in the plan view.

Setting the UCS ICON system variable (Not available in LT)
The **UCSVP** system variable controls the UCS icon display configuration.

0 The UCS icon, in each viewport, is relative to each other. If you change one, they all change.

1 The UCS icon, in each viewport is independent. Each viewport may have it's own UCS orientation. (This is the default setting)

I find it less confusing, for students new to 3D, to use the "0" setting.
If you change one, you change them all.
*So I want you to change the **UCSVP** variable to "0".*

1. ***Type: ucsvp<enter>***
2. ***Type: 0 <enter>***

You should change this system variable on all of your master setup drawings then resave the drawing. Then you do not have to worry about this until you decide you want to change it to "1".
(Do this now so you don't forget, but I will remind you for awhile)

HOW TO SPLIT THE SCREEN INTO 2 VIEWPORTS

1. Open EX-14A. (Select Model tab.)
2. Check the **ucsvp** setting. (I will continue to remind you)
3. Select **VIEW / VIEWPORTS / 2 VIEWPORTS**
4. Enter a configuration option [Horizontal/Vertical] <Vertical>: ***press <enter>***
5. Activate the right hand viewport. (Click in it)
6. Select **SE Isometric** view. (Notice the UCS icon location)
7. Activate the left hand viewport (Click in it)
8. Select the **VIEW / 3D VIEWS / TOP**

If your origin icon is not attached to the object, zoom out a little.

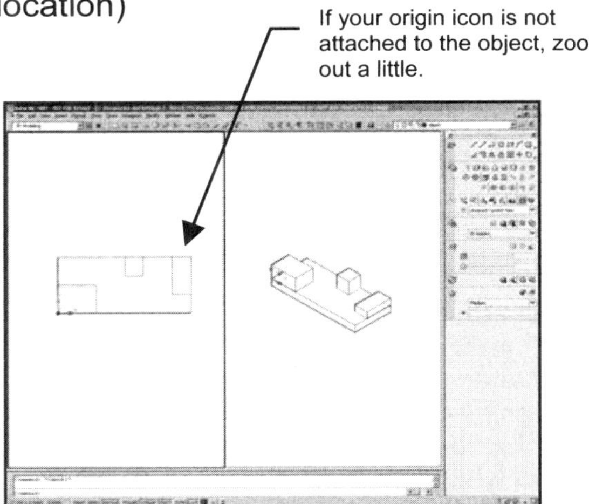

Now think about this. With the left viewport still active, select some of the other views such as, Bottom or Side. Watch the UCS icon move around on the isometric view to match your selection in the left viewport.
If you move the UCS icon in the right hand viewport (isometric view) it will also reflect the change within the left hand viewport. They are relative to each other.

To return to 1 viewport, select VIEW / VIEWPORTS / 1 VIEWPORT.

On the next page I will discuss how this will assist you when constructing a 3D model.

PLAN VIEW

Now that you have learned to split the screen into 2 viewports, let's see how you can use this to assist you with constructing a 3D model.

Remember, you always draw on the XY plane.

HOW TO USE THE "PLAN" VIEW

1. Set **UCSVP** to "**0**" (refer to page 15-7)
2. Split the screen into 2 viewports (refer to page 15-7)
3. Activate the right hand viewport. (isometric view)
4. Move the UCS icon to the plane that you wish to draw on.
5. Activate the left hand viewport
6. Type: ***PLAN <enter>***
   Enter an option [Current ucs/Ucs/World] <Current>: ***<enter>***

Use 3D Wireframe to see through the surfaces.
The left hand viewport will display the current XY plane.
You can now draw on the XY plane as if you were drawing a 2D drawing.
(Note: Each viewport may be zoomed and panned independently.)

EXAMPLES

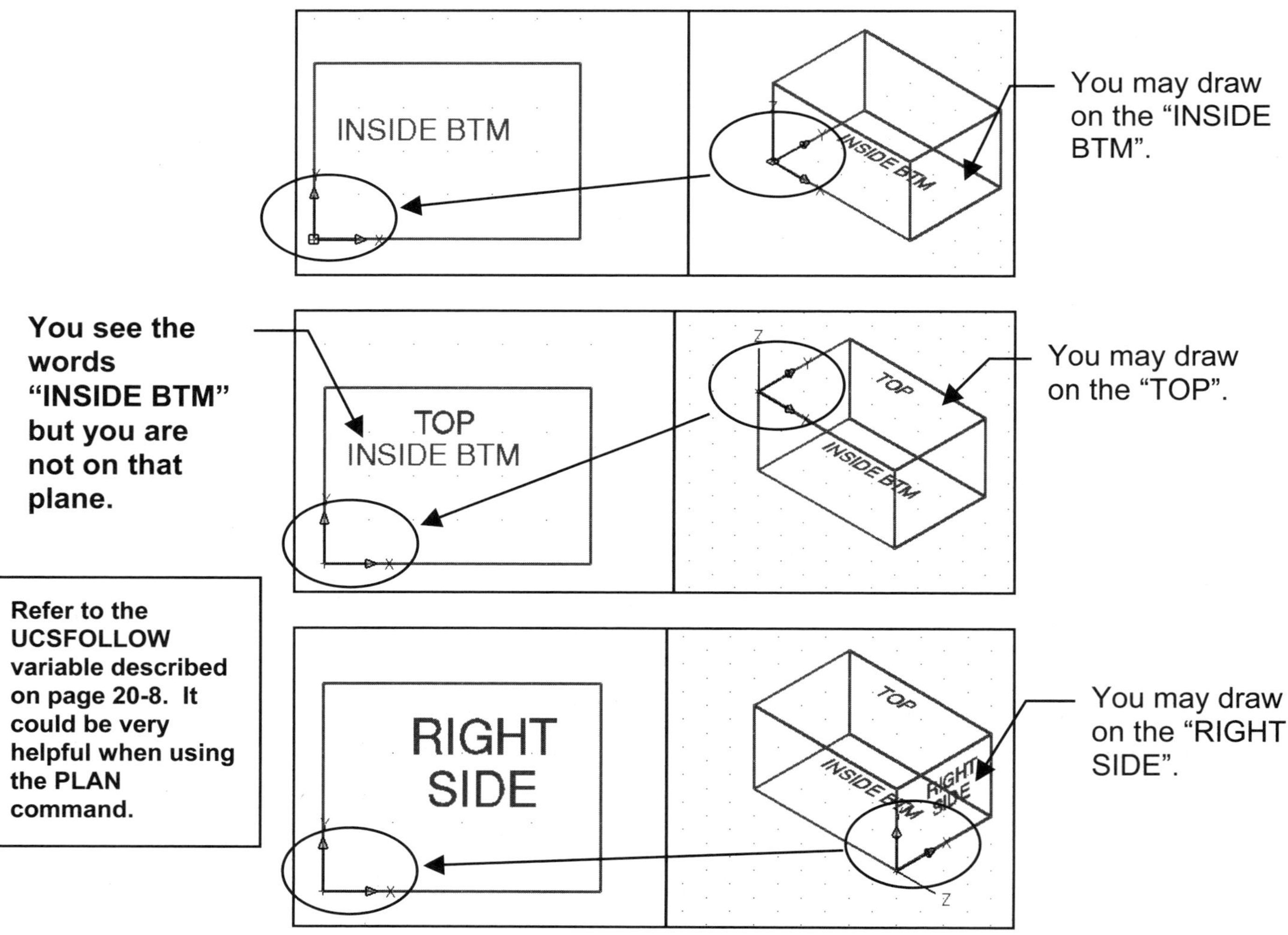

BOOLEAN OPERATIONS (Sorry, not available in version LT)

Solid objects can be combined using Boolean operations to create ***composite solids.*** AutoCAD's **Boolean** operations are **Union, Subtract** and **Intersect.**

Boolean is a math term which means using logical functions, such as addition or subtraction on objects. George Boole (1815-1864) was an English mathematician. He developed a system of mathematical logic where all variables have the value of either one or zero. Boole's two-value logic, or binary algebra, is the basis for the mathematical calculations used by computers and is required in the construction of composite solids.

UNION

The **Union** command creates one solid object from 2 or more solid objects.

1. Select the **Union** command using one of the following:

   TYPE = UNI or UNION
   PULLDOWN = MODIFY / SOLID EDITING / UNION
   3D MAKE CONTROL PANEL = SOLIDS EDITING

2. Select objects: ***select the solids to be combined <enter>***

BEFORE UNION – 5 separate objects

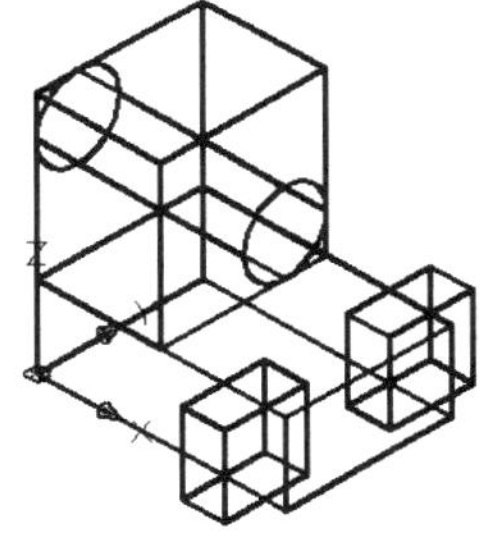

SE Isometric

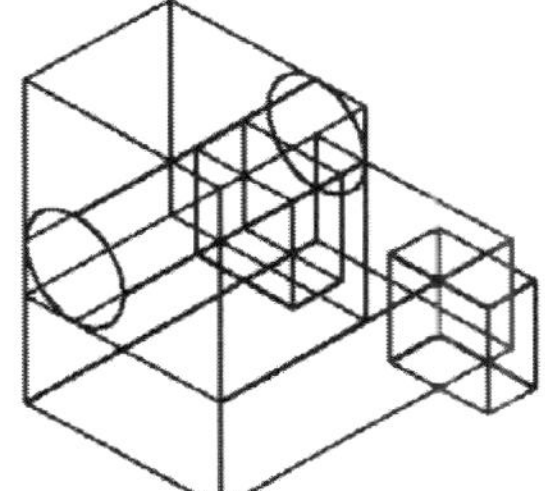

SW Isometric

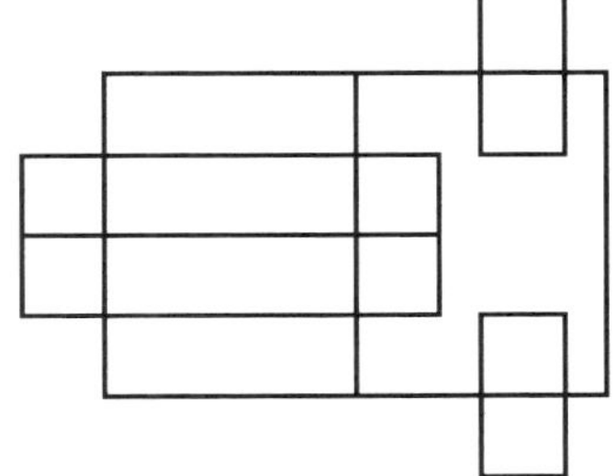

Top View

AFTER UNION – 1 object

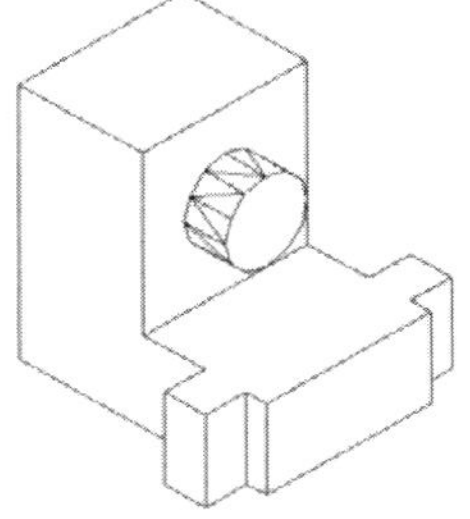

SE Isometric

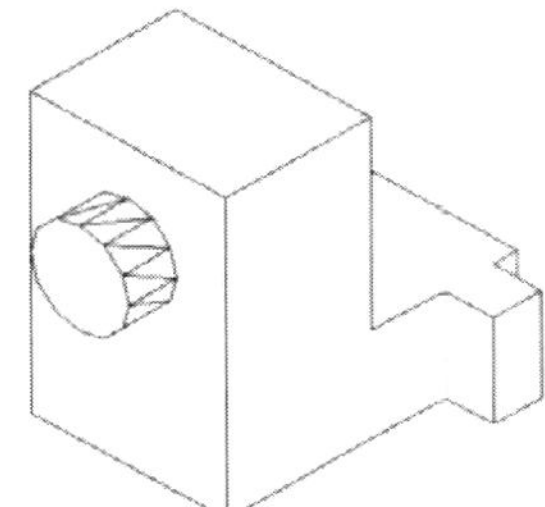

SW Isometric

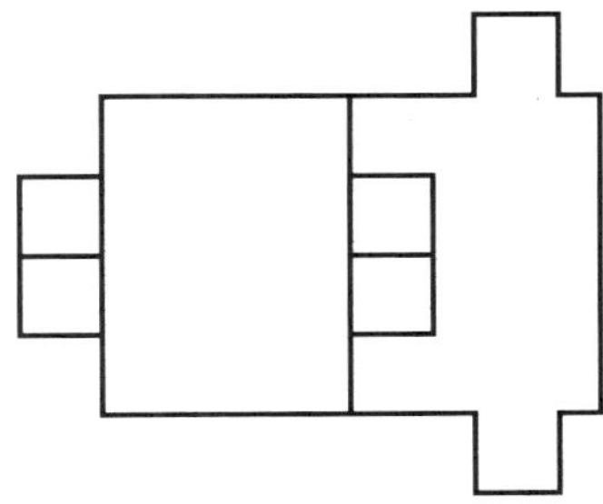

Top View

SUBTRACT

The **Subtract** command subtracts one solid from another solid.

1. Select the **Subtract** command using one of the following:

   TYPE = SU or SUBTRACT
   PULLDOWN = MODIFY / SOLID EDITING / SUBTRACT
   3D MAKE CONTROL PANEL = SOLIDS EDITING

2. Select solids and regions to subtract from…
   Select objects: ***select the solid object to subtract <u>from</u>***
   Select objects: ***<enter>***

3. Select solids and regions to subtract…
   Select objects: ***select the solid object to subtract***
   Select objects: ***<enter>***

BEFORE SUBTRACT – 5 separate objects

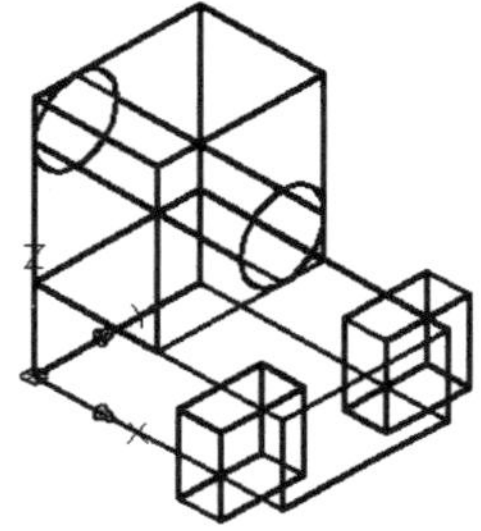

SE Isometric

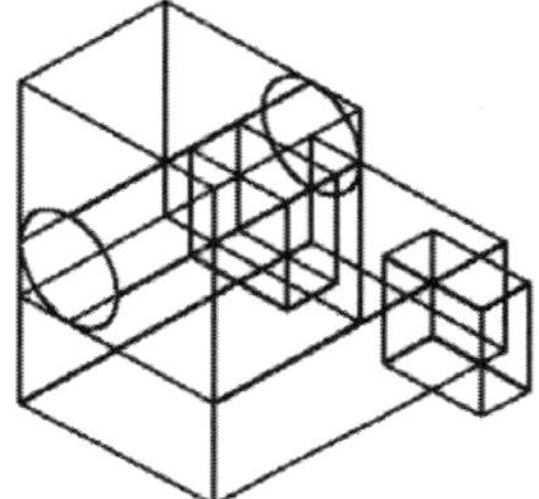

SW Isometric

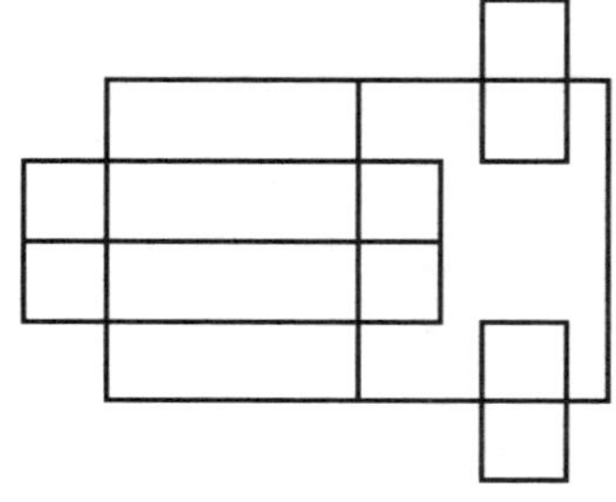

Top View

AFTER SUBTRACT – 1 object

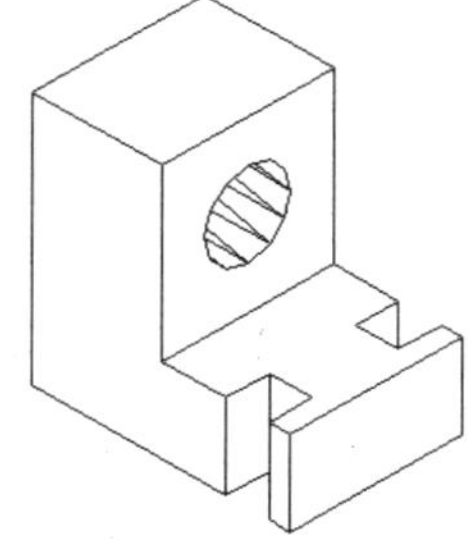

SE Isometric

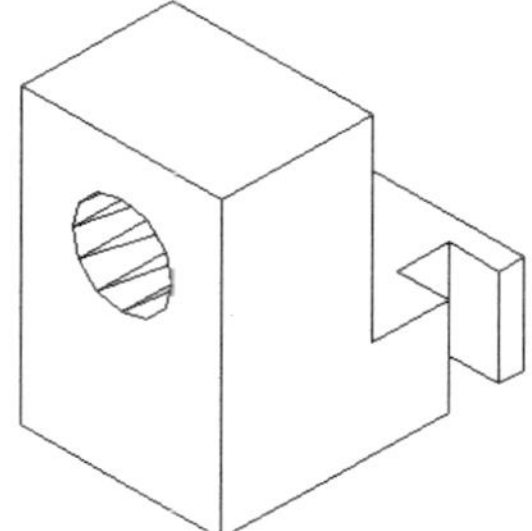

SW Isometric

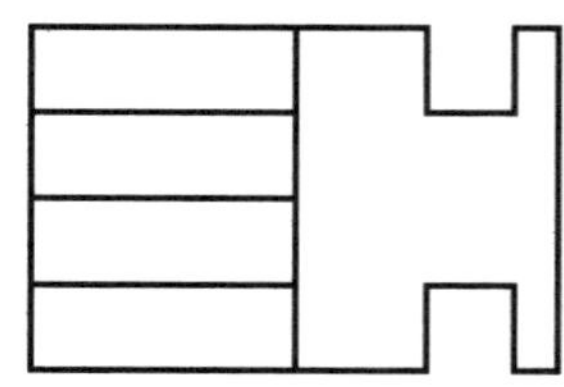

Top View

INTERSECTION

If solid objects intersect, they share a space. This shared space is called an Intersection. The **Intersection** command allows you to create a solid from this shared space.

1. Select the **Intersection** command using one of the following:

    TYPE = IN or INTERSECT
    PULLDOWN = MODIFY / SOLID EDITING / INTERSECT
    3D MAKE CONTROL PANEL = SOLIDS EDITING

2. Select objects: ***select the solid objects that form the intersection***
    Select objects: ***<enter>***

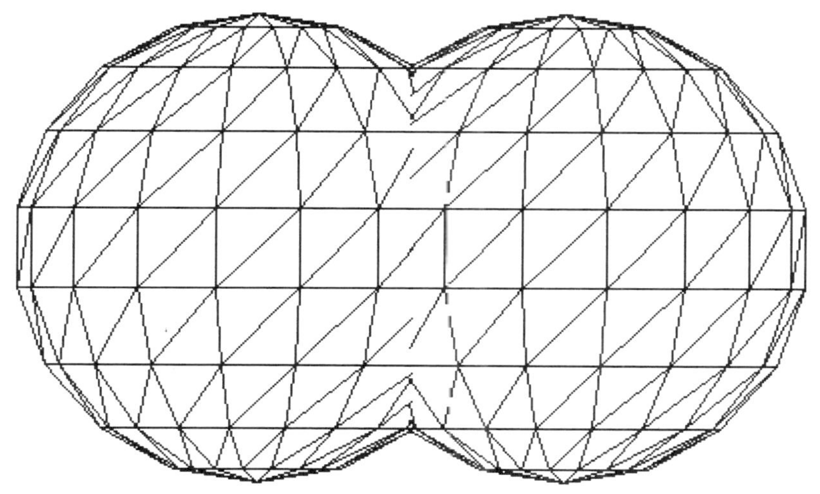

BEFORE INTERSECTION

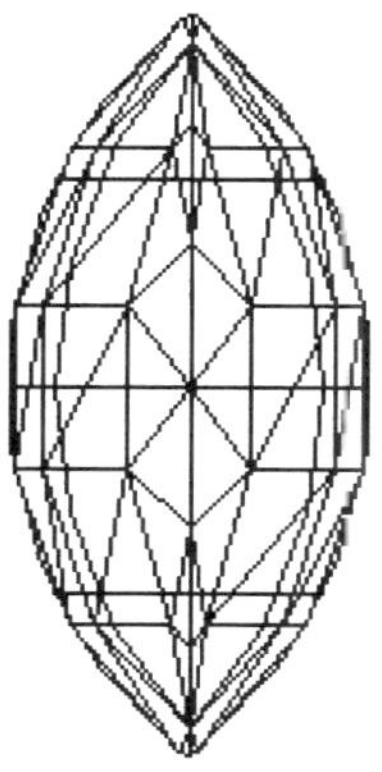

AFTER INTERSECTION

EXERCISE 15A

Subtract

1. **Select File / New and select Acad3d.dwt.**

2. Select the **SE Isometric view.**

3. Check the **UCSVP** setting. It should be "0"
   (Very important if you are using multiple viewports)

4. Draw the solid object shown below.

5. Save as **EX-15A.**

6. **Do not Dimension.**

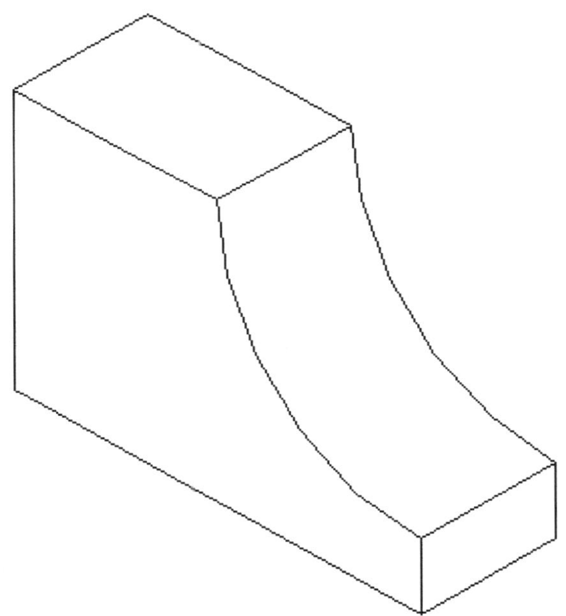

Step by step instructions and dimensions are shown on the next page.

Note: There are many different methods to create the object above.
The steps on the next page are designed to make you think about UCS positioning and rotating and the positive and negative inputs.
Can you think of other methods?

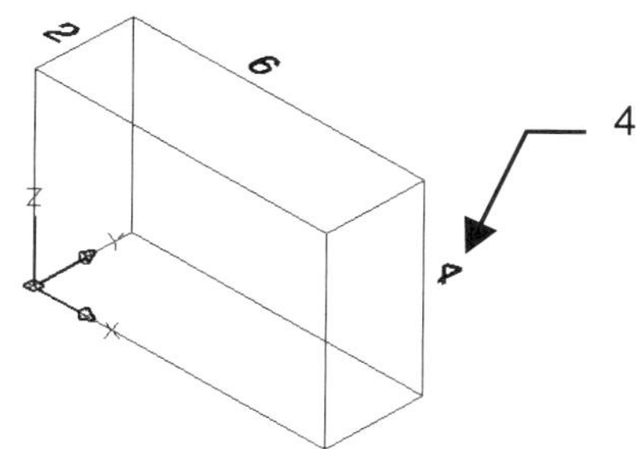

Step 1.
Create a solid box.

15-4

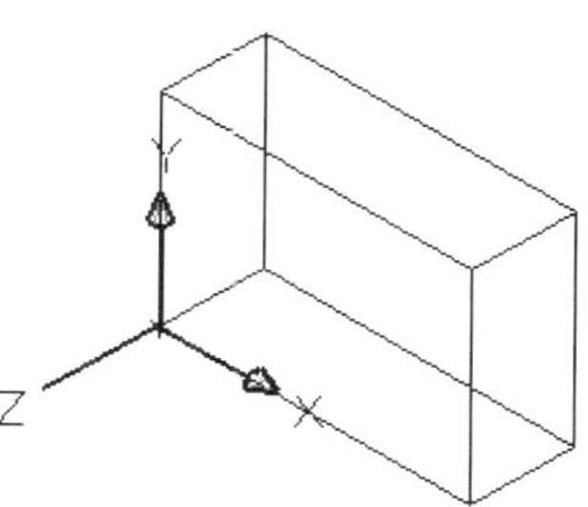

Step 2.
Rotate the UCS icon (pg. 15-5)
or 3point (pg. 15-3) or Dynamic (pg

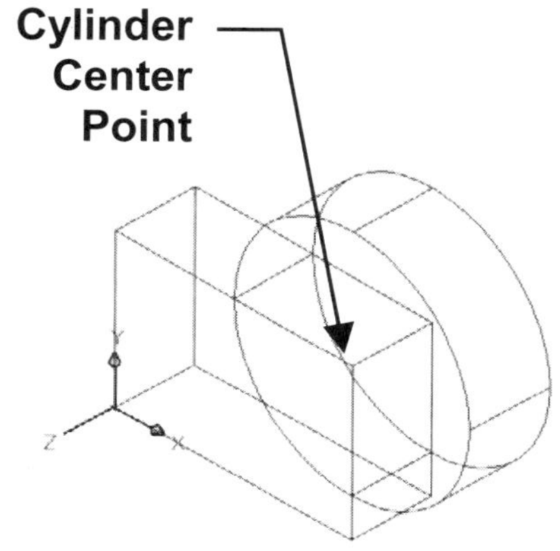

Step 3.
Draw a 3" Radius Cylinder
(Note: Ht = negative 2)

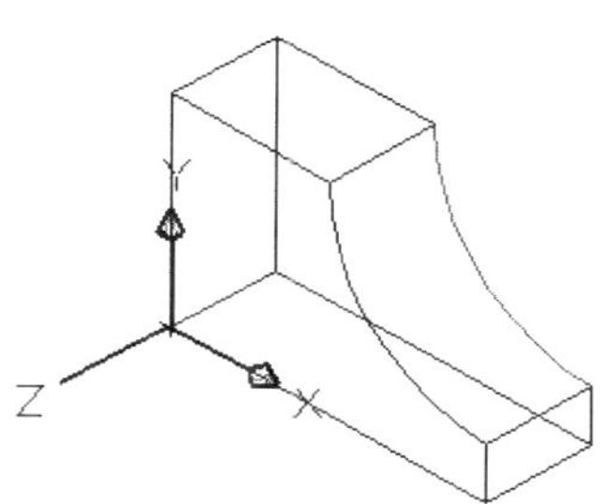

Step 4.
Subtract the Cylinder from the box

EXERCISE 15B

Union and Subtract

1. Open **EX-15A**
2. Check the **UCSVP** setting. It should be "0"
   (very important if you are using multiple viewports)
3. Add and subtract from the existing model to form the solid object below.
4. Save as **EX-15B**
5. **Do not Dimension**

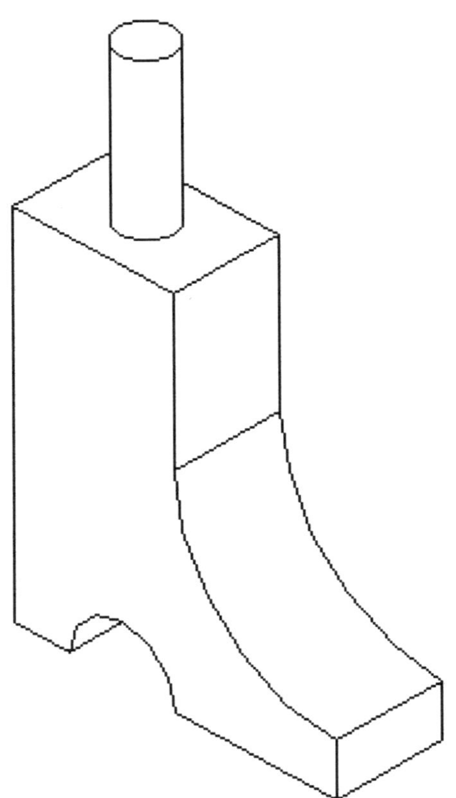

A few construction hints and dimensions are shown on the next page.

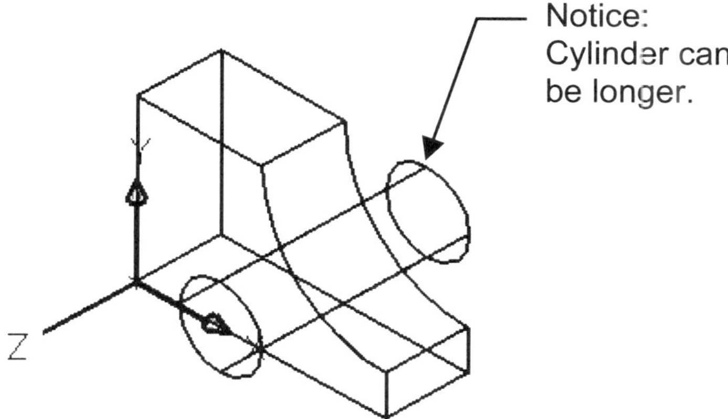

Step 1.
Subtract a Cylinder

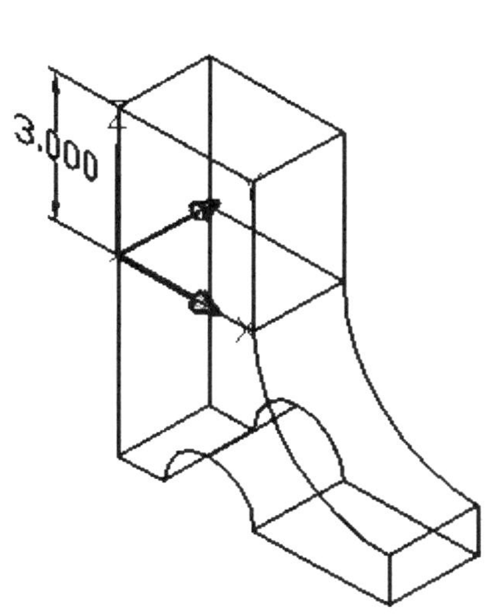

Step 2.
Add a solid Box.

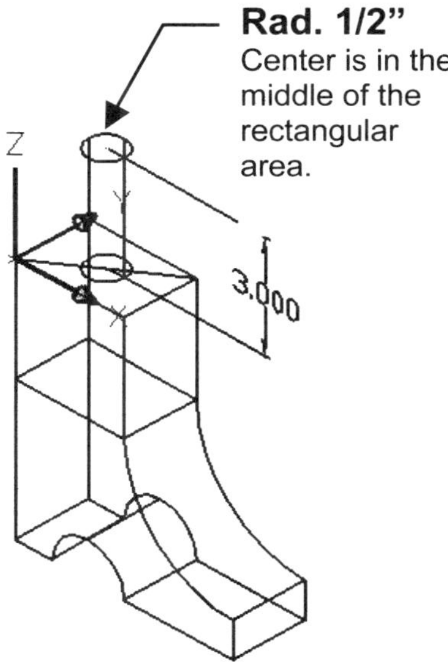

Step 3.
Add a Cylinder.

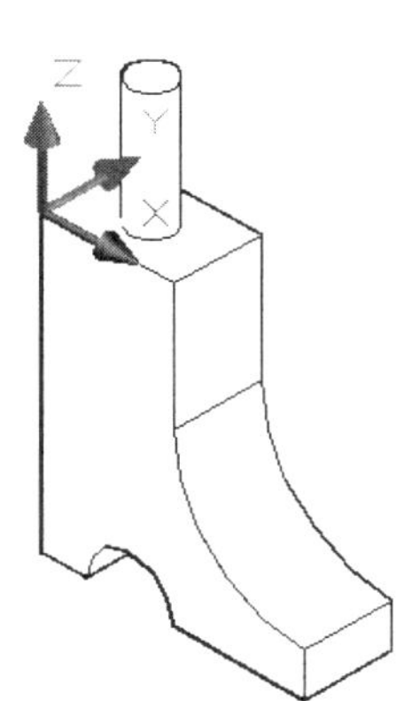

Step 4.
Union

EXERCISE 15C

Moving the UCS

1. **Select File / New and select Acad3d.dwt.**

2. Check the **UCSVP** setting. It should be "0"
   (very important if you are using multiple viewports)

3. Draw the 6" solid cube shown below

4. Add the text (Height = .50) on the outside of each side approximately as shown.

5. View as a 3D Wireframe to see if you did it correctly.

6. Save as **EX-15C**

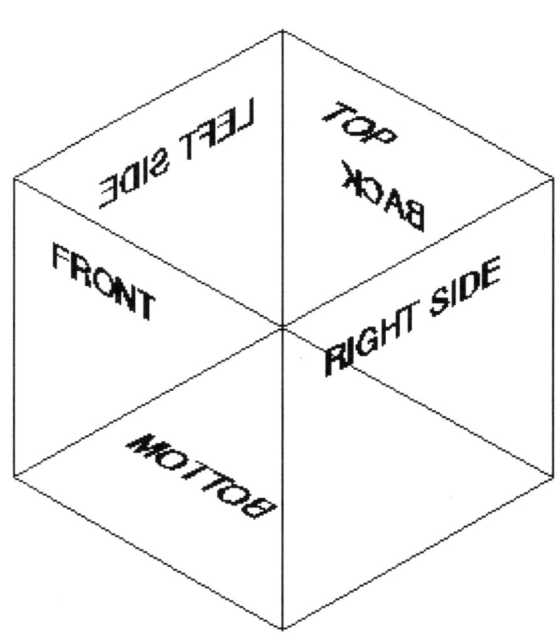

**This is a perfect exercise for using 2 viewports and the "Plan" view.
Use Isometric views or 3D Orbit to spin the cube around.
Use "3point" to move the Origin to reposition the XY plane or
consider using Dynamic UCS.
Oh so many choices!**

EXERCISE 15D

Assembling 3D solids

1. **Select File / New and select Acad3d.dwt.**
2. Change the **Units** to Architectural and the **Limits** to 60, 60 (inches).
3. Check the **UCSVP** setting. It should be "0"
   (very important if you are using multiple viewports)
4. Draw the table shown below
5. Save as **EX-15D**

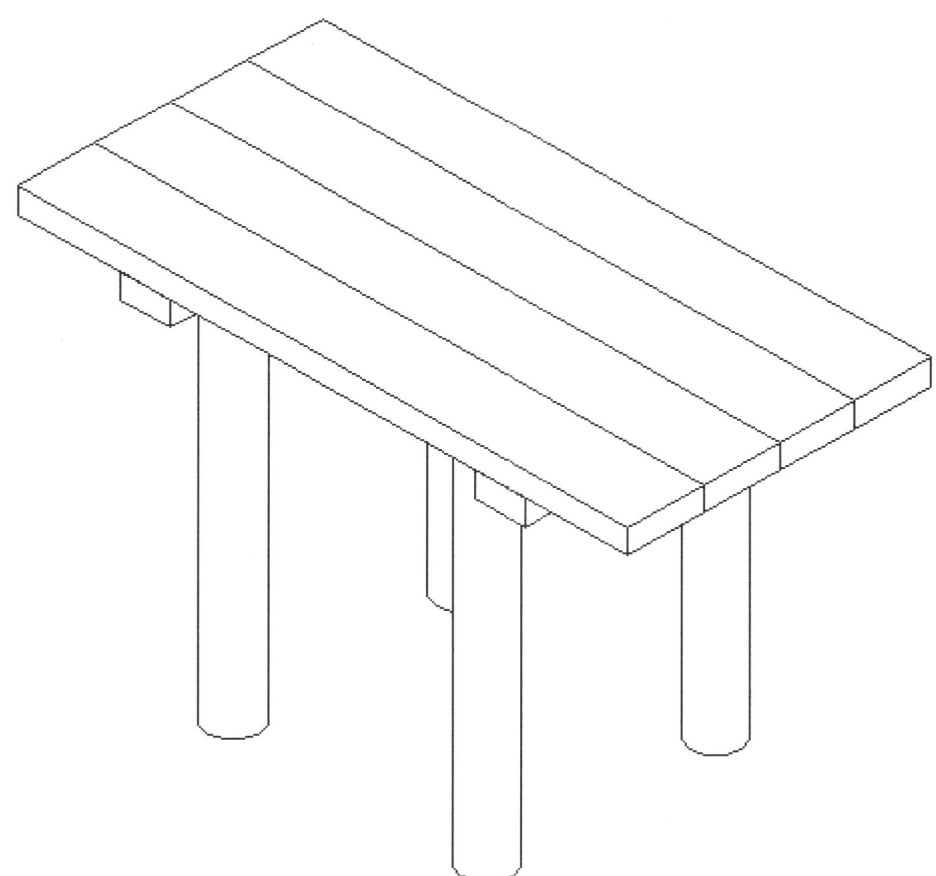

Dimensions and construction hints on the next page.
Have some fun with this.
If you have time try to add some benches.

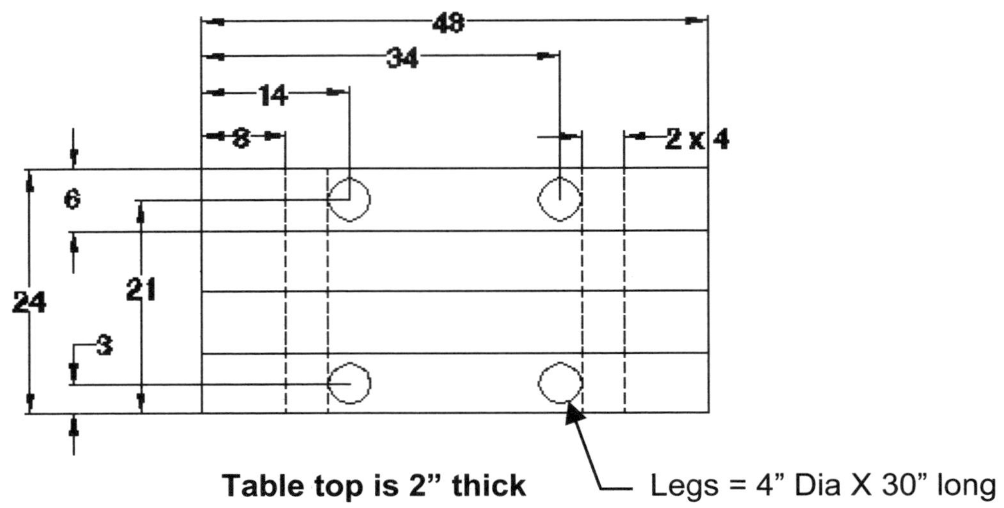

Table top is 2" thick

Legs = 4" Dia X 30" long

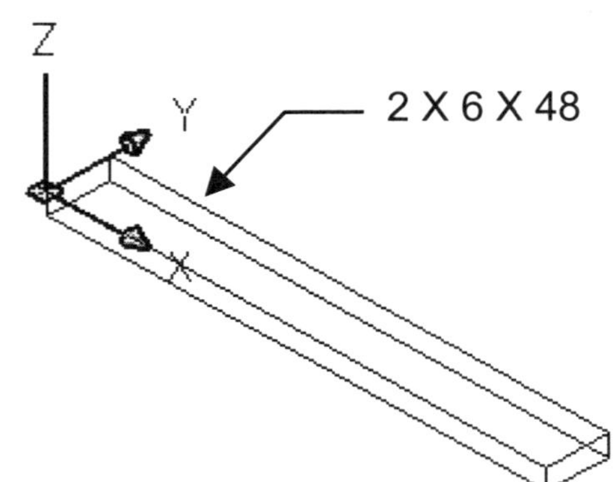

Start with a plank.

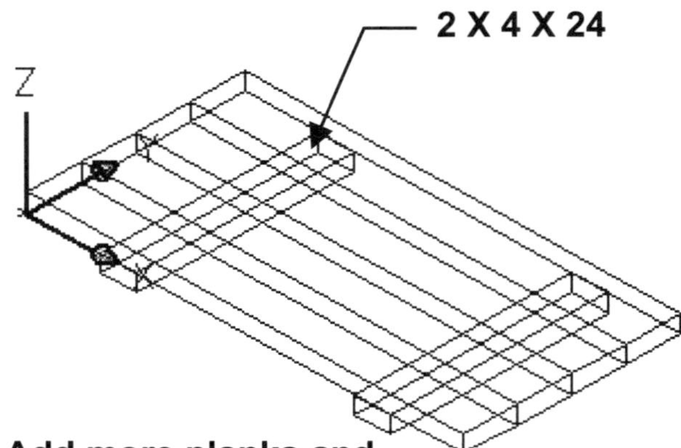

Add more planks and
the 2 x 4's (Notice the UCS position)

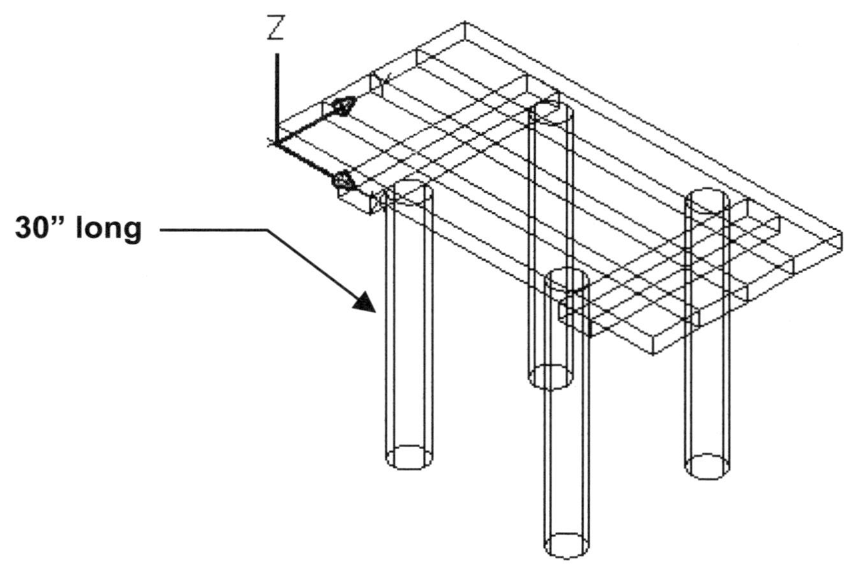

Add the <u>30" long</u> legs
Remember, to make the legs go in the negative Z axis

LEARNING OBJECTIVES

After completing this lesson, you will be able to:

1. Extrude solid surfaces.
2. Create a Region.
3. Review Polyline "Join" command.
4. Use Presspull command.

Sorry LT users, you do not have this option.

LESSON 16

EXTRUDE

AutoCAD's solid objects are helpful in constructing 3D solids quickly. But what if you want to create a solid object that has a more complex shape, such as the one shown here.

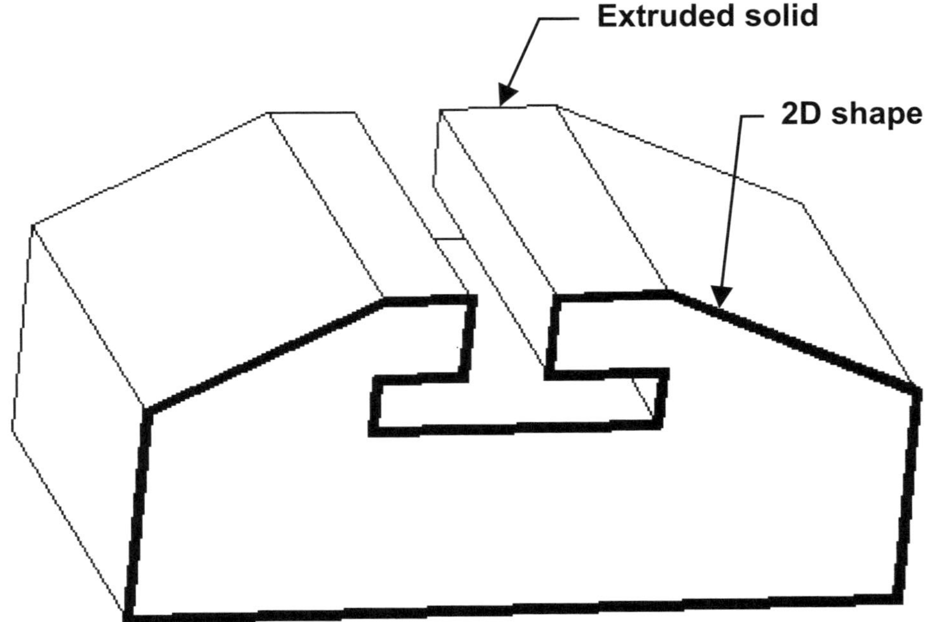

To create the solid shown above you first need to draw a **closed 2D shape** and then **EXTRUDE** it.

The **EXTRUDE** command allows you to take a 2D shape and extrude it into a solid. It extrudes along the Z-Axis. Extrusions can be created along a straight line or along a curved path. The extrusion may also have a taper.

You may EXTRUDE Polylines, Arcs, Circle, Polygon, Rectangle, Ellipses, Donuts, and Regions. (Splines can also be extruded but we are not discussing those in this workbook)

There are 4 methods for extruding a 2D shape:
1. Perpendicular to the 2D shape with straight sides.
2. Perpendicular to the 2D shape with tapered side.
3. Along a Path
4. Extrude a Region.

HOW TO SELECT THE EXTRUDE COMMAND

TYPE = EXT or EXTRUDE
PULLDOWN = DRAW / MODELING / EXTRUDE
3D MAKE CONTROL PANEL =

EXTRUDE - PERPENDICULAR WITH STRAIGHT SIDES

1. Select the **SE Isometric view.**
2. Draw a 2D shape on the XY plane.
3. Select the **EXTRUDE** command using one of the methods shown on page 16-2.
4. Select the objects: ***select the objects to extrude <enter>***
5. Select the objects: ***<enter>***
6. Specify height of extrusion or [Direction/Path/Taper angle]: ***type the height <enter>***

   (Note: a **positive** value extrudes **above the XY plane** and **negative** extrudes **below the XY plane,** along the Z axis.)

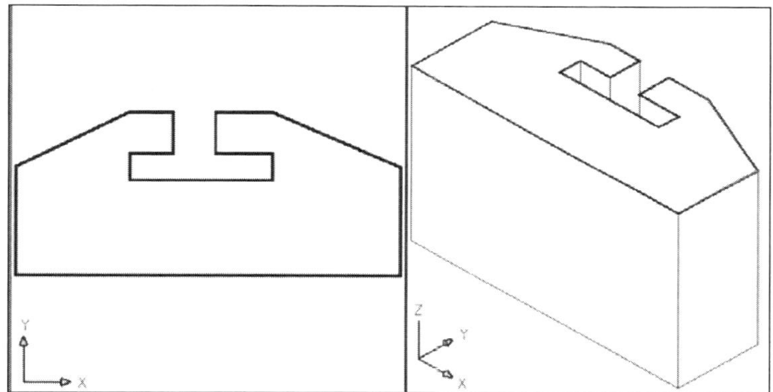

*This is a great time to use the **PLAN** method.*

Before drawing the 2D shape do the following:

1. Set the UCSVP setting to "0".
2. Split the screen into 2 viewports
3. Set the right hand viewport to **SE Isometric** view.
4. Rotate the UCS if necessary so Z axis is pointing in the correct direction if you are typing the height.
5. Set the left hand viewport to **PLAN** view.
6. Draw the 2D shape in the left hand viewport.
7. **Extrude**

EXTRUDE - <u>PERPENDICULAR WITH TAPERED SIDES</u>

1. Draw a 2D shape on the XY plane.
2. Select the **EXTRUDE** command
3. Select the objects: ***select the objects to extrude <enter>***
4. Select the objects: ***<enter>***
5. Specify height of extrusion or [Direction/Path/Taper angle]: ***T <enter>***
6. Specify angle of taper for extrusion <0>: ***enter taper angle <enter>***
7. Specify height of extrusion or [Direction/Path/Taper angle]: ***enter height <enter>***

How to control the taper direction
If you enter a positive angle the resulting extruded solid will taper inwards.
If you enter a negative angle the resulting extruded solid will taper outwards.

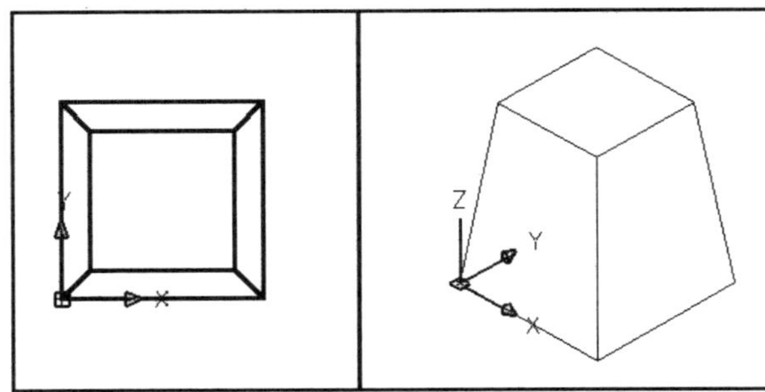

Positive angle

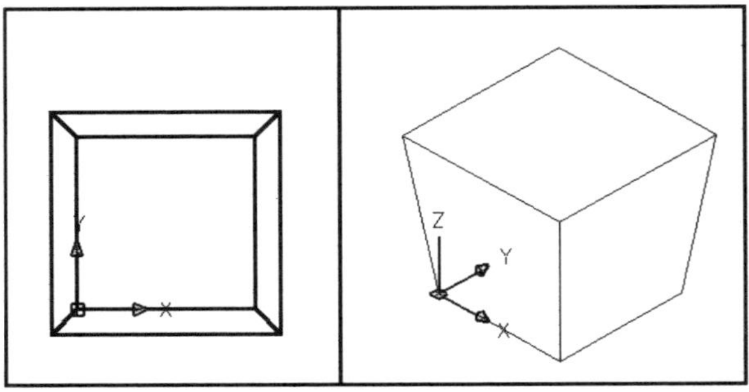

Negative angle

EXTRUDE - <u>ALONG A PATH</u>

Important: The path should be drawn perpendicular to the 2D shape.

1. Draw a 2D shape on the XY plane.
2. Draw the path perpendicular to the shape.
   (hint: rotate the UCS 90 degrees)
3. Select the **EXTRUDE** command.
4. Select the objects: ***select the objects to extrude <enter>***
5. Specify height of extrusion or [Direction/Path/Taper angle]: ***P <enter>***
6. Select extrusion path or [Taper angle]:***select the path (object)***

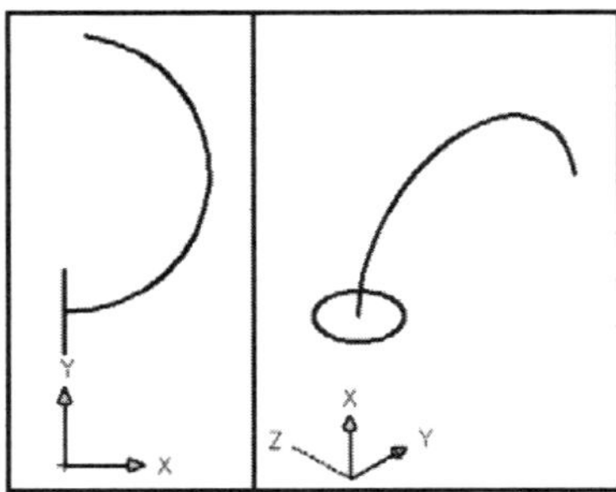

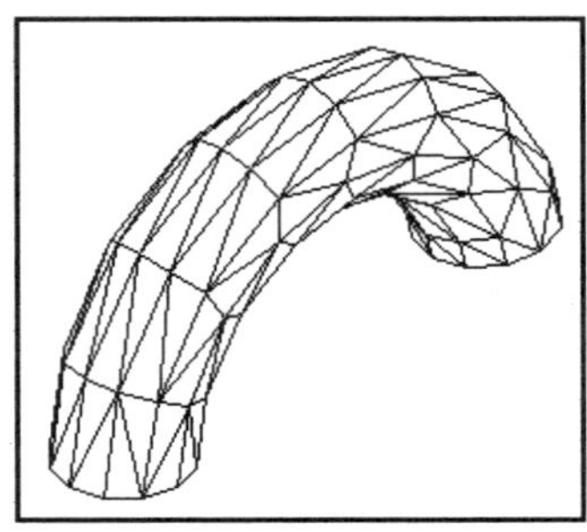

REGION

A **REGION** is a solid with no thickness. Think of a piece of paper. Thin but it is a solid. You can use all the Boolean operations, Union, Subtract and Intersect, on the region and you can extrude it.

Example:

1. You have a **2D** drawing of a flat plate with circles on it. You would like to extrude this plate and you want the circles to be actual holes.

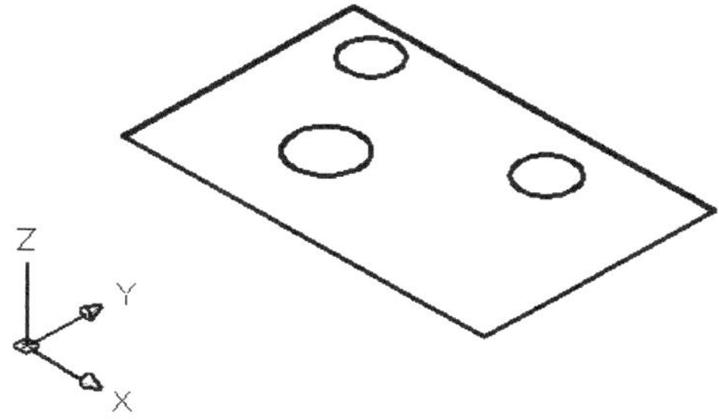

2. Use the **REGION** command to transform these objects into a solid with holes.

    a. Select the **REGION** command using one of the following:
        TYPE = REG or **REGION**
        PULLDOWN = DRAW / REGION
        TOOLBAR = DRAW

    b. Select objects: ***select the objects <enter>***
        Select objects: ***<enter>***
        4 loops extracted.
        4 Regions created.

3. Now subtract the circles from the rectangle, so they are actually "holes".

4. Extrude the region.

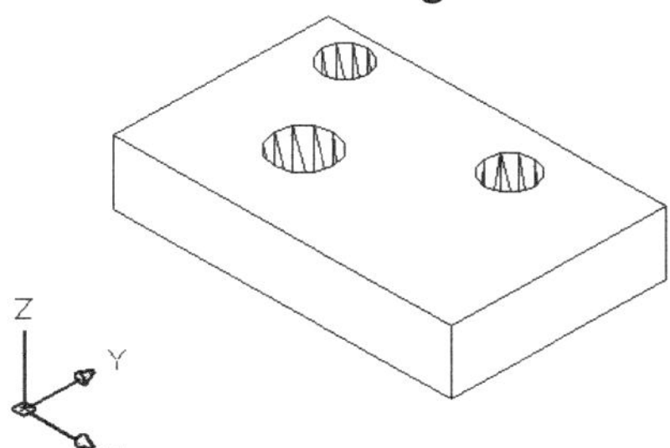

Which is more efficient?

1. Draw a 2D drawing, create a region and then extrude.

or

2. Draw a box and 3 cylinders then subtract cylinders from box.

Polyline's "JOIN" option

You learned how to edit Polylines in Lesson 24 of the Exercise Workbook for ***Beginning*** *AutoCAD. But you may not remember Polyline's "****JOIN****" option. This option is very useful when attempting to extrude a 2D shape, so I would like to discuss it again.*

The **JOIN** option allows you to join multiple polyline segments into one polyline.
The JOIN option also will transform a **LINE** into a Polyline. Touching Line segments are easier to control if you Extrude them separately or JOIN them into one polyline and Extrude all in one operation.

HOW TO USE THE JOIN COMMAND

1. Use the "Line" command to draw the shape below. You can even use "close" but it is still not considered a "closed object".

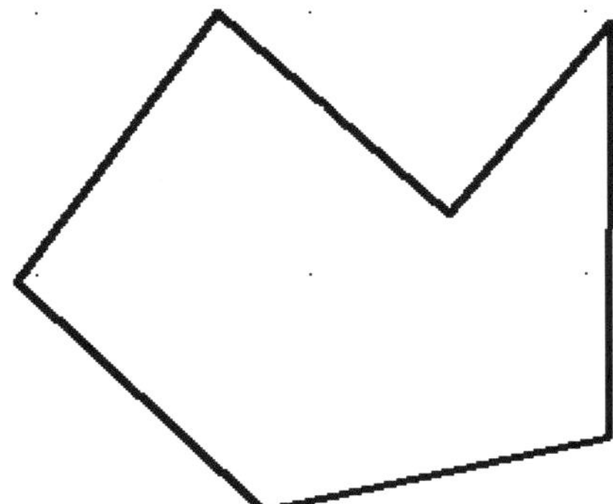

2. Select the Polyedit command using one of the following:

   TYPE = PEDIT
   PULLDOWN = MODIFY / OBJECT / POLYLINE
   TOOLBAR = MODIFY II

3. Command: _pedit Select polyline or [Multiple]: ***select ONE line segment.***
   Object selected is not a polyline (***if the object is not a polyline this appears)***
   Do you want to turn it into one? <Y> ***Y <enter>***

4. Enter an option [Close/Join/Width/Edit vertex/Fit/Spline/Decurve/Ltype gen/Undo]:***J <enter>***

5. Select objects: ***select first corner of window***

6. Specify opposite corner: ***select opposite corner of window***

7. Select objects: **<enter>**
   5 segments added to polyline

8. Enter an option [Open/Join/Width/Edit vertex/Fit/Spline/Decurve/Ltype gen/Undo]: ***<enter>***

Now you have transformed 6 LINES into a closed polyline. You are now able to extrude this shape more easily.

PRESSPULL COMMAND

The **PRESSPULL** command allows you to create a Region and Extrude a closed shape all in one operation. You may also use it to modify the height of an already extruded object. The area must be closed such as a circle, rectangle, polygon or closed polylines. (That is another reason that the JOIN command is important)
You may even select an area created by two overlapping objects.

1. Draw a shape.

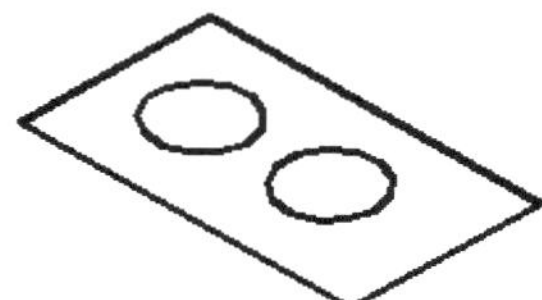

2. Move the UCS to the plane to extrude. (The plane to be extruded must be on the XY plane.

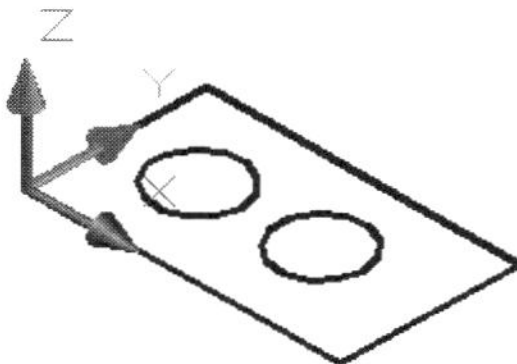

3. Select the Presspull command using:

   TYPE = PRESSPULL
   PULLDOWN = None
   3D Make Control Panel =

4. Click in the boundary just as you would to select a hatch boundary.

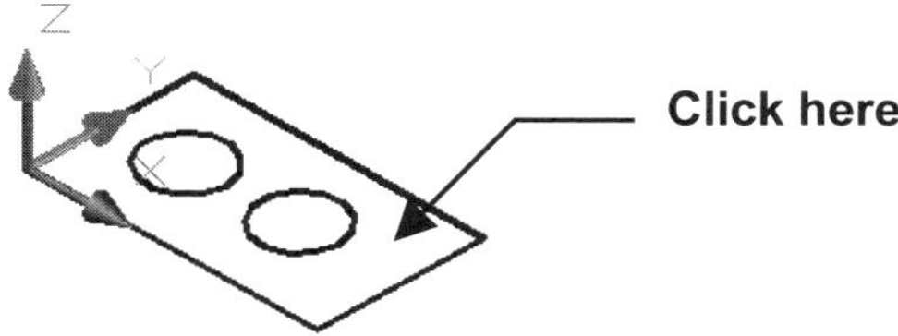

5. Move the cursor up or down in either direction to press or pull the region. You may click to determine the height or enter a value.

EXERCISE 16A

Extrude

1. **Select File / New and select <u>Acad3d.dwt</u>.**
2. Draw the 2D drawing approximately as shown below to form a closed Polyline. (Fig. 1)
   (Exact size is not important.)
3. Use Extrude or Presspull to form the solid object. Height = 4" (Fig. 2)
4. Select the **SE Isometric** view.
6. Save as **EX-16A**

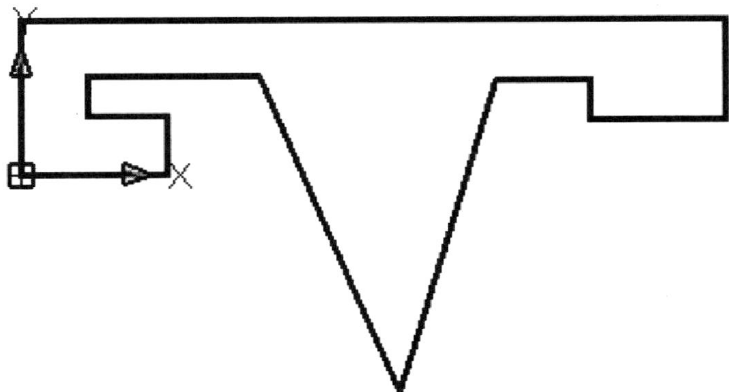

Figure 1

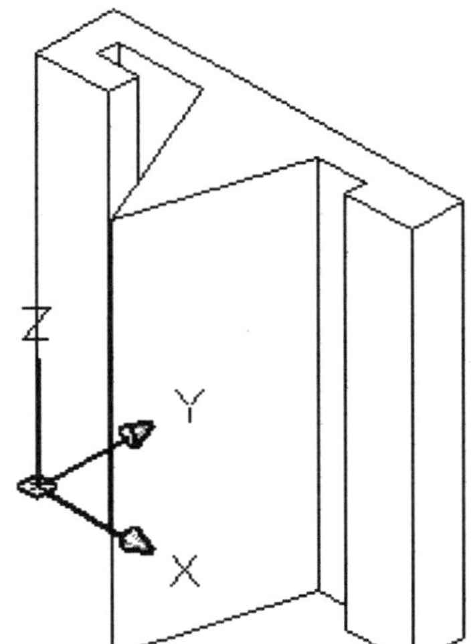

Figure 2

EXERCISE 16B

Extrude with Taper

1. **Select File / New and select <u>Acad3d.dwt</u>.**
2. Draw a 6 sided Polygon with a radius of 2 (Fig. 1)
3. Extrude to form the solid object. Ht = 4 Taper = 5 (Fig. 2)
4. Add the letter & Number shapes (Use polyline, do not use text).
   Extrude each. Ht = 1 (Fig. 3)
   Don't forget to use subtract on the shapes "4" , "6" and "B".

   Refer to the next page for construction suggestions

5. Select the **SE Isometric** view.
7. Save as **EX-16B**

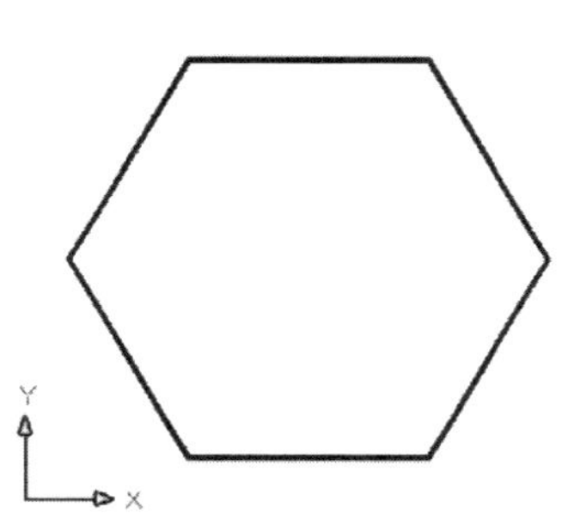

Figure 1

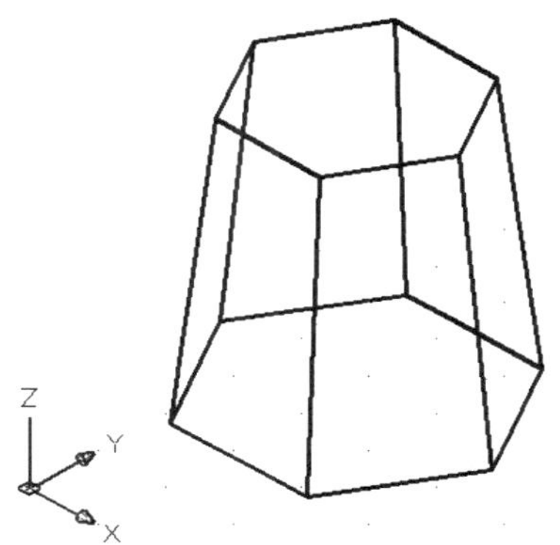

Figure 2

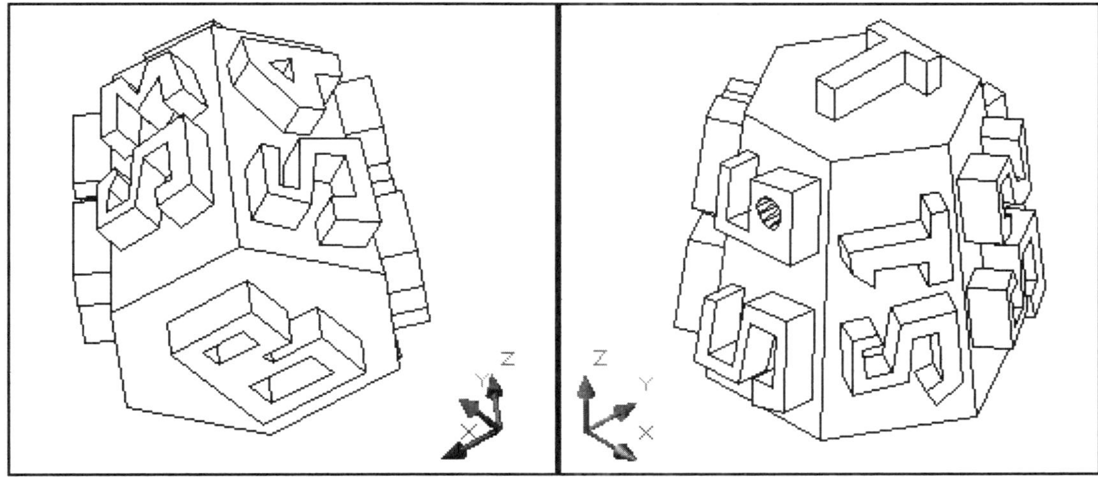

Figure 3

Construction suggestion:

a. After constructing the shape shown below, move the UCS, using 3 Point (pg 15-3), to select the correct surface to draw on.

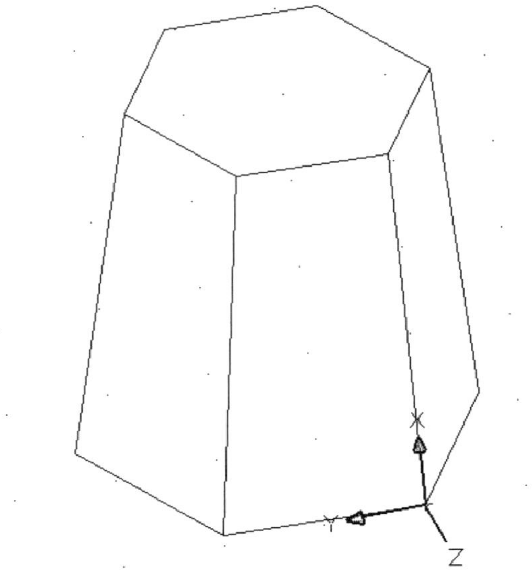

b. View the **current Plan view** to draw the text.
Type **Plan <enter> <enter>**

c. Extrude the polyline letter shapes.

d. Select **SE Isometric view** and then use **3D Orbit** to rotate around to the next surface.

EXERCISE 16C

Extrude along a Path

1. **Select File / New and select <u>Acad3d.dwt</u>.**
2. Draw a 6 sided Inscribed Polygon with a radius of 2 (Fig. 1)
3. Rotate the UCS, around the Y axis, -90. (Perpendicular to current UCS.)
4. Draw a polyline path perpendicular to the Polygon approximately as shown (Fig. 2)
5. Extrude the Polygon along the path (Fig. 3)
6. Select the **SE Isometric** view.
7. Save as **EX-16C**

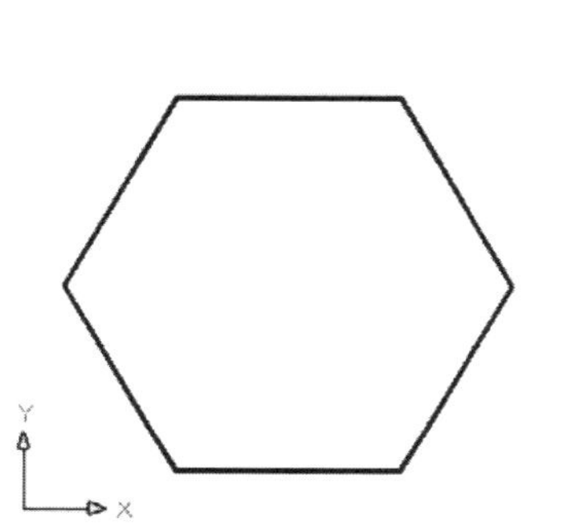

Figure 1

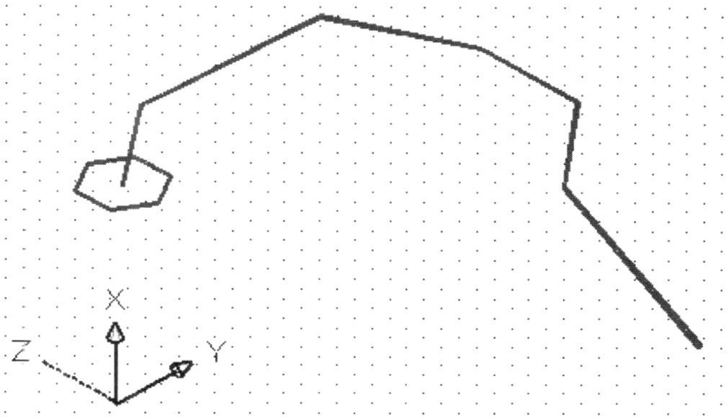

Figure 2

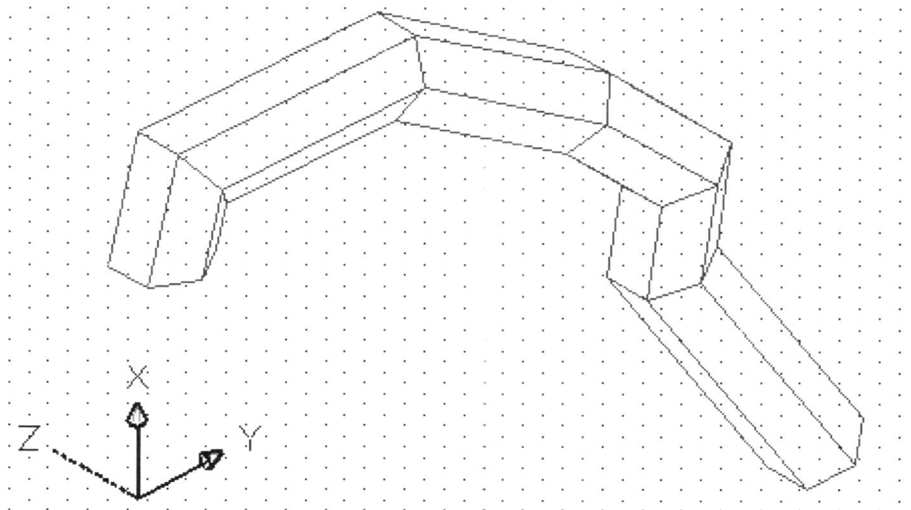

Figure 3

EXERCISE 16D

Extrude or Presspull a Region

1. **Select File / New and select Acad3d.dwt.**
2. Draw the 2D drawing shown below. (Fig. 1)
   Note: Orientation must be correct so you may have to rotate the UCS.
4. Create a “Region” (pg. 16-5) and Extrude or use “Presspull” command to form the solid object (Fig. 2). Height = 4”
5. Save as **EX-16D**

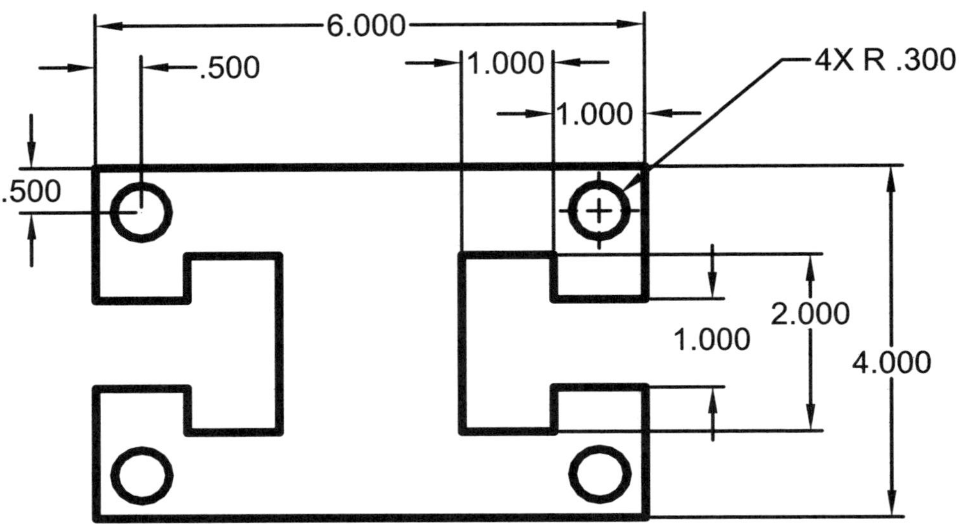

Figure 1

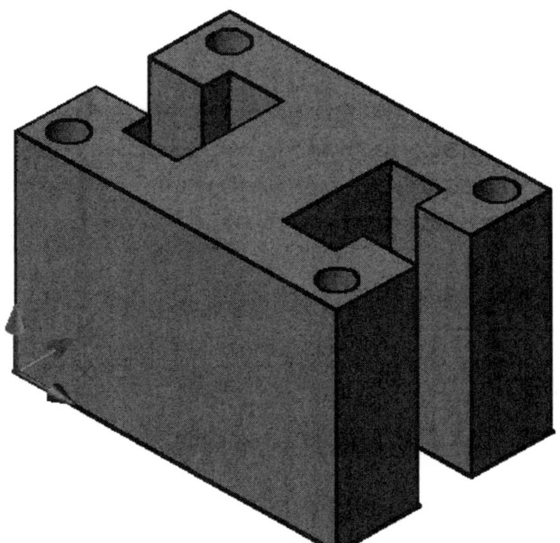

Figure 2

EXERCISE 16E

Extrude or Presspull a Region

1. Open **EX-5D**
2. **Freeze** all layers except “Walls”. (Hopefully you used the correct layers when drawing 5D.)
3. Select the **SE Isometric** View.
4. Use the “**REGION**” command (refer to page 16-5) to create 4 regions.

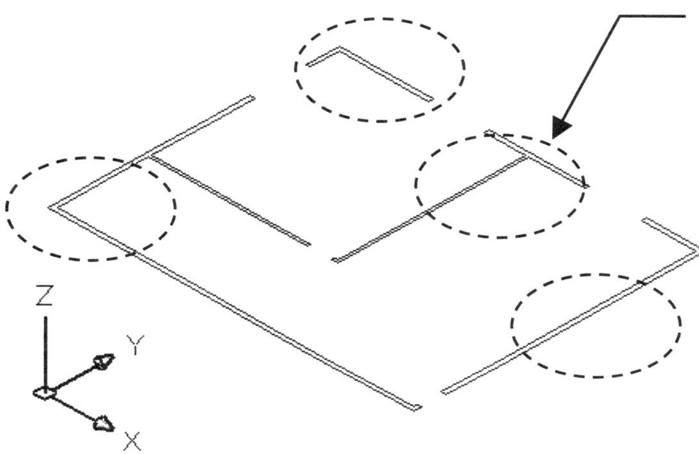

Note: The missing doors and windows cause a separation in the walls. You must use “Join” or “Region” on each wall section.

5. EXTRUDE or Presspull the walls to a height of 8 feet.
6. Save as **EX-16E**

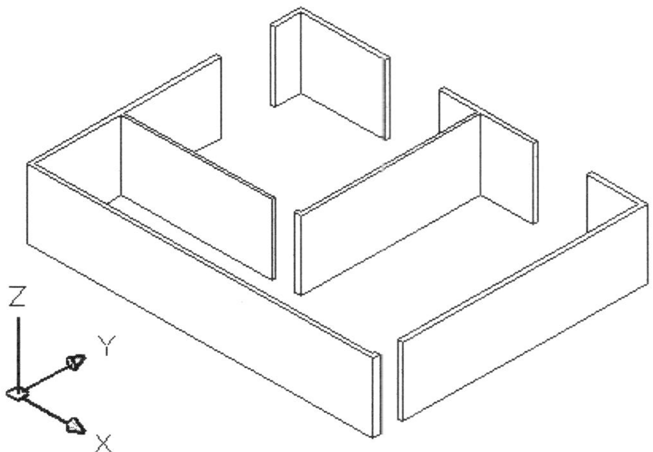

To add walls or windows you can use the Union or Subtract.
Have some fun with this one. Try adding the furniture.

DELOBJ SYSTEM VARIABLE

The **DELOBJ** system variable determines whether the object used when extruding, revolving, sweeping or presspulling, is retained or deleted. For example, when you extrude a circle to make a cylinder, what should happen to the original circle? Do you want to delete it or do you want to keep it?

Set the **DELOBJ** system variable before using the Modeling command.

Options:

0 = Retains all source objects.

1 = This is the default setting. Deletes profile curves. For example, if you use a circle to create a cylinder, the circle will be deleted. Cross sections and guides used with the SWEEP command are also deleted. However, if you extrude along a path, the path is not deleted.

2 = Deletes all defining objects, including paths

-1 = Just like "1" except you get a prompt to allow you to choose.

-2 = Just like "2" except you get a prompt to allow you to choose.

LEARNING OBJECTIVES

After completing this lesson, you will be able to:

1. Understand 3D Operations.
   Mirror 3D, 3D Rotate, 3D Align and 3D Array

LESSON 17

3D OPERATIONS

3D Operations allow you to **3D Mirror, 3D Rotate, 3D Align** and **3D Array** a solid. The methods are almost identical to the 2D commands with the exception of defining the plane. <u>You do not have to move the UCS.</u>

Note: You may use the equivalent 2D commands but they will only work in the XY plane. <u>It may be necessary to move the UCS.</u>

THE FOLLOWING ARE EXPLANATIONS FOR THE <u>3D COMMANDS.</u>

3D MIRROR

You must define the mirror plane.

1. Select the **3D MIRROR** command using one of the following:

    TYPE = **3DMIRROR**
    PULLDOWN = **MODIFY / 3D OPERATIONS / 3D MIRROR**
    TOOLBAR = **NONE**

2. Select objects: ***select the solid to be mirrored***

3. Select objects: ***<enter>***

4. Specify first point of mirror plane (3 points) or [Object/Last/Zaxis/View/XY/YZ/ZX/3points] <3points>: ***<enter>***

5. Specify first point on mirror plane: ***select first point on mirror plane (P1)***

6. Specify second point on mirror plane: ***select second point on mirror plane (P2)***

7. Specify third point on mirror plane: ***select third point on mirror plane (P3)***

8. Delete source objects? [Yes/No] <N>:***select Y or N <enter> (the example is NO)***

EXAMPLE:

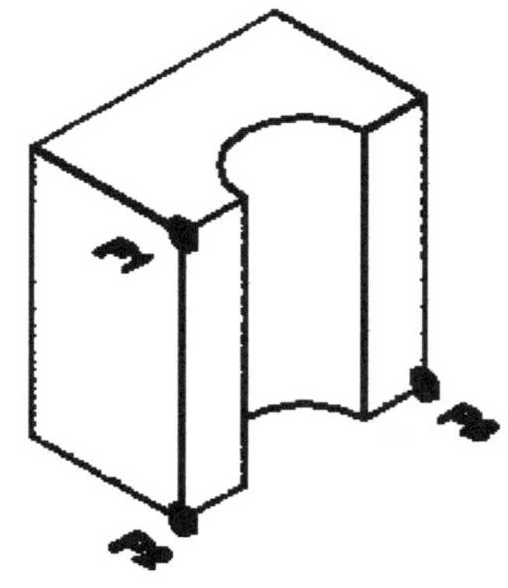

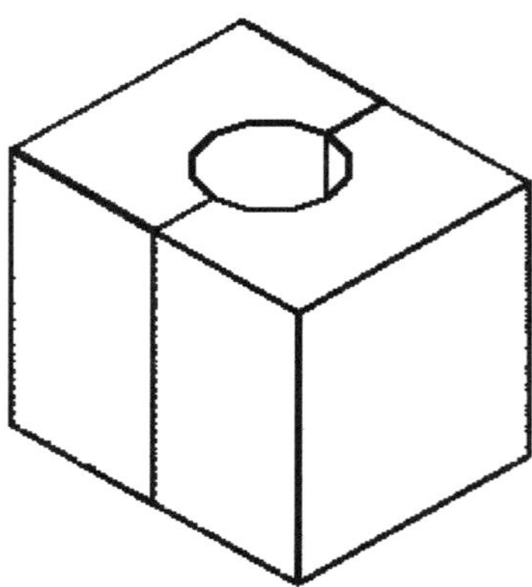

3D ROTATE

You must pick 2 points to define the axis of rotation and the rotation angle. To determine the rotation angle you must look down the axis from the second point.

If you use my stabbing arrow analogy, you will put the positive end of the second point selected, in your chest and rotate the other two axes. <u>*Positive input is counterclockwise and Negative input is clockwise.*</u>

1. Select the **3D ROTATE** command using one of the following:

    TYPE = **3DROTATE**
    PULLDOWN = **MODIFY / 3D OPERATIONS / 3D ROTATE**
    3D MAKE CONTROL PANEL =

2. Current positive angle: ANGDIR=counterclockwise ANGBASE=0
   Select objects: ***select the solid to be rotated***

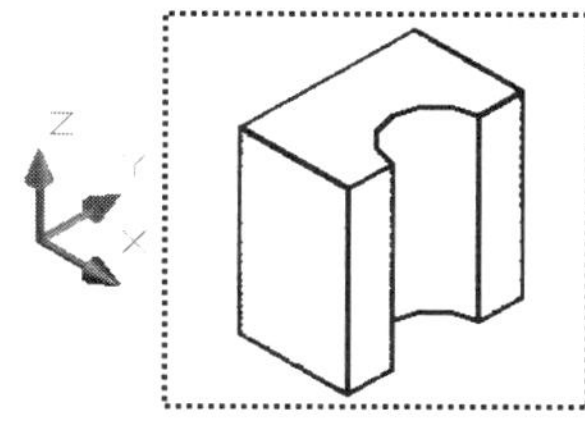

3. Select objects: ***<enter> to stop***

4. Specify Base Point: ***Snap to the base point of the rotation***

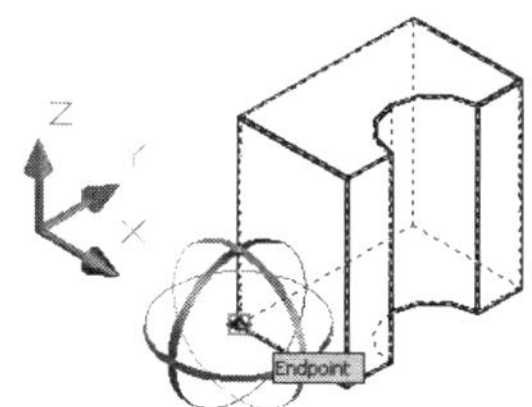

5. Pick a Rotation axis: ***Touch the ribbons until the correct line appears then click***

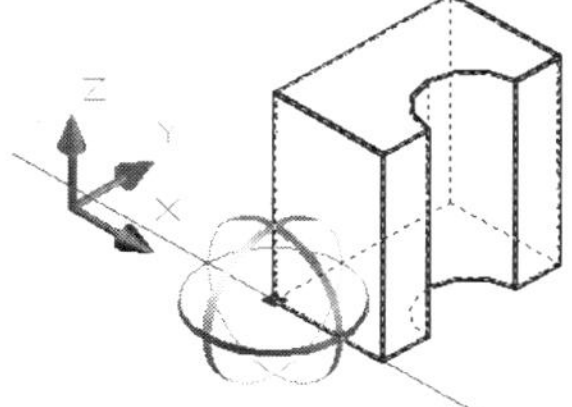

6. Specify angle start point: ***type rotation angle <enter>***

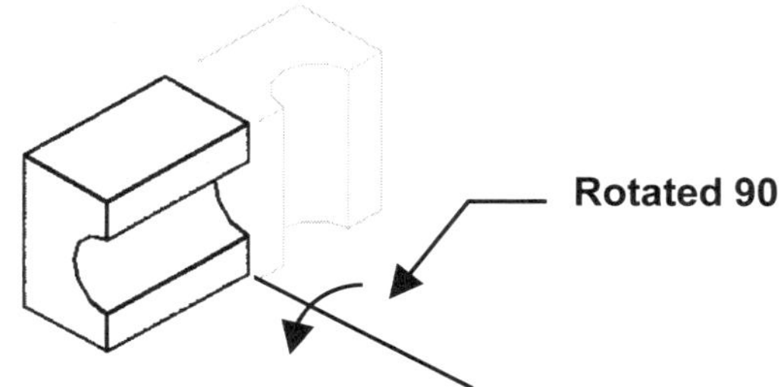

3D ALIGN

The **3D ALIGN** command allows you to MOVE and ROTATE an object from its existing location to a new location.

You define the two or three points on the **SOURCE** object and then the points on the **DESTINATION** location.

1. Select the **3D ALIGN** command using one of the following:

    TYPE = **3DALIGN**
    PULLDOWN = **MODIFY / 3D OPERATIONS / 3D ALIGN**
    TOOLBAR = **NONE**

2. Select objects: ***select the object to be aligned***

3. Select objects: ***<enter>***

4. Specify source plane and orientation ...
   Specify base point or [Copy]:: ***select the 1st source (S1)***

5. Specify second point or [Continue] <C>: ***select the 2nd source (S2)***

6. Specify third point or [Continue] <C>: ***select the 3rd source (S3)***

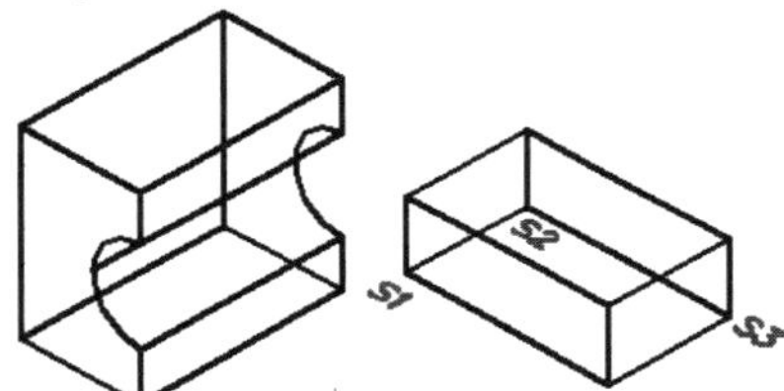

7. Specify destination plane and orientation ...
   Specify first destination point: ***select the 1st destination point (D1)***

8. Specify second destination point or [eXit] <X>: ***select the 2nd destination point (D2)***

9. Specify third destination point or [eXit] <X>: ***select the 3rd destination point (D3)***

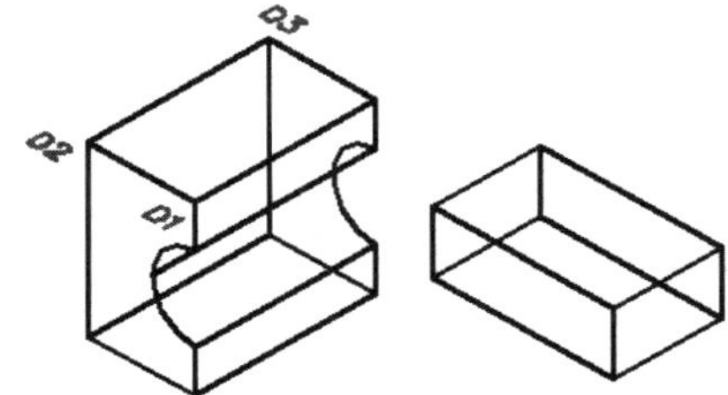

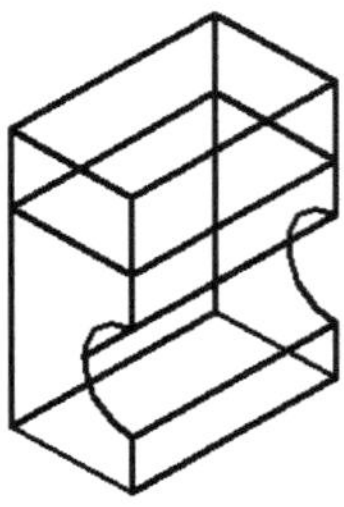

Completed alignment

3D ARRAY (Rectangular and Polar)

RECTANGULAR ARRAY

You can array any object on the current XY Plane.
You define the following:

ROWS = number of copies needed in the **Y** direction
COLUMNS = number of copies needed in the **X** direction
LEVELS = number of copies needed in the **Z** direction
DISTANCE BETWEEN ROWS, COLUMNS and LEVELS

1. Select the **3DARRAY** command using one of the following:

   TYPE = **3A** or **3DARRAY**
   PULLDOWN = **MODIFY / 3D OPERATIONS / 3D ARRAY**
   TOOLBAR = **NONE**

2. Select objects: ***select object to be arrayed***

3. Enter the type of array [Rectangular/Polar] <R>: ***R <enter>***

5. Enter the number of rows (---) <1>: ***type number of rows (Ydirection) <enter>***

6. Enter the number of columns (|||) <1>: ***type number of columns (X direction) <enter>***

7. Enter the number of levels (...) <1>: ***type number of levels (Z direction) <enter>***

8. Specify the distance between rows (---): ***type distance between rows<enter>***

9. Specify the distance between columns (|||): ***type distance between columns<enter>***

10. Specify the distance between levels (...): ***type distance between levels <enter>***

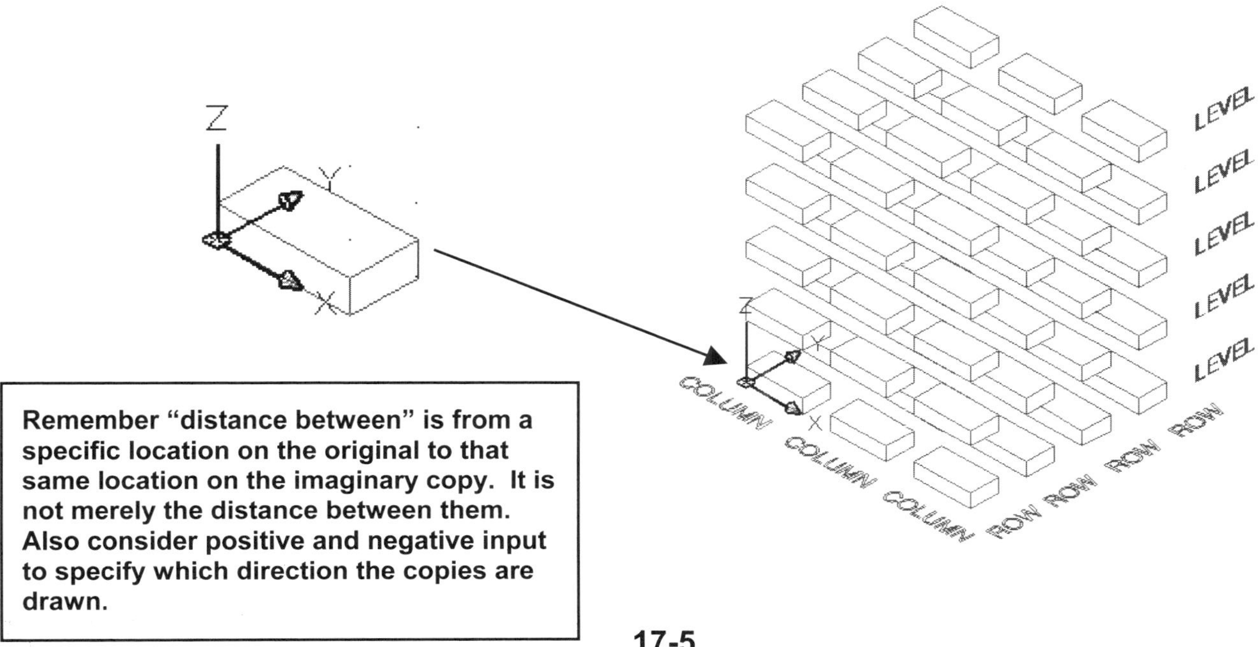

Remember "distance between" is from a specific location on the original to that same location on the imaginary copy. It is not merely the distance between them. Also consider positive and negative input to specify which direction the copies are drawn.

POLAR ARRAY

The object is arrayed around an entire axis rather than just a point.

You define the following:

NUMBER OF COPIES
ANGLE TO FILL
ROTATION DIRECTION
ROTATION AXIS ENDPOINTS

1. Select the **3D ARRAY** command using one of the following:

   TYPE = **3A** or **3DARRAY**
   PULLDOWN = **MODIFY / 3D OPERATIONS / 3D ARRAY**
   TOOLBAR = **NONE**

2. Select objects: ***select object to be arrayed***
3. Select objects: **<enter>**
4. Enter the type of array [Rectangular/Polar] <R>: ***P <enter>***
5. Enter the number of items in the array: ***type the number of copies (include original)***
6. Specify the angle to fill (+=ccw, -=cw) <360>: ***enter angle for copies to fill***
7. Rotate arrayed objects? [Yes/No] <Y>: ***copies rotated, yes or no?***
8. Specify center point of array: <Osnap on>***snap to first endpoint of axis (P1)***
9. Specify second point on axis of rotation:***snap to second endpoint of axis (P2)***

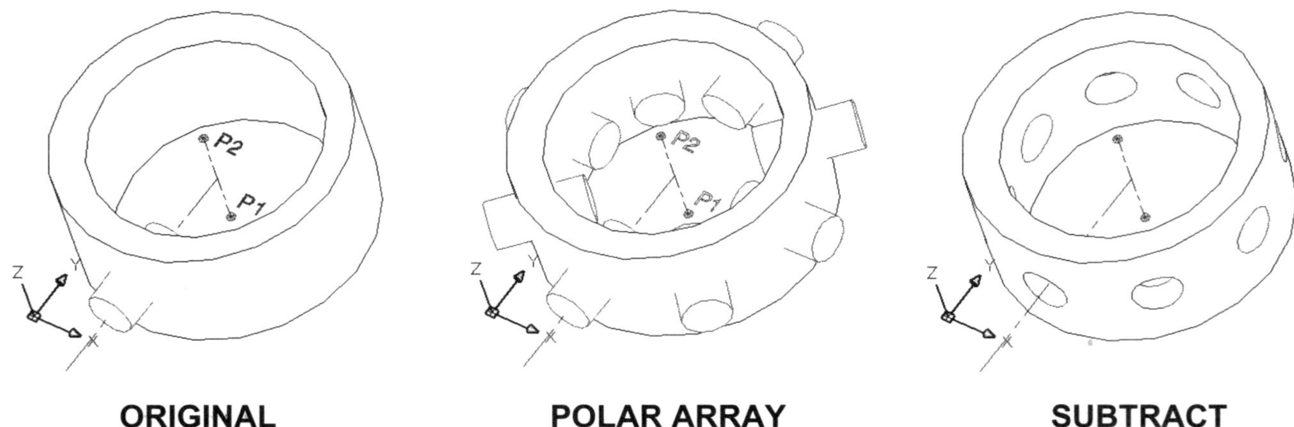

ORIGINAL **POLAR ARRAY** **SUBTRACT**

Note: To select the 1st (P1) and 2nd (P2) points on the axis you could use object snap "center" to snap to the center of the top circle and then the bottom circle. Just want to keep you thinking.

EXERCISE 17A

3D MIRROR

1. **Select File / New and select <u>Acad3d.dwt</u>.**
2. Draw the solid object below.
3. Dimensions and construction hints on page 17-8.
4. Save as **EX-17A**

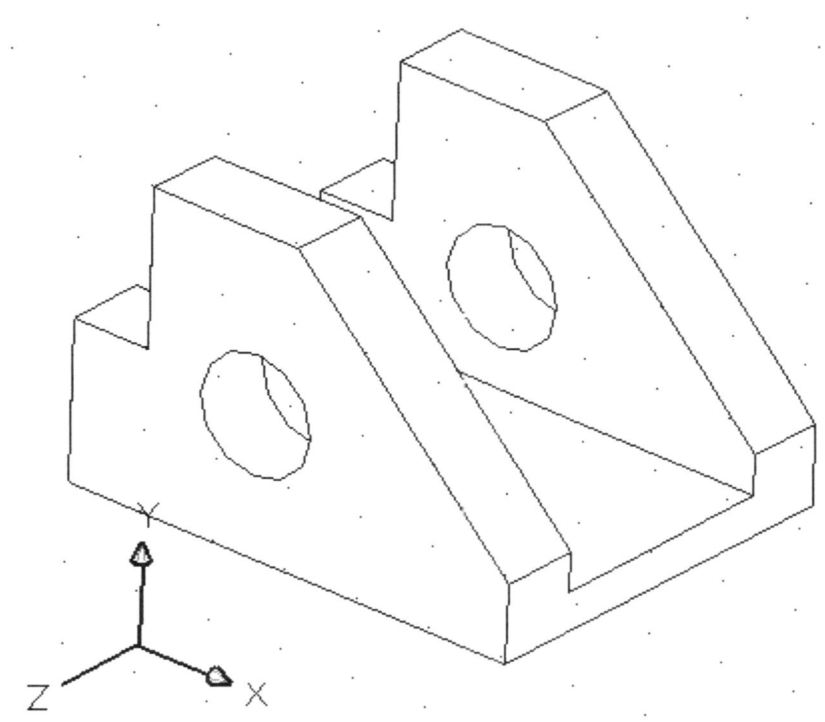

Drawing hints on the next page.

Step 1. Draw the side profile using Plan view.

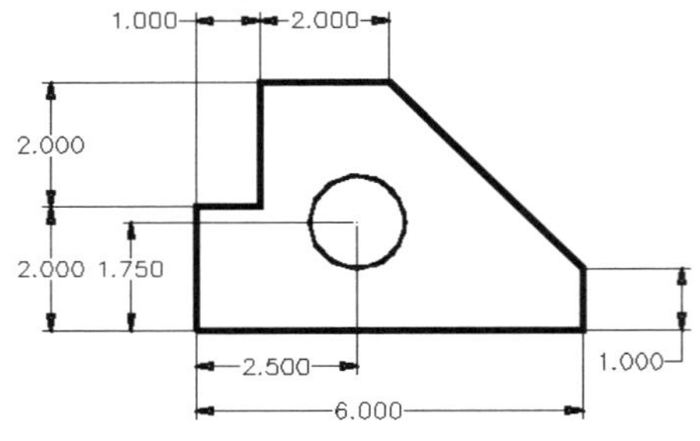

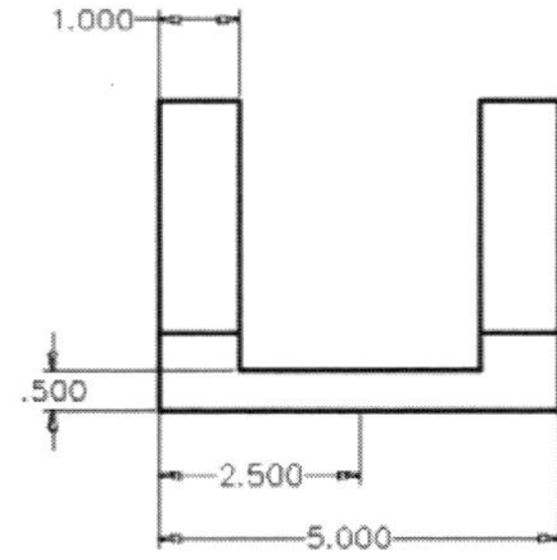

Step 2. Presspull

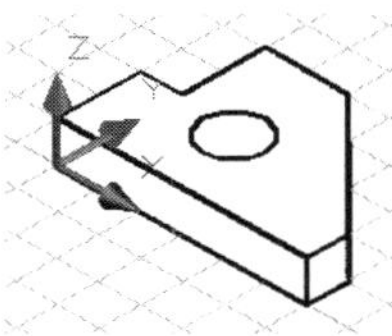

Step 3. Add center divider.

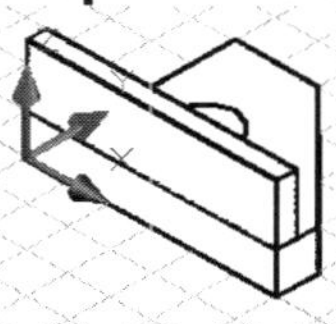

Step 4. Rotate

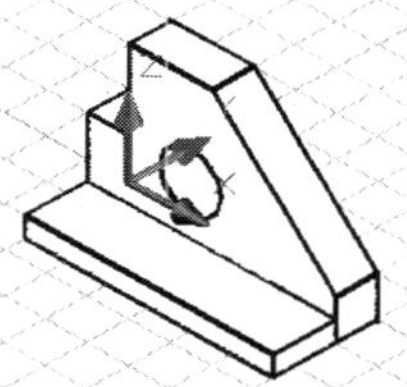

Step 5. Mirror

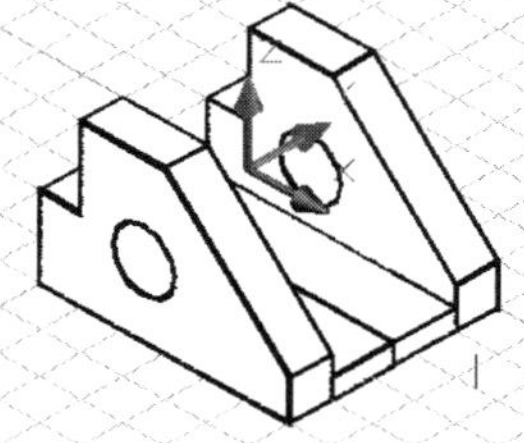

Step 6. Union

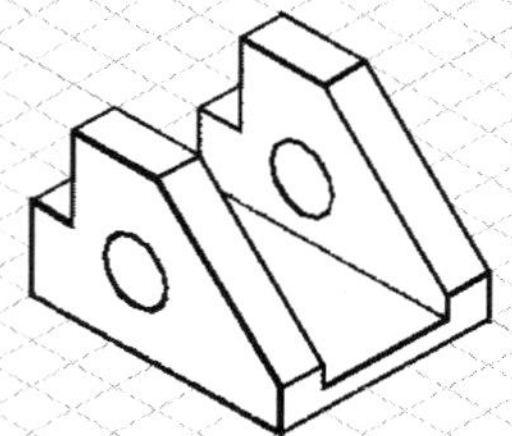

EXERCISE 17B

3D ROTATE

1. Open **EX-17A** (Figure 1)
2. Rotate the solid model 180 degrees as shown in Figure 2
3. Save as **EX-17B**

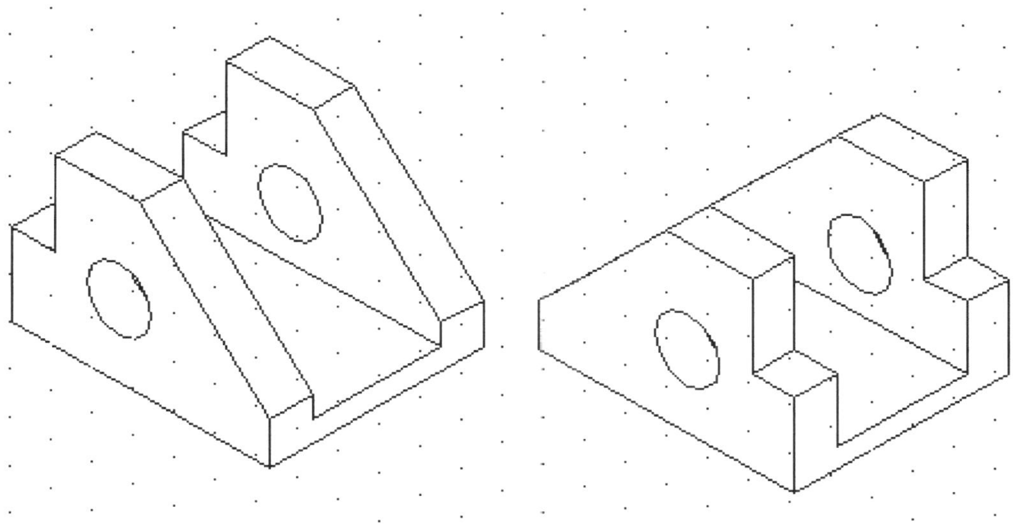

Figure 1 **Figure 2**

EXERCISE 17C

3D ALIGN

1. **Select File / New and select <u>Acad3d.dwt</u>.**
2. Draw the 3 solid objects shown below in Figure 1 (Use the dimensions shown)
3. Using the **3D ALIGN** command, assemble the objects as shown in Figure 2.
4. Save as **EX-17C**

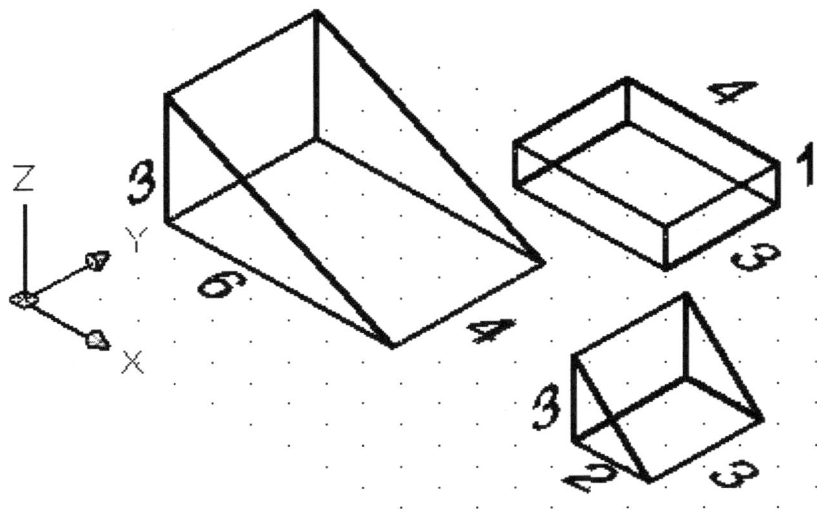

Figure 1

Note: It will help if you display 3D Wireframe while selecting source points.

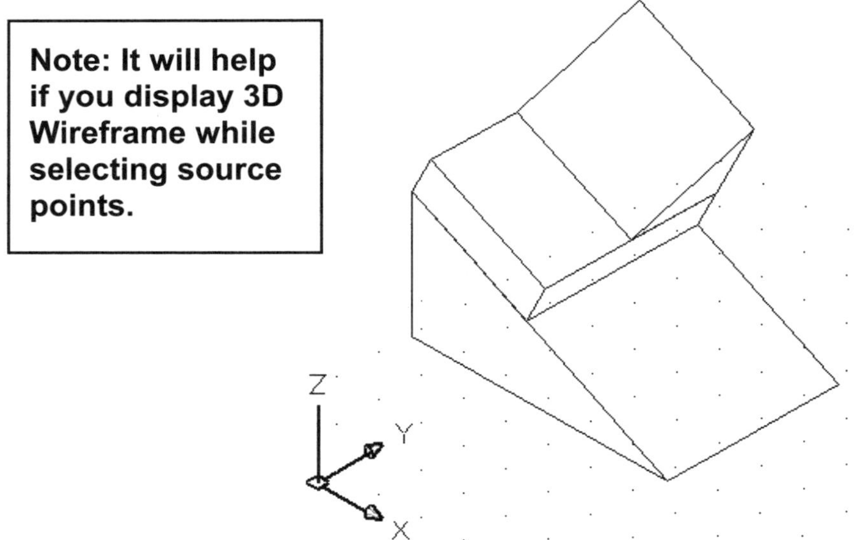

Figure 2

EXERCISE 17D

2D ARRAY

1. **Select File / New and select <u>Acad3d.dwt</u>.**
2. Draw the box shown below in Figure 1 (Use the dimensions in Figure 2).
3. Try the **2D** Array command to draw the holes.
4. Create a Region, Subtract the circles and Extrude. (Refer to lesson 16)
   Or select any method you prefer.
5. Save as **EX-17D.**

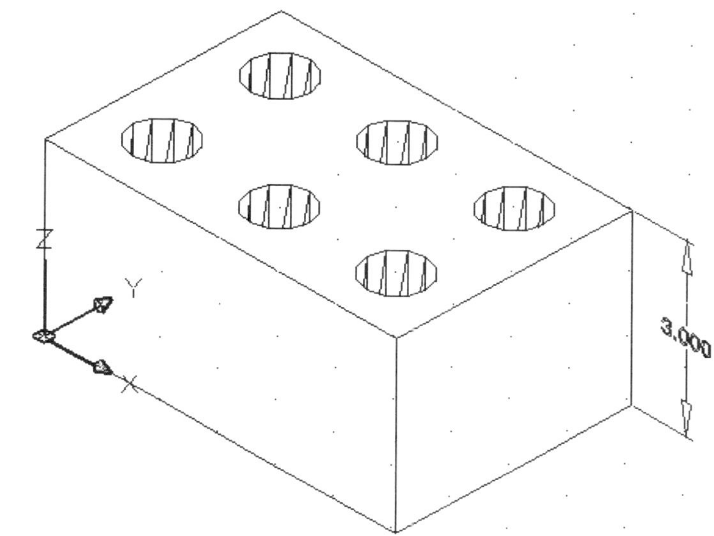

Figure 1

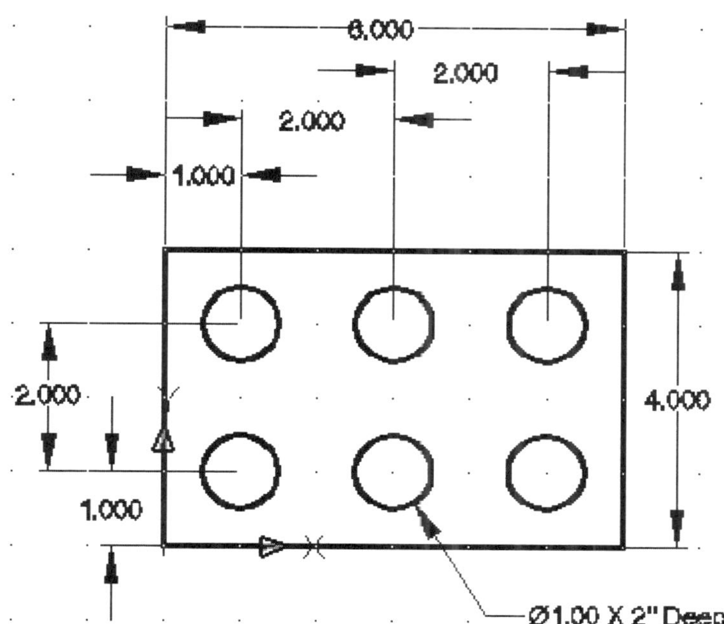

Figure 2

EXERCISE 17E

3D ARRAY - RECTANGULAR

1. **Select File / New and select Acad3d.dwt.**
2. Draw the solid objects shown below in Figure 1 (Use the dimensions shown at the bottom of the page)
3. Try the **3D Array** command to create the solid model in Figure 2.
4. Save as **EX-17E**

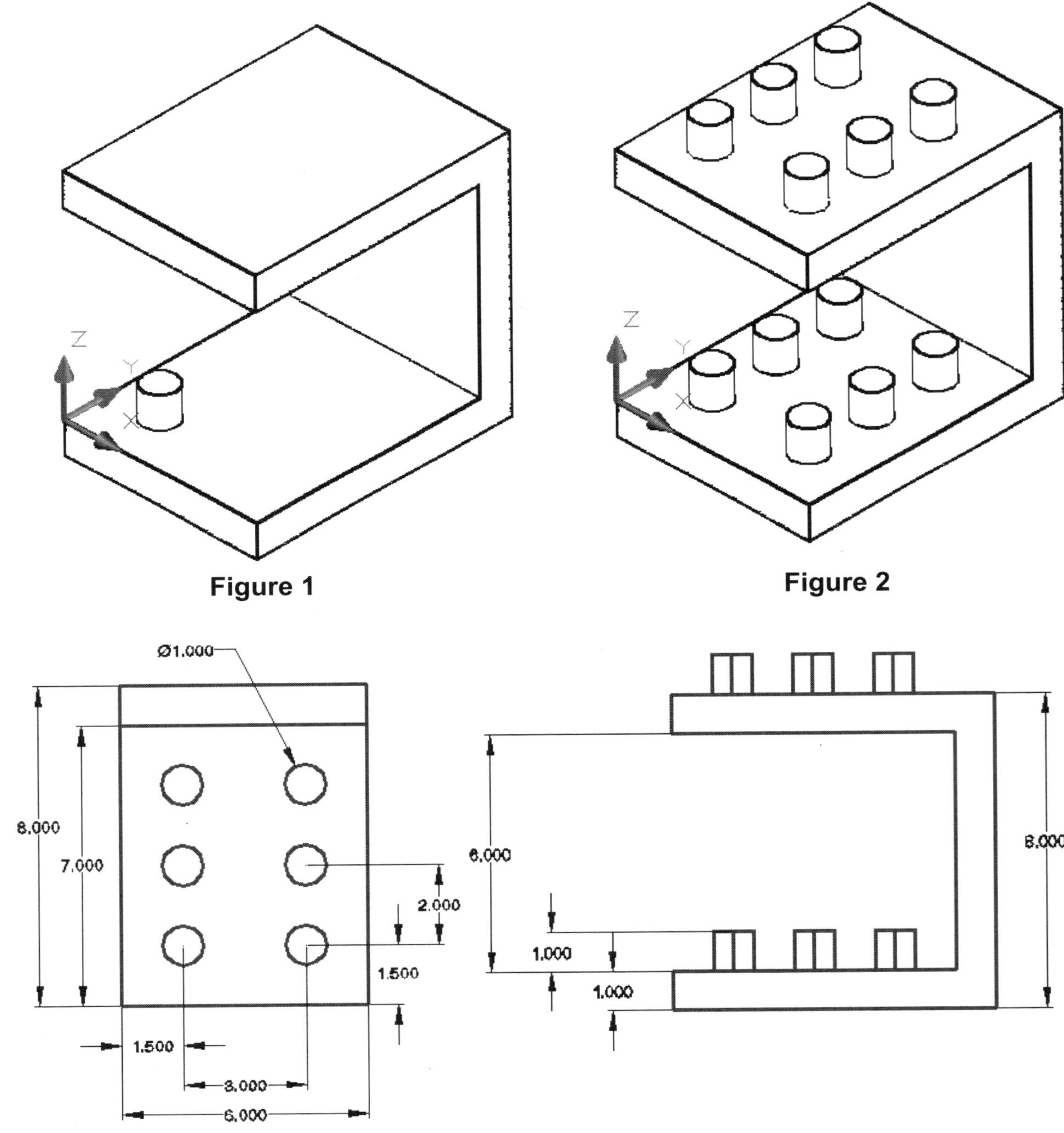

EXERCISE 17F

3D ARRAY - POLAR

1. **Select File / New and select <u>Acad3d.dwt</u>.**
2. Create the solid cylinder with 8 holes shown below
   (Dimensions and construction hints on page the next page)
3. Save as **EX-17F**

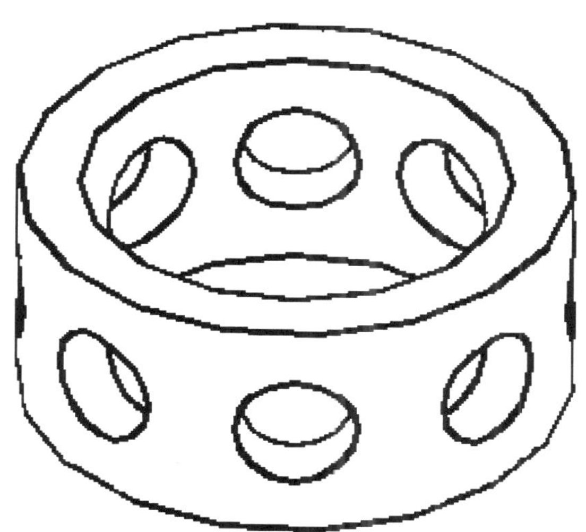

Drawing hints on the next page and refer to page 17-6

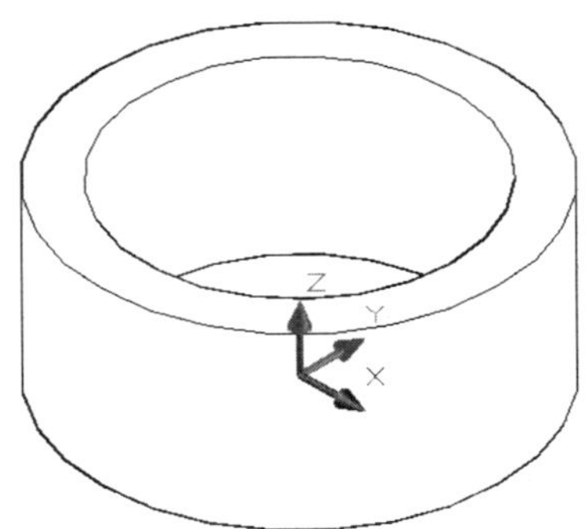

Step 1. Draw 2 Cylinders. 9" Dia and 7" Dia. Height = 4"
Subtract the small from the large.

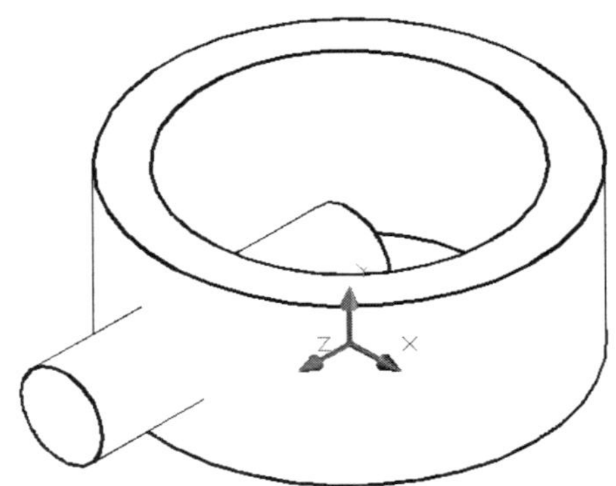

a. Rotate the UCS
b. Draw Cylinder 2" up from the bottom.
Height = 7"
(You really have to think about this)

Step 2. Draw a 2" Dia Cylinder

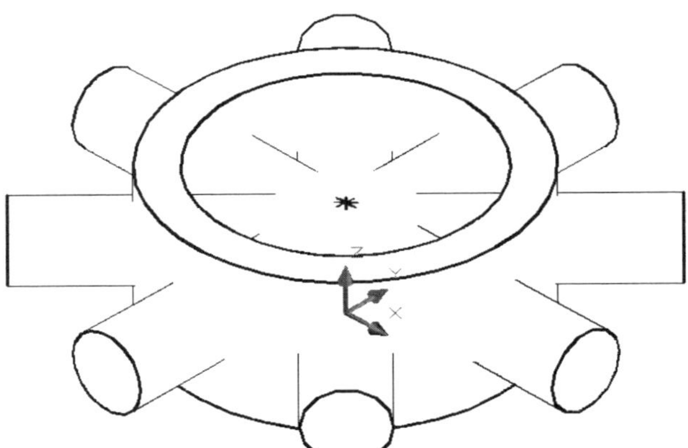

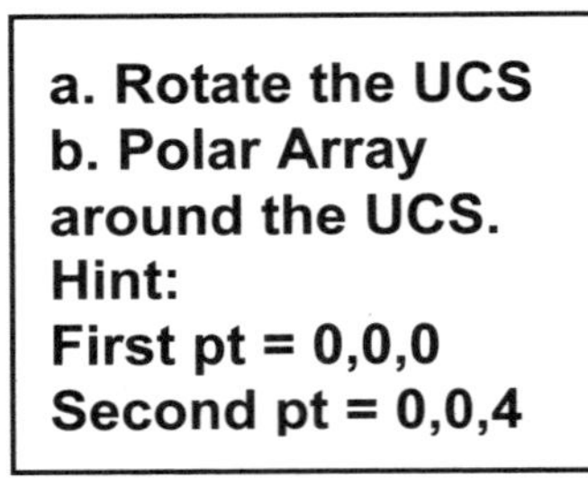

a. Rotate the UCS
b. Polar Array around the UCS.
Hint:
First pt = 0,0,0
Second pt = 0,0,4

Step 3. Polar Array 8 Cylinders

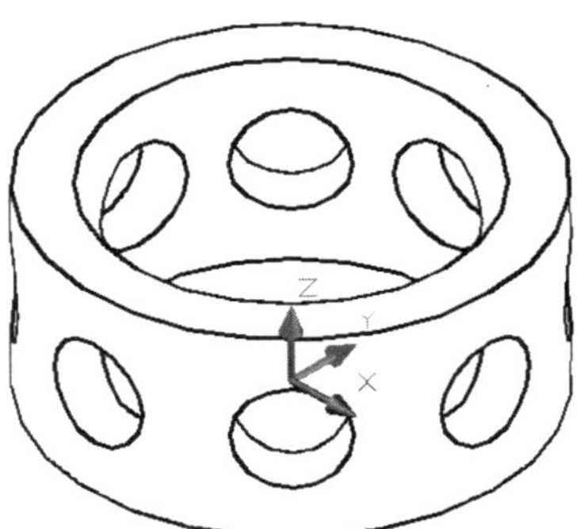

Step 4. Subtract the small cylinders from the large Cylinder.

LEARNING OBJECTIVES

After completing this lesson, you will be able to:

1. Understand 4 Solidedit commands.
   Extrude faces, Move faces, Offset faces and Delete faces

LESSON 18

SOLID EDITING

AutoCAD has many editing commands that allow you to edit existing solids. In this Lesson four of the more frequently used editing commands will be discussed. They are: **Extrude faces, Move faces, Offset faces** and **Delete faces**.

Before you can learn how to use these editing commands, you must first understand how to select the surfaces and how to add or subtract material.

Selecting a face to edit.
It is easiest to select the correct face or hole if the model is displayed in 2D wireframe. To select a face, place your cursor on an open area of the face (surface). Do not select an edge. (Consider turning off Osnap. It may prevent you from placing the cursor in the correct location)
To select a hole, place the cursor on the height line, inside the hole.
If you select too many, hold the shift key down while clicking on the faces you did not want.

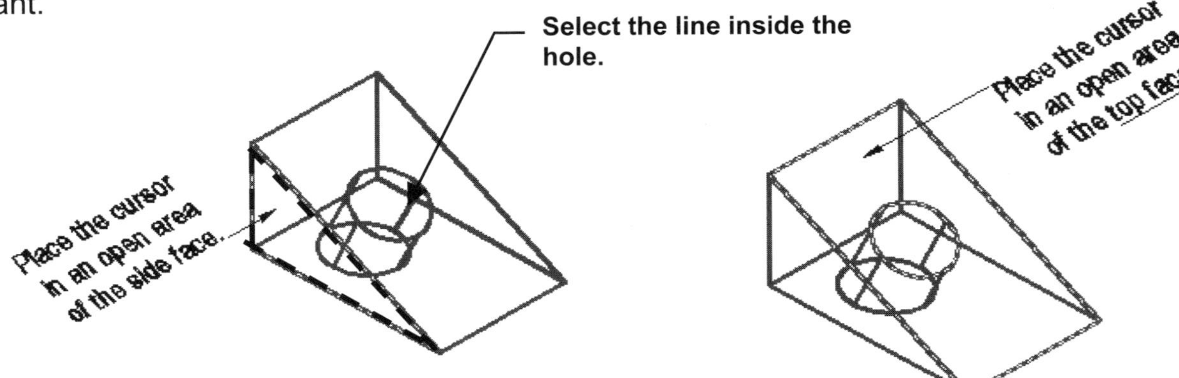

Positive vs. Negative value
Positive value adds material. Negative value subtracts material.
This sounds basic but you need to think about this. For example, if you need to increase the size of a hole you will enter a negative value because when the hole increases in size material is subtracted from the solid model.

ERROR MESSAGES
Sometimes, when using these commands, AutoCAD gets a little confused. If AutoCAD gets confused it stops working. If this happens you usually have to reboot the computer. So it is always a good idea to "**Save your drawing**" before attempting the Solidedit commands.

Also, AutoCAD may display an error message. The following are explanations for a few of the AutoCAD error messages.

"Invalid face/edge (or edge/edge, or face/face, etc. intersection"
You have selected a face that may interfere with another face after the operation.

"Gap cannot be filled"
If you try to delete a large face use **Union** instead if you get this message.

"Modeling error: [some command] will fail to produce a valid ACIS solid
AutoCAD is really confused and will not complete the command. Give up and find another method to edit your solid.

EXTRUDE FACES

The **Extrude faces** command adds or subtracts material perpendicular to the selected face. You will be prompted for a **height**. The height will always be perpendicular to the face and is not affected by the position of the UCS.
A Positive value will add material and a **Negative** value will remove material.

Note: This command has basically been replaced by the Presspull command.

1. Select the **Extrude Faces** command using one of the following:

   TYPE = solidedit
   PULLDOWN = MODIFY / SOLID EDITING / EXTRUDE FACES
   TOOLBAR = SOLID EDITING **(Note: the button looks just like the Extrude command button but it is on a different toolbar)**

2. Select faces or [Undo/Remove]: ***select a face***

3. Select faces or [Undo/Remove/ALL]: ***select more faces or Remove to remove faces that were selected at the same time or <enter> to stop***

4. Specify height of extrusion or [Path]: ***type height (positive or negative)<enter>***

5. Specify angle of taper for extrusion <0>: ***<enter>***

   Solid validation started.
   Solid validation completed.

6. Enter a face editing option
   [Extrude/Move/Rotate/Offset/Taper/Delete/Copy/coLor/Undo/eXit] <eXit>:***<enter>***

   Solids editing automatic checking: SOLIDCHECK=1

7. Enter a solids editing option [Face/Edge/Body/Undo/eXit] <eXit>: ***<enter>***

Example:

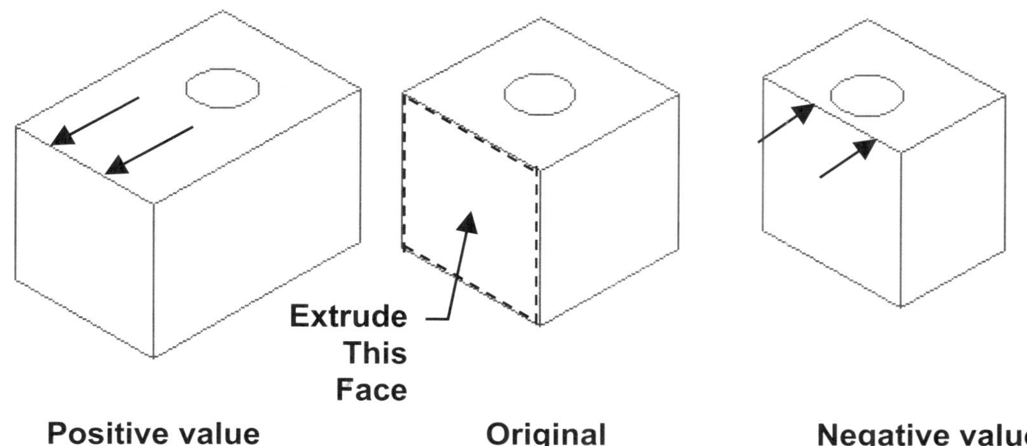

MOVE FACES

The **Move Faces** command allows you to move a negative space such as a hole. This command is very useful and easy to use.

1. Select the **Move Faces** command using one of the following:

   TYPE = solidedit
   PULLDOWN = MODIFY / SOLID EDITING / MOVE FACES
   TOOLBAR = SOLID EDITING

2. Select faces or [Undo/Remove]: ***select the hole***

3. Select faces or [Undo/Remove/ALL]: ***select other holes or remove faces that were selected at the same time or <enter> to stop.***

4. Specify a base point or displacement: ***select a base point***

5. Specify a second point of displacement: ***type coordinates or DDE***

   Solid validation started.
   Solid validation completed.

6. Enter a face editing option
   [Extrude/Move/Rotate/Offset/Taper/Delete/Copy/coLor/Undo/eXit] <eXit>: ***<enter>***

7. Solids editing automatic checking: SOLIDCHECK=1
   Enter a solids editing option [Face/Edge/Body/Undo/eXit] <eXit>:***<enter>***

EXAMPLE:

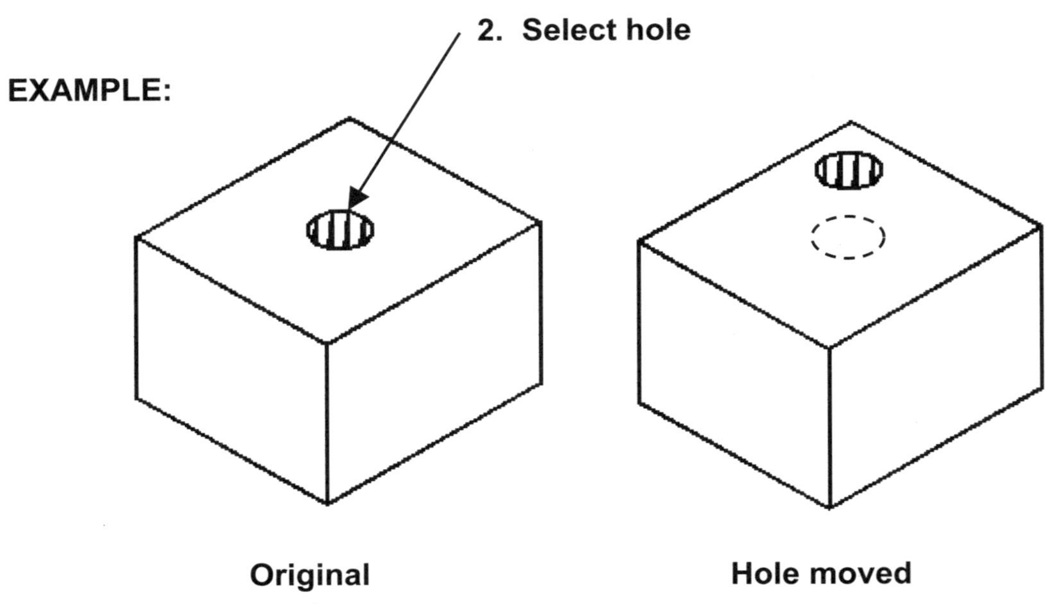

Original **Hole moved**

OFFSET FACES

The **Offset Faces** command adds or removes material also. It is especially helpful when editing the size of holes. If the object is a rectangular shape Offset faces add or removes material equally by the distance entered. (Think about it.)

Remember, a Negative value removes material from the solid, so the hole will get larger. A Positive value adds material to the solid, so the hole will get smaller.

1. Select the **Offset Faces** command using one of the following:

    TYPE = solidedit
    PULLDOWN = MODIFY / SOLID EDITING / OFFSET FACES
    TOOLBAR = SOLID EDITING

2. Select faces or [Undo/Remove]:***select a hole***

3. Select faces or [Undo/Remove/ALL]:***select more holes or <enter>***

4. Specify the offset distance: ***type the offset distance***

    Solid validation started.
    Solid validation completed.

5. Enter a face editing option
    [Extrude/Move/Rotate/Offset/Taper/Delete/Copy/coLor/Undo/eXit] <eXit>: ***<enter>***

    Solids editing automatic checking: SOLIDCHECK=1

6. Enter a solids editing option [Face/Edge/Body/Undo/eXit] <eXit>:***<enter>***

EXAMPLE:

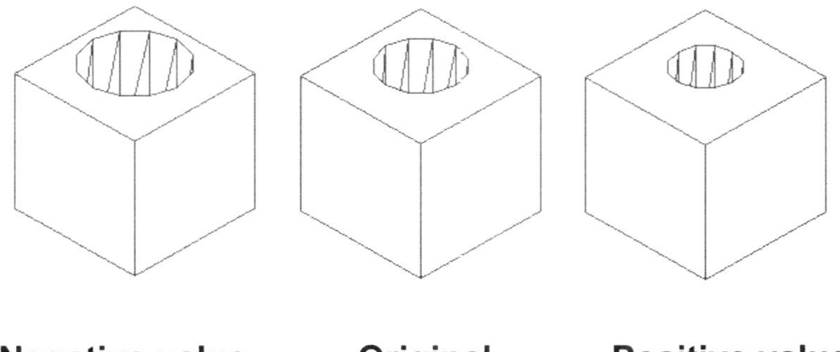

Negative value **Original** **Positive value**

DELETE FACES

The **Delete Faces** command removes a face. Use this command to delete holes.

1. Select the **Delete Faces** command using one of the following:

   TYPE = solidedit
   PULLDOWN = MODIFY / SOLID EDITING / DELETE FACES
   TOOLBAR = SOLID EDITING

2. Select faces or [Undo/Remove]: ***select a face (hole)***

3. Select faces or [Undo/Remove/ALL]: ***select another face or <enter>***

   Solid validation started.
   Solid validation completed.

4. Enter a face editing option
   [Extrude/Move/Rotate/Offset/Taper/Delete/Copy/coLor/Undo/eXit] <eXit>: ***<enter>***

   Solids editing automatic checking: SOLIDCHECK=1

5. Enter a solids editing option [Face/Edge/Body/Undo/eXit] <eXit>:***<enter>***

EXAMPLE:

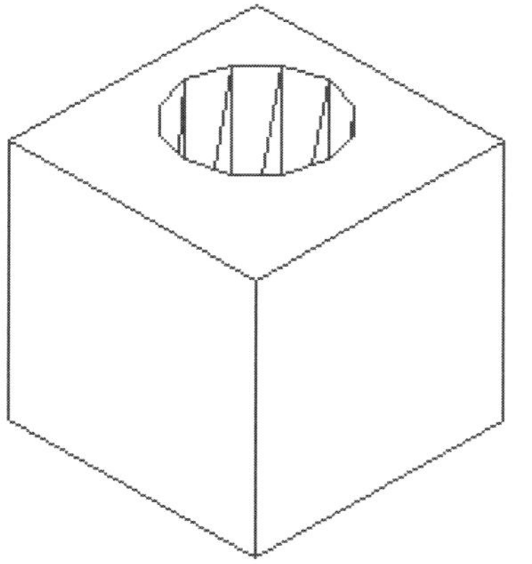

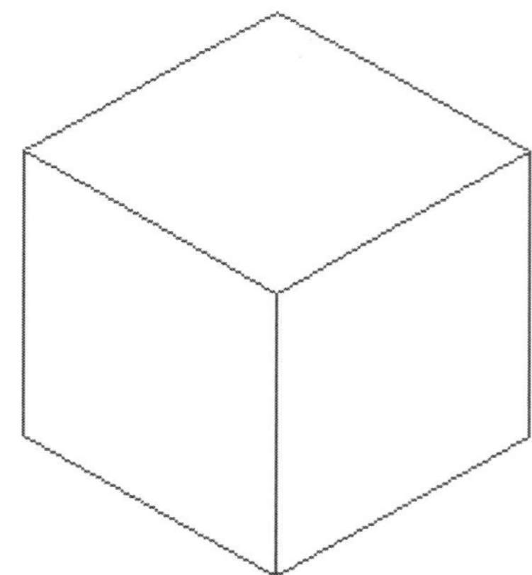

Original with hole

Hole deleted

EXERCISE 18A

EXTRUDE FACES or Presspull

1. Open **EX-17A**
2. Select the **Model tab**
3. Edit the solid model as shown.
4. Save as **EX-18A**

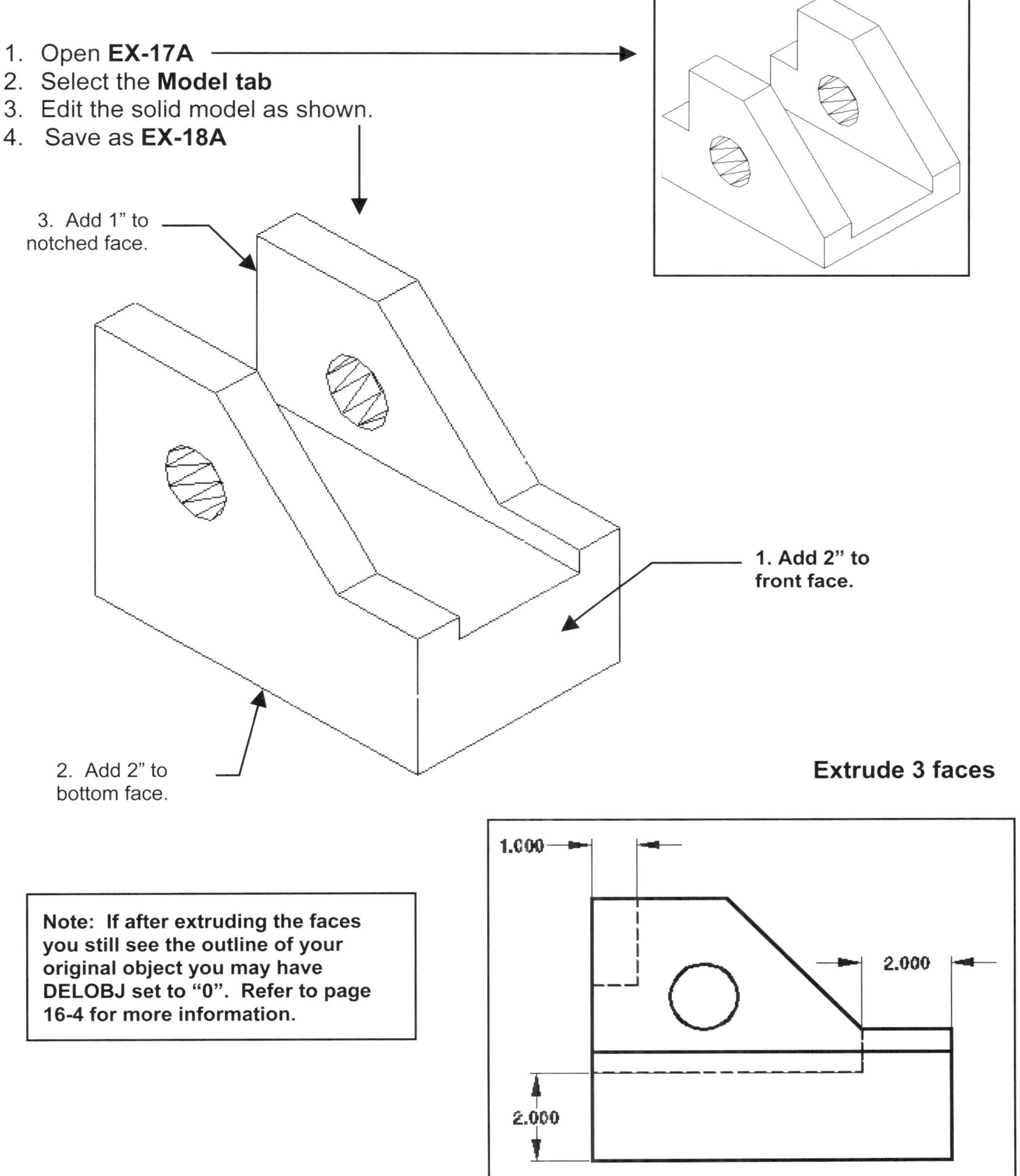

Note: If after extruding the faces you still see the outline of your original object you may have DELOBJ set to "0". Refer to page 16-4 for more information.

EXERCISE 18B

MOVE FACES

1. Open **EX-16E**
2. Edit the solid model as shown below.
3. Refer to page 18-9 for instructions for removing an opening.
4. Save as **EX-18B**

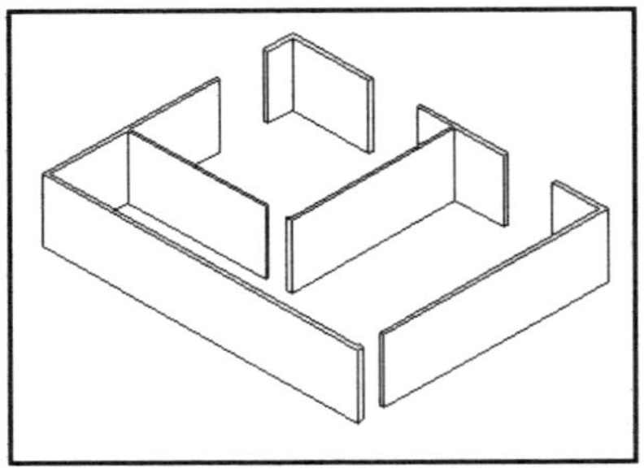

EX-16E

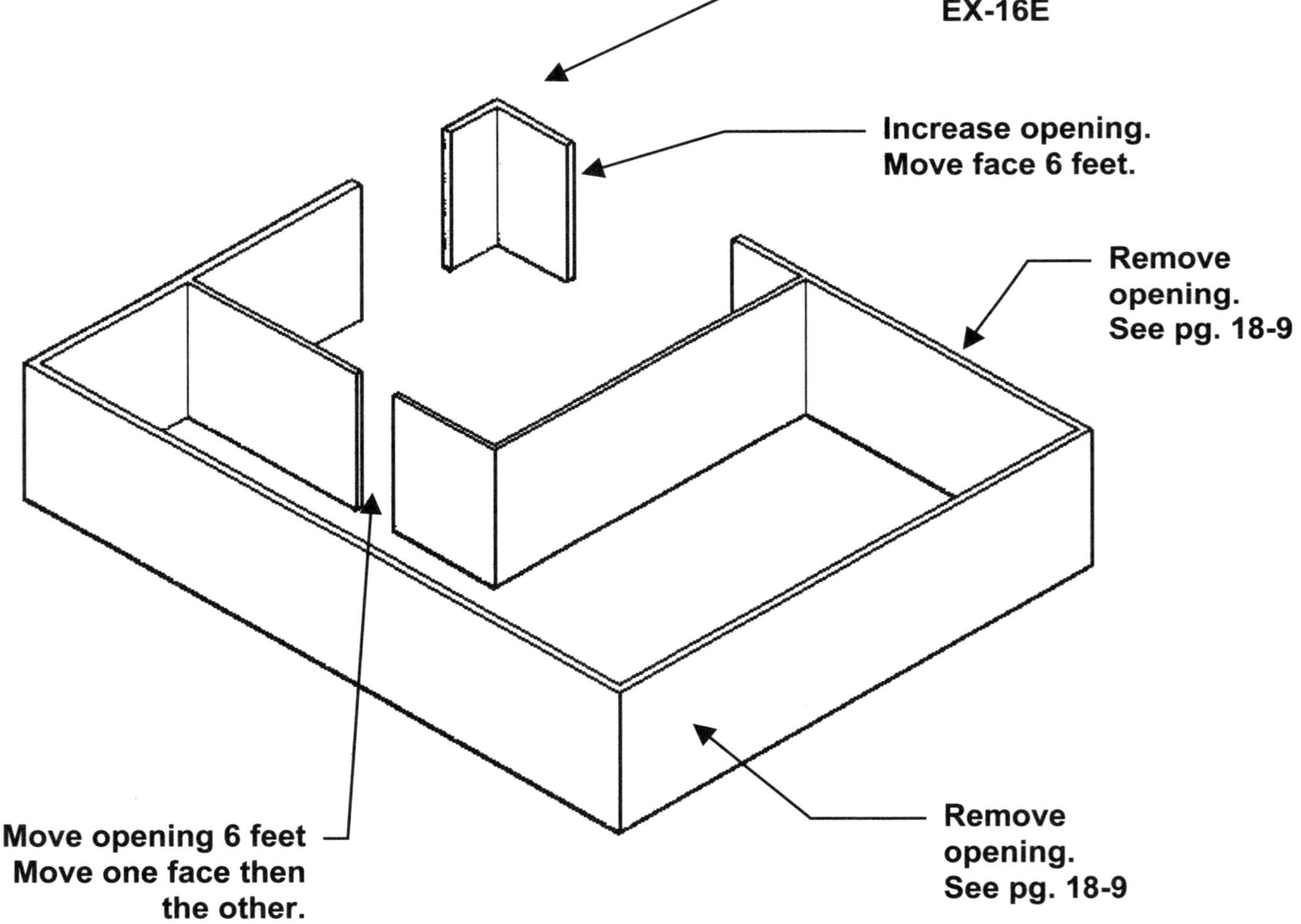

Increase opening. Move face 6 feet.

Remove opening. See pg. 18-9

Move opening 6 feet Move one face then the other.

Remove opening. See pg. 18-9

HOW TO REMOVE AN OPENING.

1. Select the **MOVE FACES** command.

2. Select the face.

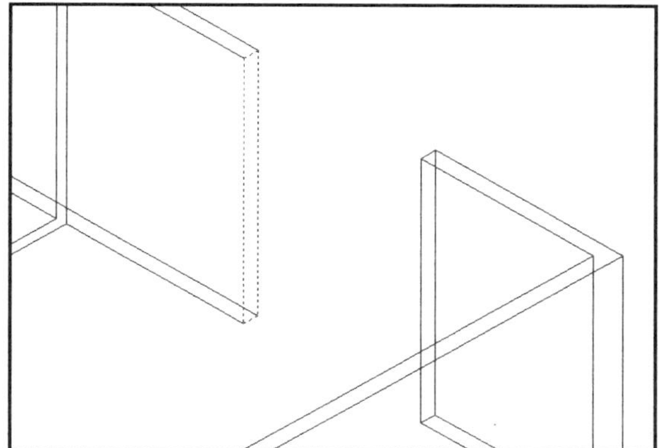

3. Specify the basepoint

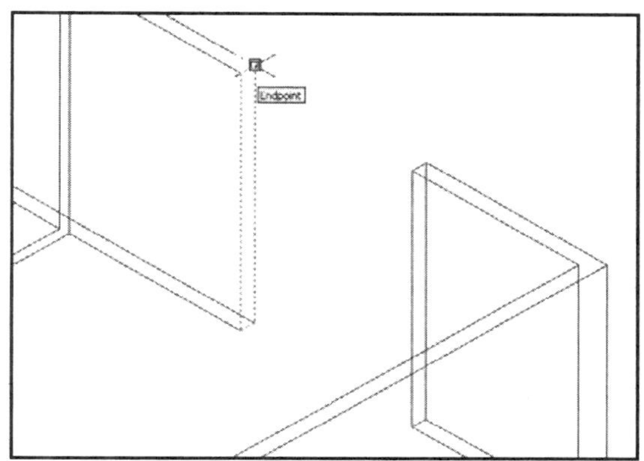

4. Specify the new location.

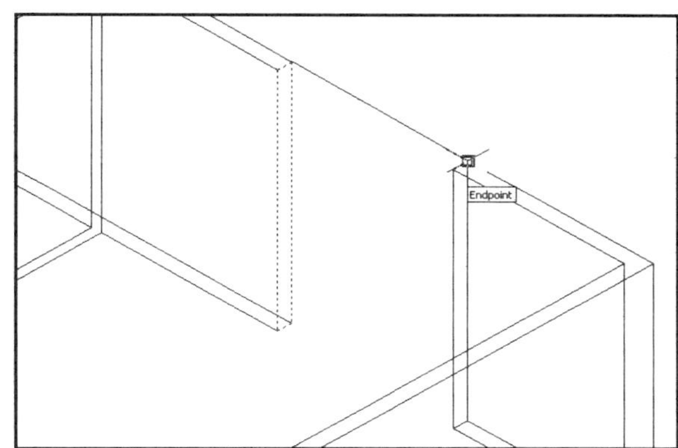

5. Union

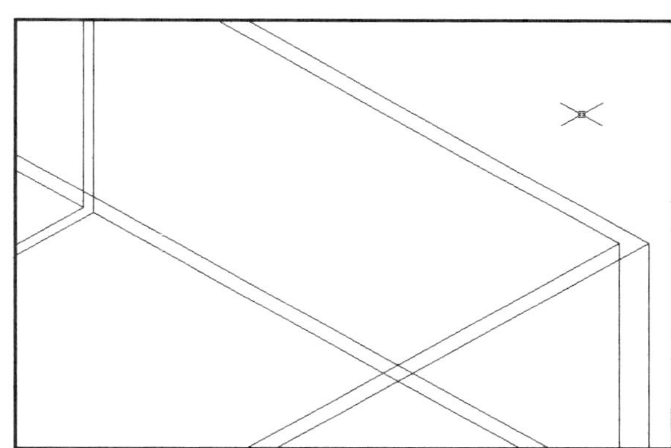

EXERCISE 18C

OFFSET FACES

1. **Select File / New and select <u>Acad3d.dwt</u>.**
2. Draw the solid shown in Figure 1. (Dimensions on page 18-11)
3. Save as **EX-18C1**

FIGURE 1

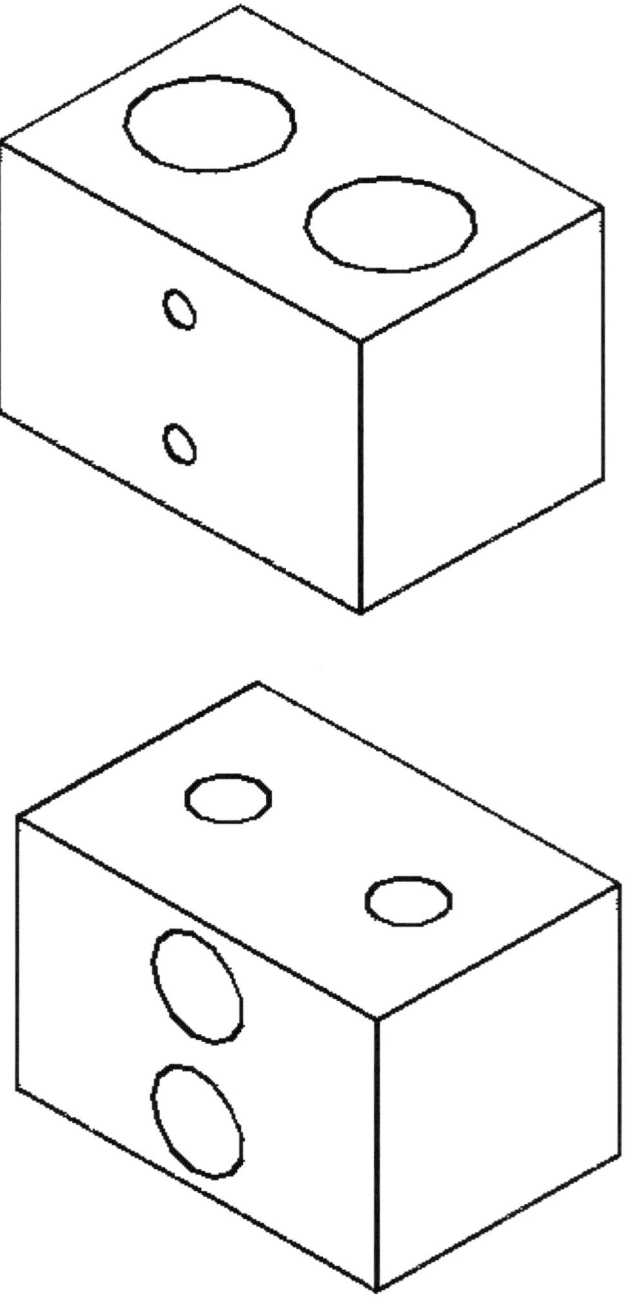

FIGURE 2

4. Edit the holes per dimensions on page 18-11.

5. Save as **EX-18C2**

DIMENSIONS FOR FIGURE 1

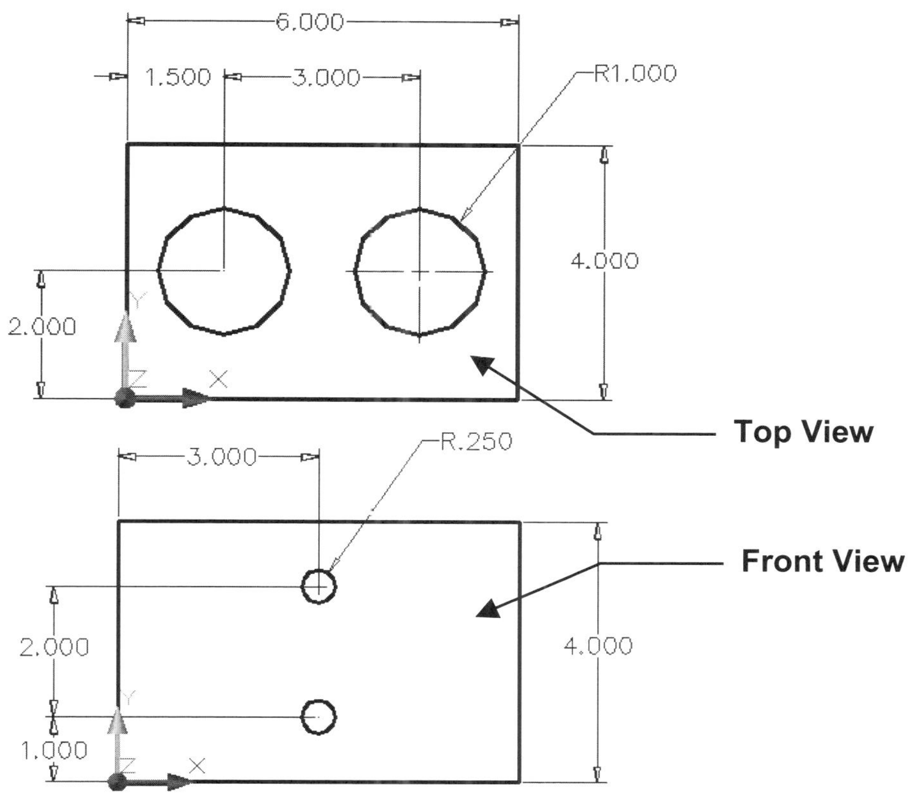

DIMENSIONS FOR FIGURE 2

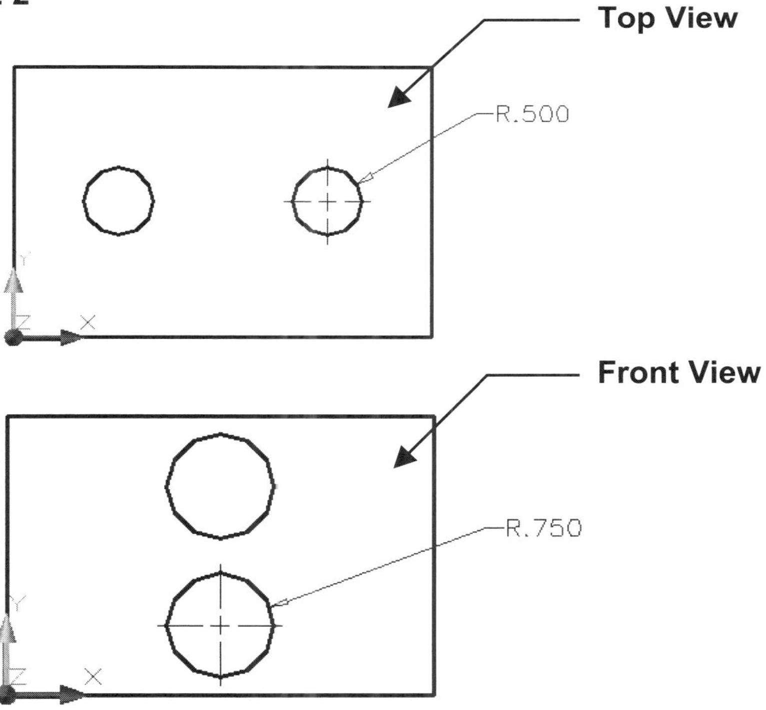

EXERCISE 18D
DELETE FACES

1. Open **EX-18A**
2. Delete the holes as shown in Figure 2.
3. Save as **EX-18D**

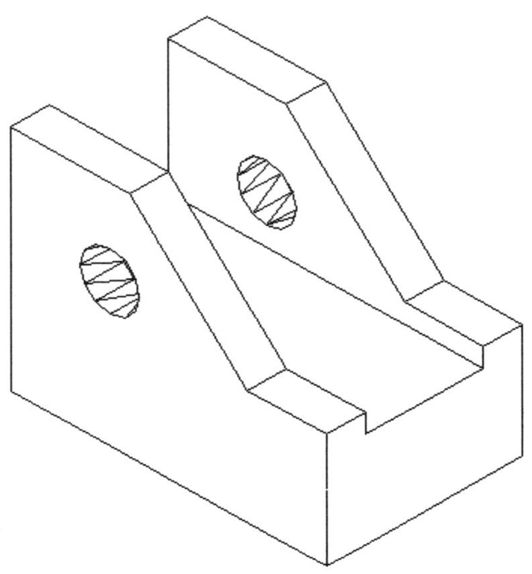

Figure 1

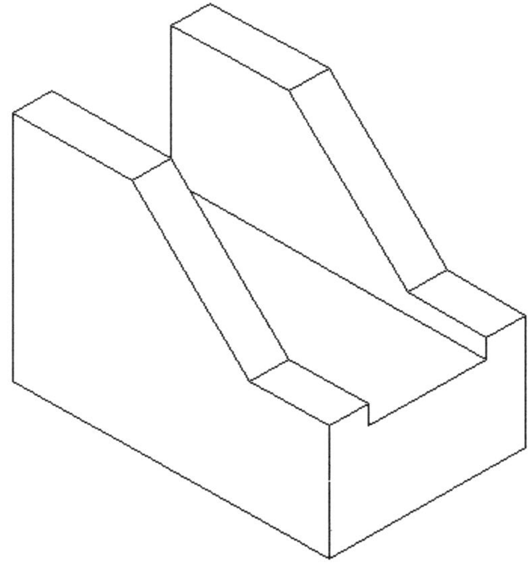

Figure 2

LEARNING OBJECTIVES

After completing this lesson, you will be able to:

1. Create a solid object by revolving a 2D shape
2. Slice a solid object into 2 segments
3. Create a section view through a solid object.
4. Create a twisted object along a path.
5. Create a Helix.

LESSON 19

REVOLVE

The **REVOLVE** command allows you to revolve a profile around an axis. If the shape is closed it will become a solid. If the shape is open it will become a surface. The shape can be revolved from 1 to 360 degrees.

You will select the <u>axis to revolve around</u>, the <u>solid to be revolved</u> and enter the <u>angle of revolution.</u>

Selecting the axis of revolution
The axis of revolution can be the X or Y axis. It may also be an object such as a line. The axis may be located on the shape or not.

HOW TO USE THE REVOLVE COMMAND

1. Select the **REVOLVE** command using one of the following:

    TYPE = REV or **REVOLVE**
    PULLDOWN = DRAW / MODELING / REVOLVE
    3DMAKE CONTROL PANEL =

2. Select objects to revolve: ***select the shape or region***
3. Select objects to revolve: ***select more or <enter> to stop***
4. Specify axis start point or define axis by [Object/X/Y/Z] <Object>: ***select option***
5. Select an object: ***select the axis of object to revolve about***
6. Specify angle of revolution or [STart angle] <360>: ***type the angle (positive ccw or negative cw determines direction of rotation.***

Below are examples of how you can make 3 different solids by selecting different "Axis of Revolution".

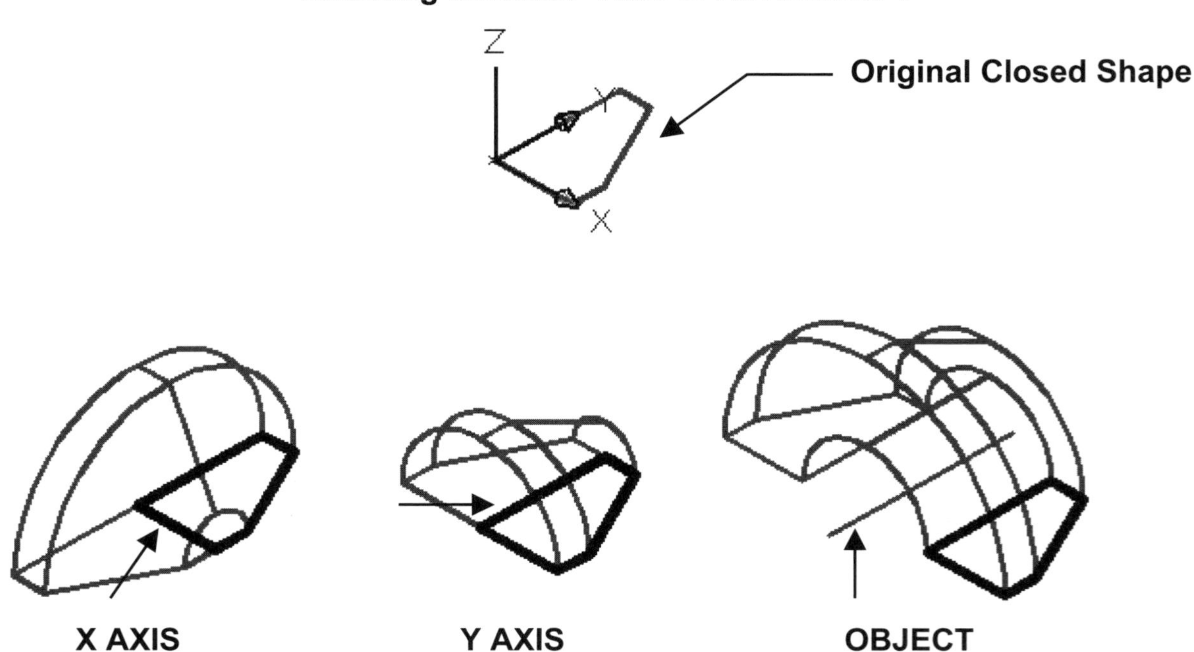

SLICE

The **SLICE** command allows you to make a knife cut through a solid. You specify the location of the slice by specifying the plane. After the solid has been sliced, you determine which portion of the slice you want removed. You may also keep both portions but they are still separate.

THERE ARE 2 EASY METHODS TO SPECIFY THE CUTTING PLANE.

Method 1.
1. Define the cutting plane by selecting 3 points on the object.
2. Click on the side you want to keep or "B" to keep both sides.

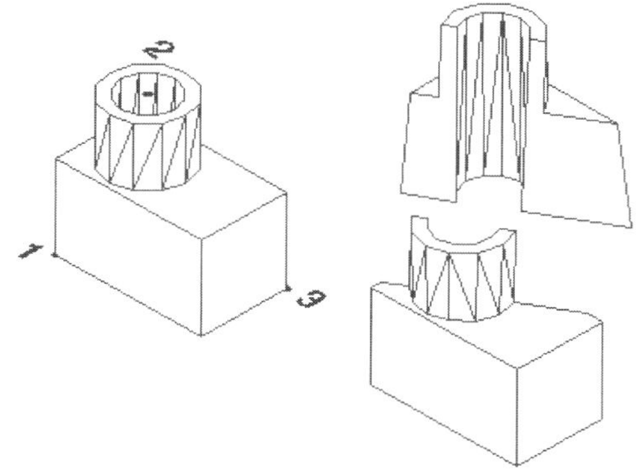

Method 2.
1. Move the UCS to the desired cutting plane location.
2. Select the XY, YZ or ZX plane. (The cut will be made along the selected plane)

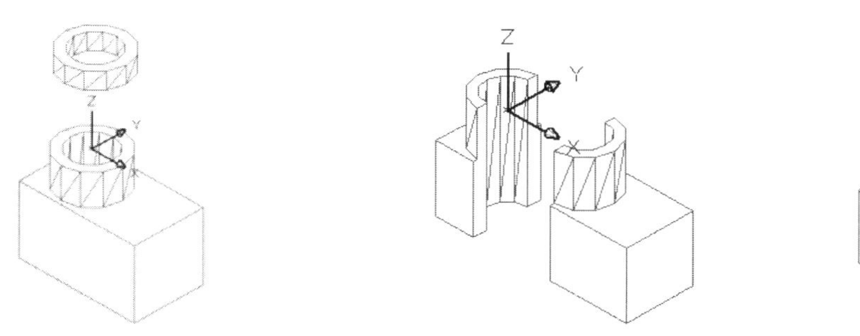

XY PLANE **YZ PLANE** **ZX PLANE**

HOW TO USE THE SLICE COMMAND

1. Select the **SLICE** command using one of the following:

    TYPE = SL or **SLICE**
    PULLDOWN = MODIFY / 3D OPERATIONS / SLICE
    3D MAKE CONTROL PANEL (EXTENDED) =

2. Select objects: ***select the solid object***
3. Select objects: ***<enter>***
4. Specify start point of slicing plane or [planer object/Surface/Zaxis/View/SY/YZ/ZX/3pts <3points>: ***select the option preferred.***

See methods above.

SECTION

The **SECTION** command process is very similar to the Slice command. But the Section command does not actually separate the solid. It creates a 2D region from the plane that you specify. The 2D region will be created on the current layer. The 2D Region can then be moved, extruded, revolved, copied, etc.

To specify the section plane refer to the Slice methods on page 19-3.

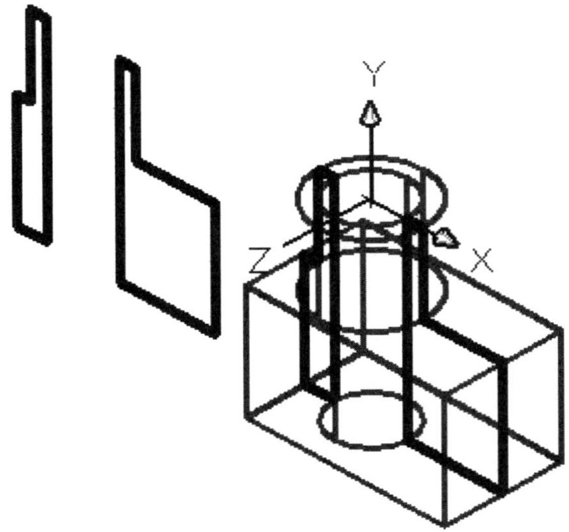

HOW TO USE THE SECTION COMMAND

1. Select the Layer you want the section to be on and the Linetype.
2. Select the **SECTION** command using one of the following:

   TYPE = SEC or **SECTION**
   PULLDOWN = NONE
   TOOLBAR = NONE

3. Select objects: ***select the solid object***
4. Select objects: ***<enter>***
5. Specify first point on section plane by [Object/Zaxis/View/XY/YZ/ZX/3points] <3points>: ***select the option preferred***

HOW TO CREATE A HATCHED SECTION VIEW

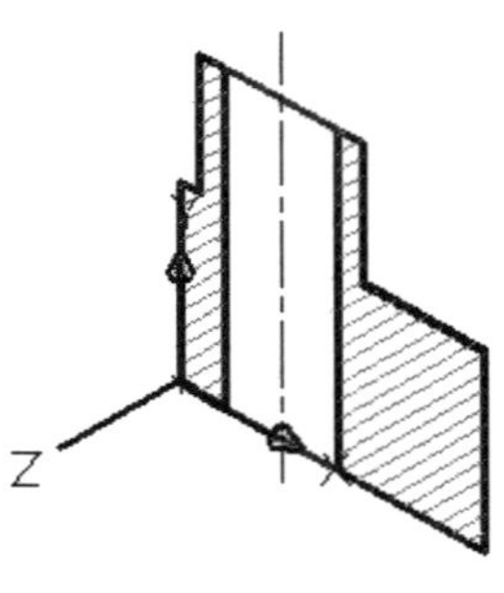

1. Create a section (2D region) as described above.
2. Move or copy the region to a new location.
3. Explode if necessary to create separate regions.
4. Use the Hatch command to create section lines.
   (Note: The area to be hatched must be parallel to the XY plane or you will receive an error message. *My suggestion is, if you intend to hatch a section, use the XY plane method for defining the plane.)*
5. Draw any connecting lines to complete the view.

SWEEP

The SWEEP command is similar to the Extrude along a path but it does much more. You draw a path and an object to sweep along that path in the same plane. It will automatically aligns the object to sweep perpendicular to the path. That makes drawing them a lot easier.

HOW TO USE THE SWEEP COMMAND

1. Select the Layer that you want the finished object on.
2. Draw the object and the path on the same plane.

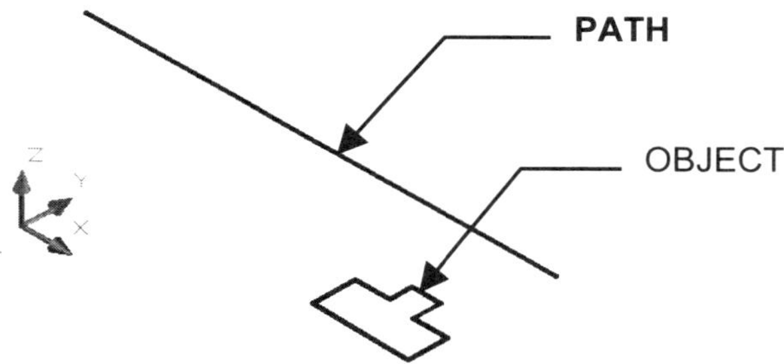

3. Select the **SWEEP** command using one of the following:

    TYPE = SWEEP
    PULLDOWN = DRAW / MODELING / SWEEP
    3D MAKE CONTROL PANEL =

4. Select objects to sweep: ***select the object <enter>***
5. Select objects to sweep: ***select more or <enter> to stop***
6. Select sweep path or [Alignment/Base Point/Scale/Twist]: ***select the path***

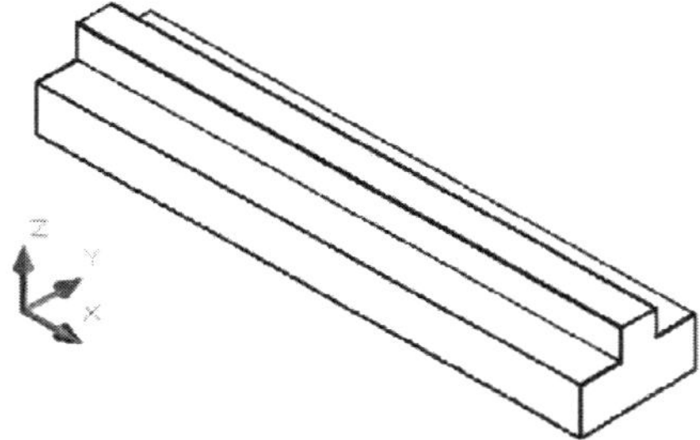

If you would like it to twist, when prompted for the path, select "Twist". Then enter the total angle that you would like it to twist along the entire path. Then select the path.

Example of 180 degree twist from the start of the path to the end.

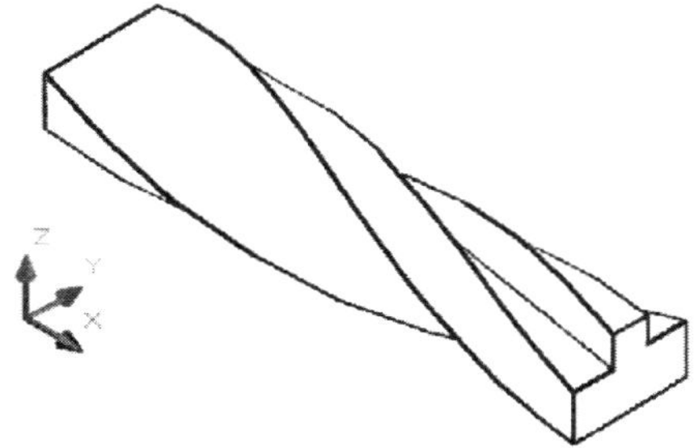

HELIX

The HELIX command allows you to easily construct a wireframe helix or spiral. You will specify the center of the bottom and top, bottom and top diameters, number of turns, turn height distance and which direction it turns. The default is counterclockwise.

HOW TO USE THE HELIX COMMAND

1. Select the Helix command using one of the following:

TYPE = HELIX
PULLDOWN = DRAW / HELIX
3D MAKE CONTROL PANEL

Command: _Helix
Number of turns = 3.000   Twist=CCW

1. Specify center point of base: 0,0: ***select the location for the center of the base***
2. Specify base radius or [Diameter] <1.000>: ***enter base radius <enter>***
3. Specify top radius or [Diameter] <2.000>: ***enter top radius <enter>***
4. Specify helix height or [Axis endpoint/Turns/turn Height/tWist] <1.000>: ***enter Axis endpoint or select one of the other options.***

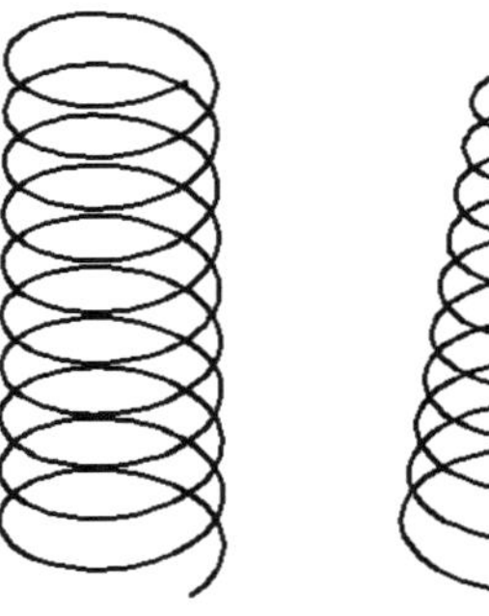

Options:

Axis endpoint = center of the top of the Helix.

Turns = How many turns do you want? (500 max) Then you will be prompted for distance between turns. AutoCAD will calculate the height from that information.

Turn Height = Distance between revolutions.

Twist = The default twist is CCW. You may change it to CW.

EXERCISE 19A

REVOLVE

1. **Select File / New and select Acad3d.dwt.**
2. Draw the 2D shape shown below.
3. Create 2 additional copies (3 total)
4. Revolve in the X axis, Y axis and around an object, as shown on the next page.
5. Save as **EX-19A**

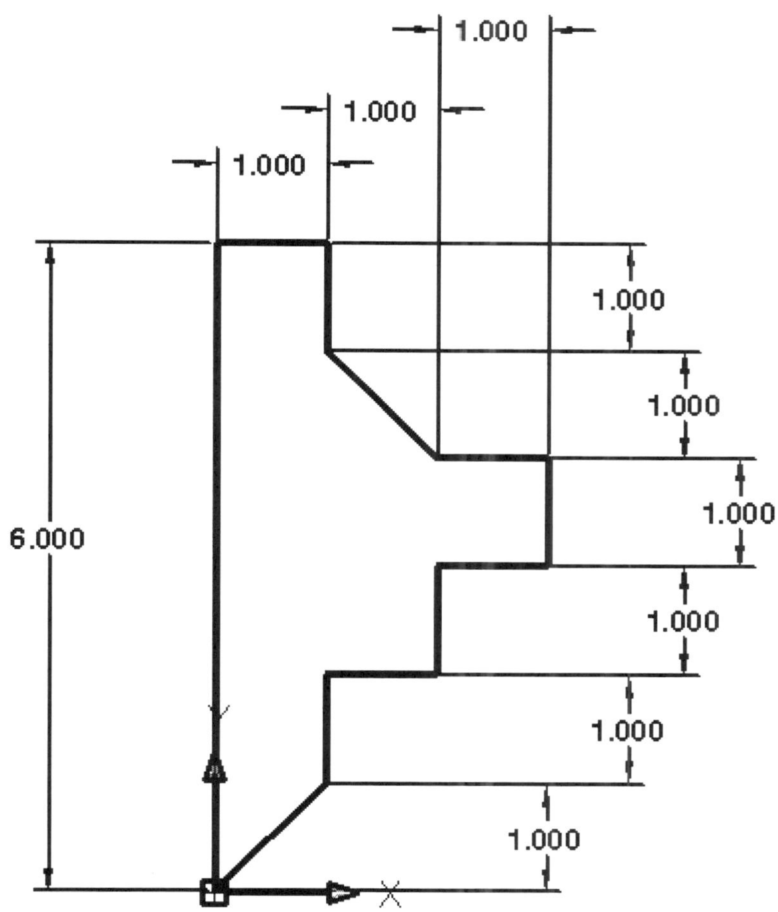

ORIGINAL SHAPE

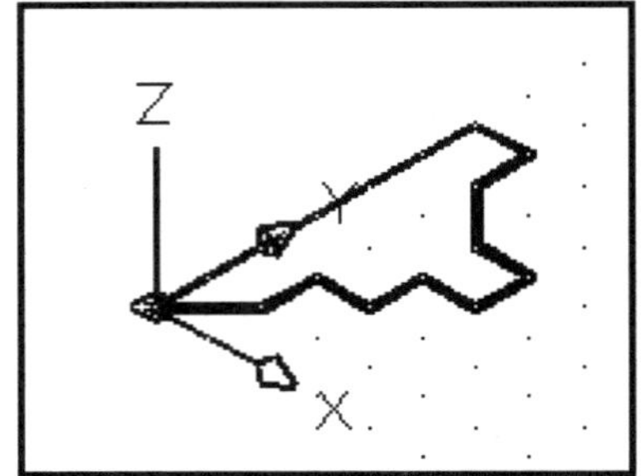

REVOLVE 180 DEGREES AROUND THE X AXIS

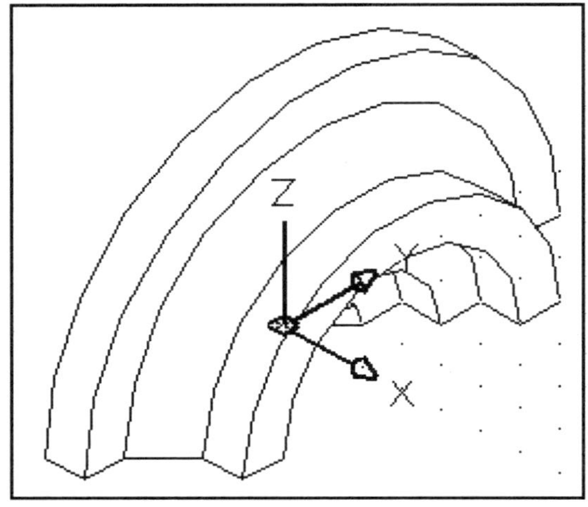

REVOLVE –90 DEGREES AROUND THE Y AXIS

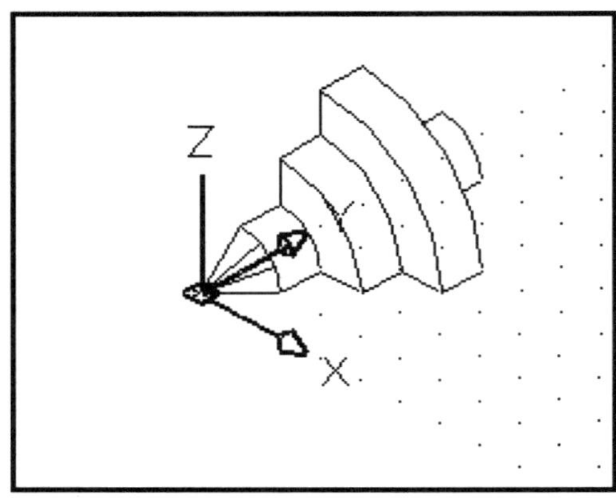

Remember, the point of "Y" is in your chest and you revolve the XZ clockwise (negative).

REVOLVE 360 DEGREES AROUND AN OBJECT

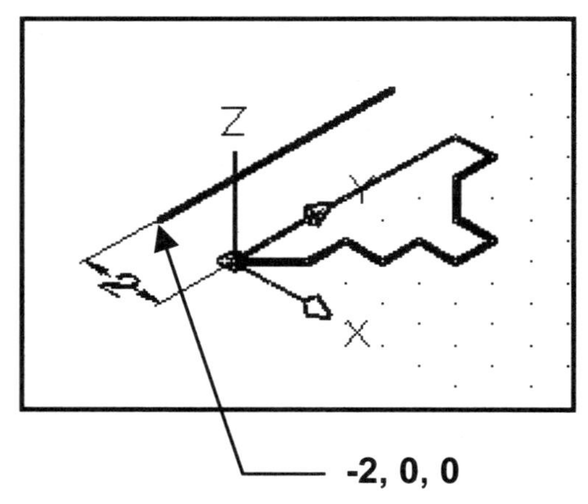

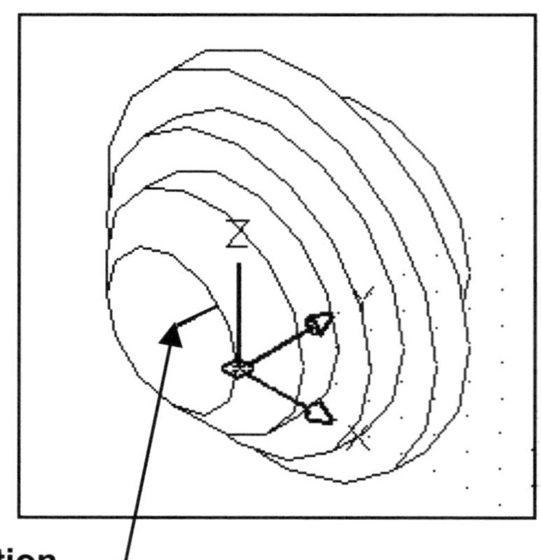

EXERCISE 19B

SLICE

1. **Select File / New and select <u>Acad3d.dwt</u>.**
2. Draw the 3D solid shown in the upper right view **FIGURE 1**.
   (The other 3 views are for dimensional information only)
3. Save as **EX-19B1 (This drawing will be used in EX-19C also.)**
4. Split the solid object as shown.
5. **SLICE** the solid as shown in **FIGURE 2.**
6. Save as **EX-19B2**

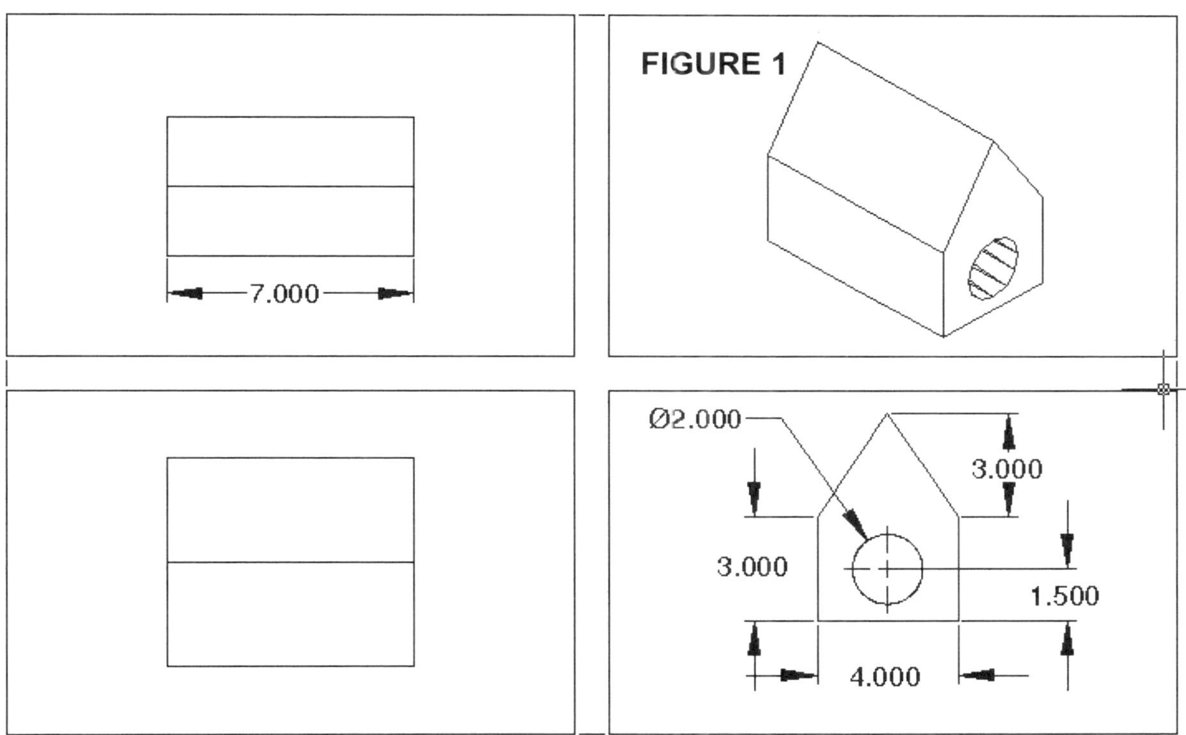

FIGURE 2

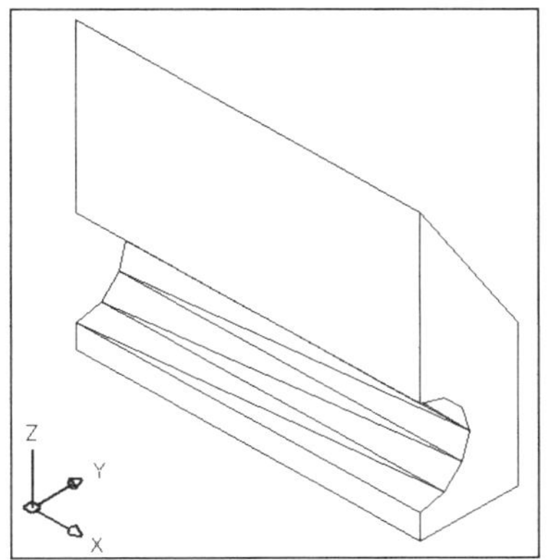

EXERCISE 19C

SECTION

1. Open **EX-19B1**
2. Create a **SECTION** as shown. (Refer to page 19-4)
3. The new section should be on the "**Section**" layer.
4. The linetype should be **Continuous**.
5. Save as **EX-19C**

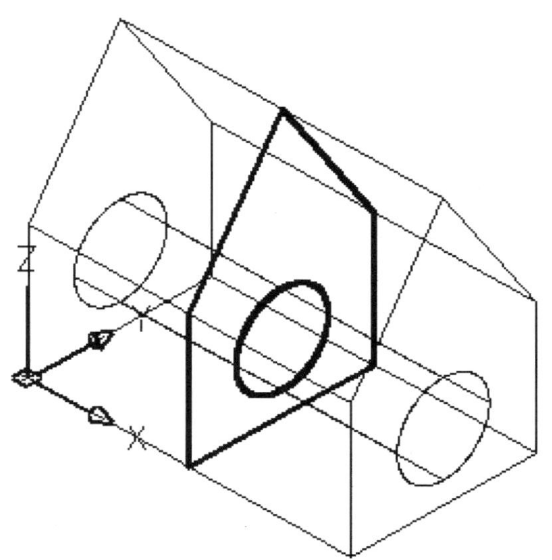

EXERCISE 19D

HATCH THE SECTION

1. Open **EX-19C**
2. Move the section (2D region) away from the solid model.
3. **HATCH** the section. (Refer to page 19-4)
4. Place the hatch lines on the "Hatch" <u>layer</u>.
5. Hatch specs: Angle = 0 Scale = 4
6. Save as **EX-19D**

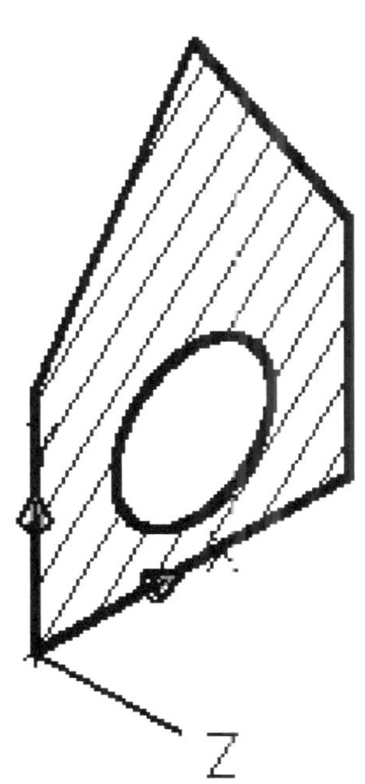

EXERCISE 19E

SWEEP

1. **Select File / New and select <u>Acad3d.dwt</u>.**
2. Draw the 2D shape shown below and the path on the XY plane.

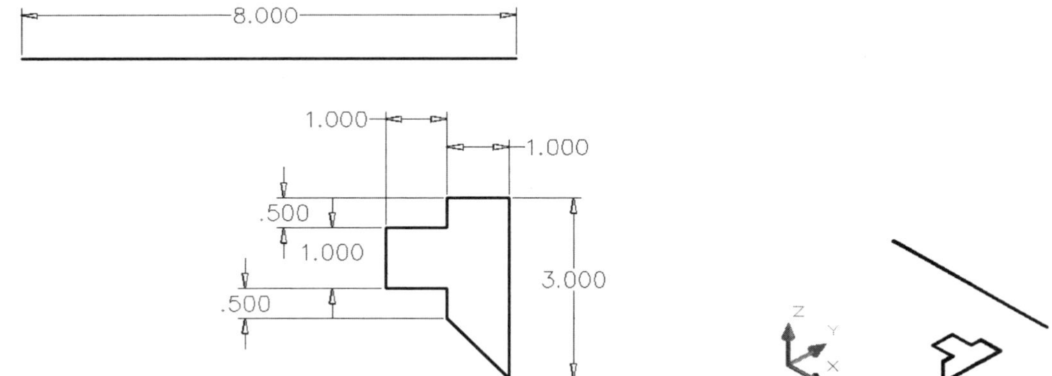

3. Make a Region of the shape.
4. Create a copy of both the Region and the Line.

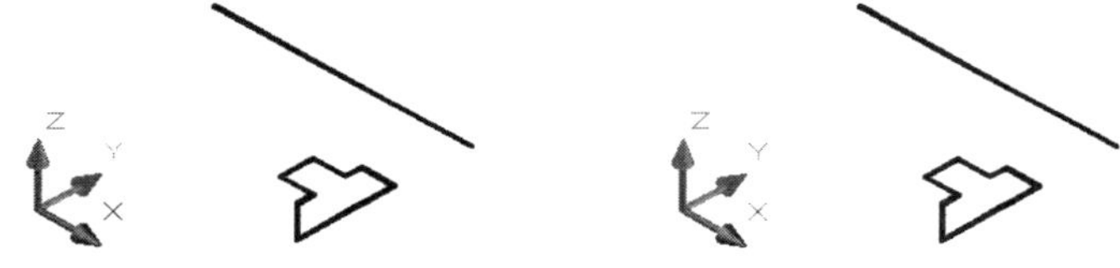

5. Select the layer where you want the sweep placed.
6. First do a straight Sweep. Then do a Sweep with a twist of 180 degrees. (Refer to page 19-5)
7. Save as **EX-19E**

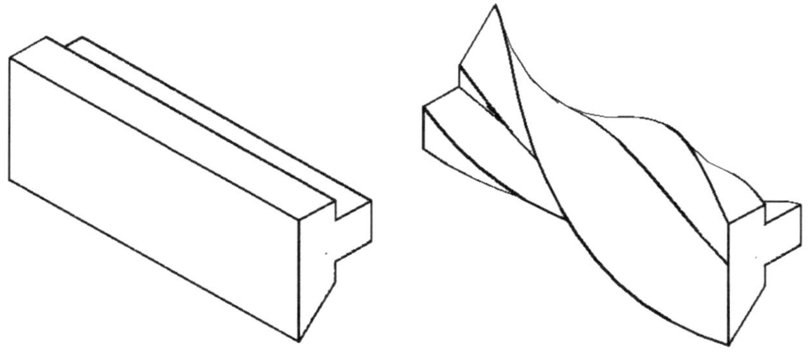

STRAIGHT SWEEP **TWIST 180 DEGREES SWEEP**

EXERCISE 19F

HELIX

1. **Select File / New and select Acad3d.dwt.**
2. Draw the 2 HELIX shown below.
3. Save as: **EX-19F**

Number of turns = 10
Twist = CCW
Base radius = 3.000
Top radius = 3.000
Axis end point = 8

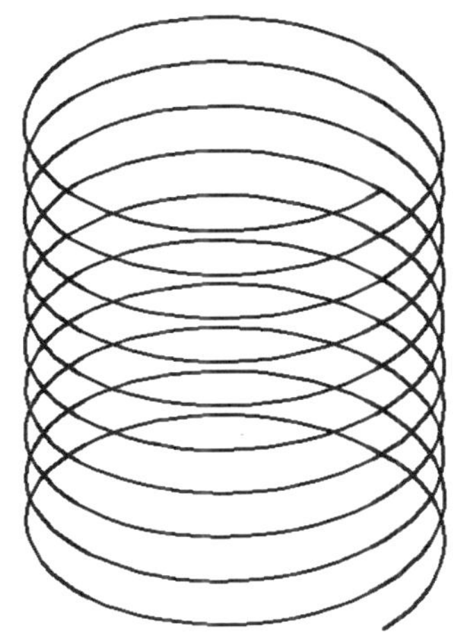

Number of turns = 5   Twist = CCW
Base radius = 6
Top radius = 2
Turns = 5
Distance between turns = 1

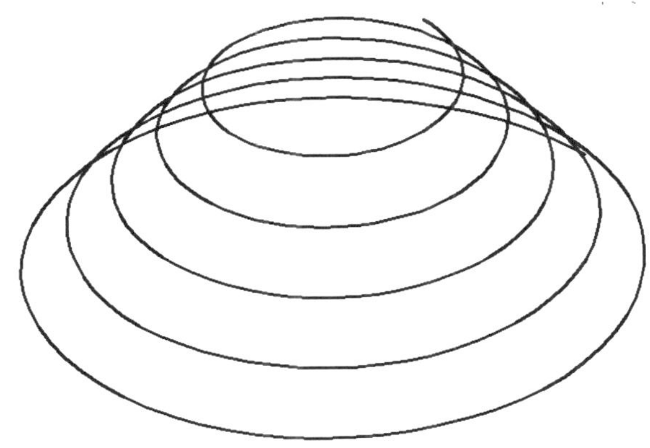

NOTES:

LEARNING OBJECTIVES

After completing this lesson, you will be able to:

1. Create a multiple view plot setup
2. Control the hidden line display for plotting
3. Shade one of the views
4. Dimension a multiple view display
5. Hollow the inside out of a solid

LESSON 20

PLOTTING MULTIPLE VIEWS

AutoCAD has many commands to create a multiview layout, such as SOLVIEW, SOLDRAW, SOFPROF and DXB plot. You may consider researching these, but in this lesson I will show you 5 quick and simple steps to develop a multiview layout to use for plotting.

STEP 1. SELECTING THE PLOTTER / PRINTER AND PAPER SIZE.

1. Change to your 2D Workspace.
2. Open a solid model drawing. (Example: 2007-3D demo.dwg)
3. Select an unused Layout tab. (If all Layout tabs have been used, create a new Layout tab.)
4. Create a new Page Setup named "Multiview". (Refer to page 3-7)
5. Select "your printer", "Plot Style Monochrome" and " Paper Size-Letter". Close and Set Current the Multiview page setup and Close again.(Refer to pg 3-8)
6. Rename the Layout tab to "Multiview". (Refer to page 3-9.)

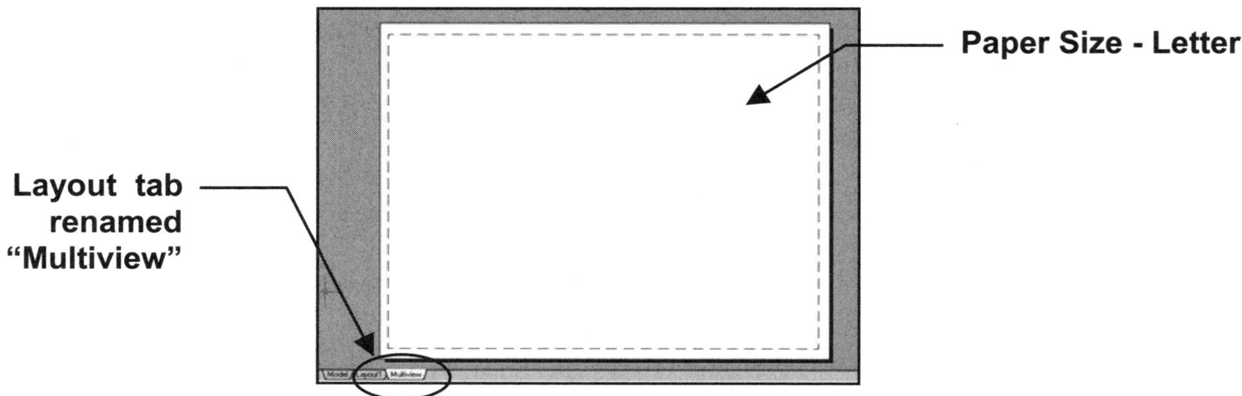

STEP 2. CREATE VIEWPORTS

1. Make the **VP** layer current.
2. Select **VIEW / VIEWPORTS / NEW VIEWPORTS**
   a. Select **FOUR EQUAL**
   b. Set **Viewport Spacing to: .50**
   c. Set **Setup to: 3D**
   d. Select **OK**
   e. Specify first corner or [Fit] <Fit>: **<enter>**

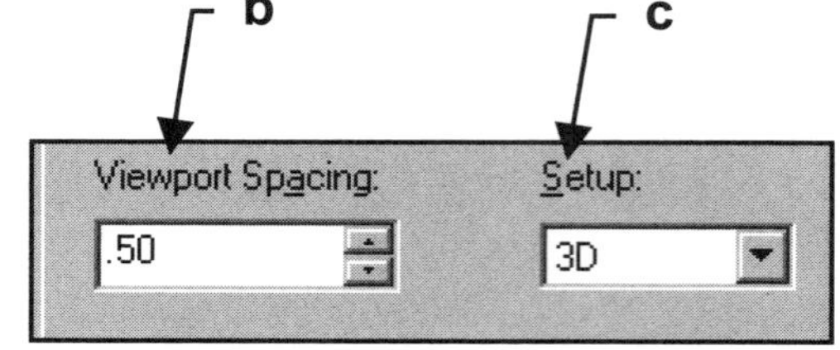

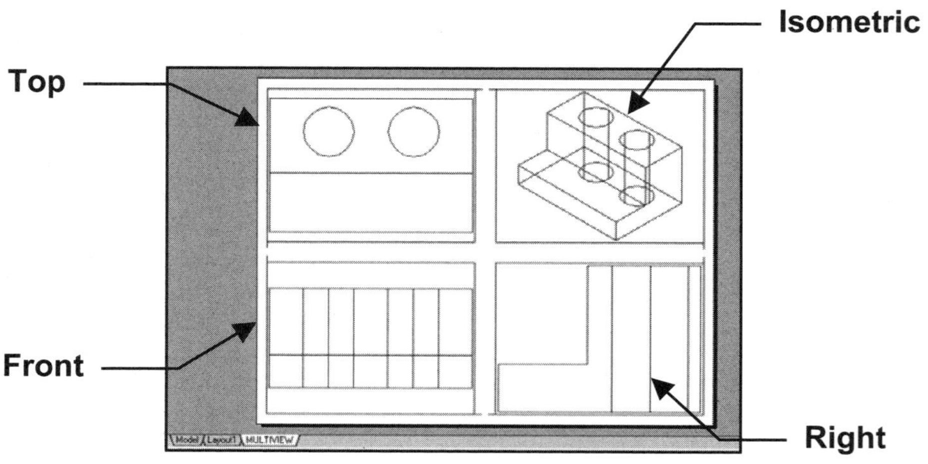

STEP 3. ADJUST VIEWPORT SCALE, PAN and LOCK

1. Open the VIEWPORT TOOLBAR and adjust the scale of each Viewport.
2. Pan the object in each viewport to the desired location. (Do not use Zoom)
3. **LOCK** each Viewport.

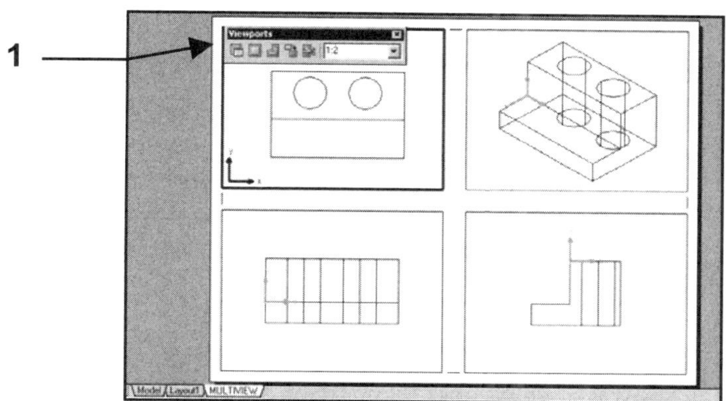

STEP 4. SELECT THE PLOT APPEARANCE as follows:

1. Make sure you are in Paperspace.
2. Select **ALL** the viewports. (Use crossing window.)
3. Right click and select "**Shade Plot / Hidden**" from the menu.

Notice that the drawing appearance has not changed. But you have specified how you want the drawing plotted. You will see when you select Preview from the Plot dialog box.

STEP 5. PLOT (Refer to pg. 3-13 or 3-30)

1. Select **FILE / PLOT.** (Notice the Page Setup name is "Multiview")
2. Select **Printer / Plotter** = yours
3. Select **Paper Size** = Letter
4. Select **Plot Area** = Layout
5. Select **Plot Offset** = 0
6. Select **Scale** = 1 : 1
7. Select **Plot Style Table** = Monochrome
8. Select the **Preview** button.

Now you should be able to see the difference.

Notice the facet lines. If you do not want those displayed, type DISPSILH <enter> and set it to 1.

Notice Hidden Lines are not displayed. This is controlled by the "Shade Plot" setting in Step 4. (Also see the next page for more options.)

How to control the hidden line plot

Option 1. Invisible hidden lines
In Step 4, on the previous page, you set the SHADE PLOT to HIDDEN. The "Hidden" setting displays the hidden lines, in the Full Preview, as "invisible" for plotting. They are not actually removed. When you return to the drawing, the hidden lines are still there.

Option 2. Dashed hidden lines
You may want to plot the drawing with hidden lines shown. To accomplish this you must change the **OBSCUREDLTYPE** variable setting.

1. On the command line type **obscuredltype <enter>**
2. Enter new value for OBSCUREDLTYPE <0>: ***type 2 <enter>***

The Obscured Linetype variable controls the appearance of the hidden lines. There are 11 choices as follows: 0=Off, 1=Solid, 2=Dashed, 3=Dotted, 4=Short Dash, 5=Medium Dash, 6=Long Dash, 7=Double Short Dash, 8=Double Medium Dash, 9=Double Long Dash, 10=Medium Long Dash and 11=Sparse Dot.

And don't forget "Dispsilh" varible to remove the facets on the internal surfaces. Refer to the previous page.

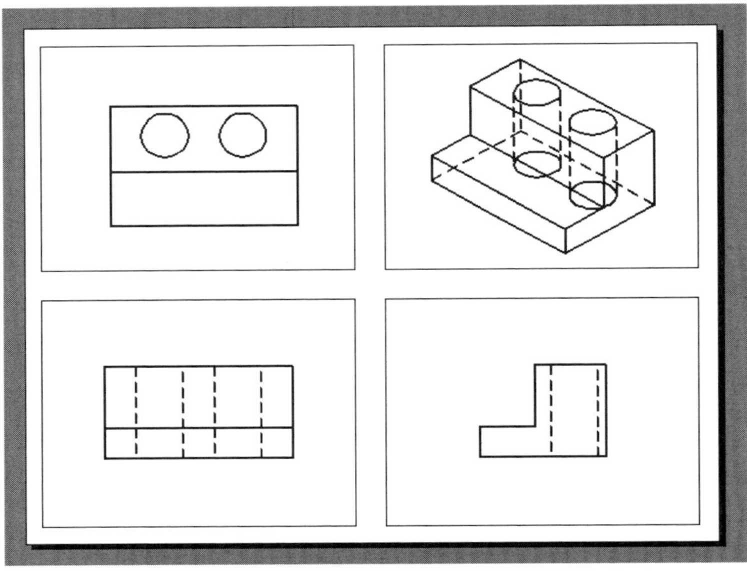

Option 3. Hidden lines a different color.
You may want the hidden lines to be a different color. To accomplish this you must change the **OBSCUREDCOLOR** setting.

1. On the command line type **obscuredcolor <enter>**
2. Enter new value for OBSCUREDCOLOR<0>: ***type the color number <enter>***

To print in color you must select a "color dependent plot style table" with color assigned. Refer to Appendix-C for instructions.

How to plot with a shaded view

1. First shade the model inside the viewport. **View / Visual Styles** Select the individual viewport and make the change.
2. Now return to Paperspace.
3. Select the shaded viewport frame. (Click once on the frame or use crossing window)
4. Right click and select **Shade Plot / As Displayed** from the menu.
5. Select **FILE / PLOT**
6. Select the Plot settings.
7. Select the **PREVIEW** button.
8. Select **OK**

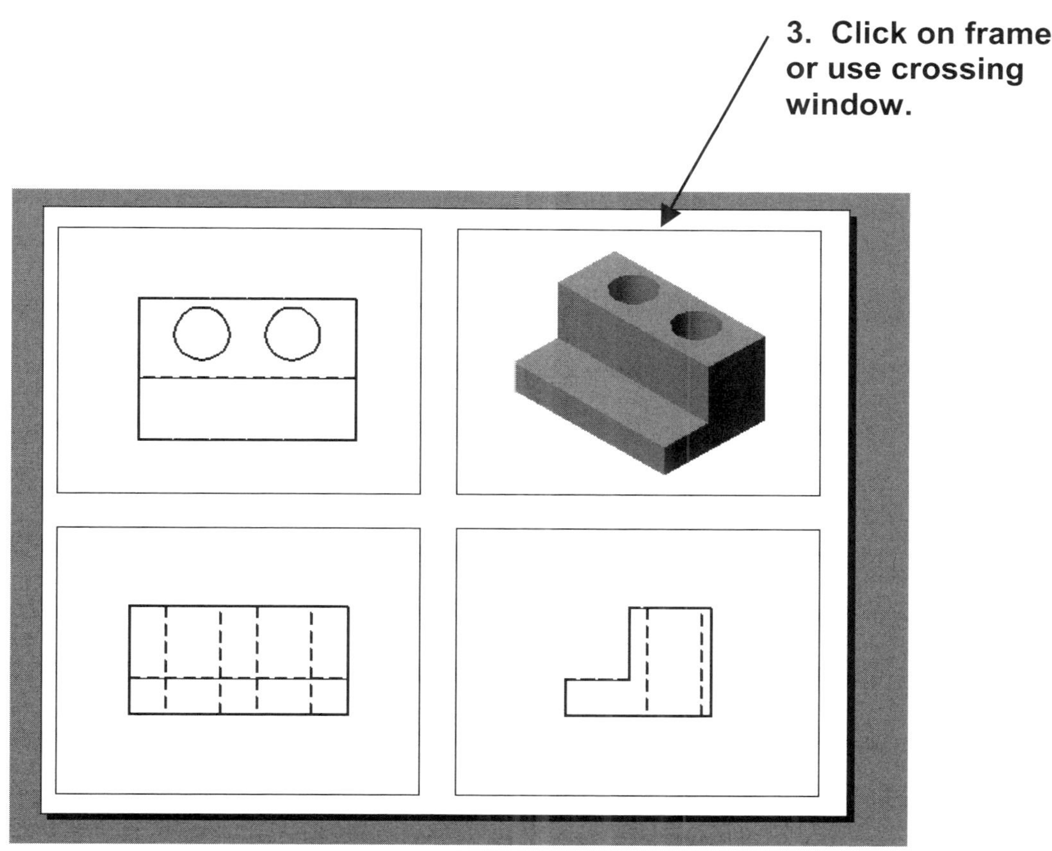

DIMENSIONING MULTIPLE VIEWS

It is important to dimension in Paperspace.

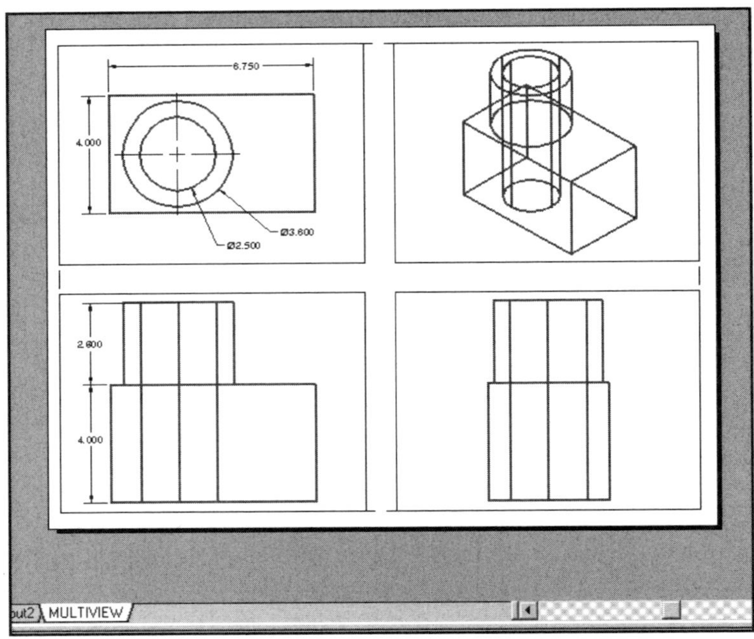

If you dimension in modelspace, each dimension will be visible in all viewports.

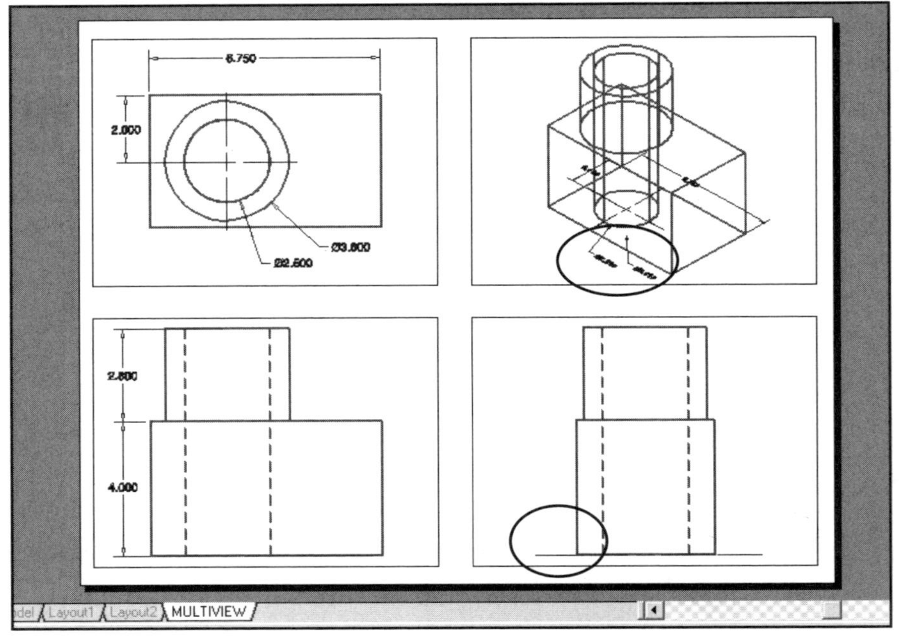

SHELL

The SHELL command allows you to hollow out the insides of a solid leaving a thin wall.

How to Shell a solid object:

1. Draw a solid object such as a box.

2. Select the Shell command using one of the following:

   TYPE = SHELL
   PULLDOWN = MODIFY / SOLID EDITING / SHELL
   SOLID EDITING TOOLBAR =

   Command: _solidedit
   Solids editing automatic checking: SOLIDCHECK=1
   Enter a solids editing option [Face/Edge/Body/Undo/eXit] <eXit>: _body
   Enter a body editing option
   [Imprint/seParate solids/Shell/cLean/Check/Undo/eXit] <eXit>: _shell

3. Select a 3D solid: ***select the solid object <enter>***

4. Remove faces or [Undo/Add/ALL]: ***select the top face***

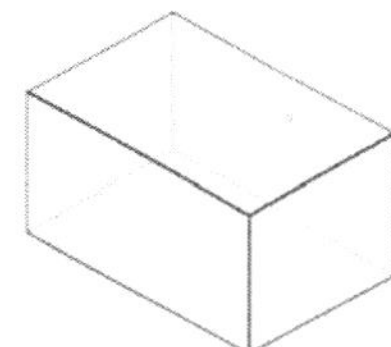

5. Remove faces or [Undo/Add/ALL]: 1 face found, 1 removed. ***(No visual notification from AutoCAD that you selected a face except this confirmation. Watch for it)***

6. Remove faces or [Undo/Add/ALL]: ***select more faces or <enter> to stop***

7. Enter the shell offset distance: ***enter the wall thickness***

   Solid validation started.
   Solid validation completed.

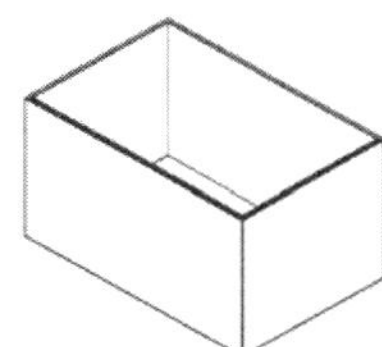

8. Enter a body editing option [Imprint/seParate solids/Shell/cLean/Check/Undo/eXit] <eXit>: ***<enter>***

   Solids editing automatic checking: SOLIDCHECK=1

9. Enter a solids editing option [Face/Edge/Body/Undo/eXit] <eXit>: ***<enter>***

UCSFOLLOW

The **UCSFOLLOW** system variable command is a very useful tool if you like using plan views to draw your 3D models.

The default setting is OFF (0). If it is ON (1) each time you change the UCS the drawing automatically displays the plan view.

You may turn this on or off at anytime.

How to set the UCSFOLLOW system variable:

1. Type on the command line: ***ucsfollow <enter>***
2. Enter new value for UCSFOLLOW <0>: ***enter "1"***

EXAMPLE:

1. Set the ucsfollow system variable to "1".
2. Move the UCS to the lower left corner.

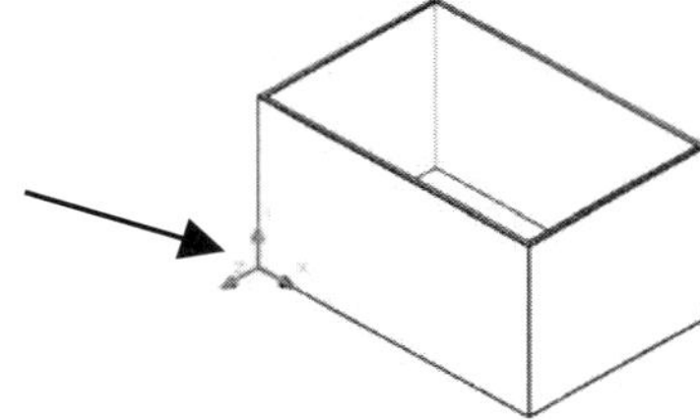

The display should have automatically displayed the current plan view.

3. Select the SE view.
4. Move the UCS to the top corner.

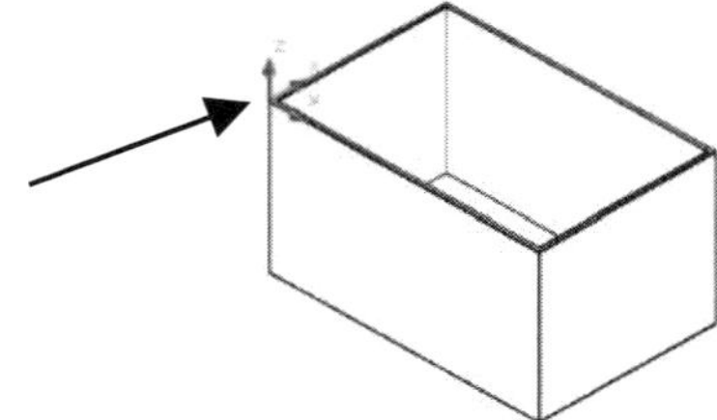

The display should have automatically displayed the current plan view.

EXERCISE 20A

MULTIVIEW PLOT SETUP

1. Open **EX-14E** in your 2D workspace.
2. Change Visual Styles to 3D Wireframe.
3. Create a **new** Layout tab.
4. Select the **HP 4MV** plotter and **11 X 17** paper. (just for this exercise)
5. Create a **multiview plot setup** as shown below.
   (Refer to the instructions on page 20-2 to 20-3)
6. Adjust the scale in each viewport to 3/8 and PAN if necessary.

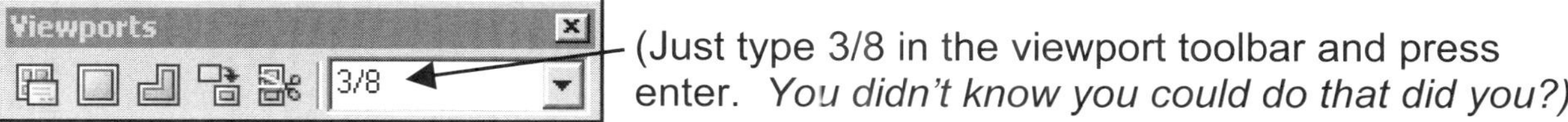

7. Save as **EX-20A**

Note: You are not plotting or previewing yet. This is just starting the page setup. Refer to Ex-20B for plotting.

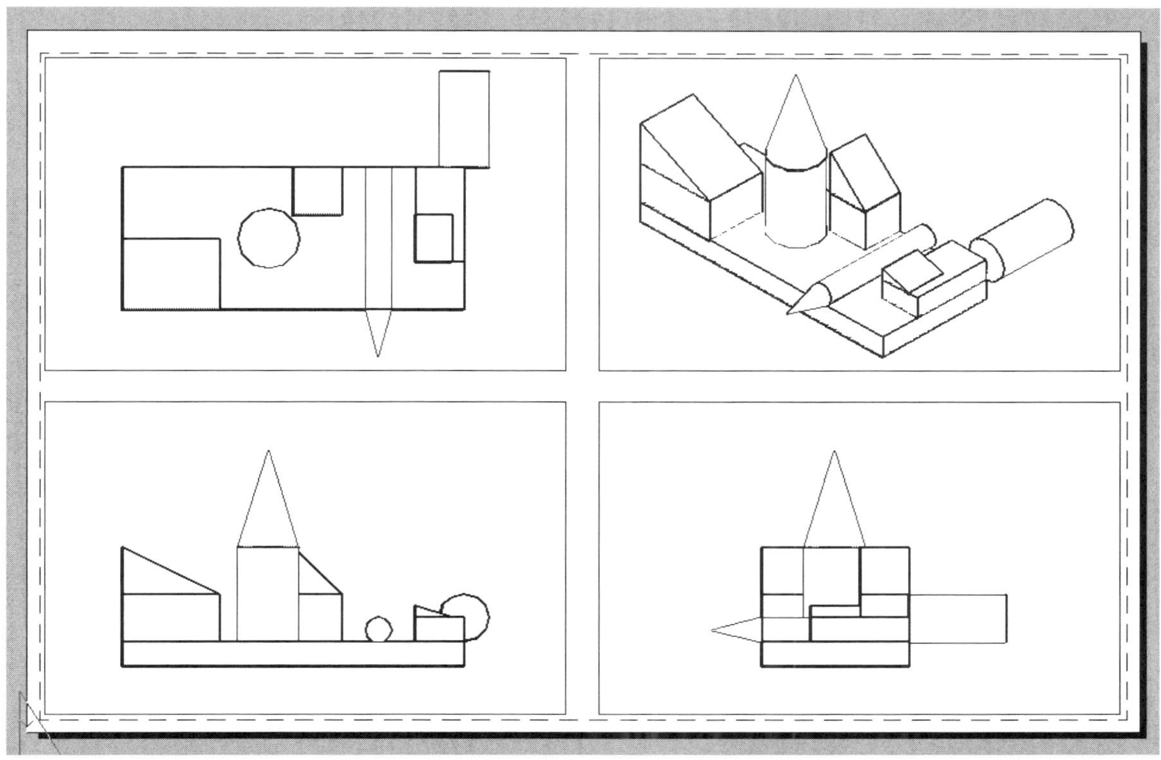

EXERCISE 20B
INVISIBLE HIDDEN LINES

1. Open **EX-20A**

2. Set the **SHADE PLOT** to **3D HIDDEN** in all of the viewports.
   (Refer to page 20-3 for instructions)

3. Select **FILE / PLOT PREVIEW**.

   Does your drawing look like the drawing below?

4. Save as **EX-20B**

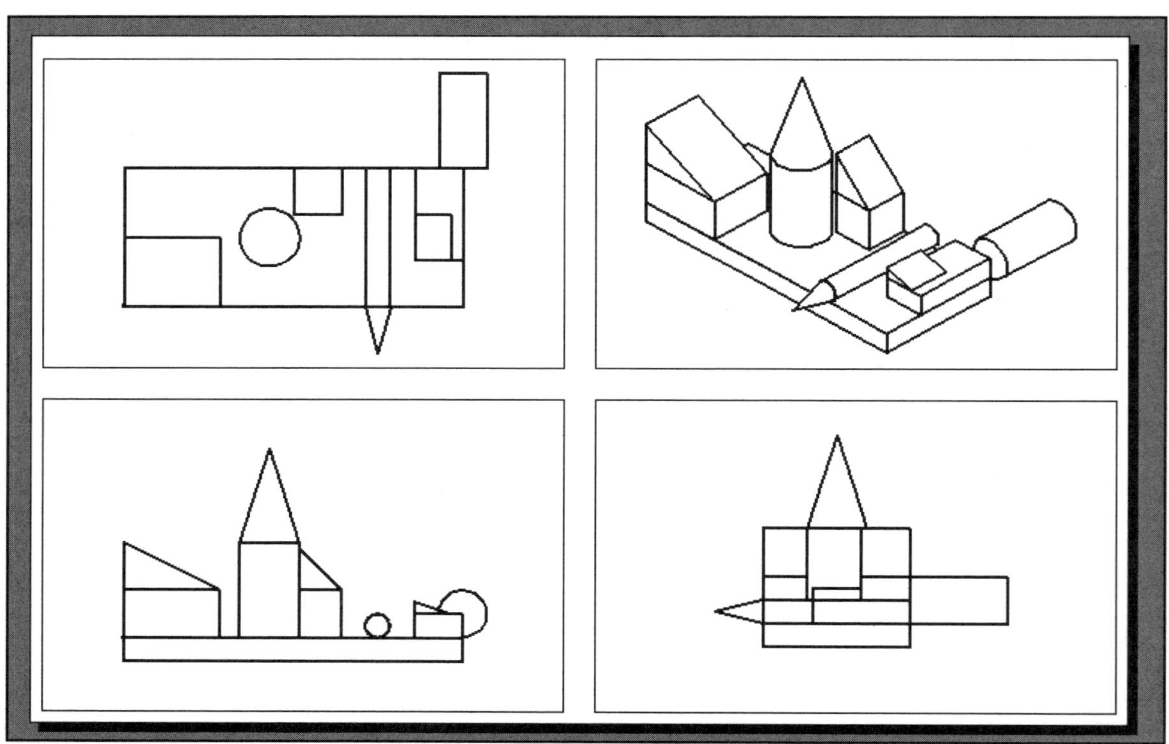

EXERCISE 20C

VISIBLE DASHED HIDDEN LINES

1. Open **EX-20B**

2. Set the **OBSCUREDLTYPE** variable to 2 to display dashed hidden lines. (Refer to page 20-4 for instructions)

3. Select **FILE / PLOT PREVIEW**.

   Does your drawing look like the drawing below?

4. Save as **EX-20C**

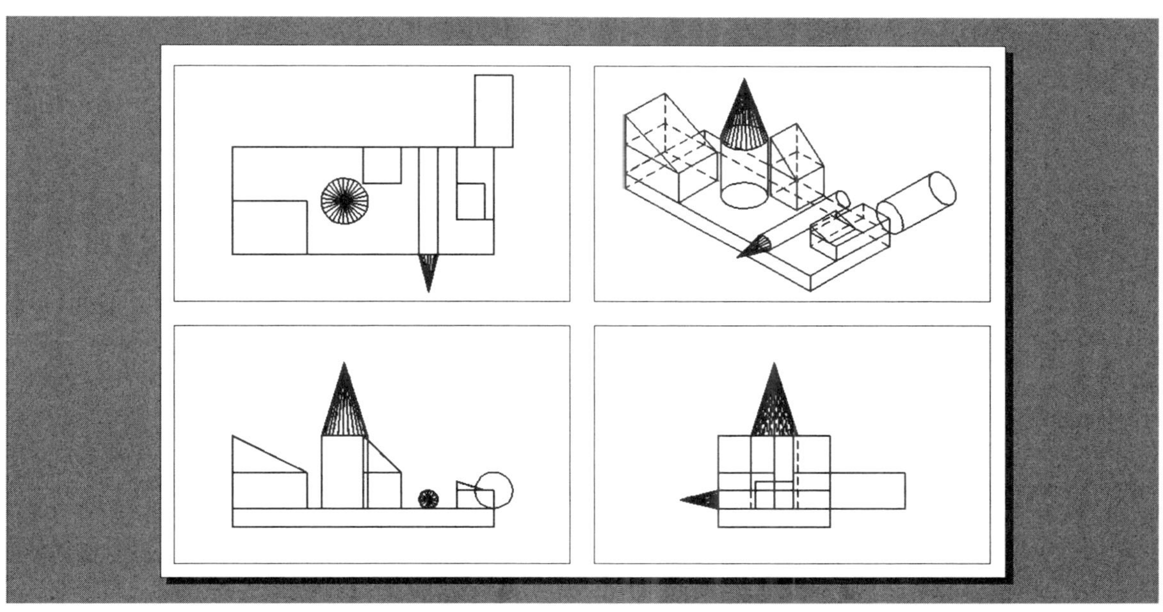

EXERCISE 20D
DISPLAY A SHADED VIEW

1. Open **EX-20C**

2. Set the **Obscuredltype** back to **"0"**

2. Shade the isometric view using Visual Styles Conceptual

3. Set the **SHADEPLOT** to "As Displayed".
   (Refer to page 20-5 for instructions)

4. Select **FILE / PLOT PREVIEW**.

   Does your drawing look like the drawing below?

5. Save as **EX-20D**

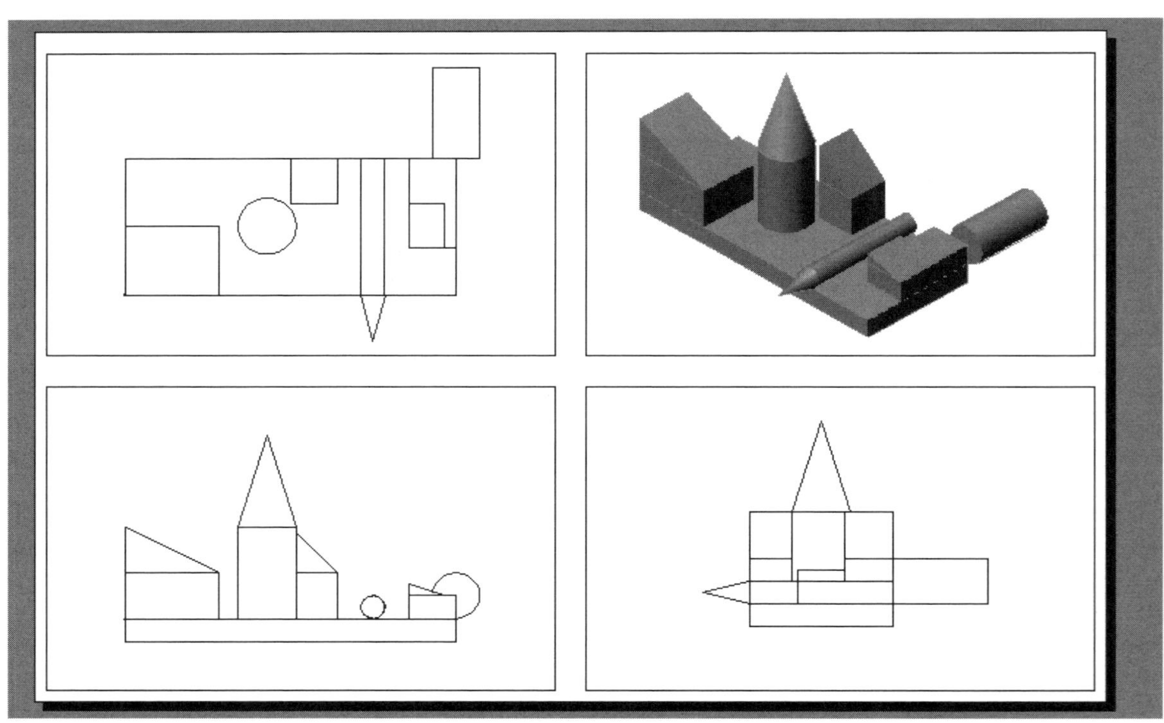

EXERCISE 20E

DIMENSIONING

1. Open **EX-20B (not 20D)**

2. Dimension as shown below.
   (Refer to page 20-6 for instructions.)

3. Save as **EX-20E**

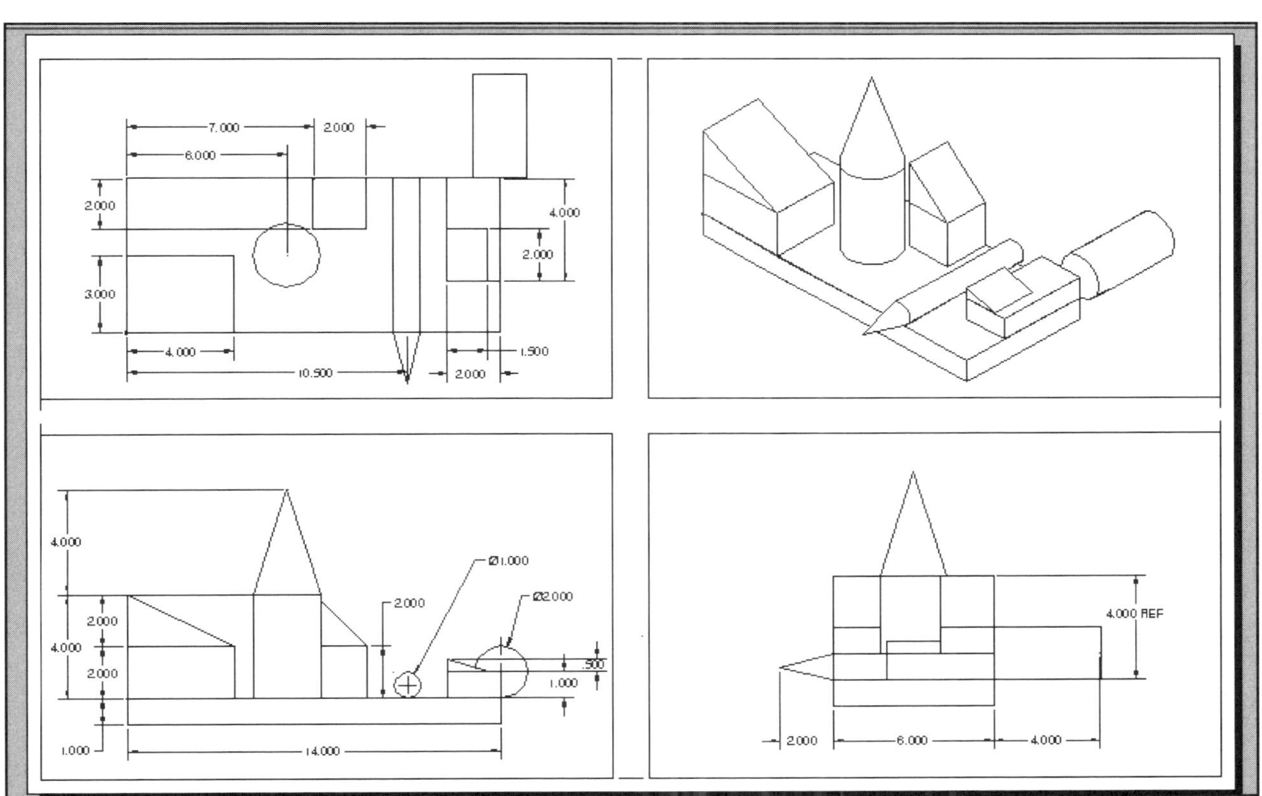

EXERCISE 20F

DIMENSIONING

1. **Select File / New and select <u>Acad3d.dwt</u>**

2. Draw the Box shown below. Place a cylinder at each end.
   Box = L = 8 W = 6 Ht = 6
   Cylinder = 2Rad. L = 3

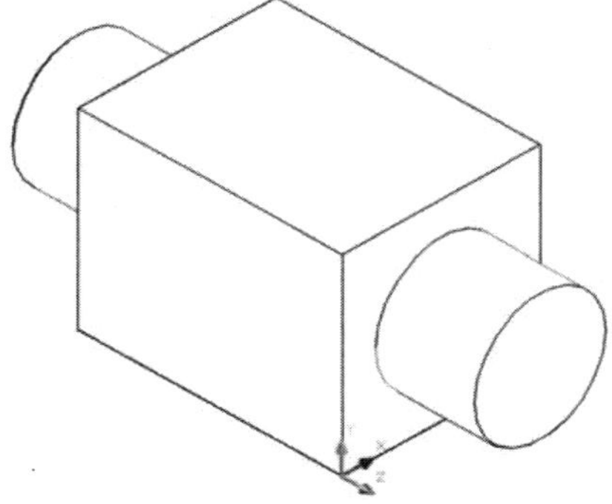

3. Union

4. Shell
   a. Remove the Top Surface
   b. Offset distance = .20

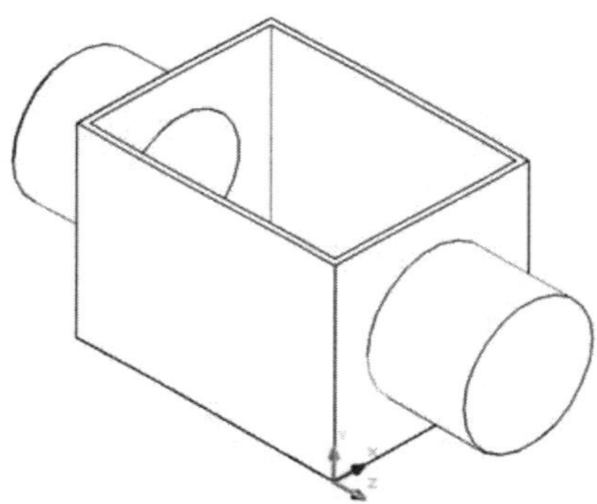

5. Slice through the center line as shown.

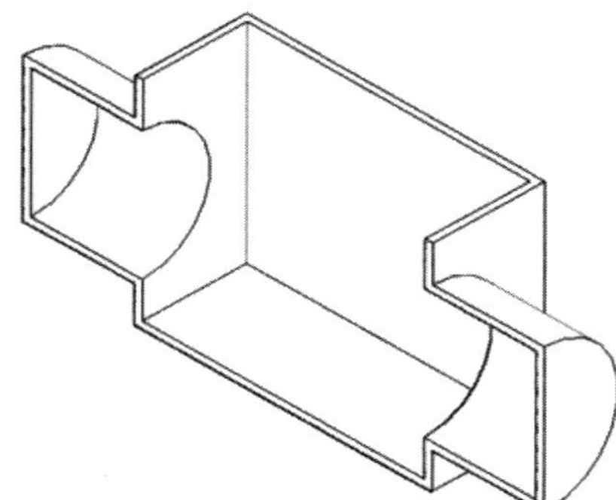

6. Save as **EX-20F**

LEARNING OBJECTIVES

After completing this lesson, you will be able to:

1. Create a Table.
2. Insert a Table.
3. Modify an existing Table.
4. Insert a Block into a Table.
5. Insert a Formula into a Table.
6. Create a Field.
7. Update a Field.
8. Edit an existing Field.

LESSON 21

TABLES

A Table is an object that contains data organized within columns and rows. AutoCAD's Table feature allows you to modify an existing Table Style or create your own Table Style and then enter text or even a block into the table cells. This is a very simple to use feature with many options.

This is an example of a Table.

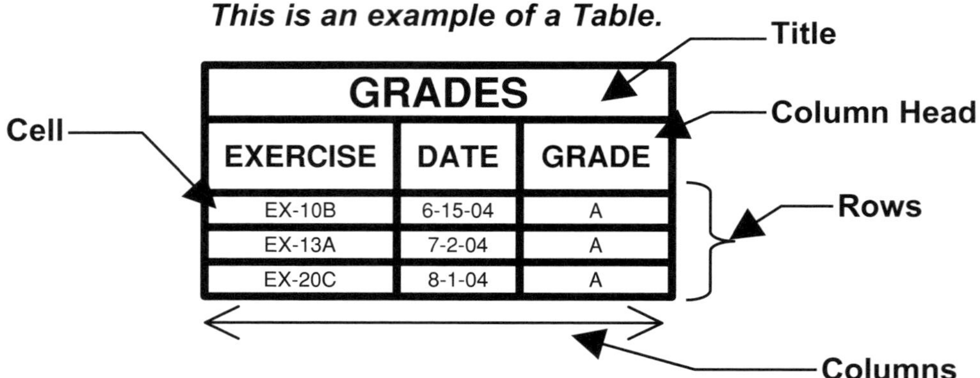

GRADES		
EXERCISE	**DATE**	**GRADE**
EX-10B	6-15-04	A
EX-13A	7-2-04	A
EX-20C	8-1-04	A

HOW TO CREATE A TABLE.

1. Select the Table Style command using one of the following:

   TYPE = tablestyle
   PULLDOWN = FORMAT / TABLE STYLE
   TOOLBAR = FORMAT

   The following dialog box will appear.

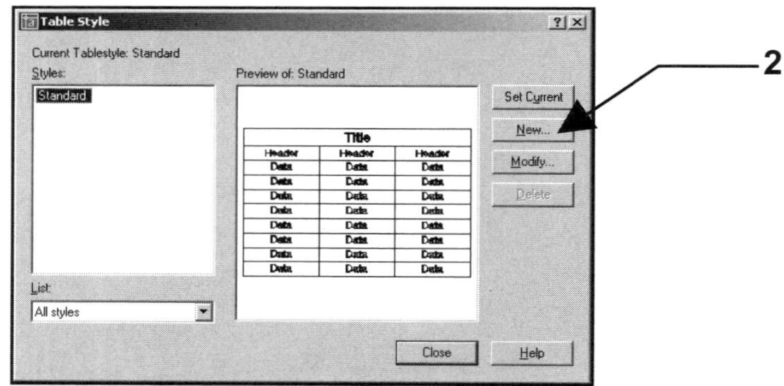

2. Select the **NEW** button. ***The following dialog will appear.***

3. Enter the new Table Style name. *Note: When you create a new table style you always "Start With" an existing style and you specify the differences.*

4. Select the **Continue** button.

5. Specify table properties for the Title, Column Heads and the Data.
   Notice the **3 tabs** in the upper left corner. The properties within each of these tabs is the same except for the following:

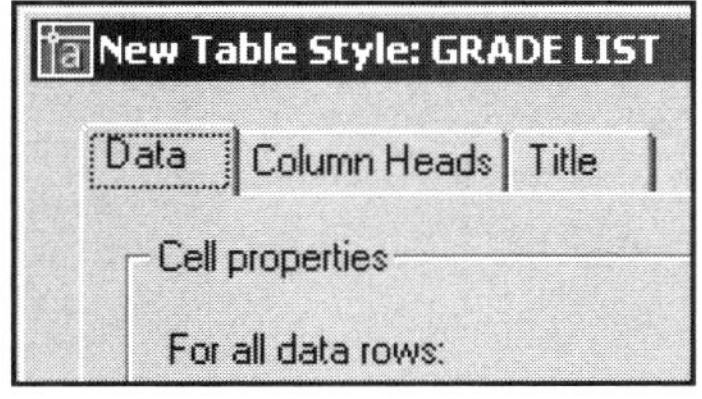

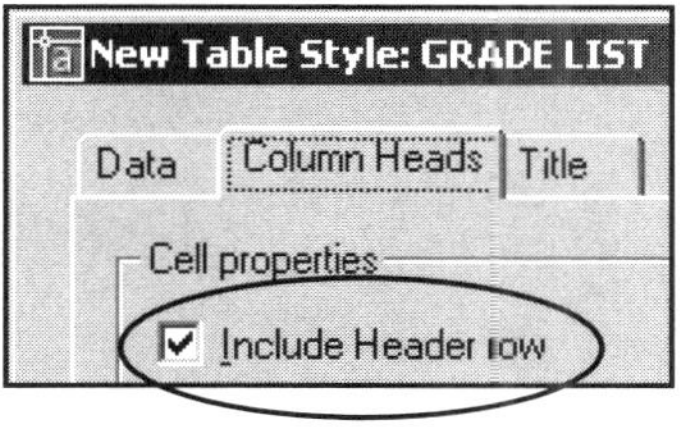

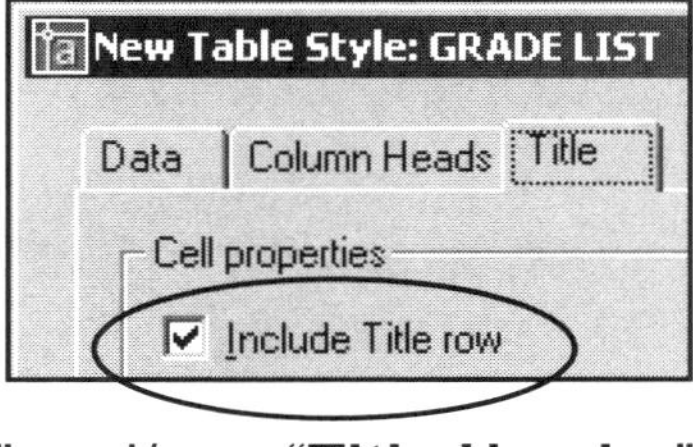

The check mark indicates that you want a "**Header for the Column**" and/or a "**Title Header**". If you do not want one or both, remove the check mark. The setting boxes on that tab will turn gray and you will not be required to enter information for that tab.

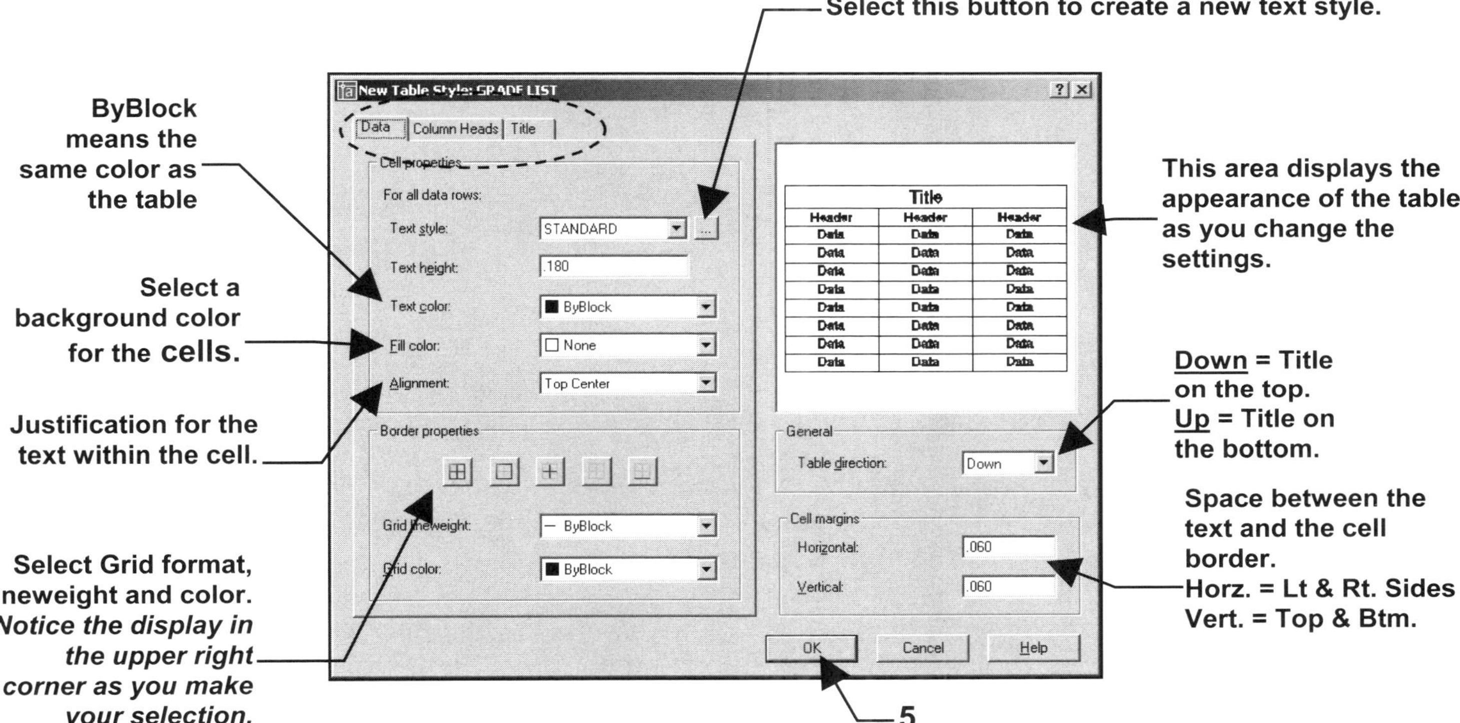

5. Select the **OK button** after you have completed all the settings necessary shown above.

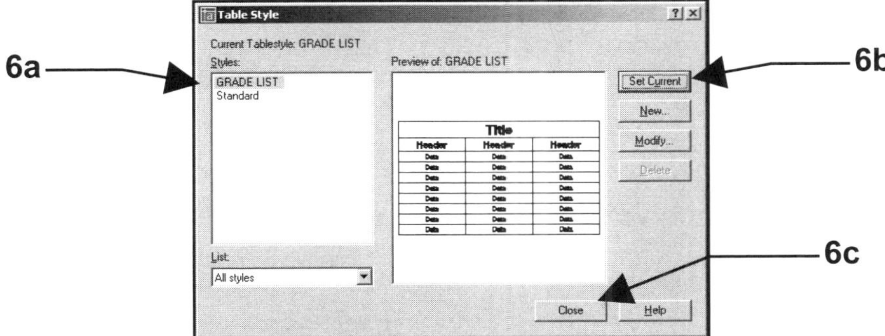

6. Select the **(a) New Table Style** and select the **(b) Set Current**" button and **(c) close**.

<u>Note: Tables are saved in the drawing. If you open another drawing this table will not be there. You must copy it using Design Center. (Refer to page 10-4)</u>

HOW TO INSERT A TABLE

1. Select **DRAW / TABLE.**

The following dialog will appear.

Click this button to create a new Table Style. Refer to page 21-2.

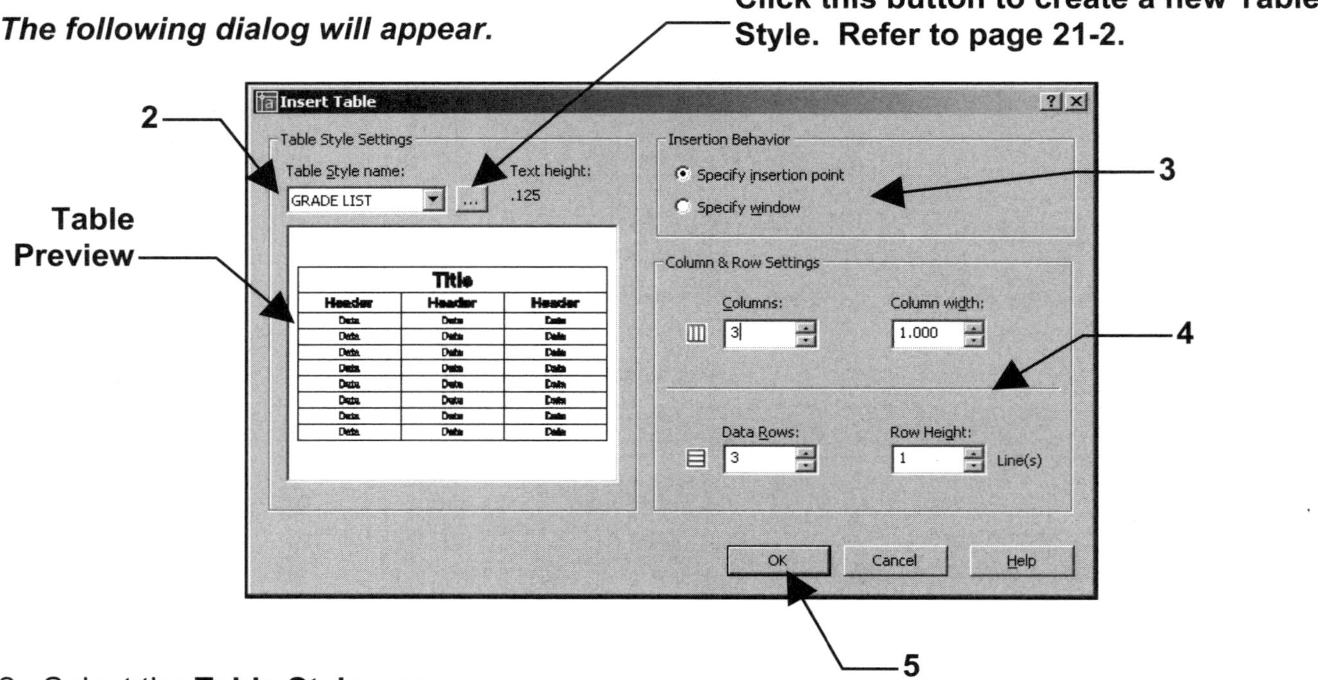

2. Select the **Table Style name**.

3. Select the Insertion Behavior.

   Specify Insertion Point: Specify the Columns, Column width, Rows and Rows Height. When you select the OK button you select the location to insert the table. The size of the table is determined by the settings you specified.

   Specify window: Specify the number of Columns and Row height. When you select the OK button you select the location for the upper left corner of the table. Then drag the cursor to specify the Column width and number of Rows, on the screen.

4. Specify the Column and Row specifications.

5. Select **OK.**

6. Place the insertion point. (Refer to #3 above.)

7. The Table is now on the screen waiting for you to fill in the data. Use the Tab or Arrow keys to move between the cells.

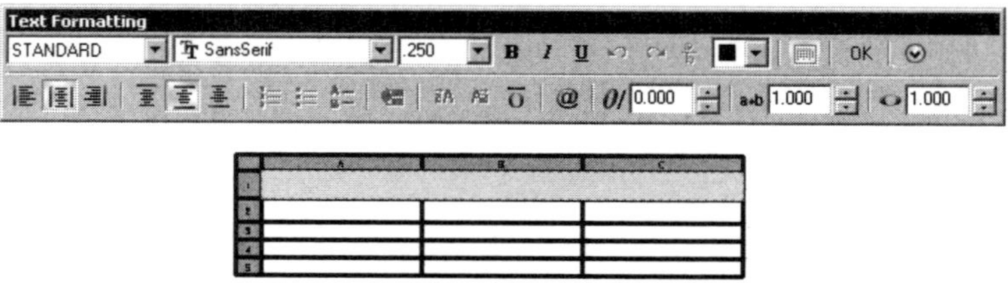

8. When you have filled all of the cells, select **OK.**

HOW TO INSERT A BLOCK INTO A TABLE CELL

1. Left click in the cell you wish to insert a block.
2. Right click to display the menu.
3. Select Insert Block from the menu.

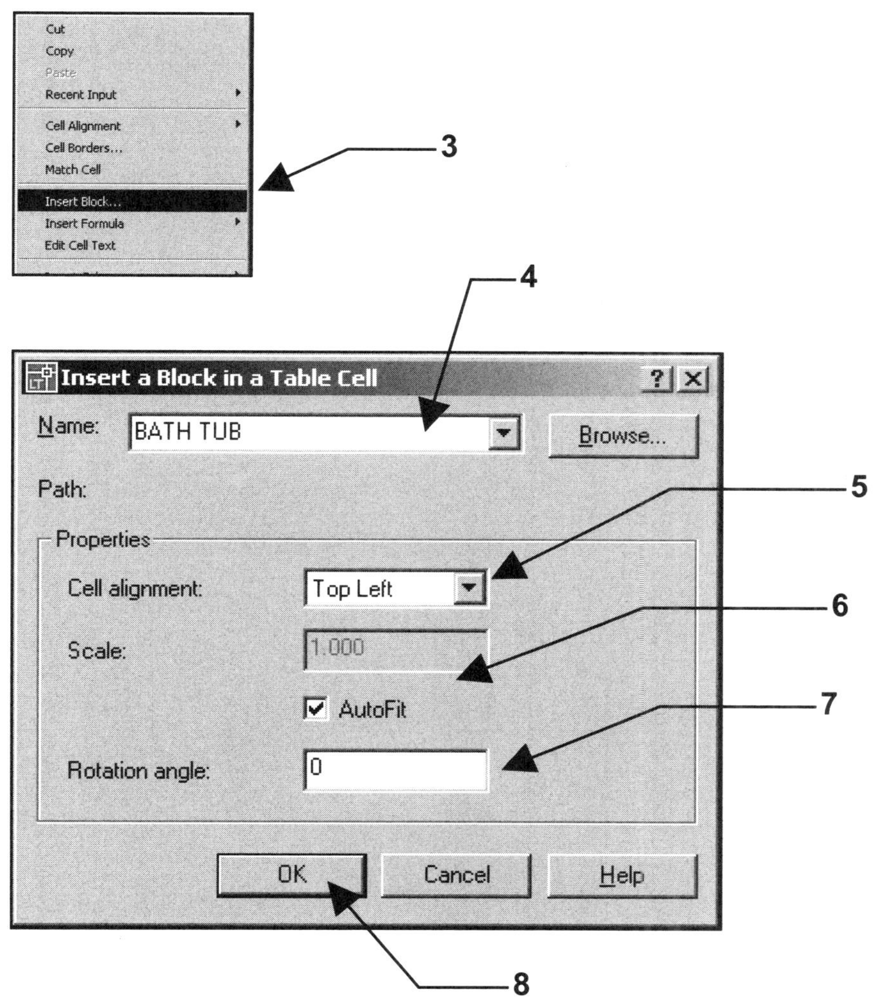

4. Select the **Block name**.
5. Select the **Cell alignment**.
6. Select the **Scale.**
   Note: AutoFit will automatically size the block to fit within the cell.
7. Select **Rotation angle**.
8. Select the **OK** button.

HOW TO INSERT A FORMULA INTO A CELL

You may apply simple numerical operations such as Sum, Average, Count, set cells equal to other cells or even add an equation of your own.

The following examples are for Sum and Average operations.

SUM

1. Click in the Cell that you wish to enter a formula.

ROOM	TABLES	CHAIRS	COST
B14	2	4	100
F22	3	6	400
G7	4	8	500
TOTAL			
AVERAGE			

1

2. Right click and select:
   Insert Formula / Sum

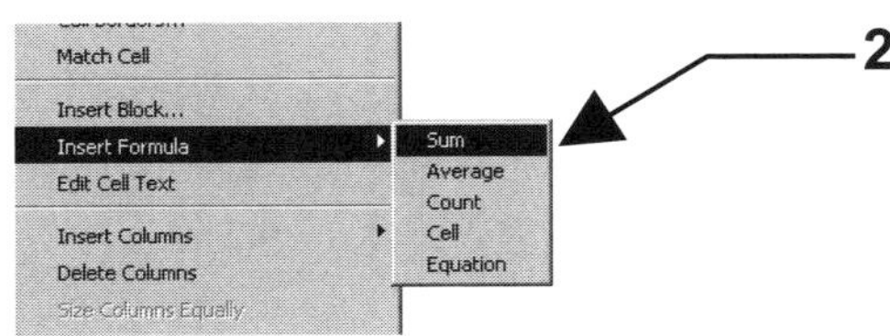

3. Select the cells content to sum using window.

ROOM	TABLES	CHAIRS	COST
B14	2	4	100
F22	3	6	400
G7	4	8	500
TOTAL			
AVERAGE			

3

4. Verify the formula and select OK.
   You may edit the formula if necessary.

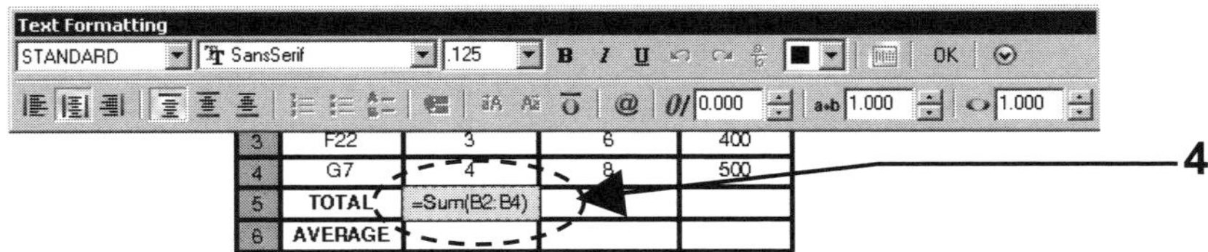

Notice the formula disappears and the sum of the cells selected has been calculated. Also the value is shaded to make you aware that this cell has a formula in it.

ROOM	TABLES	CHAIRS	COST
B14	2	4	100
F22	3	6	400
G7	4	8	500
TOTAL	9		
AVERAGE			

Sum

AVERAGE

1. Set the precision in **Format / Units**.
   Note: It is important to set the precision before you start the operation. This is only necessary for the Average operation. Each cell that you assign the Average formula to may have a different precision.

2. Click in the Cell that you wish to enter a formula.

ROOM	TABLES	CHAIRS	COST
B14	2	4	100
F22	3	6	400
G7	4	8	500
TOTAL	9		
AVERAGE			

2

3. Right click and select:
   Insert Formula / Average

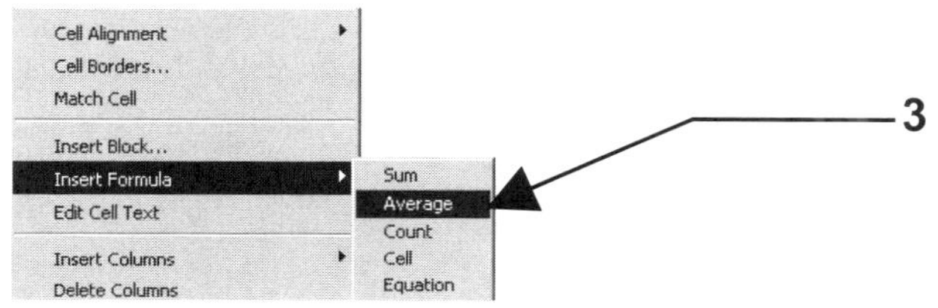

4. Select the cells content to average using a window.

ROOM	TABLES	CHAIRS	COST
B14	2	4	100
F22	3	6	400
G7	4	8	500
TOTAL	9		
AVERAGE			

4

5. Verify the formula and select OK.
   You may edit the formula if necessary.

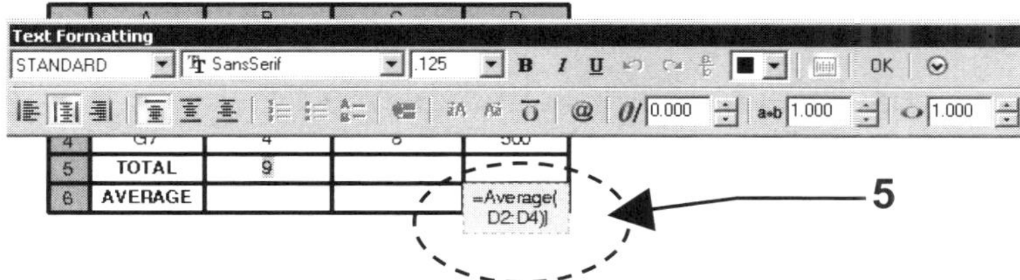

Notice the formula disappears and the average of the cells selected has been calculated.

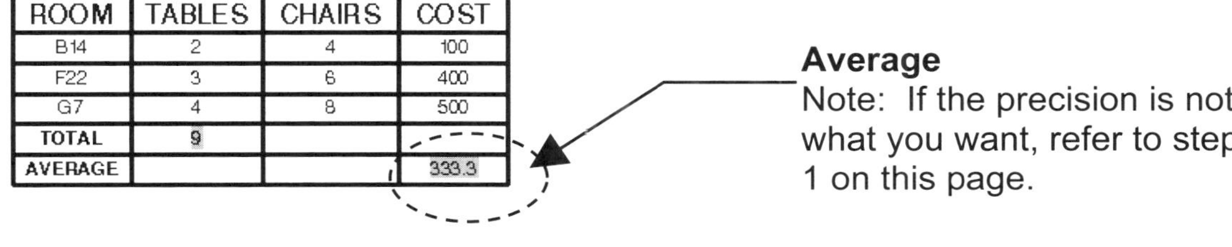

ROOM	TABLES	CHAIRS	COST
B14	2	4	100
F22	3	6	400
G7	4	8	500
TOTAL	9		
AVERAGE			333.3

Average
Note: If the precision is not what you want, refer to step 1 on this page.

HOW TO MODIFY AN EXISTING TABLE

EDIT TEXT WITHIN A CELL.

1. Double click inside the cell. (Be careful not to click on a border line.)
   The text formatting dialog box will appear.

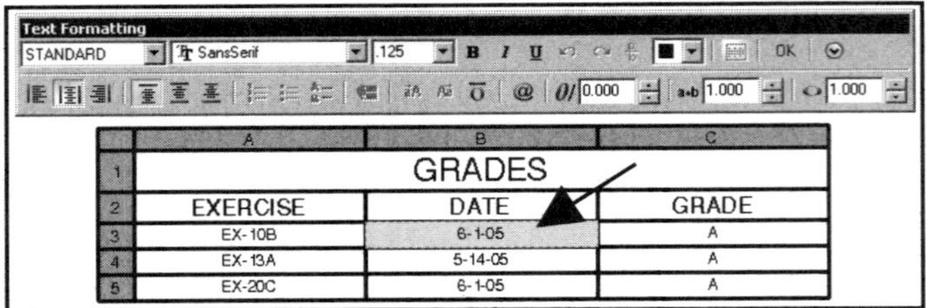

2. Make the text changes.
3. Select the OK button.

CHANGE THE COLUMN WIDTH OR ROW HEIGHT.

1. Click once inside the cell or row that you wish to change.
2. Right click and select "Properties" from the menu.

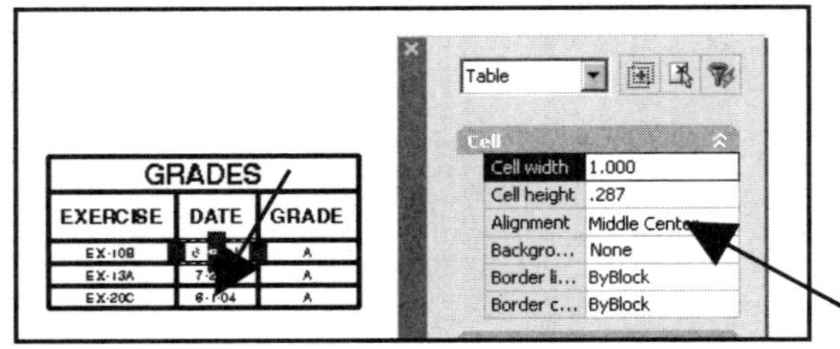

3. In the Properties Palette under Cell, click the cell width or the cell height value and enter the new value.

ADD A COLUMN OR ROW.

1. Click once inside a cell where you wish to add a column or a row.
2. Right click and select one of the following options from the menu:

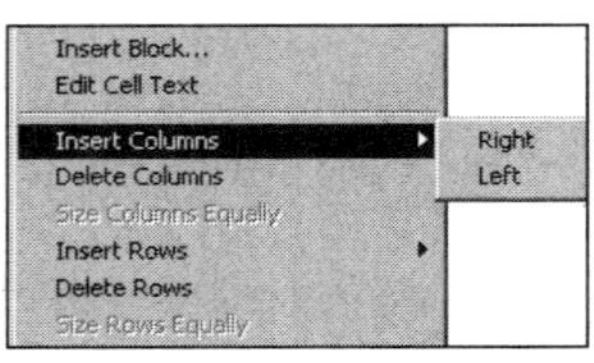

COLUMNS:
Right: Inserts a column to the right of the selected cell.
Left: Inserts a column to the left of the selected cell.

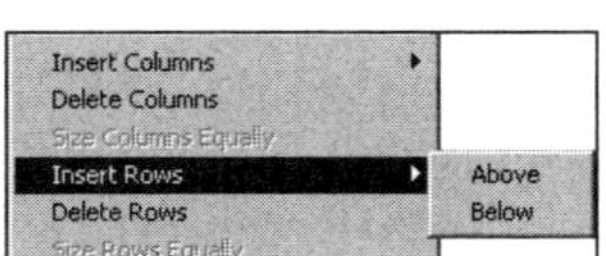

ROWS:
Above: Inserts a row above the selected cell.
Below: Inserts a row below the selected cell.

DELETE A COLUMN OR ROW.

1. Click once inside one of the cells in the column or row that you wish to delete.
2. Right click and select Delete Columns or Delete Rows from the menu.

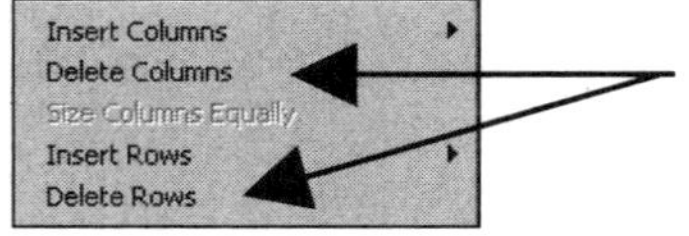

MODIFY A TABLE USING GRIPS

You may also modify tables using Grips. When editing with Grips, the left edge of the table remains stationary but the right edge can move. The upper left Grip is the Base Point for the table.

To use Grips, click on a table border line. The Grips should appear. Each Grip has a specific duty, shown below. To use a Grip, click on the Grip and it will change to red. Now click and drag it to the desired location.

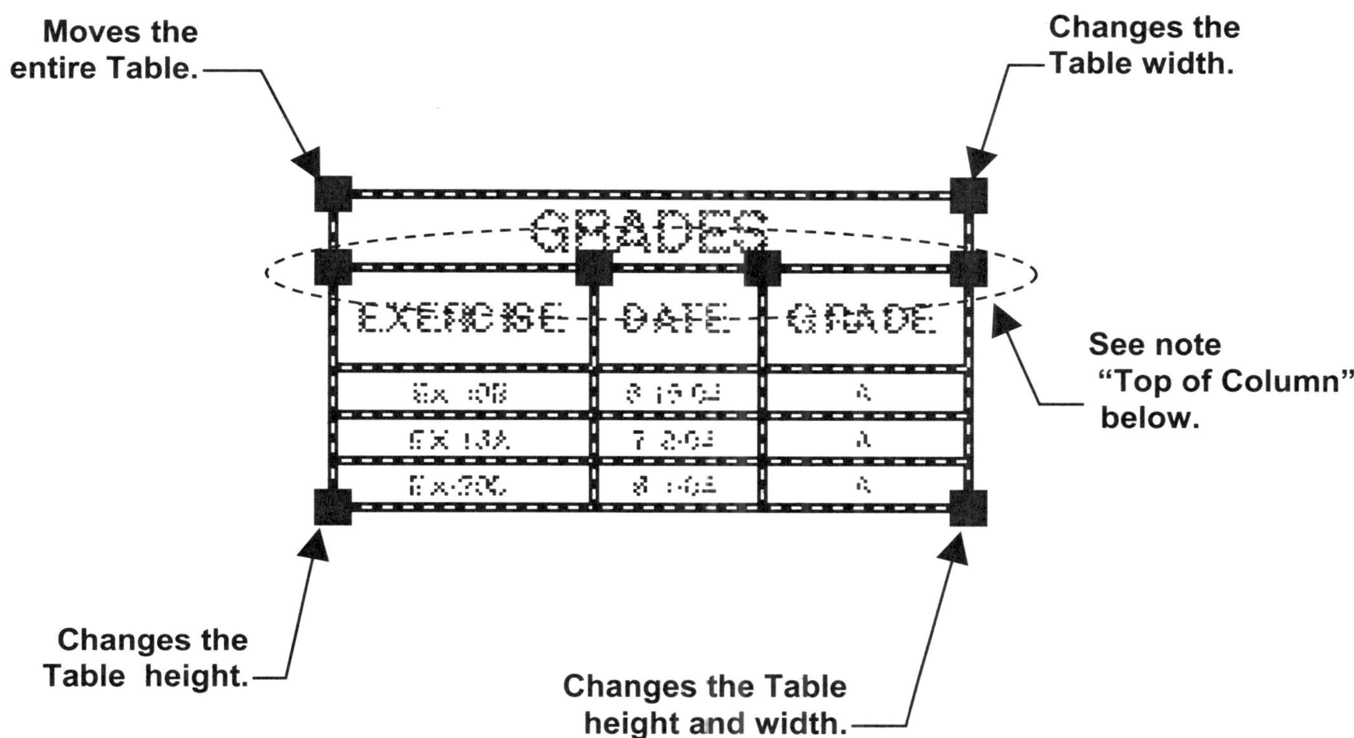

Top of Column
There is a Grip located at the top of each column line. These Grips adjust the width of the column <u>to the left</u> of the Grip. The entire Table adjusts at the same time. If you hold the CTRL key down while moving a column Grip, column will change but the width of the entire Table remains unchanged.

FIELDS (Not available in LT version)

A **Field** is a string of text that has been set up to display data that it gets from another source. For example, you may create a field that will display the Circumference of a specific Circle within your drawing. If you changed the diameter of that Circle you could "update" the field and it would display the new Circumference.

Fields can be used in many different ways. After you have read through the example below you will understand the steps required to create and update a Field. Then you should experiment with some of the other Field Categories to see if they would be useful to you. Consider adding Fields to a cell within a Table.

CREATE A FIELD

1. Draw a 2" Diameter Circle and place it anywhere in the drawing area.
2. Select "**Insert / Field...**".

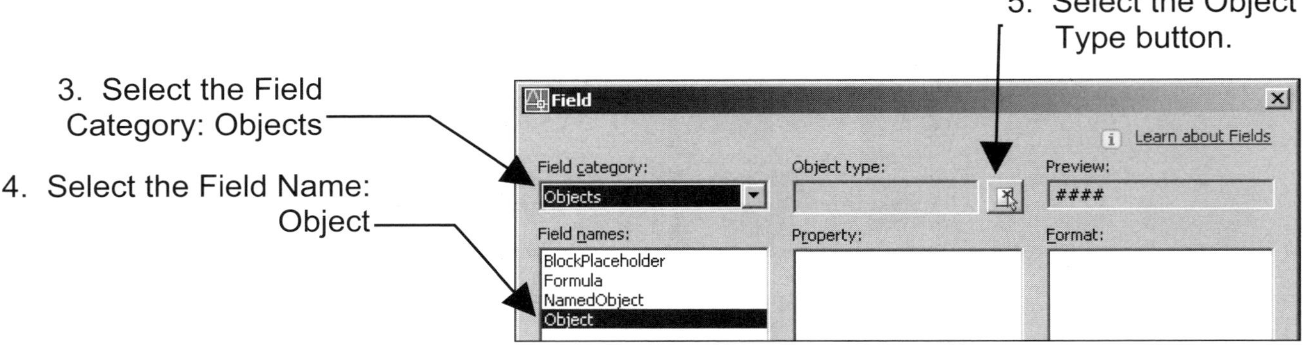

6. Select the Circle that you just drew.

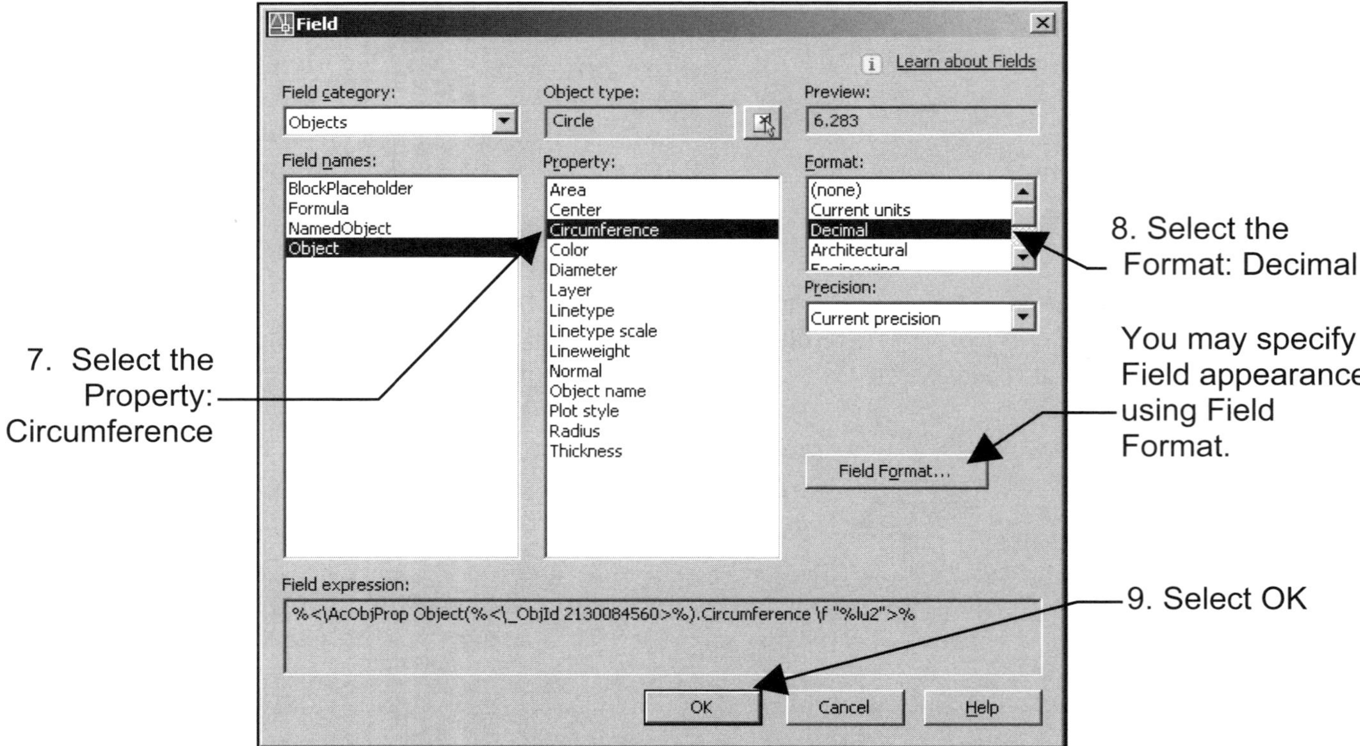

10. Place the "Field" inside the Circle.

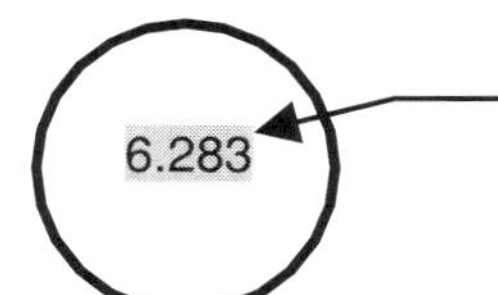

Note: Watch the command line closely. You may change the "Height" and "Justification" of the text before you place it.

Notice that the Field appears with a gray background. This background will not plot. The background can be turned off but I find it helpful to be able to visually distinguish a Field from plain text. If you wish to turn it off use: Tools / Options / User Preferences tab. In the Fields section, uncheck the "Display background of Fields" box.

UPDATE A FIELD

Now let's change the diameter of that Circle and see what happens to the Field.

1. Change the size of the Circle.
   a. Select the Circle.
   b. Right click and select **Properties.**
   c. Change the Diameter to 4.
   d. Close the Properties Palette.
   e. Press ESC key to clear the Grips.

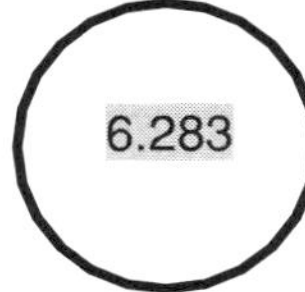

Notice that the Field has not changed yet.
In order to see the new Circumference value, you must "Update the Field".

2. Select **Tools / Update Fields** or **View / Regen.**

3. Select the Field and <enter>.

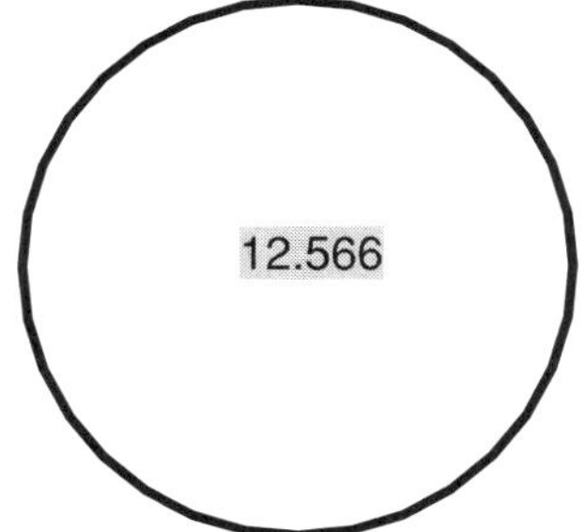

The Field updated to the new Circumference value.

Note: The Field will update automatically each time you Save, Plot or Regen the drawing.

Editing FIELDS

Editing a Field is very easy. The process is basically the same as creating a Field.

EDIT A FIELD

1. Double click the **Field text**. *The Multiline Text Editor will open.*
2. Right click on the **Field text**.
3. Select **Edit Field** from the menu.
4. Make the changes.
5. Select **OK**.
6. You may also add more text to the Field text but you may not change the text within the Field.
   Example: Add the words "Circle Circumference" under the Field text on the previous pages.
7. Select **OK** to exit the Multiline Text Editor.

ADD A FIELD TO A TABLE CELL

1. Select the cell.
2. Right click.
3. Select "Insert Field" from the menu.
4. Create a Field as described on page 21-10.

Questions about Fields

1. What happens when you Explode a Field?

The Field will convert to normal text and will no longer update.

2. What happens if someone opens the drawing in another version like AutoCAD 2000 or LT?

The last value displayed will appear but it will not be a Field.

3. Will all of my Fields be ruined now?

No. When you open the drawing in AutoCAD 2006 all the Fields should return.

EXERCISE 21A

Create a Table

1. Open **My Decimal Setup.**

2. Create a Table. (Refer to page 21-2)

    a. Name = CIRCLE INFORMATION.
    b. Start With = “Standard” table.
    c. TITLE: Text Style = Class Text Text Ht = .250 Fill color = Green
    d. COLUMN HEAD: Text Style = Class Text Text Ht = .180 Fill color = Yellow
    e. DATA: Text Style = Class Text Text Ht = .125 Fill color = Cyan

3. Set it Current.

4. Save the drawing as: **EX-21A.**

EXERCISE 21B

Draw a Table

1. Open **EX-21A.**

2. Draw the Table shown below. (Refer to page 21-4)

   Columns = 4 Column Width = 1.00

   Data Rows = 3 Row Height = 1

3. Enter the Title and Column Heads.

4. Save as: **EX-21B**

CIRCLE INFORMATION			
ITEM	DIA	CIR	AREA

Note: If your model space display color is "White", when you enter text into the table, the background for each cell may be Black. If the text is also black it will be difficult to see the text as you type. You may find it necessary to change the display color of model space to Black temporarily while entering data to tables. Refer to page 2-17.

EXERCISE 21C

Modify an existing Table

1. Open **EX-21B.**

2. Modify the Column Head "CIR" as shown below.

   a. Change the Cell Width to: 2.750 (Refer to page 21-8)
   b. Change the Column Head text to: CIRCUMFERENCE

3. Add the data for "ITEM" and "DIA" columns.

4. Save as: **EX-21C**

CIRCLE INFORMATION			
ITEM	**DIA**	**CIRCUMFERENCE**	**AREA**
1	2		
2	3		
3	4		

EXERCISE 21D

Add Fields to an existing Table

1. Open **EX-21C.**

2. Draw 3 Circles and place their item number in the middle, as shown below:

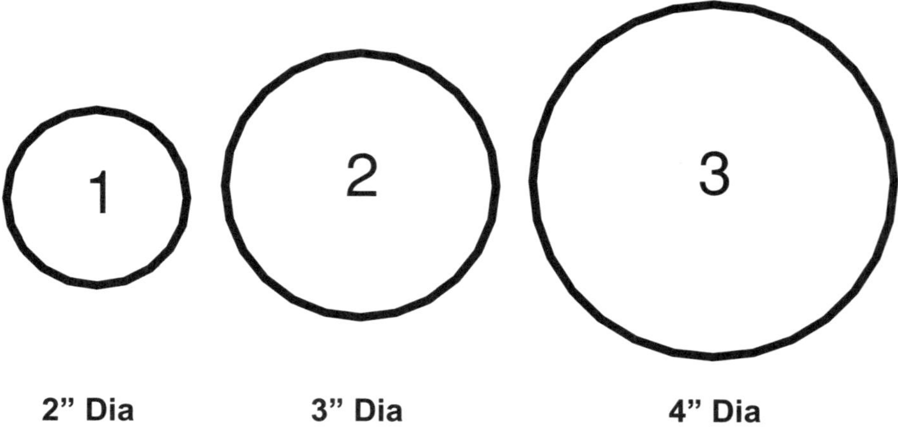

3. Add the FIELDS for CIRCUMFERENCE and AREA in the appropriate Data Cells within the Circle Information Table shown below. (Refer to page 21-12)

Text Ht. = .125

CIRCLE INFORMATION			
ITEM	DIA	CIRCUMFERENCE	AREA
1	2	6.283	3.142
2	3	9.425	7.069
3	4	12.566	12.566

4. Save as: **EX-21D**

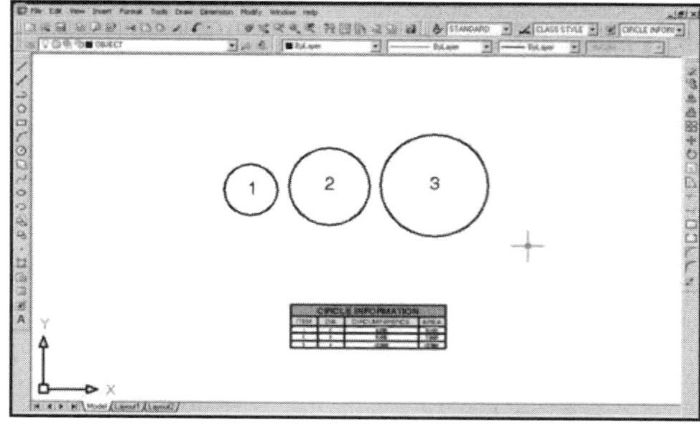

EXERCISE 21E

Update a Field

1. Open **EX-21D.**

2. Modify the Diameters of the 3 Circles using the Properties Palette.

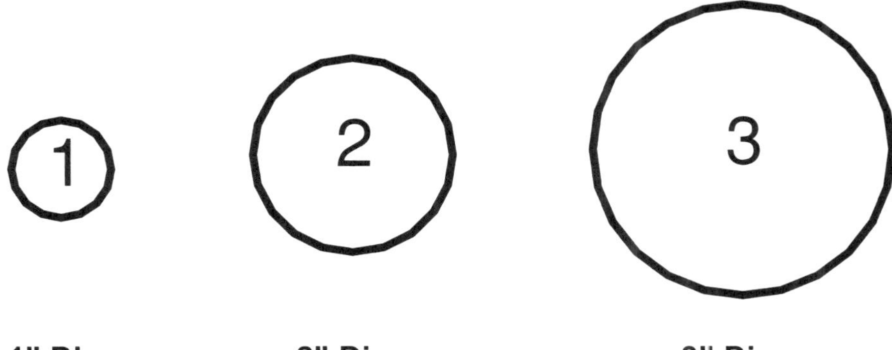

Change to: **1" Dia** **2" Dia** **3" Dia**

3. Edit the "Dia" Column to reflect the modification.

4. Update the Fields. (Refer to page 21-11)

CIRCLE INFORMATION			
ITEM	**DIA**	**CIRCUMFERENCE**	**AREA**
1	1	3.142	0.785
2	2	6.283	3.142
3	3	9.425	7.069

3

4

5. Save as: **EX-21E**

NOTES:

ARCHITECTURE

ARCHITECTURAL SYMBOL LIBRARY

When you are using a CAD system, you should make an effort to only draw an object once. If you need to duplicate the object, use a command such as: Copy, Array, Mirror or Block. This will make drawing with CAD more efficient.

In the following exercise, you will create a file full of architectural symbols that you will use often when creating an architectural drawing. You will create them once and then merely drag and drop them, from the DesignCenter, when needed. This will save you many hours in the future.

Save this library file as **Library** so it will be easy to find when using the DesignCenter or create a Library Palette.

1. Open **My Feet-Inches Setup**
2. Select the **Qtr equals foot** tab.
3. Draw each of the Symbol objects, shown on the following pages, actual size. Do not scale them.
4. Create an individual Block for each one using the **BLOCK** command.
   Use the number as the name. The actual name is too long.
5. Save this drawing as: **Library**
6. Plot the drawing, of your library symbols, for reference.
   The format is your choice.
   Use Page setup: **24 X 18 ARCH**

Consider creating a library palette. Refer to page 10-11.

SYMBOLS		INSTRUCTIONS
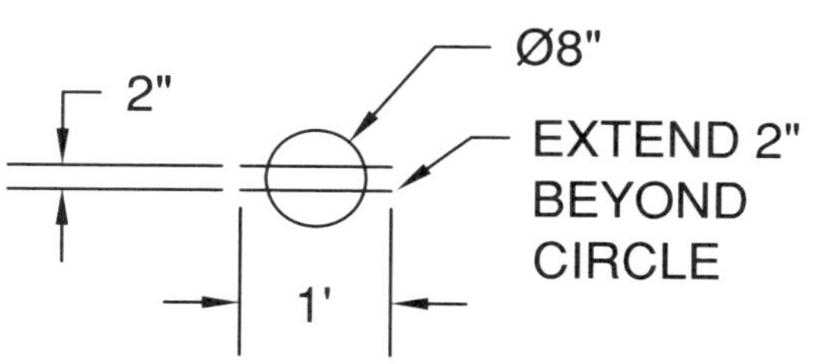 Ø8" · 2" · EXTEND 2" BEYOND CIRCLE · 1' DUPLEX CONVENIENCE OUTLET	1	LAYER = ELECTRICAL
 GFI GROUNDED DUPLEX OUTLET	2	LAYER = ELECTRICAL GFI TEXT HEIGHT= 3" POSITION GFI TEXT APPROXIMATELY AS SHOWN
 GFI · WP GROUNDED WEATHER PROOF OUTLET	3	
 220V 220 VOLT OUTLET	4	
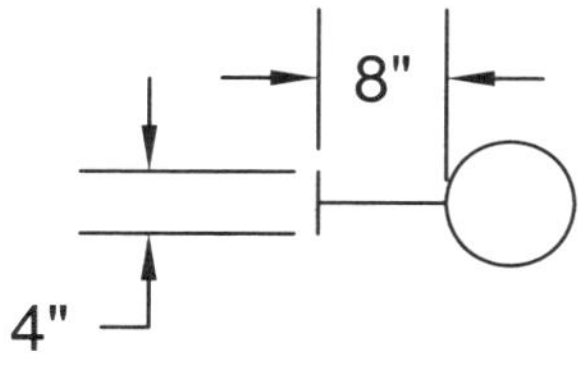 8" · 4" WALL MOUNTED FIXTURE W/INCANDESCENT LAMP	5	DIMENSIONS NOT SHOWN ARE THE SAME AS SYMBOL NUMBER 1. LAYER = ELECTRICAL
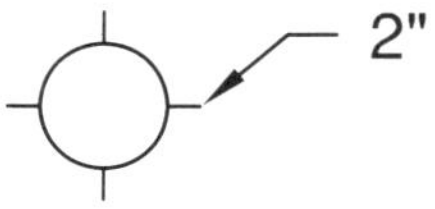 2" CEILING MOUNTED FIXTURE W/FLUORESCENT LAMP	6	
SUSPENDED FIXTURE W/INCANDESCENT LAMP	7	
2" RECESSED FIXTURE W/INCANDESCENT LAMP	8	

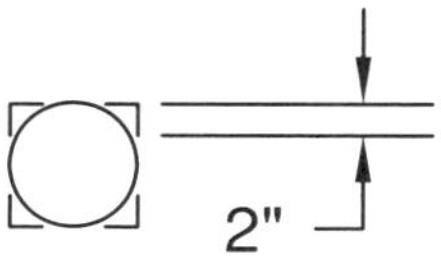

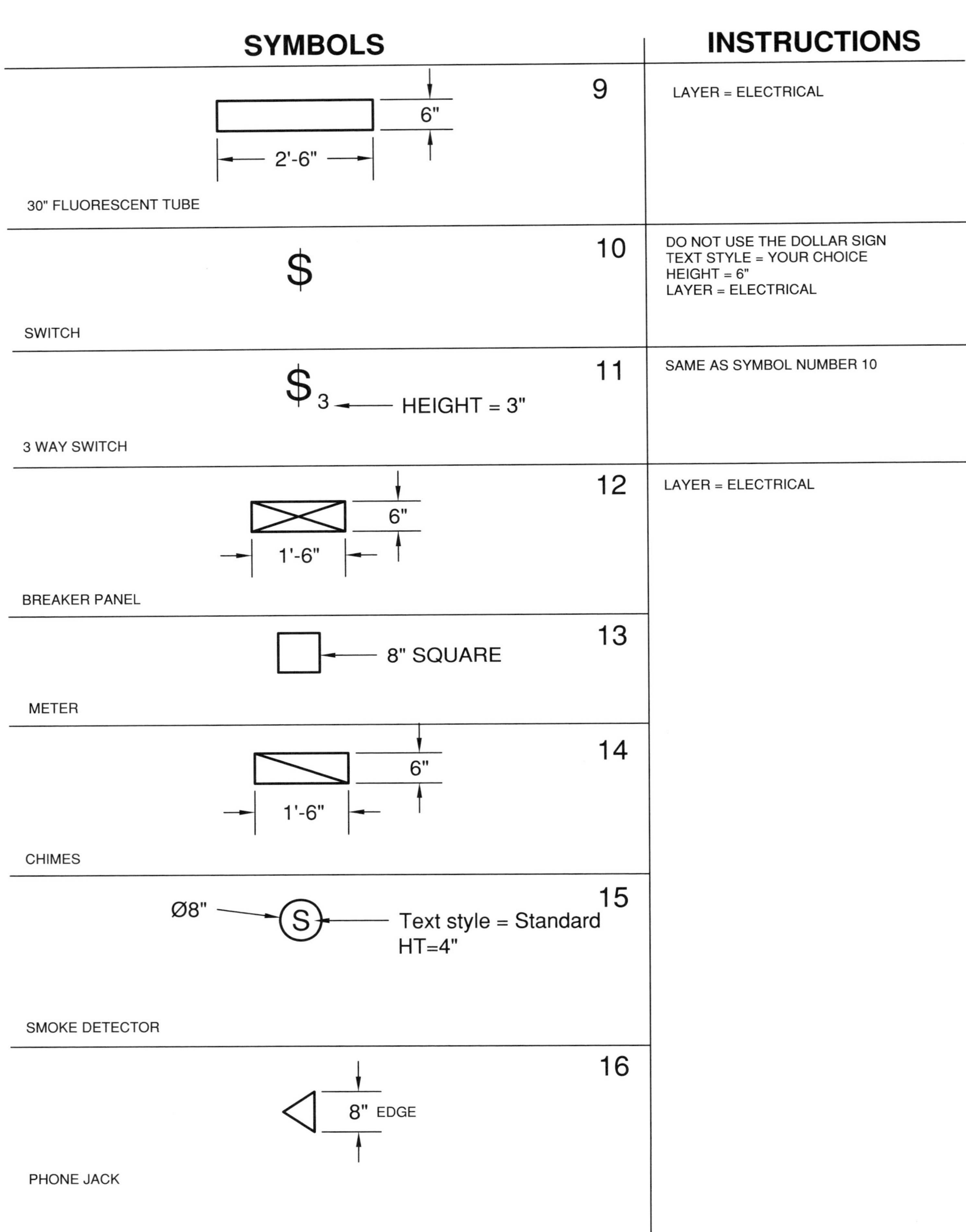
SYMBOLS
INSTRUCTIONS
9
6"
2'-6"
30" FLUORESCENT TUBE
LAYER = ELECTRICAL
10
$
SWITCH
DO NOT USE THE DOLLAR SIGN
TEXT STYLE = YOUR CHOICE
HEIGHT = 6"
LAYER = ELECTRICAL
11
$ 3
HEIGHT = 3"
3 WAY SWITCH
SAME AS SYMBOL NUMBER 10
12
6"
1'-6"
BREAKER PANEL
LAYER = ELECTRICAL
13
8" SQUARE
METER
14
6"
1'-6"
CHIMES
15
Ø8"
S
Text style = Standard
HT=4"
SMOKE DETECTOR
16
8" EDGE
PHONE JACK

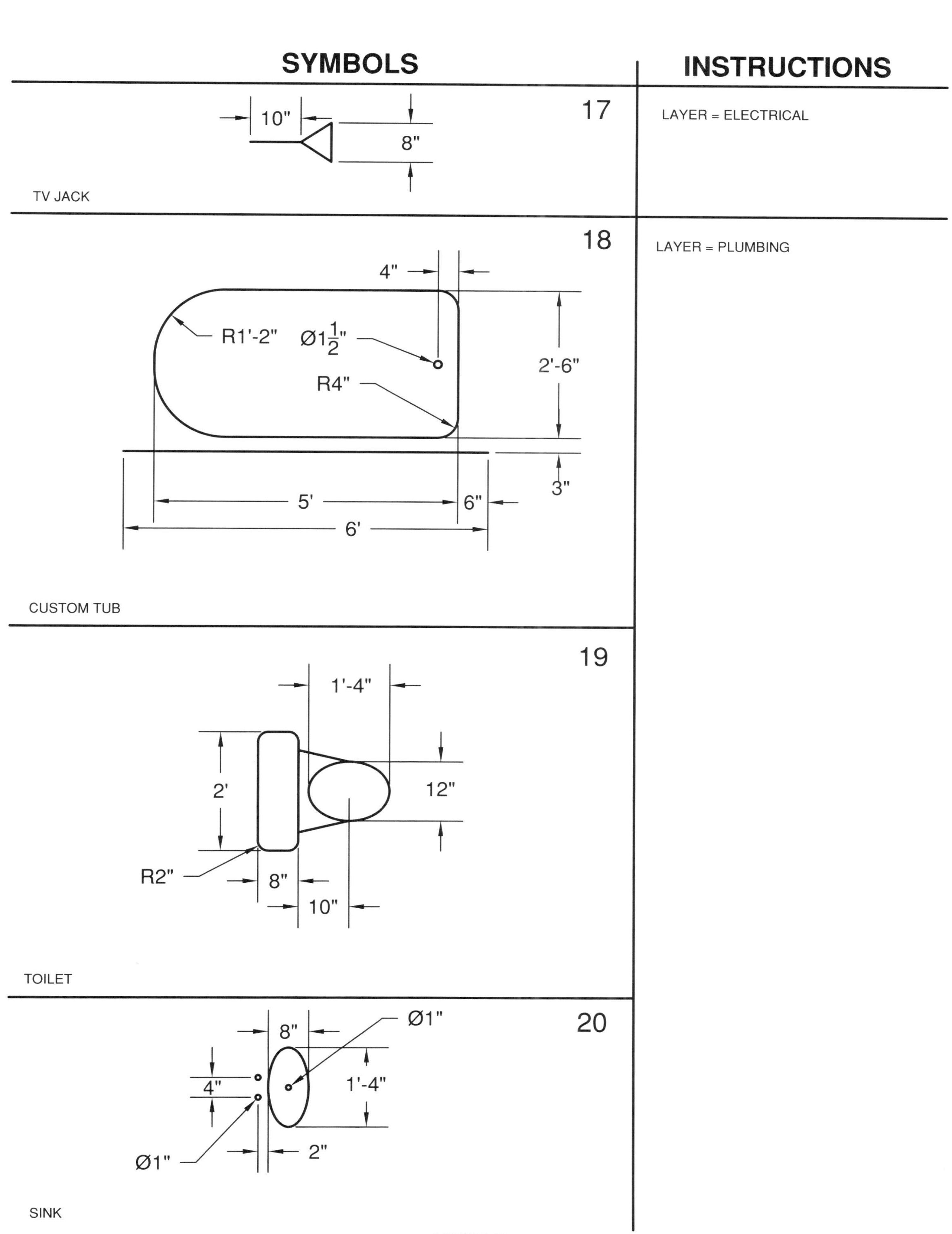
SYMBOLS
INSTRUCTIONS
17
10"
8"
TV JACK
LAYER = ELECTRICAL
18
4"
R1'-2"
Ø1½"
R4"
2'-6"
3"
5'
6"
6'
CUSTOM TUB
LAYER = PLUMBING
19
1'-4"
2'
12"
R2"
8"
10"
TOILET
20
8"
Ø1"
4"
1'-4"
Ø1"
2"
SINK

SYMBOLS	INSTRUCTIONS
21 3' R3' 3' 36 inch DOOR	LAYER = DOOR
22 2'-6" R2'-6" 2'-6" 30 inch DOOR	
23 6" 5' 5 FOOT WINDOW	LAYER = WINDOW You could experiment with Dynamic Blocks and only make one block that would dynamically change when you insert it. Refer to Lesson 8
24 6" 3' 3 FOOT WINDOW	
25 6" 2' 2 FOOT WINDOW	

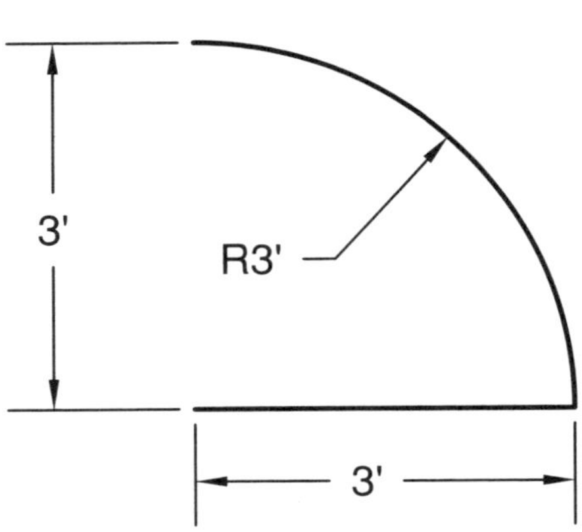

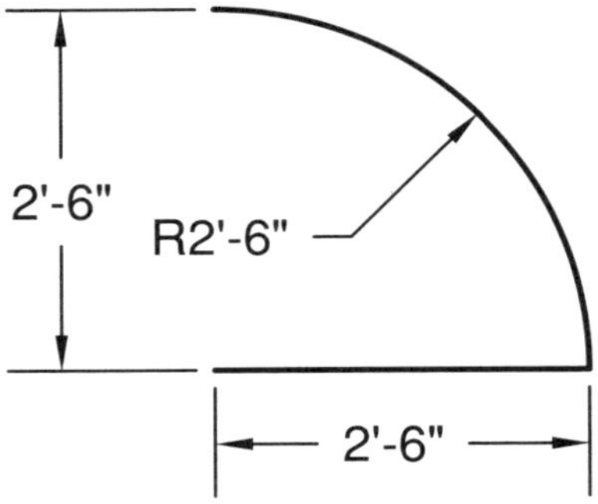

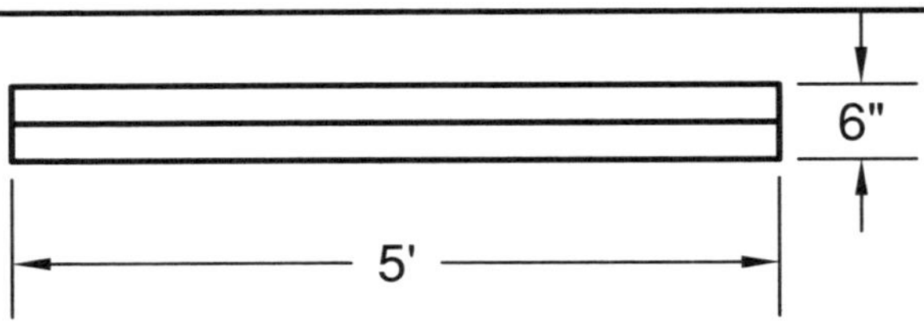

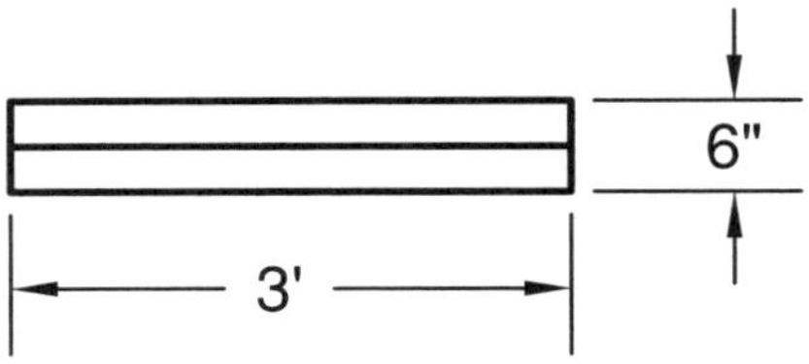

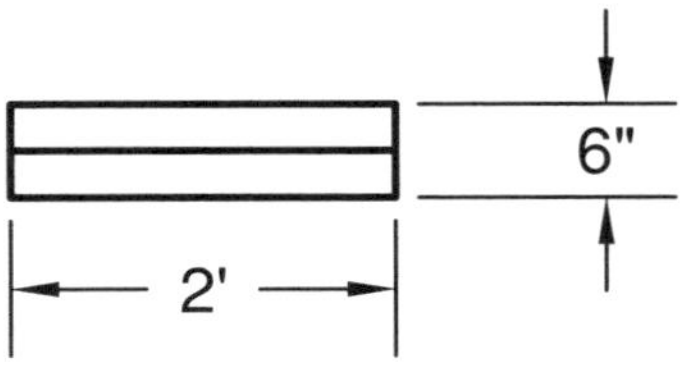

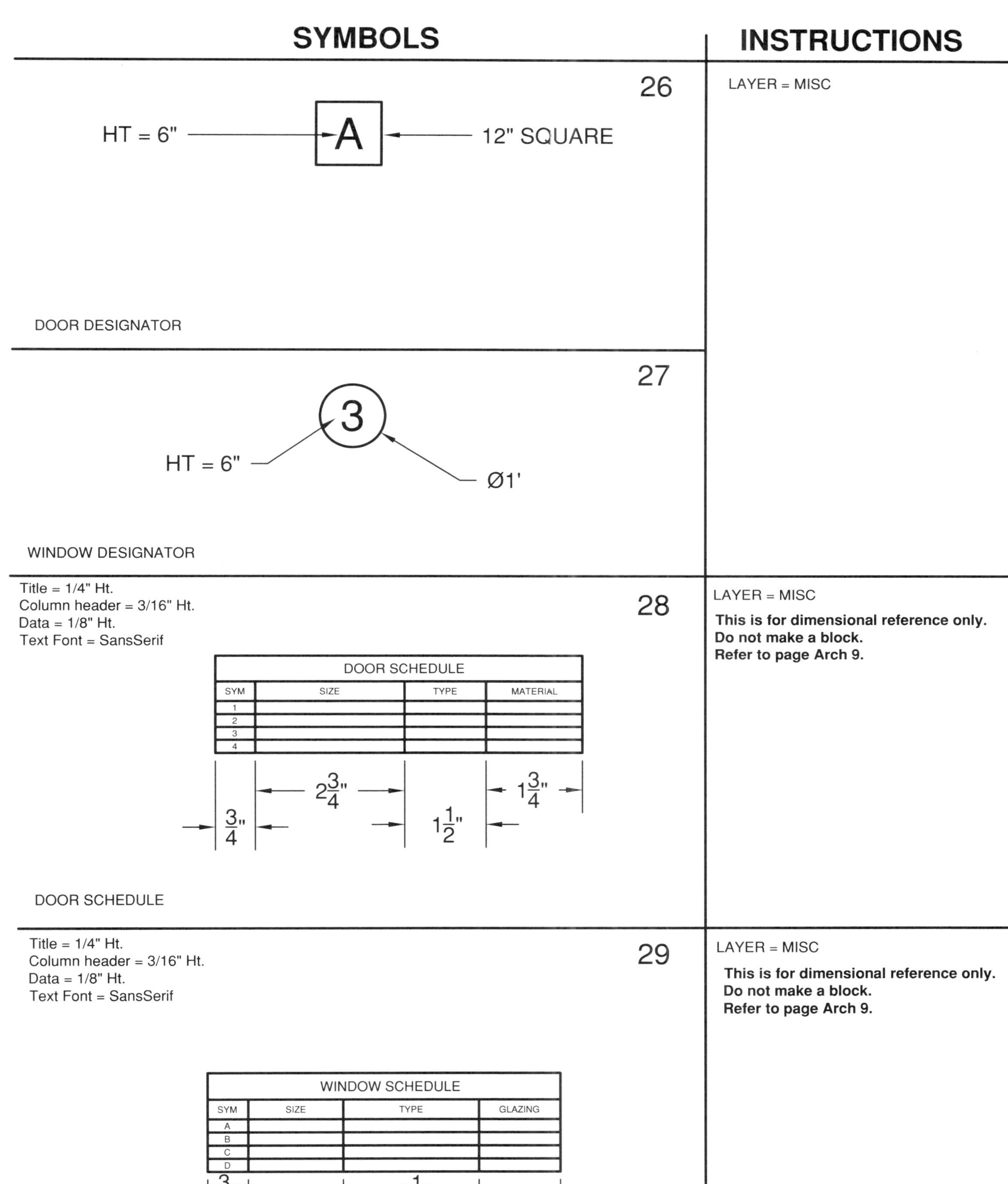

WINDOW SCHEDULE

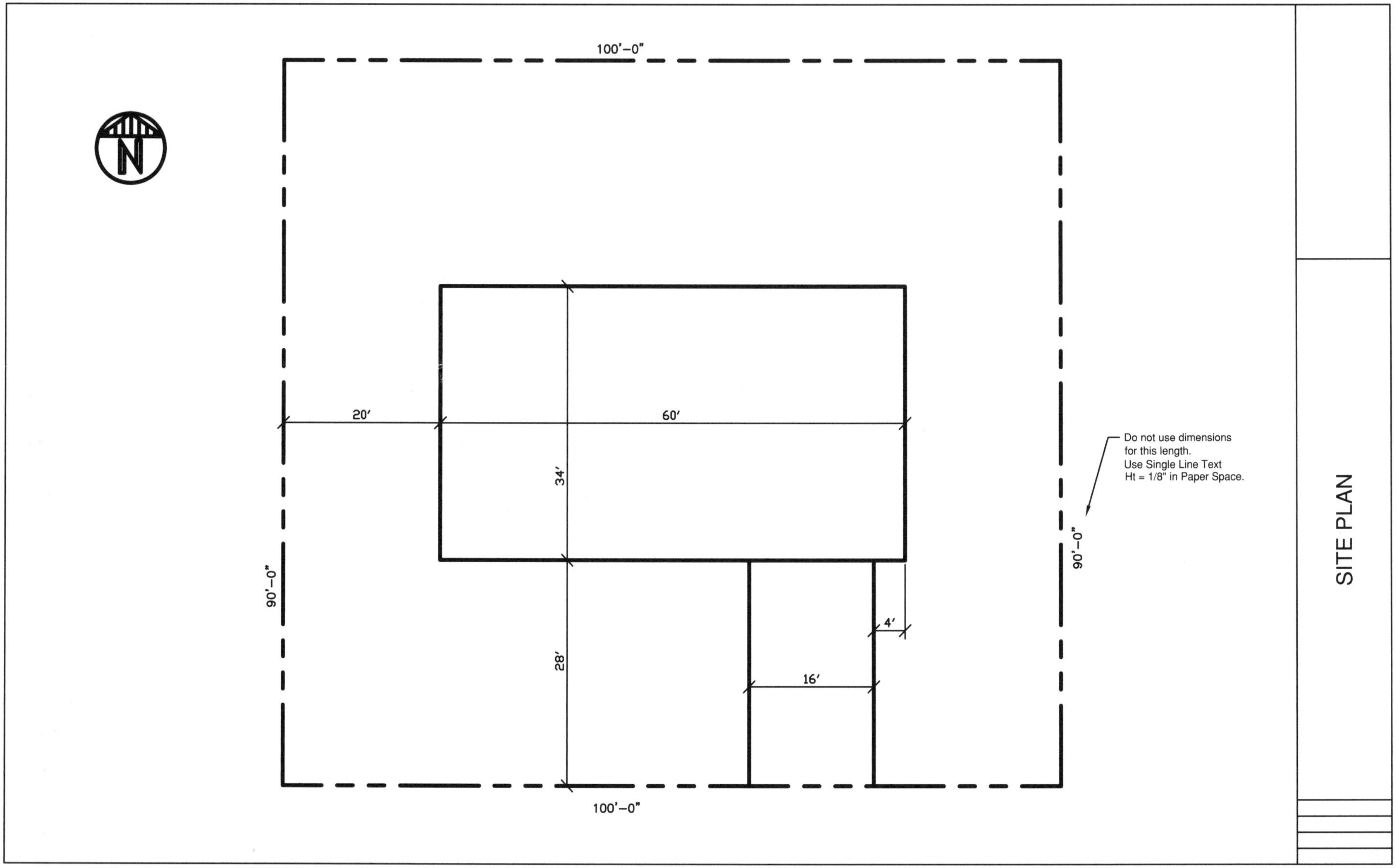

100'–0"

20'

60'

34'

28'

16'

4'

90'–0"

90'–0"

Do not use dimensions
for this length.
Use Single Line Text
Ht = 1/8" in Paper Space.

100'–0"

N

SITE PLAN

INSTRUCTIONS:

1. Open **MY Feet-Inches SETUP**
2. Select the **24 X 18 (EIGHTH - FT)** tab.
3. Draw the site plan above, in model space, full scale.
4. Dimension in Paper Space. Make sure True Associative dimensioning is On. (Dimassoc=2)
5. Design your own North symbol.
6. Fill in the information in the title block area.
7. Save as: **EX-158-1**
8. Plot using Page Setup: **24 X18 ARCH**

LOT 2 BLOCK A
PINE HILLS SUBDIVISION
SANTA ANA, CALIFORNIA

SCALE = 1/8" = 1'

EXERCISE 158-1

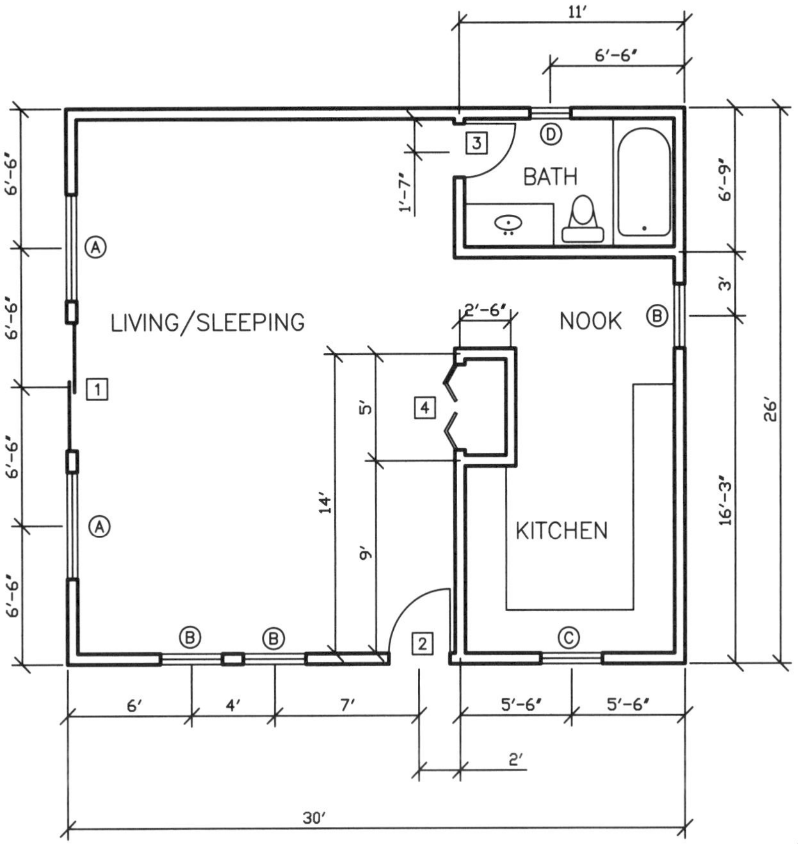

LIVING/SLEEPING

BATH

NOOK

KITCHEN

FLOOR PLAN

DRAW TABLE IN PAPER SPACE
ON LAYER SYMBOL
REFER TO ARCH-7 FOR DIMENSIONS

WINDOW SCHEDULE			
SYM	SIZE	TYPE	GLAZING
A	5'-0" X 4'-0"	WOOD FIXED	3/16" SHEET
B	3'-0" X 4'-0"	WOOD FIXED	3/16" SHEET
C	3'-0" X 3'-0"	WOOD FIXED	3/16" SHEET
D	2'-0" X 3'-0"	ALUMINUM SLIDER	3/16" SHEET

DOOR SCHEDULE			
SYM	SIZE	TYPE	MATERIAL
1	6'-0" X 6'-8"	WOOD SLIDER	1/4" POL PL
2	3'-0" X 6'-8"	PANEL	STAIN GRADE
3	2'-6" X 6'-8"	H.C. SLAB	STAIN GRADE
4	4'-0" X 6'-8"	H.C. SLAB	STAIN GRADE

INSTRUCTIONS:

1. Open **MY FEET-INCHES SETUP**
2. Select the **24 X 18(QTR-FT)** tab.
3. Draw the floor plan above, in model space, full scale. (Dimension in Paper Space, Dimassoc = 2)
4. Room title Ht. = 6" Walls = 6" wide Cabinets = 24" deep.
5. Create Tables for the SCHEDULES (ON LAYER "SYMBOLS")
6. Fill in the information in the title block area.
7. Save as: **EX-158-2**
8. Plot using Page Setup: **24 X18 ARCH.**

LOT 2 BLOCK A
PINE HILLS SUBDIVISION
SANTA ANA, CALIFORNIA

SCALE = 1/4" = 1'

EXERCISE 158-2

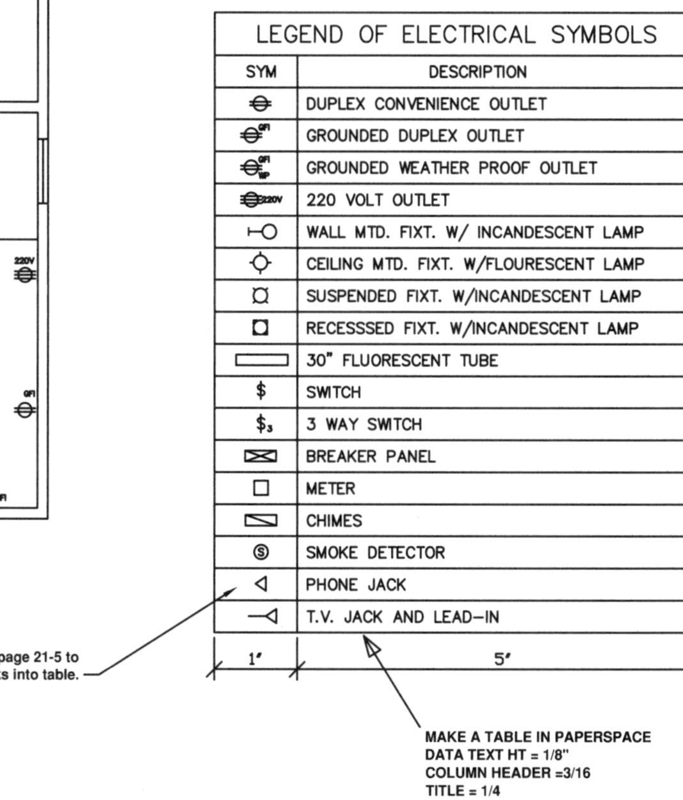

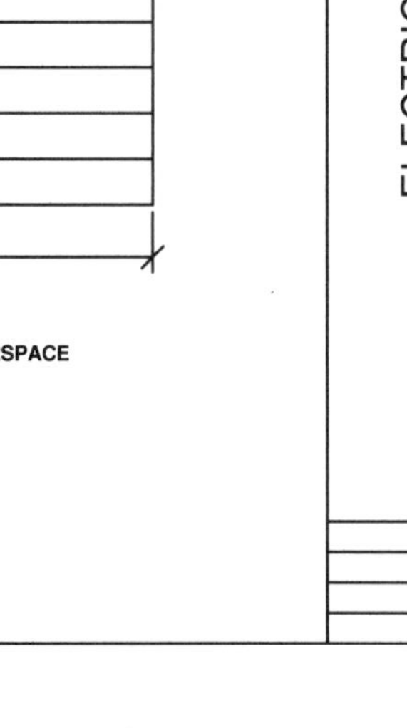

LEGEND OF ELECTRICAL SYMBOLS	
SYM	DESCRIPTION
	DUPLEX CONVENIENCE OUTLET
	GROUNDED DUPLEX OUTLET
	GROUNDED WEATHER PROOF OUTLET
	220 VOLT OUTLET
	WALL MTD. FIXT. W/ INCANDESCENT LAMP
	CEILING MTD. FIXT. W/FLOURESCENT LAMP
	SUSPENDED FIXT. W/INCANDESCENT LAMP
	RECESSSED FIXT. W/INCANDESCENT LAMP
	30" FLUORESCENT TUBE
$	SWITCH
$3	3 WAY SWITCH
	BREAKER PANEL
	METER
	CHIMES
	SMOKE DETECTOR
	PHONE JACK
	T.V. JACK AND LEAD–IN
1'	5'

Refer to page 21-5 to insert Blocks into table.

MAKE A TABLE IN PAPERSPACE
DATA TEXT HT = 1/8"
COLUMN HEADER =3/16
TITLE = 1/4

ELECTRICAL PLAN

INSTRUCTIONS:

1. Open **158-2** and immediately save it as **158-3** for safety. Select the **24 X 18(QTR-FT)** tab.
2. Freeze Layers: Dimension, Misc, Symbols and Plumbing.
3. Change the "WALLS" layer Lineweight to "default" and "Wiring" layer to .031". This will emphasis the Wiring because it will be bolder.
4. Now insert the Electrical symbols (on layer Electrical)
5. Use "Tables" to create the "Legend Of Electrical Symbols" on Layer TEXT-LIT in paperspace.
6. Fill in the information in the title block area.
7. Save as: **EX-158-3** and Plot using Page Setup: **24 x 18 ARCH.**

LOT 2 BLOCK A
PINE HILLS SUBDIVISION
SANTA ANA, CALIFORNIA

SCALE = 1/4" = 1'

EXERCISE 158-3

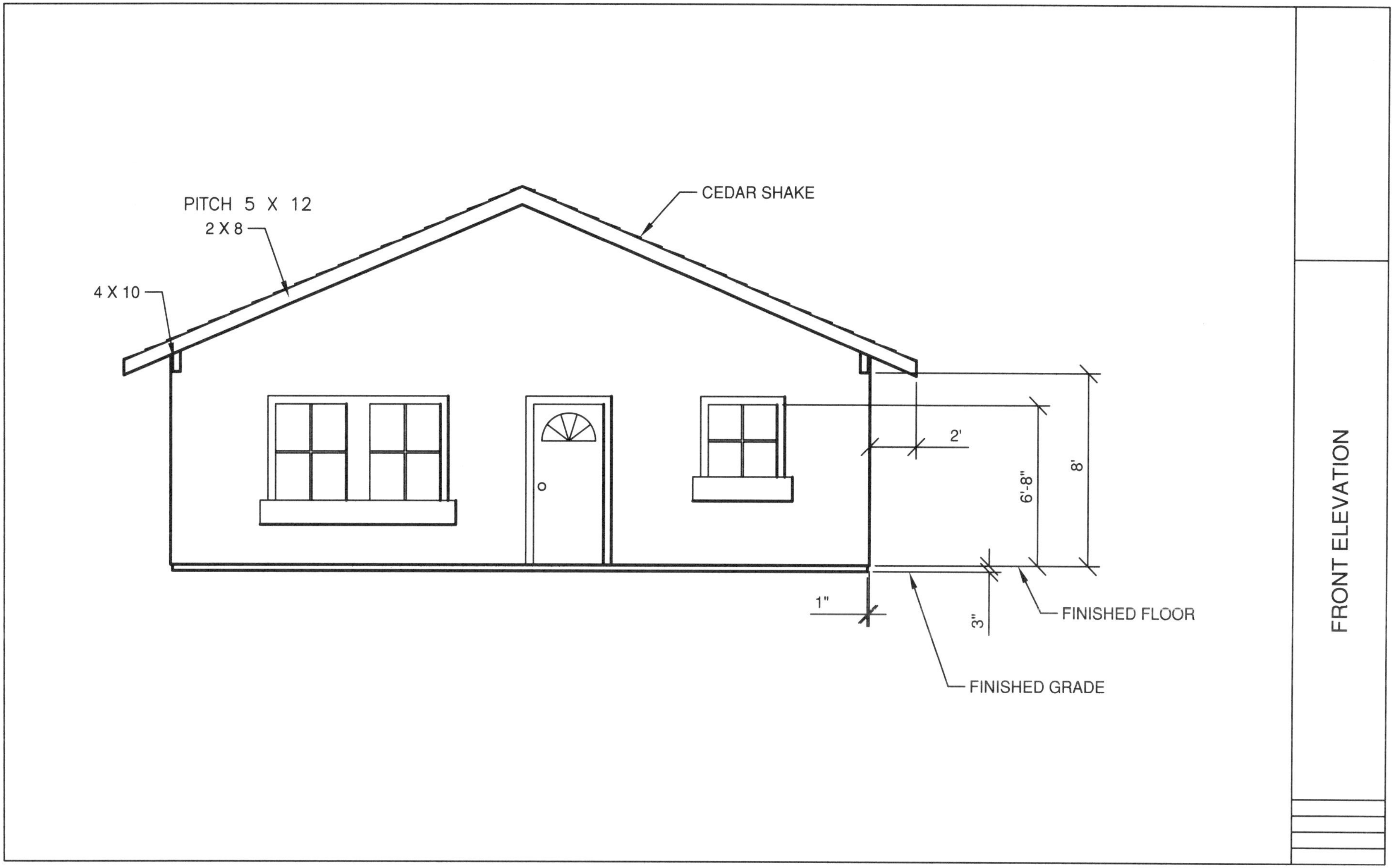

INSTRUCTIONS:

1. Open **My FEET-INCHES Setup**
2. Select the **24 X 18(QTR-FT)** tab.
3. Draw the Front Elevation shown above. Refer to 158-2 for dimensions.
4. Refer to "How to draw shingles" on the next page.
5. Use appropriate layers but Lineweights are your choice.
6. Design is your choice.
7. Fill in the information in the title block area.
8. Adjust the scale of Model Space to **3/8"=1'**
9. Save as: **EX-158-4** and Plot using Page Setup: **24 x18 ARCH**

LOT 2 BLOCK A
PINE HILLS SUBDIVISION
SANTA ANA, CALIFORNIA

SCALE = 3/8" = 1'

EXERCISE 158-4

HOW TO DRAW SHINGLES

The following is an example of how to draw shingles using the ARRAY command.

1. Draw 2 Shingles as shown below.

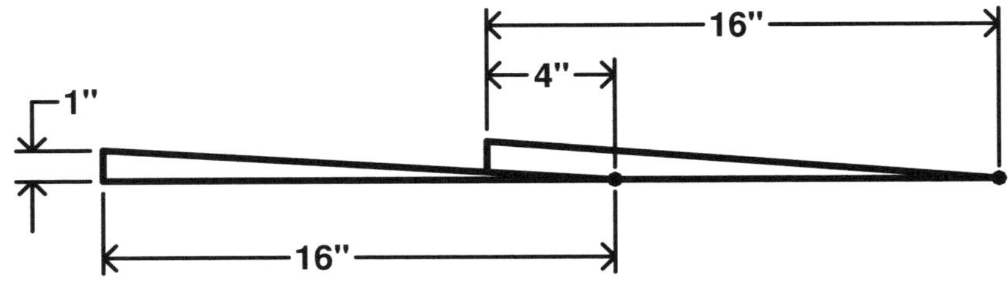

2. ARRAY the 2nd shingle as follows:
   a. Select MODIFY / ARRAY / RECTANGULAR
   b. Select the 2nd Shingle ONLY
   c. Rows = 1 Columns = 17
   d. Distance between columns:
      Snap to 1st point, then snap to 2nd point as shown below.

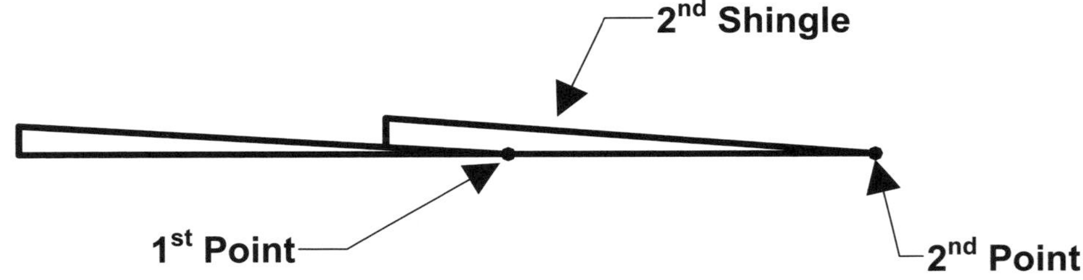

3. ROTATE the Roof Line and the Shingles 23 degrees.

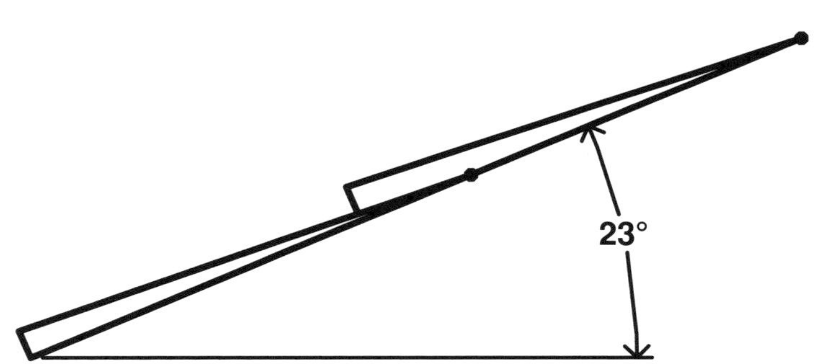

EXERCISE 158-5

WALL DETAIL

The following exercise requires that you create a new layer with a new linetype.

A. Open **My FEET-INCHES Setup**

B. Select the **24 X 18 (1 TO 1)** tab.

C. Adjust the scale to : 1" = 1'-0" and **Lock** the Viewport

D. Create a new layer
Name = Insulation   Color = magenta   Linetype = batting (LW = default)

Now experiment drawing the insulation as follows:

1. Select the Layer Insulation
2. Draw a line

Note: The insulation linetype is displayed only when you are in a layout tab.

E. Change the **Linetype scale** as follows:

1. At the command line type: **LTS <enter>**
2. Type: **.35 <enter>**
   (This will scale the "batting" linetype to be a little bit smaller than a 2 X 4)

F. Draw the Wall Detail on the next page.

1. Include the leader call outs.
2. Use Hatch Pattern "AR-CONC" for footing   Scale = .5
3. Use User defined for the Finished and Subfloor.
   Angle =45 and 135 Spacing = 3"
4. Layers and colors, your choice.

G. Save as **158-5**

H. Plot using Page Set up: **24 X 18 ARCH.**

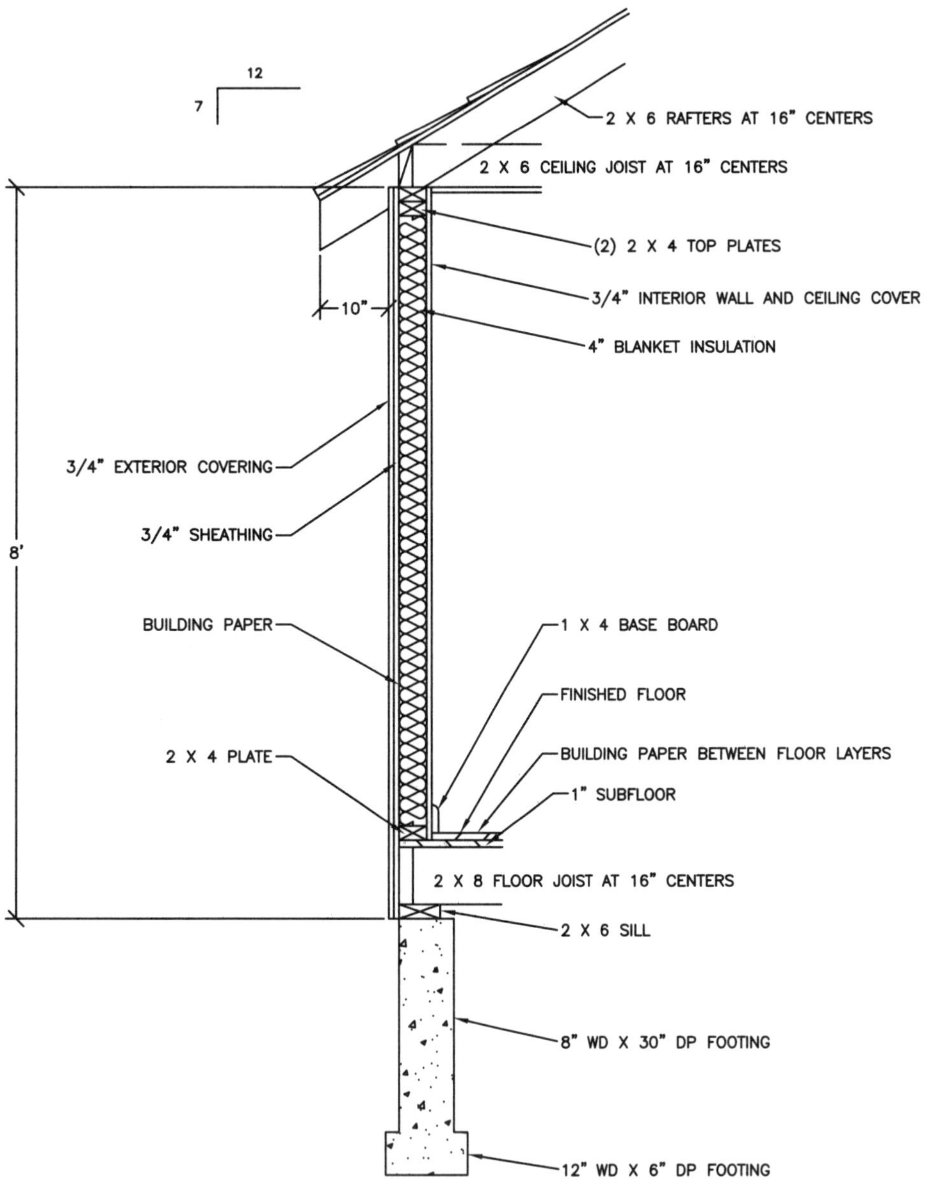

EXERCISE 158-5

ELECTRO MECH

ELECTRO-MECHANICAL SYMBOL LIBRARY

When you are using a CAD system, you should make an effort to ONLY draw an object once. If you have to duplicate the object, use a command such as: Copy, Array, Mirror or Block. Remember, this will make drawing with CAD more efficient.

In the following exercise, you will create a file full of electronic symbols that you consistently use when creating an architectural drawing. You will create them once and then merely drag and drop them, from the DesignCenter, when needed. This will save you many hours in the future.

Save this library file as **Library** so it will be easy to find when using the DesignCenter or create a Library Palette.

1. Open **My Decimal Setup**
2. Add the following layers to your **MY Decimal set up** drawing.

CIRCUIT	GREEN	CONTINUOUS
CIRCUIT2	RED	CONTINUOUS
CORNERMARK	9	CONTINUOUS
DESIGNATOR	BLUE	CONTINUOUS
MISC	CYAN	CONTINUOUS
PADS	GREEN	CONTINUOUS
PCB	WHITE	CONTINUOUS
COMPONENTS	RED	CONTINUOUS

3. **Save the My Decimal Set up**.
4. Select the **11 x 17 (1 to 1)** tab.
5. Draw each of the Symbol objects, shown on the next page, actual size. Do not scale the objects.
6. Create an individual Block for each one using the **BLOCK** command.
7. Save this drawing as: **Library**
8. Plot a drawing of your library symbols for reference. The format is your choice.
9. Plot using Page Set up: **11 x 17 (1 to 1) MONO**

SYMBOLS

1

.20
.10
.42
.20

ANTENNA

2

.10
.20
.10
.05
.35

BATTERY

3

.05
.05
.05

RESISTOR

4

.20
.20
.05

CAPACITOR

5

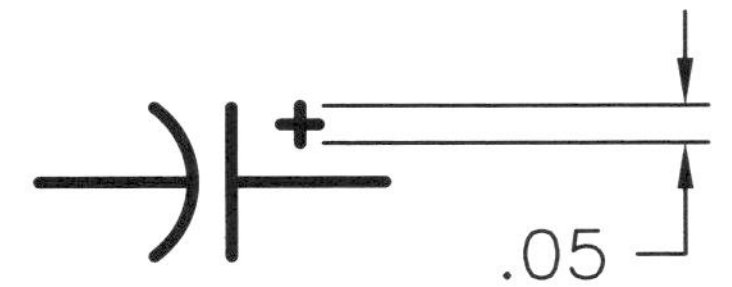

CAPACITOR-POLARIZED

INSTRUCTIONS

LAYER = COMPONENTS

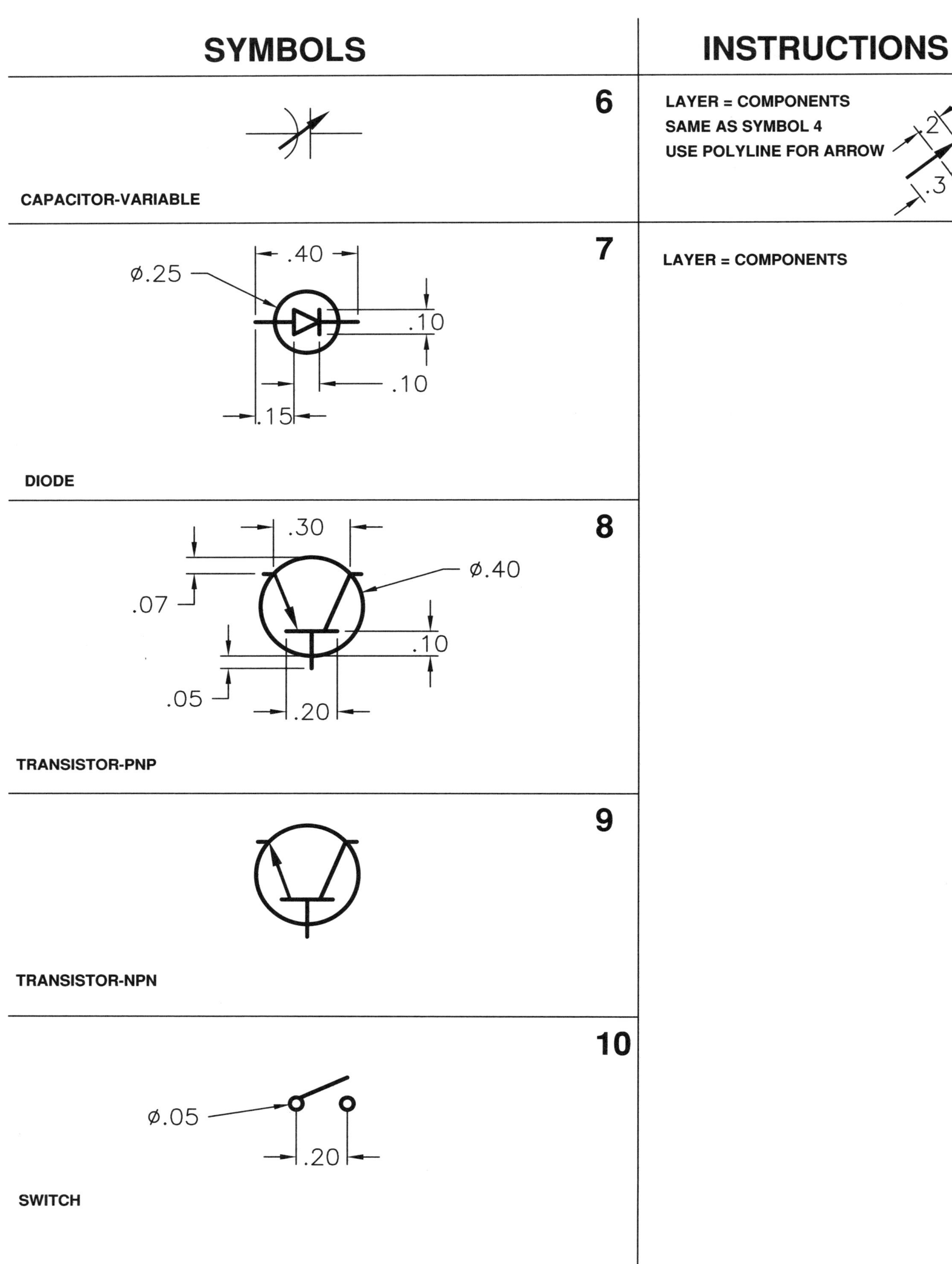
SYMBOLS
INSTRUCTIONS
6
CAPACITOR-VARIABLE
LAYER = COMPONENTS
SAME AS SYMBOL 4
USE POLYLINE FOR ARROW
.2
.3
7
ø.25
.40
.10
.10
.15
DIODE
LAYER = COMPONENTS
8
.30
ø.40
.07
.10
.05
.20
TRANSISTOR-PNP
9
TRANSISTOR-NPN
10
ø.05
.20
SWITCH

SYMBOLS		INSTRUCTIONS
.20 .07 .10 SWITCH-PUSH BUTTON	11	LAYER = COMPONENTS
.10 .10 .20 PLUG	12	
JACK - FEMALE	13	
.10 .20 .10 .30 .15 .10 SPEAKER	14	
.05 .05 .40 .10 .15 .05 TRANSFORMER	15	

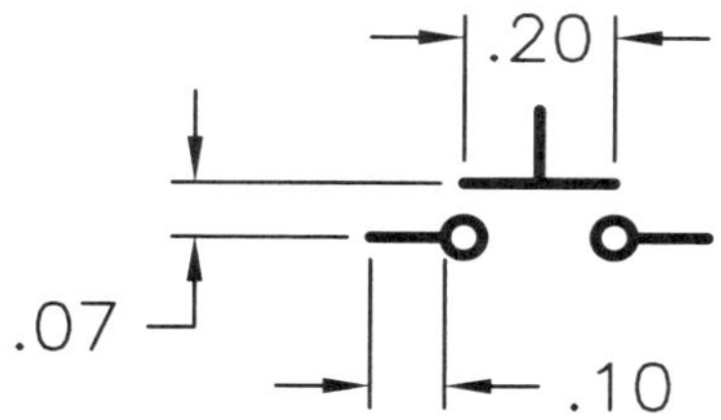

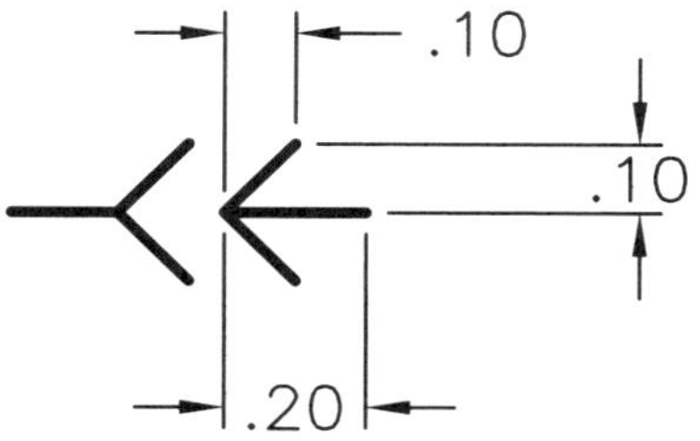

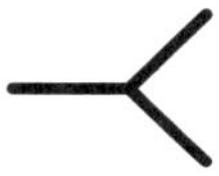

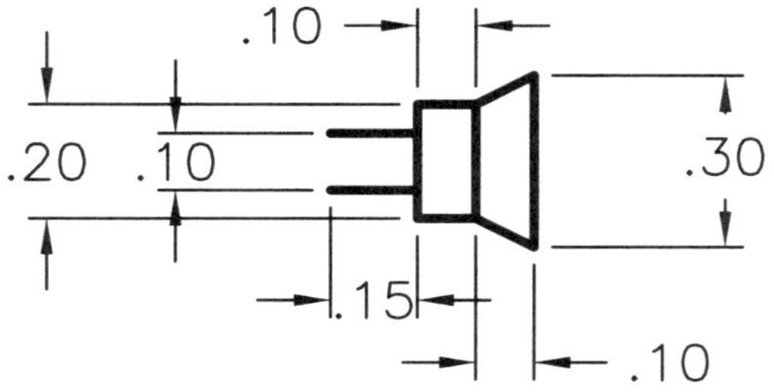

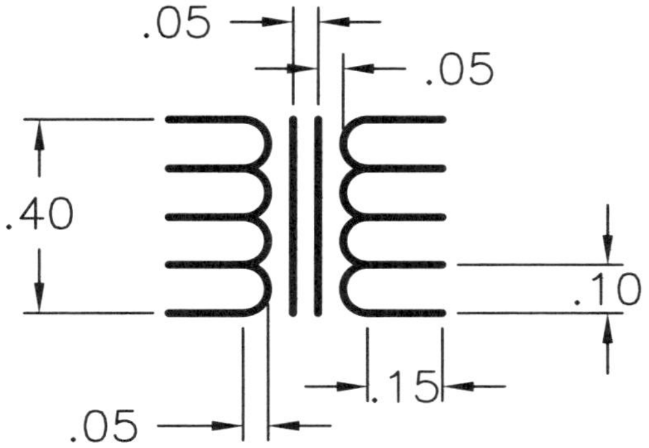

SYMBOLS		INSTRUCTIONS
.05 / .10 / .05 / .22 INDUCTOR - VARIABLE	16	LAYER = COMPONENTS
.20 / .10 / .10 / .05 CHASSIS GROUND	17	
.40 / .10 / .10 RESISTOR (1/4W)	18	
.40 / .15 / .10 RESISTOR (1/2W)	19	
.70 / .20 / .15 RESISTOR (2W)	20	

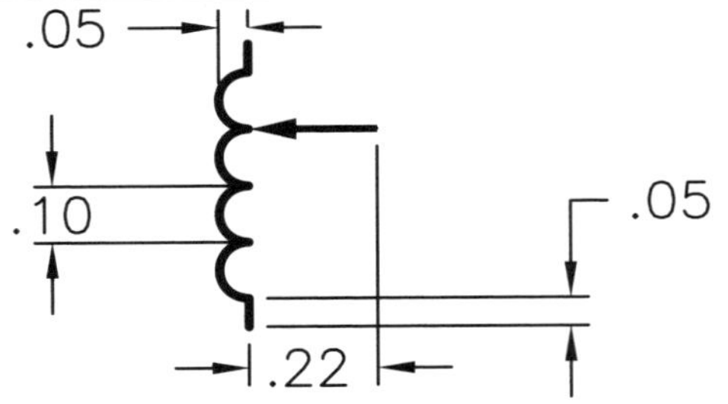

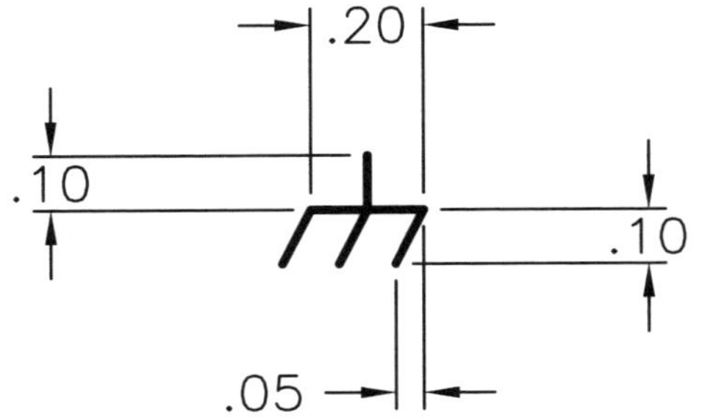

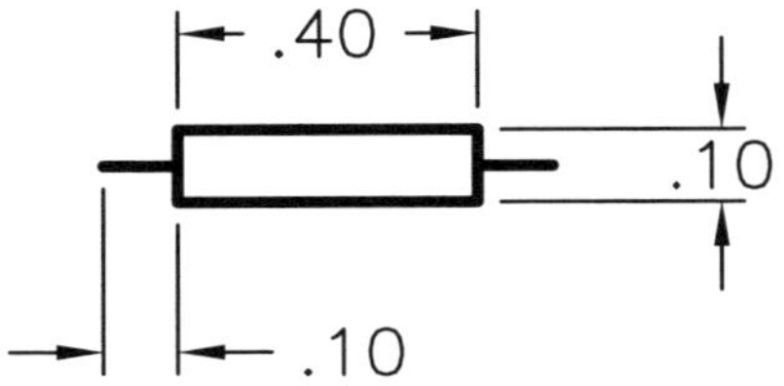

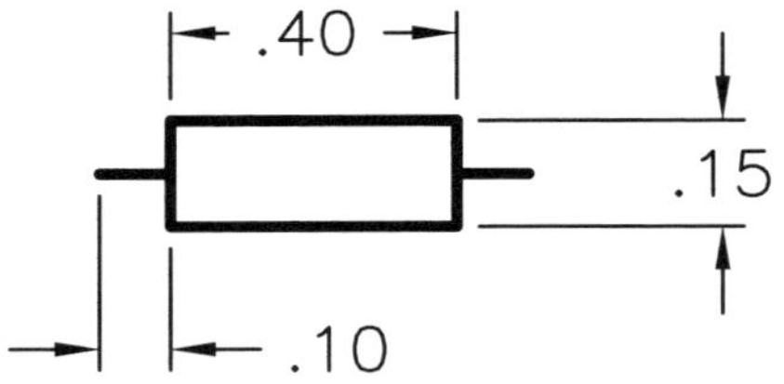

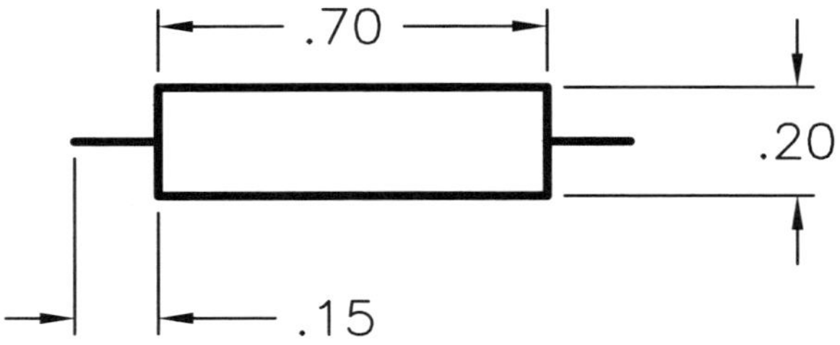

SYMBOLS	INSTRUCTIONS
21 .50 .20 .15 CAPACITOR (22mf)	LAYER = COMPONENTS
22 .30 .10 CAPACITOR (.01mf)	
23 .15 .10 .10 DIODE (IN914)	
24 .40 .15 .10 DIODE (IN4001)	
25 .25 .30 .20 ø.40 2N2646 UJT (USE FOR SCHEMATIC ONLY)	

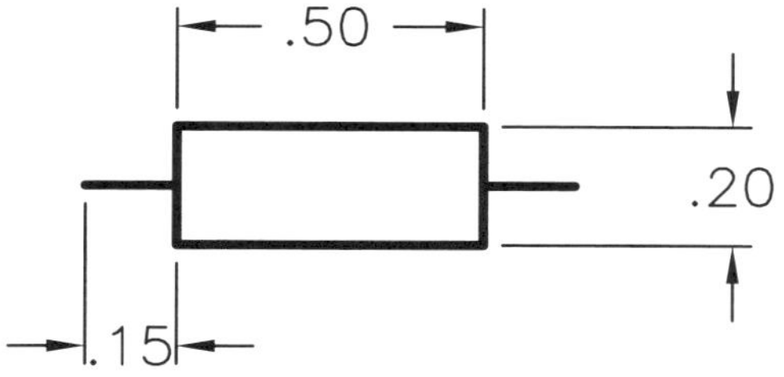

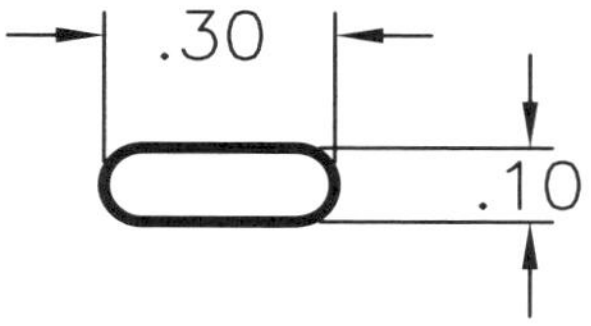

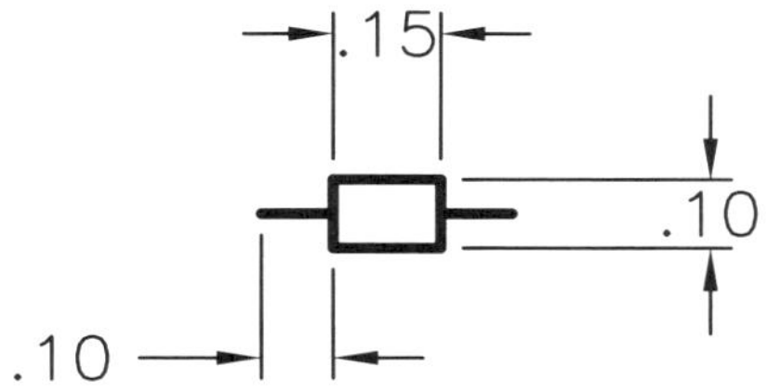

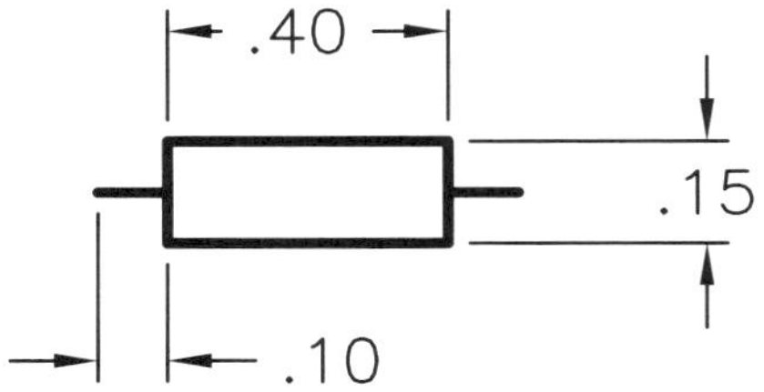

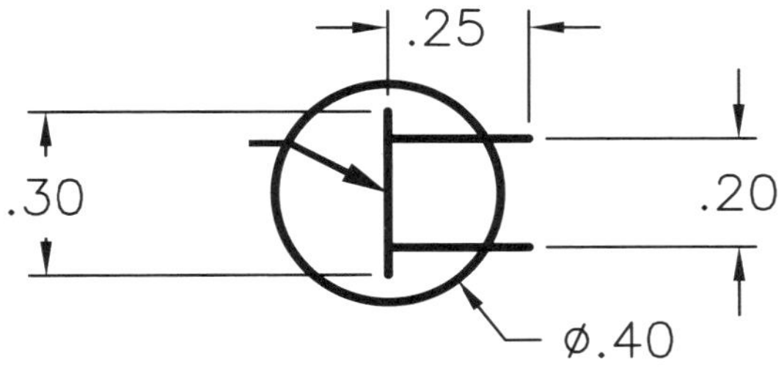

SYMBOLS

26

.15
.30
.10

TIP29A

27

.10 .03 .05
.40
.05 .05 .03
.25

TIMER-NE555

28

.50
.20
.10

CAPACITOR

29

.20
.10
.10

PLUG-MALE

30

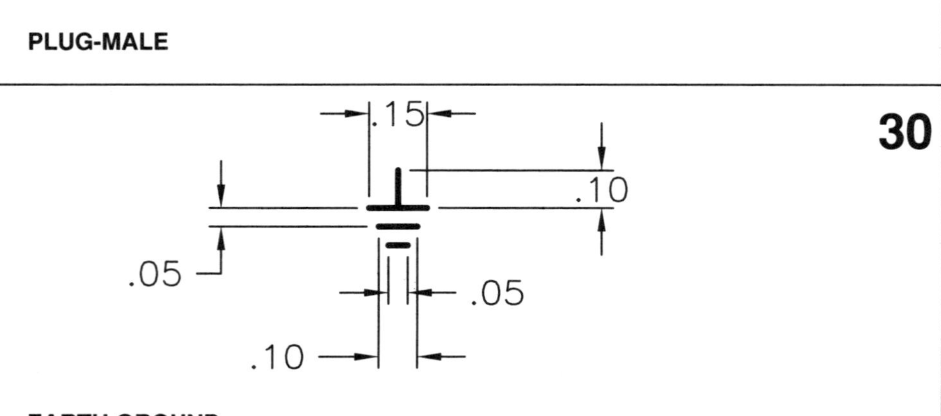

EARTH GROUND

INSTRUCTIONS

LAYER = COMPONENTS

SYMBOLS		INSTRUCTIONS
TRANSISTOR-TO-3	31	LAYER = COMPONENTS
2N2646 UJT (USE AS COMPONENT ONLY)	32	
TRIM POTENTIOMETER	33	
PAD - ROUND	34	LAYER = CIRCUIT USE "DONUT" USE THE PROPERTIES PALETTE TO CHANGE THE LINEWEIGHT TO "DEFAULT" FOR THIS OBJECT ONLY.
PAD - FINGER	35	LAYER = CIRCUIT USE "SOLID" HATCH TO FILL. USE "FILLET" TO ROUND ENDS USE THE PROPERTIES PALETTE TO CHANGE THE LINEWEIGHT TO "DEFAULT" FOR THIS OBJECT ONLY.
PAD - RECTANGULAR	36	

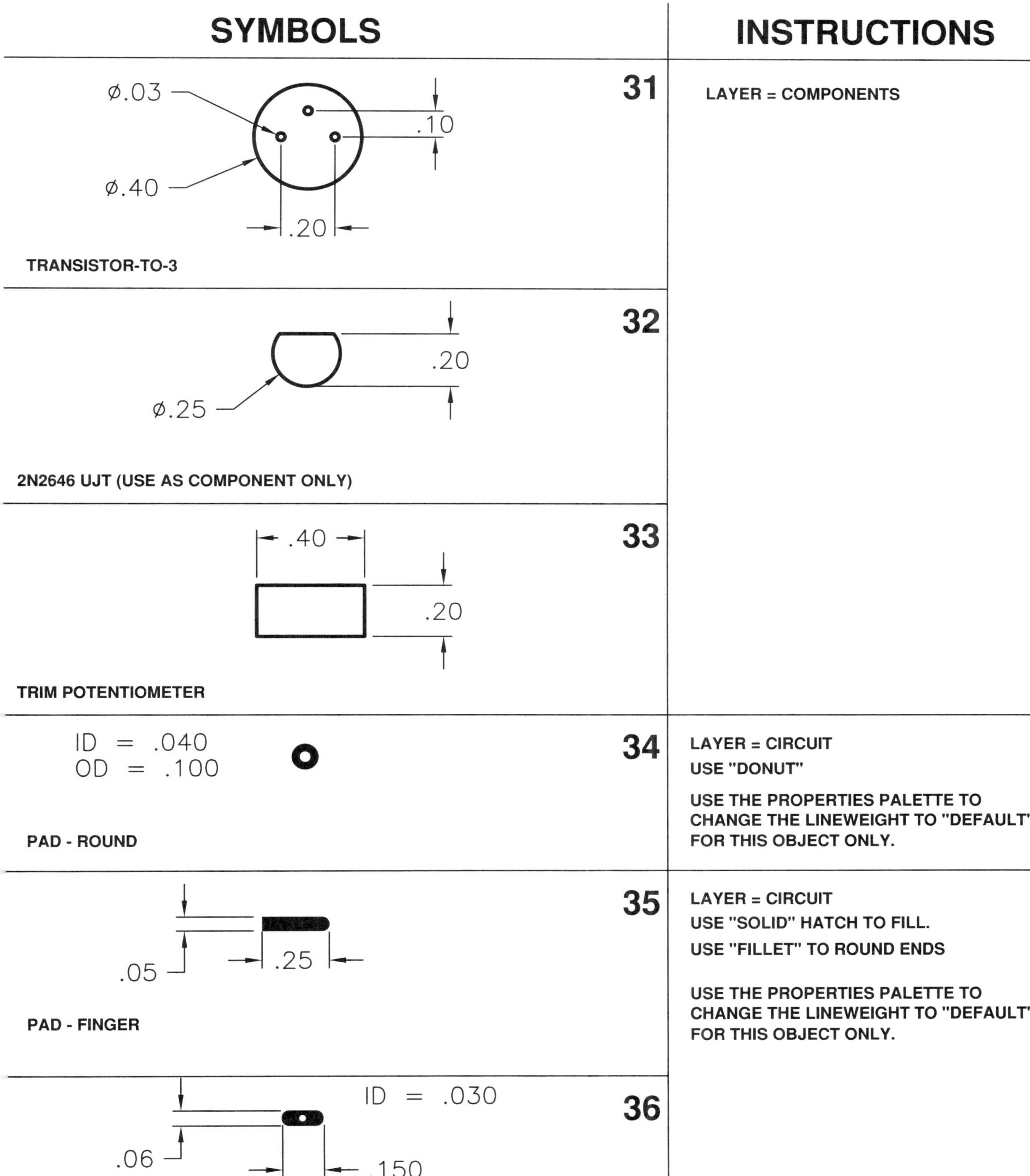

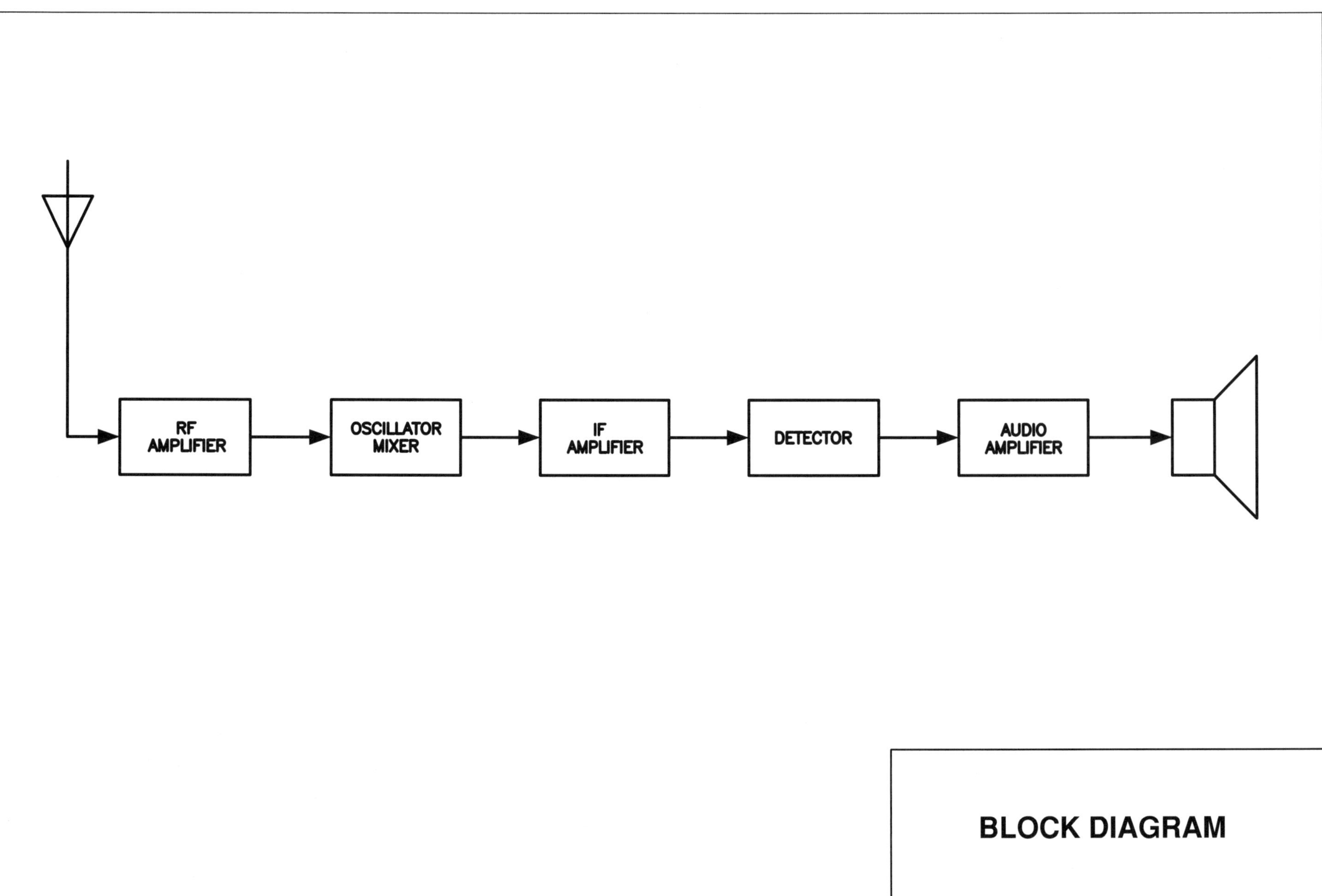

INSTRUCTIONS:

1. Open **MY DECIMAL SET UP**
2. Use Layer: **TEXT LIGHT** for text. Use Layer **OBJECT** for objects and symbols.
3. Size and proportions are your choice.
4. Save as: **EX-156-1**
5. Plot using Page Set up: **11 X 17 MONO.**

EXERCISE 156-1

EXERCISE 156-2

The following is an example of a how you might construct a schematic.

A. Open **MY DECIMAL SETUP**

B. Select the **11 X 17 (1 to 1)** tab.

C. Set: SNAP = .100 GRID = .100

D. Draw the schematic in Model Space.
The size is not critical but maintain good proportions and drawing balance.

SUGGESTION:

1. Use layer **CONSTRUCTION** to roughly layout the circuit lines
2. Change to Layer **Circuit** before inserting symbols.
3. **Insert** symbols and locate as shown. (Drag and drop from the DesignCenter, Library drawing
4. Draw the **actual** schematic lines on top of the Construction Lines skipping over the symbols. Use layer **circuit2**
5. Now **FREEZE** layer **Construction**

(Now you don't have to trim the lines under the symbols.)

E. Draw the solder points using **Donuts** .00 I.D. and .100 O.D.
Use layer **Circuit**

F. Add the designators, use layer **Designator.**
Text Ht. = .125

G. Add the Parts List (In Paperspace) Rows = .25 Text Ht. = .125
(Try "Tables")

H. Edit the title block

I. Save as **EX-156-2**

J. Plot using Page Set up: **11 X 17 MONO.**

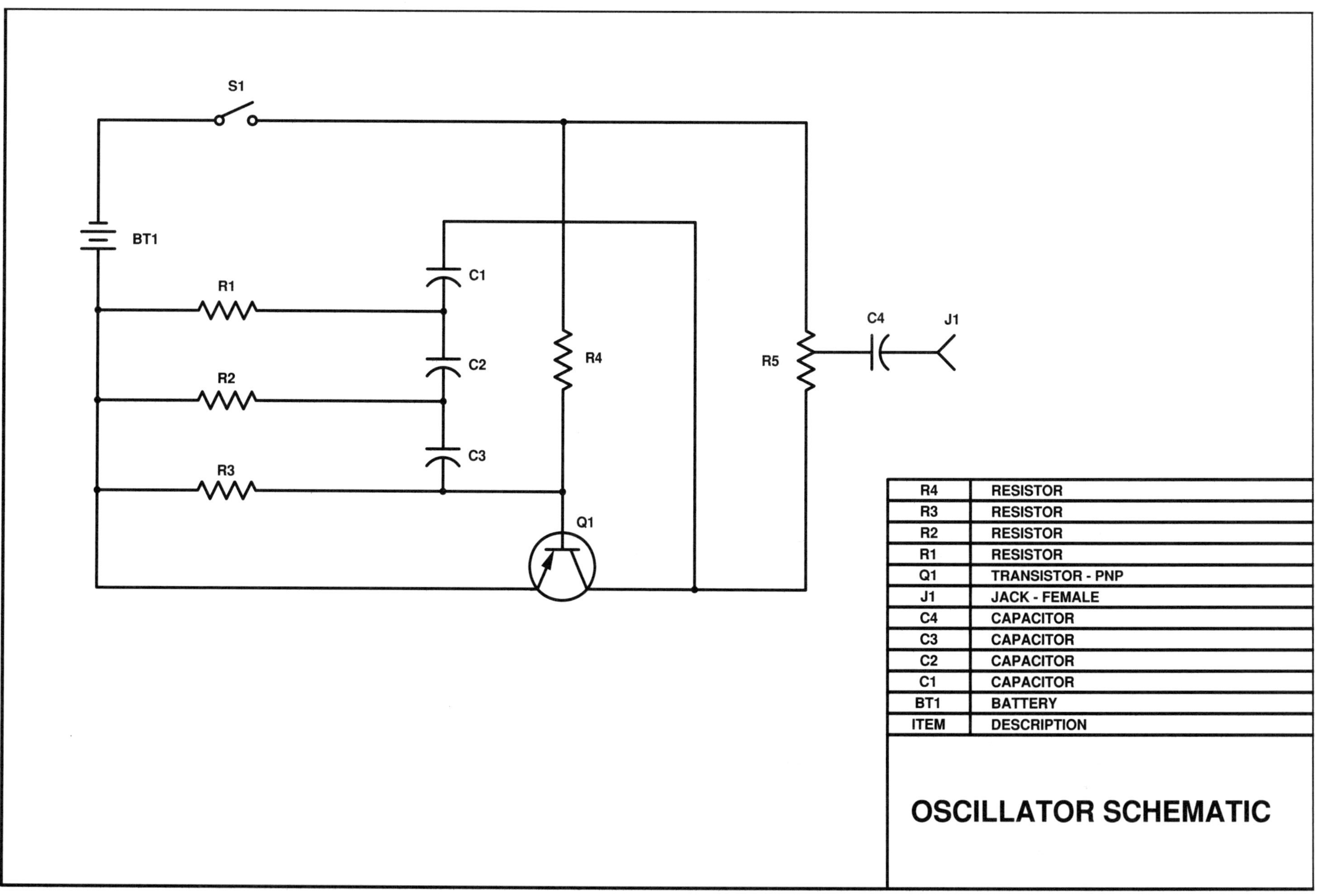

ITEM	DESCRIPTION
R4	RESISTOR
R3	RESISTOR
R2	RESISTOR
R1	RESISTOR
Q1	TRANSISTOR - PNP
J1	JACK - FEMALE
C4	CAPACITOR
C3	CAPACITOR
C2	CAPACITOR
C1	CAPACITOR
BT1	BATTERY

OSCILLATOR SCHEMATIC

EXERCISE 156-2

EXERCISE 156-3

This drawing is only a template. It will be used as a template for the following 4 drawings. If you follow the instructions and draw this template correctly, the next 4 drawings will be very easy.

A. Open **MY DECIMAL SETUP**

B. Select the **11 X 17 (1 to 1)** tab and unlock the viewport.

C. Adjust the scale of model space to 2 : 1

D. Set: SNAP = .100 GRID = .100

E. Draw the board outline (on layer PCB) Full Scale (2.50 x 2.00).
(Note: it will appear larger because you adjusted the scale of model space to 2 :1)
Printed Circuit Boards are generally drawn 2, 10 or even 100 times larger than their actual size.

F. Draw the **Circuit** on layer **CONSTRUCTION.**

G. **INSERT** symbols 19, 28, and 31 on Layer **Components** (Refer to Library)

H. Draw the designators on Layer **DESIGNATOR**. Text Ht = .060

I. Draw dimensions, in paper space, on Layer **DIMENSION**.

IMPORTANT: Make sure that True Associative Dimensioning is On:

Follow the instructions below:

1. Type **Dimassoc <enter>**
2. Type **2 <enter>**

J. **Edit** the title block

K. Save as **EX-156-3 (Do not plot)**

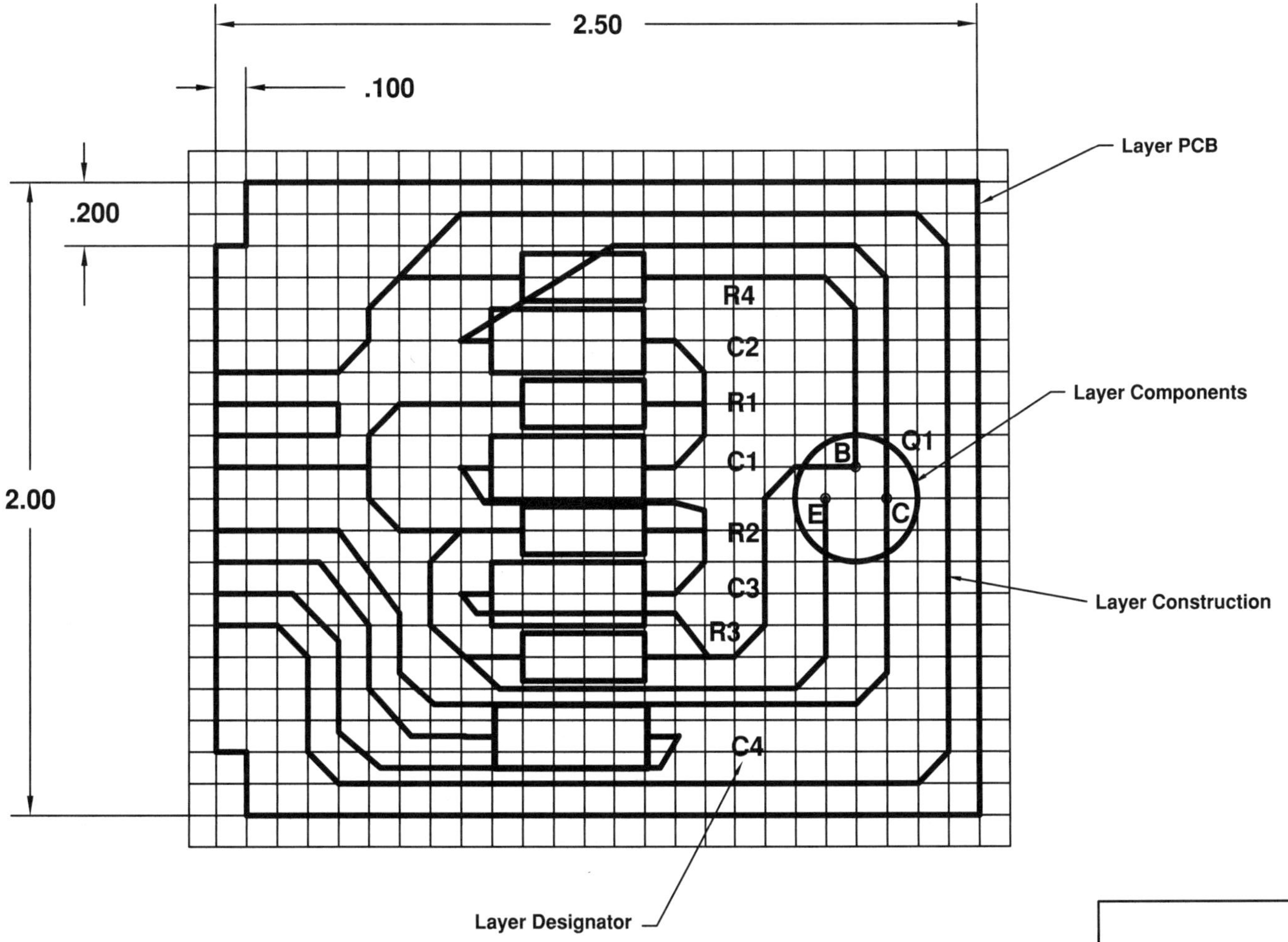

2.50

.100

.200

2.00

Layer PCB

Layer Components

Layer Construction

Layer Designator

R4

C2

R1

C1

R2

C3

R3

C4

Q1

B

E

C

GRID LAYOUT

SCALE = 2:1

Note: Do not draw the grid.
It is there only to help you
estimate positions.

EXERCISE 156-3

EXERCISE 156-4

The following is an example of a how you might illustrate the ARTWORK for the circuit on the PCB.

A. Open **EX-156-3**

B. **FREEZE** Layers **DESIGNATOR, DIMENSION** and **COMPONENTS**.
(Remember, if you were not careful when you created EX-156-3 the wrong objects may disappear. You may have to move objects to the correct layer)

C. Draw the heavy lines outside the corners of the PCB. (This defines the edges of the board)
Use layer Cornermark.
Use Polyline, width .100

Note: Consider using Offset to create a guideline, .05 from the edge of the board, for the polyline. Then just snap to the intersections.

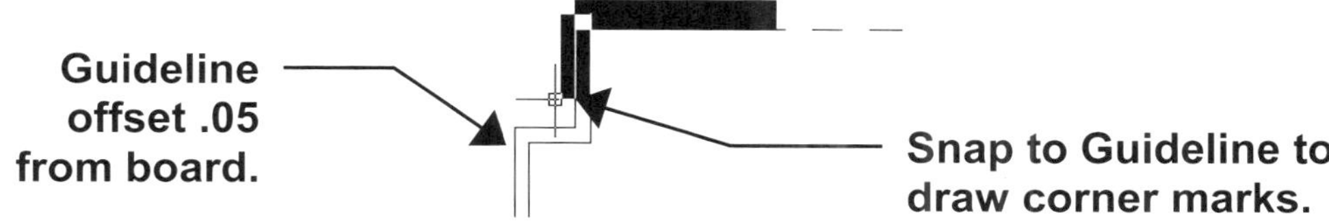

D. **Insert** symbols **34** and **35** on layer **CIRCUIT.**

E. Draw the circuit lines **on top** of the construction lines.
Use Layer Circuit
Use Polyline, width .025

F. Turn **OFF** Layers, **CONSTRUCTION** and **PCB**.

G. Save as **EX-156-4**

H. **Plot** using Page Setup: **11 X 17 MONO.**

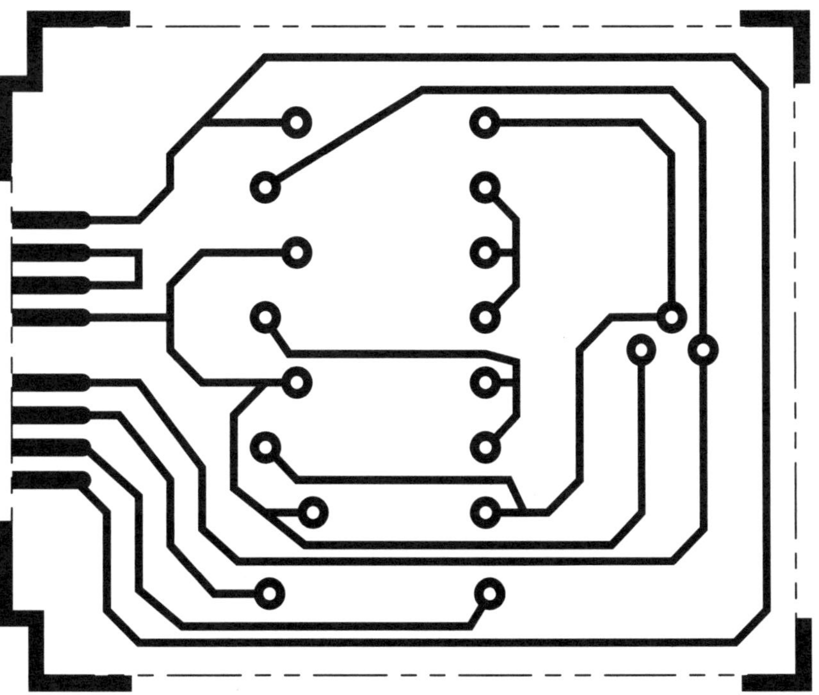

ARTWORK

SCALE = 2:1

EXERCISE 156-4

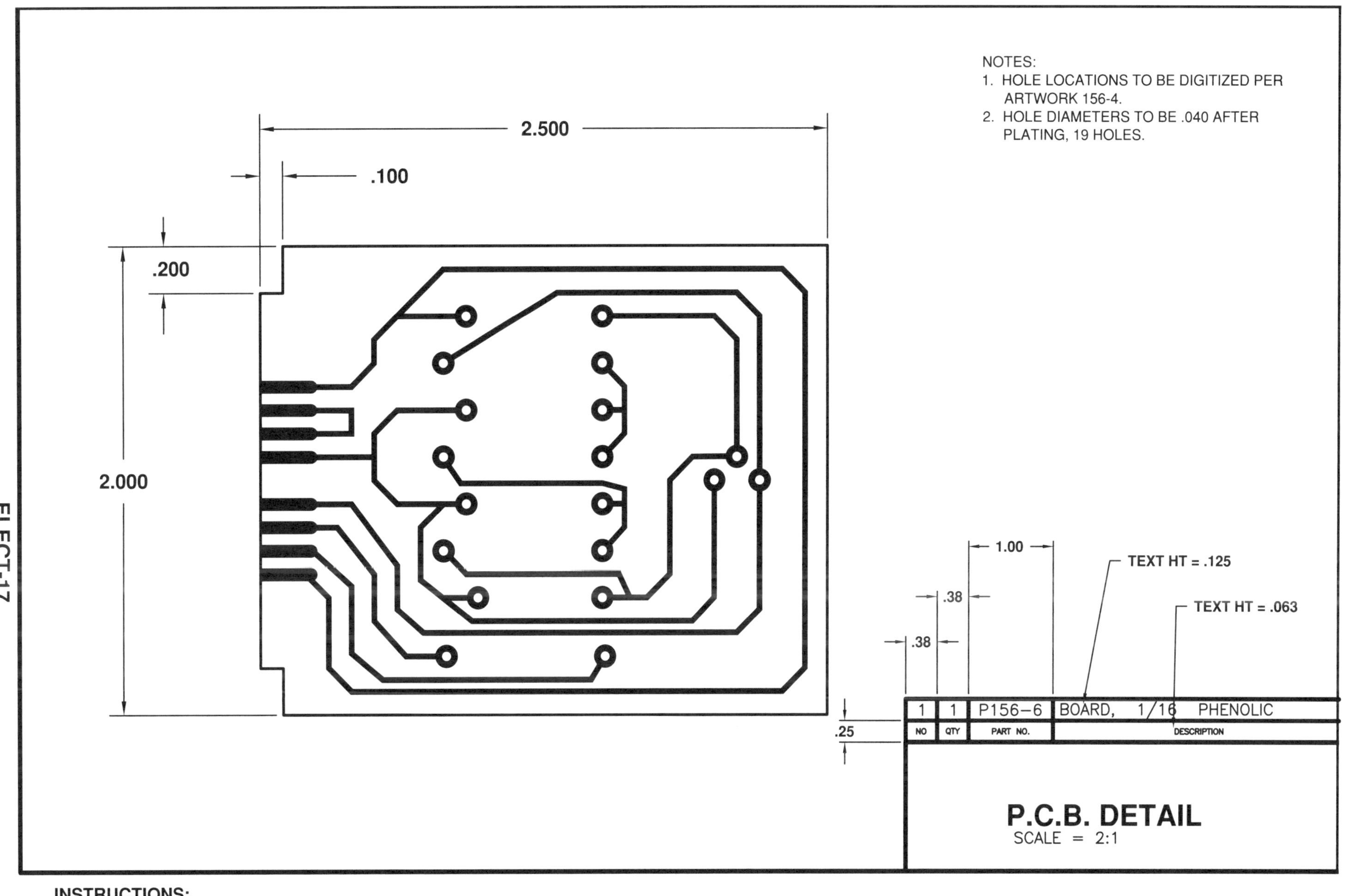

INSTRUCTIONS:

1. Open EX-156-4
2. Freeze Layer Cornermark and Thaw Layers PCB and Dimension
3. Draw the Parts List in Paperspace. Use Layer: Border and Text Light.
4. Add the notes on Layer Text Light. Ht = .125
5. Save as EX-156-5 and Plot using Page Set up: 11 X 17 MONO.

EXERCISE 156-5

EXERCISE 156-6

The following is an example of a how you might illustrate the **ARTWORK** for the **COMPONENTS**.

A. Open **EX-156-3**

B. **FREEZE** Layers **CONSTRUCTION and DIMENSION..**
(Remember, if you were not careful when you created EX-156-3 the wrong objects may disappear. You may have to move objects to the correct layer)

C. **MIRROR** the entire board and its components.

WHY? *A printed circuit board has the circuit on one side and the components on the other. We need to show the opposite side of the board to place the components..*

First, lets review the MIRRTEXT command you learned in the Beg. workbook.

1. At the command line type: **MIRRTEXT <enter>**
2. Type: **0 <enter>** (0 means OFF, 1 means ON)

The **MIRRTEXT** command controls whether the TEXT will mirror or not. It will change positions with the object but you can control the "Right Reading". In this case, we do not want the text to be shown reversed, so we set the MIRRTEXT command to "0" OFF.

D. ALIGN the designators.

E. Add the Note, in paper space, on layer TEXT LIGHT. Text ht = .125

F. Save as: **EX-156-6**

G. **Plot** using Page Setup: **11 X 17 MONO.**

NOTES:
1. SILKSCREEN TO BE PAINT, WHITE.

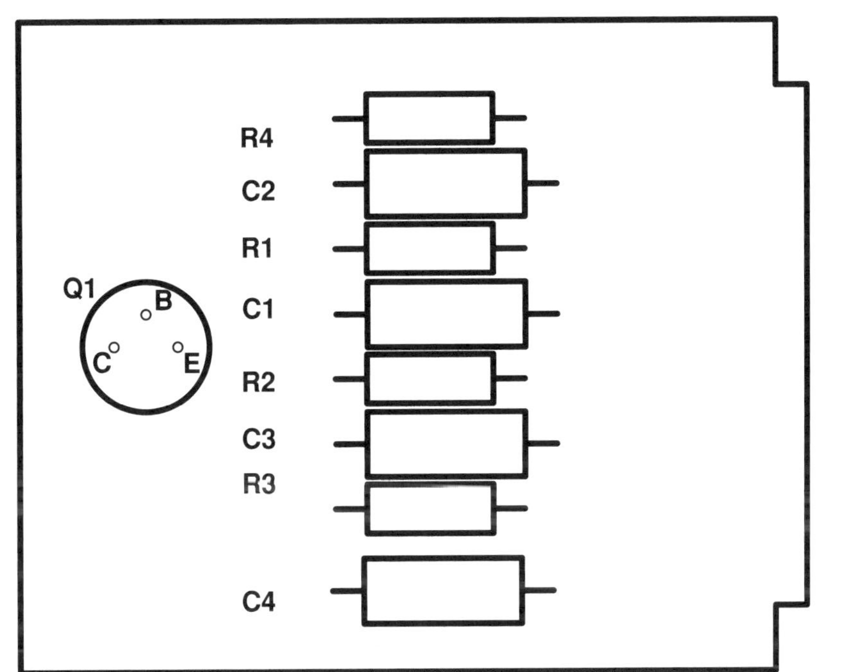

COMPONENT ARTWORK

EXERCISE 156-6

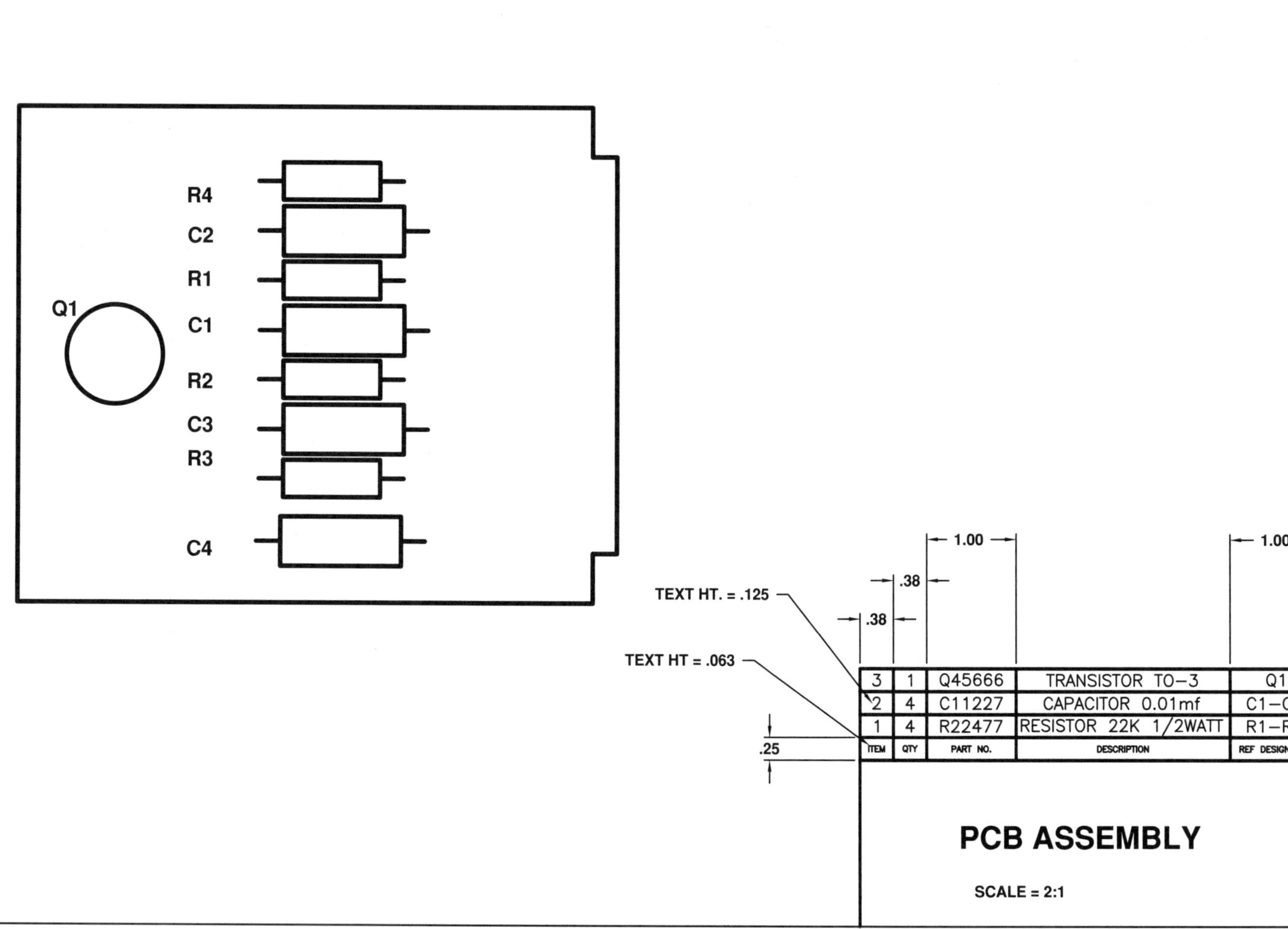

INSTRUCTIONS:

1. Open EX-156-6
2. Make the modification to the drawing. (These are the actual components shown in place)
3. Delete the Note.
4. Draw the Parts list, in paperspace.
5. Save as EX-156-7 and Plot using Page Setup: 11 X 17 MONO.

EXERCISE 156-7

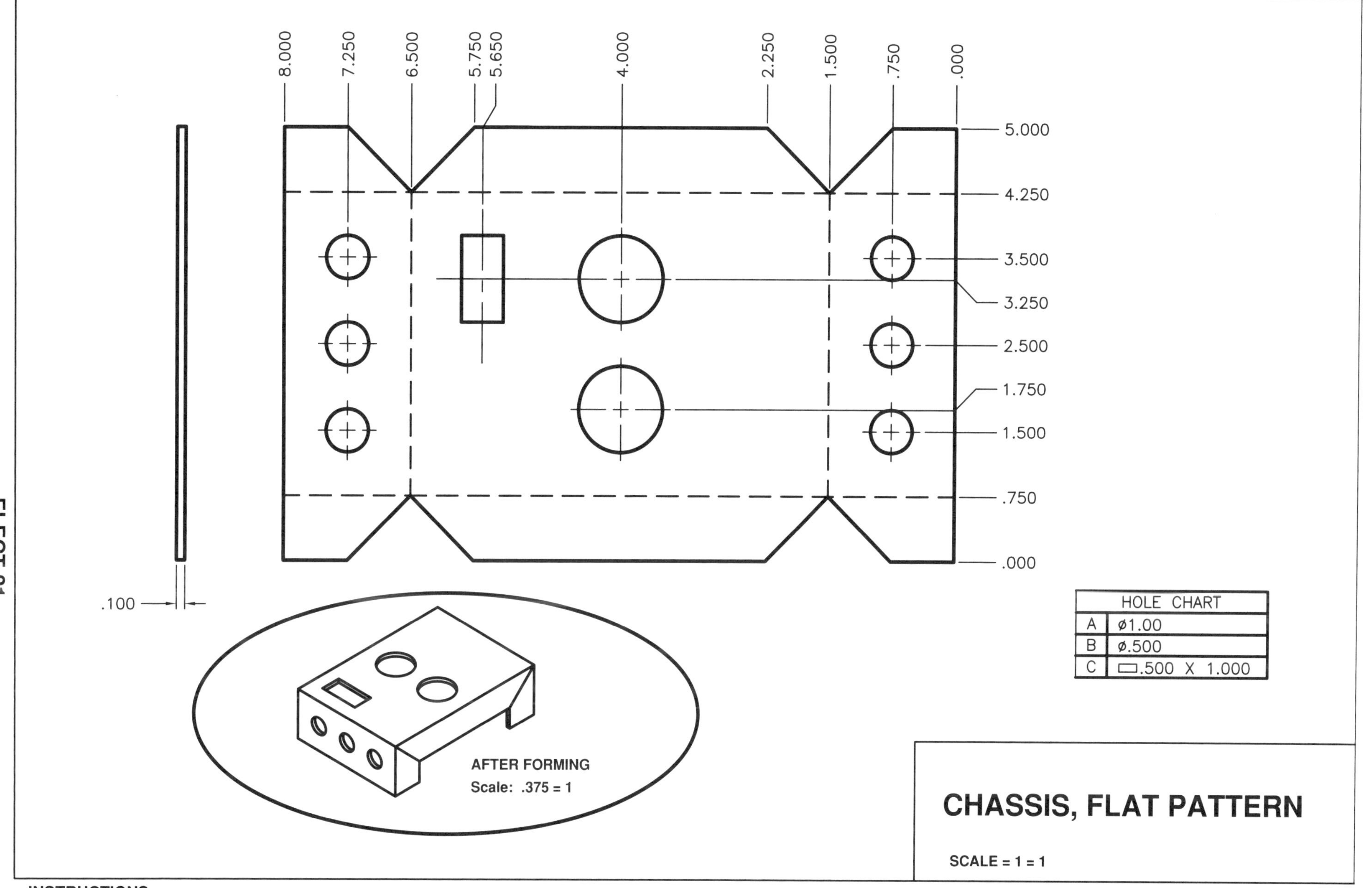

HOLE CHART	
A	ø1.00
B	ø.500
C	▭.500 X 1.000

INSTRUCTIONS:

1. Open **MY DECIMAL SET UP**
2. Adjust model space scale to 1 : 1
3. Draw the "Flat Pattern" as shown, FULL SCALE.
4. Dimension, in paper space, using ORDINATE dimensioning.
5. Draw the "Isometric" view to illustrate the "After forming" appearance. (Cut another Viewport-SC: **3/8** : 1)
6. Draw the "Hole chart" in paper space. Size is your choice. (Great for Tables)
7. Save as: **EX-156-8** and Plot using Page Set up: **11 X 17 MONO.**

EXERCISE 156-8

NOTES:

MECHANICAL

MECHANICAL SYMBOL LIBRARY

When you are using a CAD system, you should make an effort to ONLY draw an object once. If you have to duplicate the object, use a command such as: Copy, Array, Mirror or Block. Remember, this will make drawing with CAD more efficient.

In the following exercise, you will create two mechanical symbols that you use frequently. You will create them once and then merely drag and drop them, from the DesignCenter, when needed. This will save you many hours in the future.

Save this library file as **Library** so it will be easy to find when using the DesignCenter or create a Library Palette.

1. Open **My Decimal Setup**
2. Select the **11 X 17 (1 to 1)** tab.
3. Draw each of the Symbol objects, shown on the next page, actual size. Do not scale.
4. Create an individual Block for each one using the **BLOCK** command.
5. Save this drawing as: **Library**
6. Plot a drawing of your library symbols for reference. The format is your choice.
7. Plot using Page Set up: **11 X 17 MONO.**

SYMBOL	INSTRUCTIONS
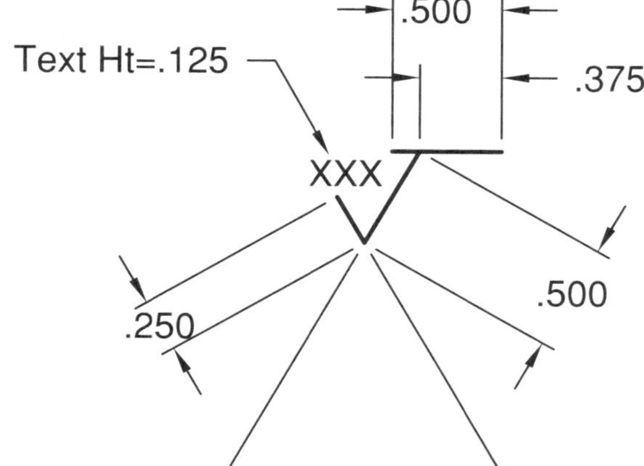 **FINISH MARK**	**Layer = Symbol** **Assign Attributes to the Text**
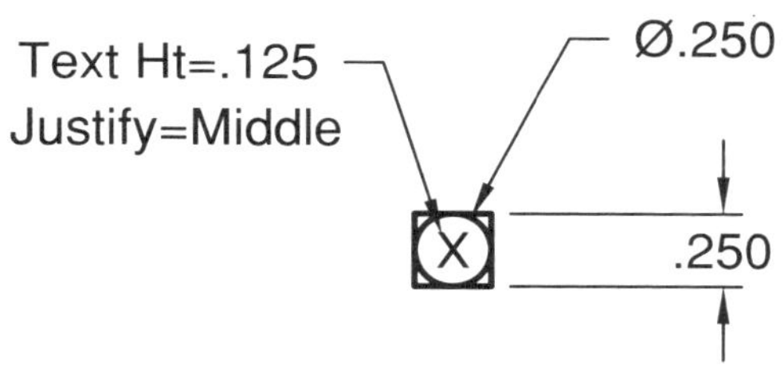 **NOTE IDENTIFIER**	

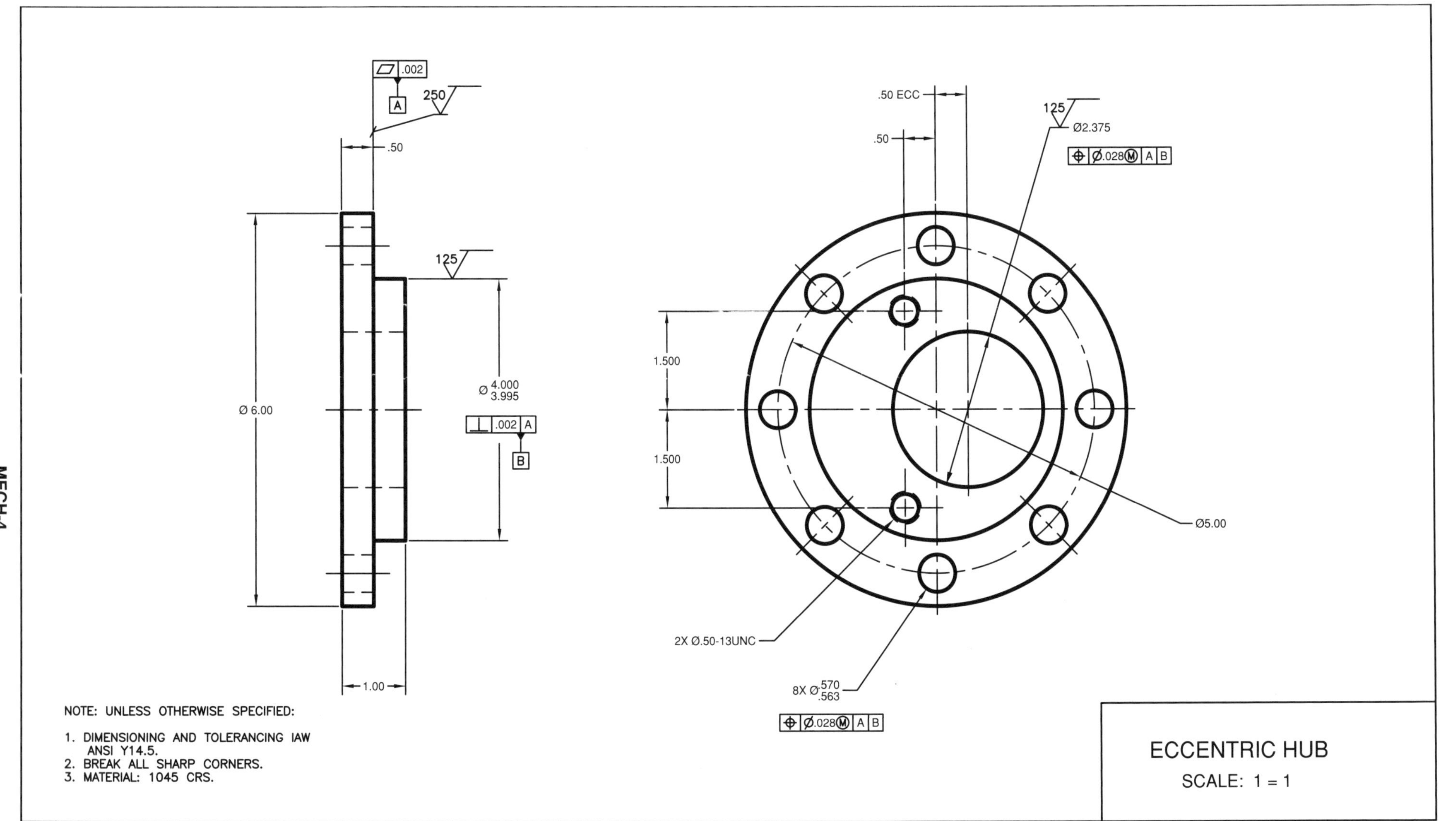

INSTRUCTIONS:
1. Open **MY DECIMAL SETUP**
2. Select the **24 X 18 (1 to 1)** tab.
3. Draw ECCENTRIC HUB above, in model space, full scale.
4. Dimension as shown. Set the Tolerance Text Ht to .125.
5. Insert the symbols, on the Symbols Layer, using DesignCenter.
6. Fill in the information in the title block area
7. Save as: **EX-157-1** and Plot using Page Setup: **24 X 18 MONO**.

EXERCISE 157-1

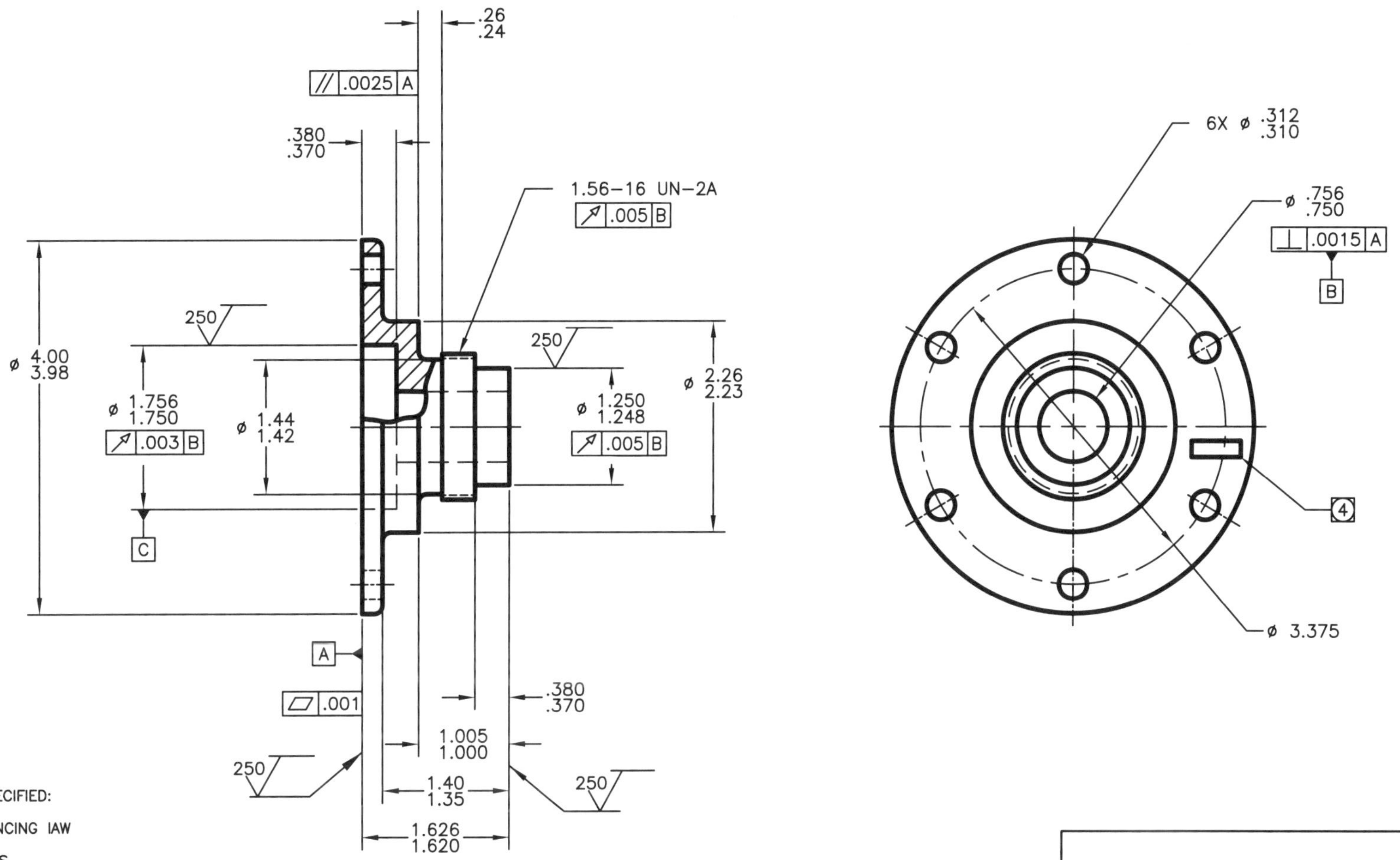

NOTE: UNLESS OTHERWISE SPECIFIED:

1. DIMENSIONING AND TOLERANCING IAW ANSI Y14.5.
2. BREAK ALL SHARP CORNERS.
3. MATERIAL: 1045 CRS.
4. RUBBER STAMP WITH .12 BLACK CHARACTERS PER MIL–STD–130.
5. ALL RADII AND FILLETS ARE R.09

SPLINE HUB

SCALE: 1 = 1

INSTRUCTIONS:

1. Open **MY DECIMAL SETUP**
2. Select the **24 X 18 (1 to 1)** tab.
3. Draw **SPLINE HUB** above, in model space, full scale. Use Hatch Pattern = ANSI 31
4. Dimension as shown. Set the Tolerance Text Ht to .125.
5. Insert the symbols, on the Symbols Layer, using DesignCenter.
6. Fill in the information in the title block area
7. Save as: **EX-157-2** and Plot using Page Setup: **24 X 18 MONO.**

EXERCISE 157-2

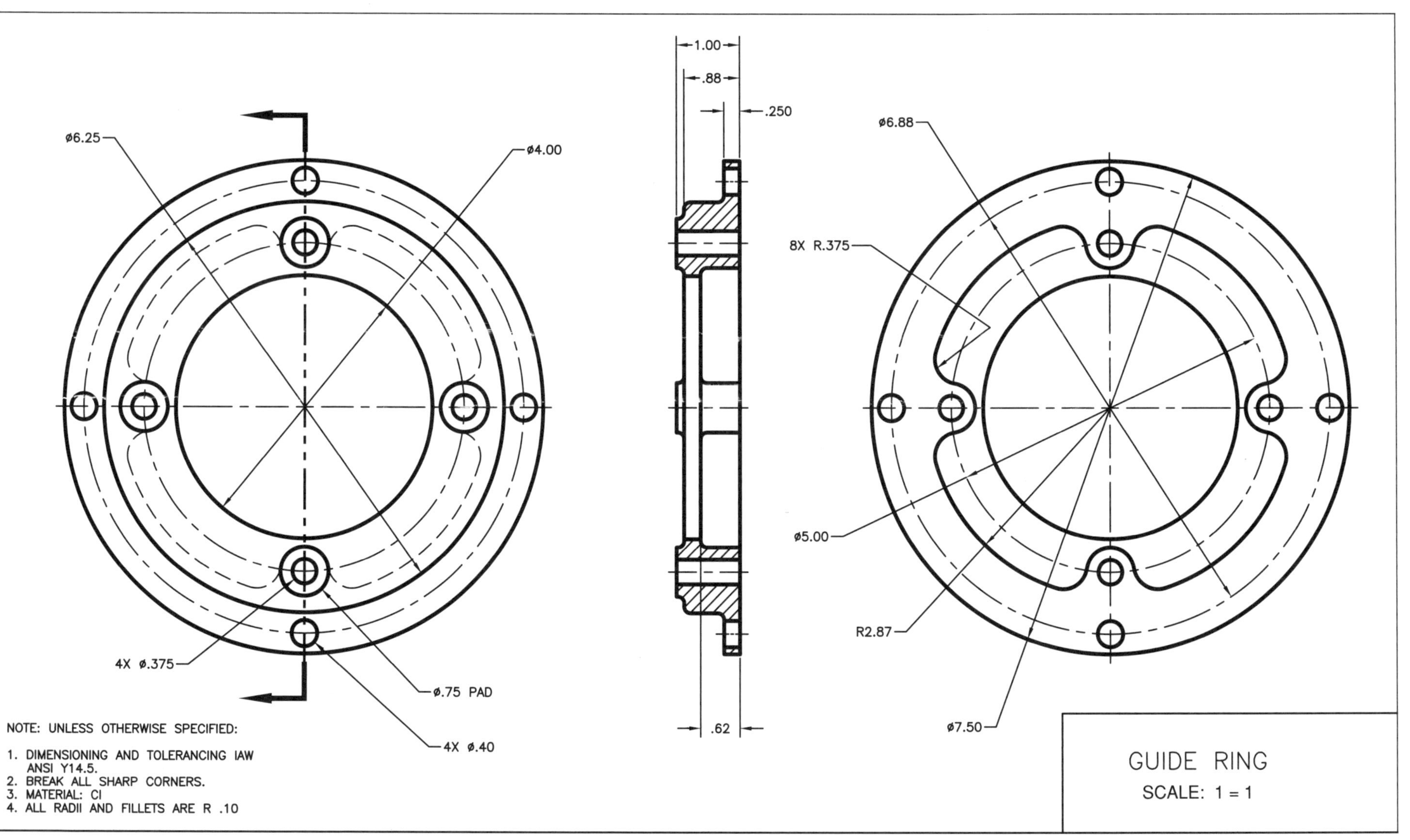

INSTRUCTIONS:

1. Open **MY DECIMAL SETUP**
2. Select the **24 X 18 (1 to 1)** tab.
3. Draw GUIDE RING above, in model space, full scale. Use POLYLINE for the Section Line. (width = .04)
4. Dimension as shown.
5. Hatch Pattern = ANSI 31.
6. Fill in the information in the title block area
7. Save as: **EX-157-3** and Plot using Page Setup: **24 X 18 MONO.**

EXERCISE 157-3

Notes:

Notes:

APPENDIX A
Add a Printer / Plotter

The following are step by step instructions on how to configure AutoCAD for your printer or plotter. These instructions assume you are a single system user. If you are networked or need more detailed information, please refer to your AutoCAD users guide.

Note: You can configure AutoCAD for multiple printers. I suggest that you configure the plotter shown below to match the exercises in this workbook.

A. Select **File / Plotter Manager**
B. Select **"Add-a-Plotter"** Wizard

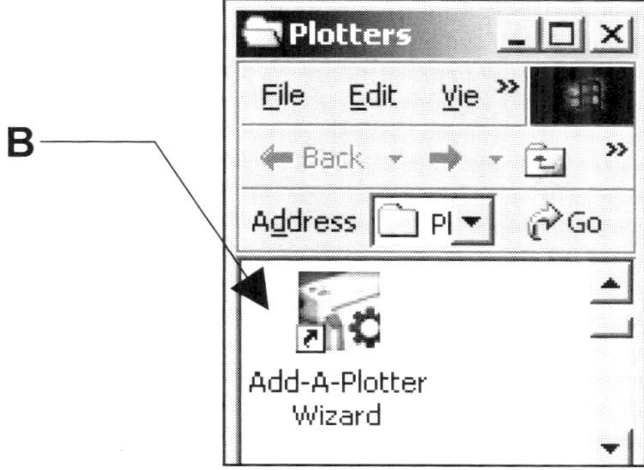

C. Select the **"Next"** button.

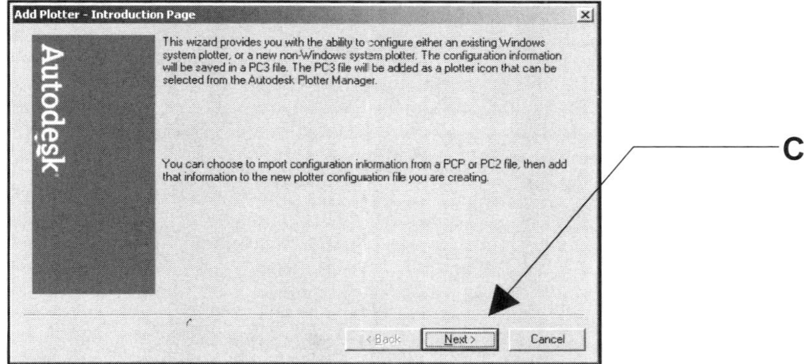

D. Select "**My Computer"** then **Next**.

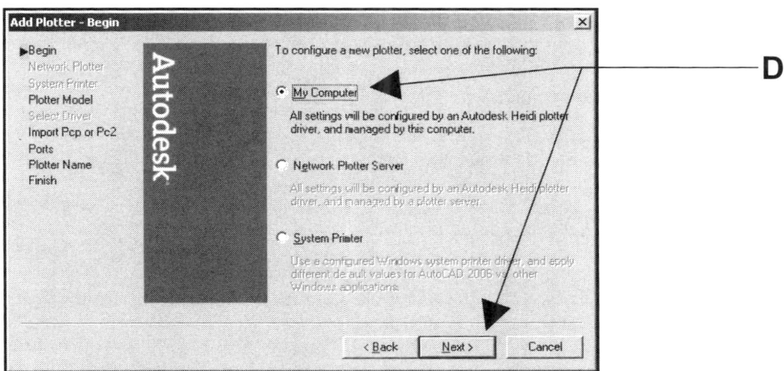

E. Select the **Manufacturer** and the specific **Model** desired then N**ext**.

(If you have a disk with the specific driver information, put the disk in the disk drive and select "Have disk" button then follow instructions.)

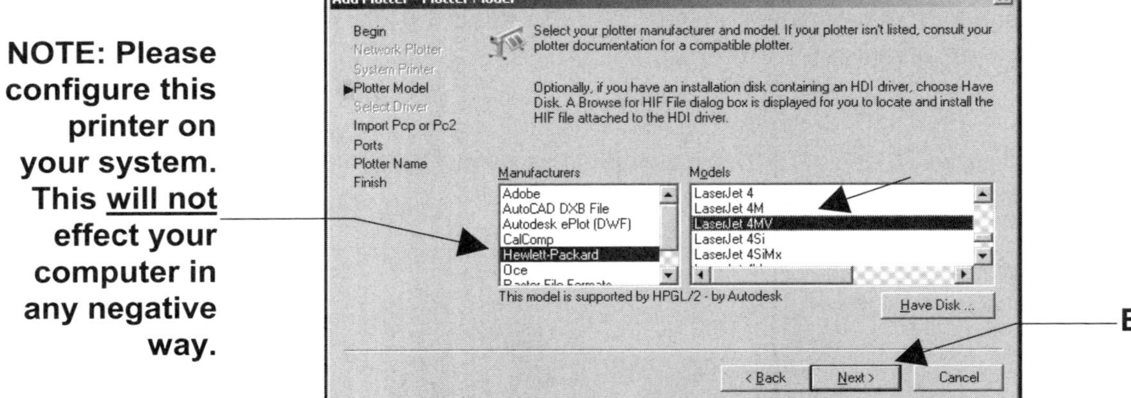

F. Select the **"Next"** box.

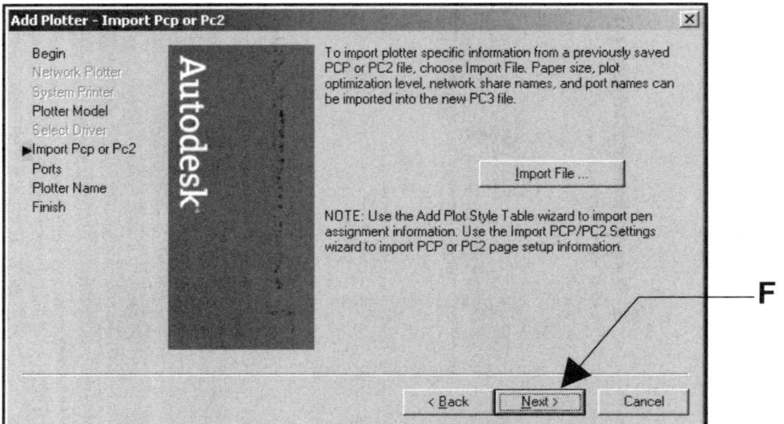

G. Select **"Plot to a port"** .
Then select **"Next".**

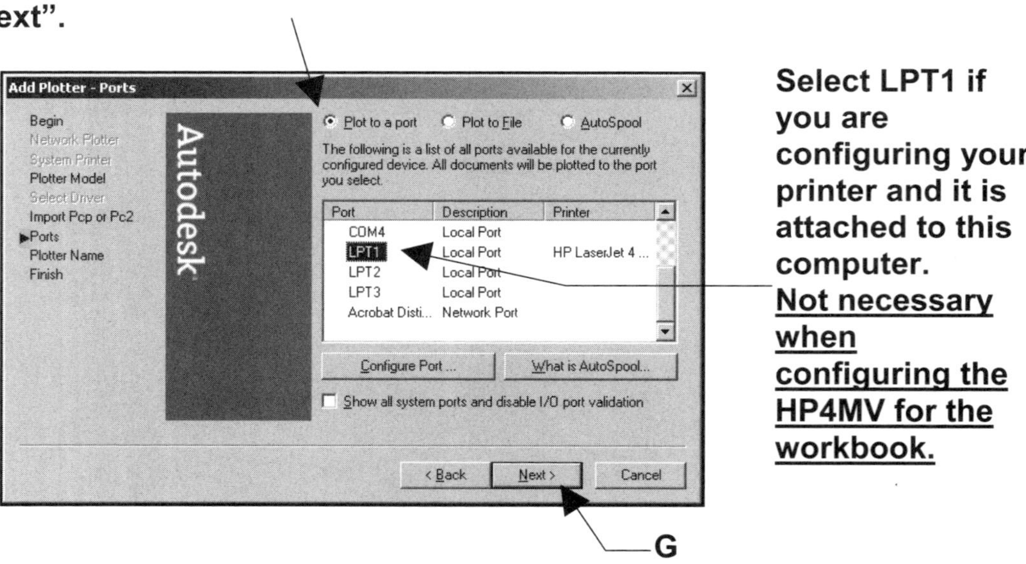

H. The Printer name, that you previously selected, should appear. Then select **“Next”**

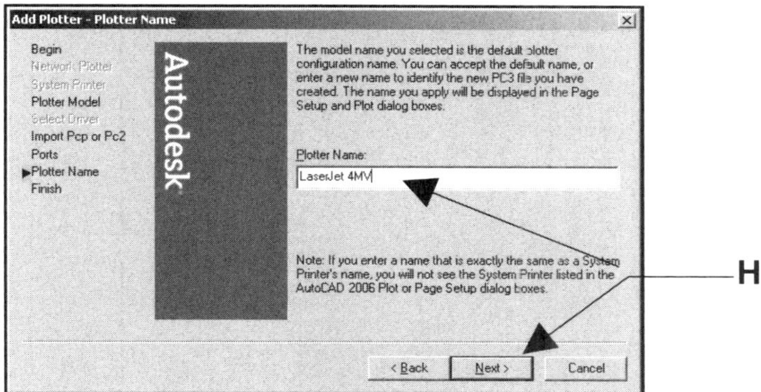

I. Select the **“Edit Plotter Configuration…”** box.

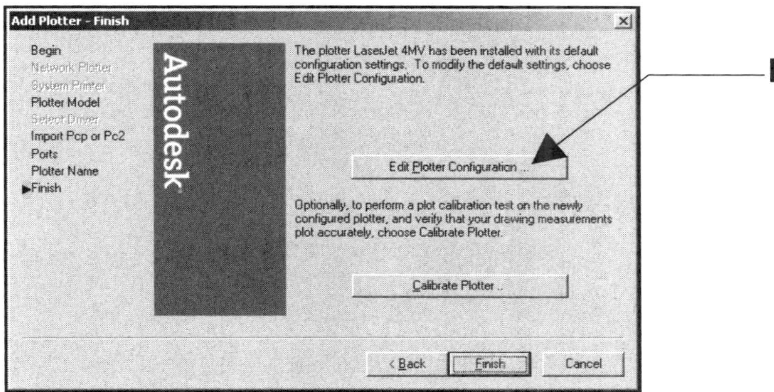

J. Select:

1. Device and Document Settings tab.
2. Media: Source and Size
3. Size: Ansi B (11 X 17 inches)
4. OK box.

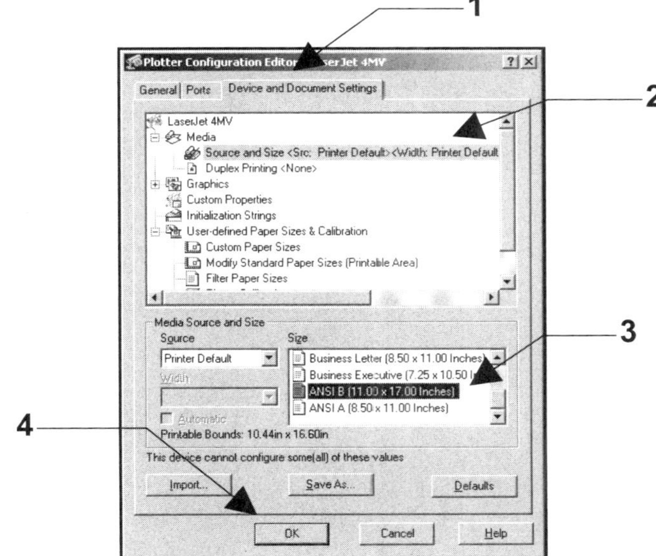

K. Select **"Finish".**

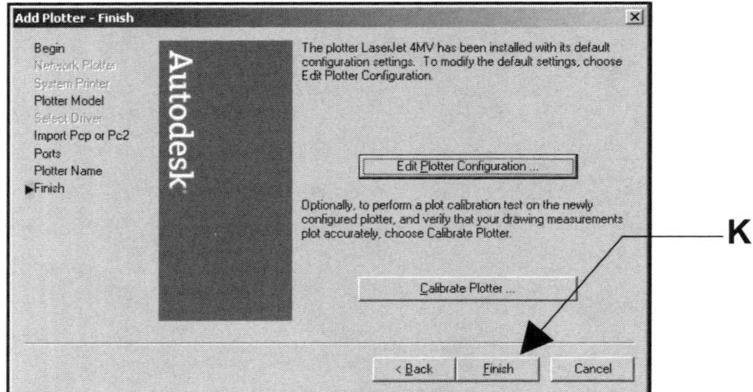

L. Now check the **File / Plotter Manager.**

Is the printer / plotter there?

APPENDIX B

PRINT A QUICK DRAFT ON A LETTER SIZE PRINTER.

Anytime you want to print a quick draft on your letter size printer do the following:

1. Open the drawing on to the screen.
2. Select **File / Plot**.
3. Select your printer. (If your printer is not listed refer to Appendix A)
4. The "AutoCAD Warning" box, shown below, will appear. Select the **OK** button.

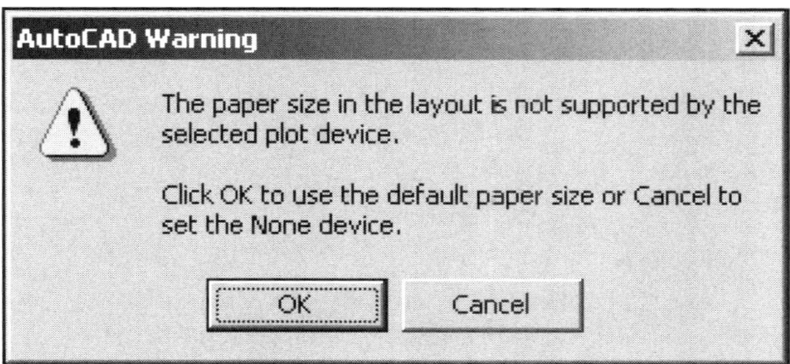

5. **Select "**Paper Size**".**
   Note: The paper size probably already changed to the default size for the printer.
6. Select "**Extents**" for the Plot Area.
7. Select "**Center** the Plot" for Plot Offset.
8. Select "**Fit to Paper**" for Plot Scale.
9. Select the "**Plot Style Table**".
   a. Black only = Monochrome
   b. Color = None (Of course your printer must be capable of printing in color.)
10. **Preview**
    a. If it looks correct press enter.
    b. If it doesn't look correct recheck 3 through 9 above.

Notes:

APPENDIX C
ASSIGN LINEWEIGHTS TO COLORS

In the 1workbook helper.dwg, lineweights have been assigned to layers. When you select the appropriate layer the lineweight is automatically drawn. For example: when you select the layer "object" the lineweight will be .031. This lineweight will be visible on the screen and will plot.

Lineweights may also be assigned to colors within the "color dependent plot style table". Some AutoCAD users prefer to assign lineweights to colors rather than layers or individual objects. As you get more familiar with Lineweights and plotting, you can determine which process you prefer.

To assign lineweights to colors, you must create a "Color dependent Plot Style Table". When this plot style table is selected, it will override the lineweights assigned within the drawing. The step by step process is described below.

1. Select **FILE / PLOT STYLE MANAGER**

2. Select "**Add-A-Plot Style Table" Wizard**

The following dialog boxes will appear.

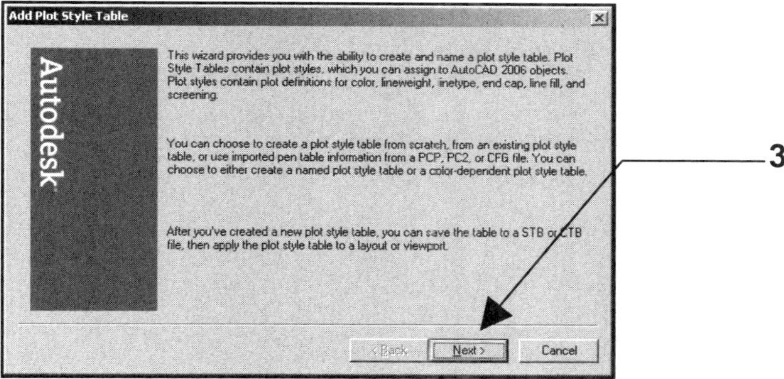

3. Select the **Next** button.

4. Select "**Start from Scratch"** then the **Next** button.

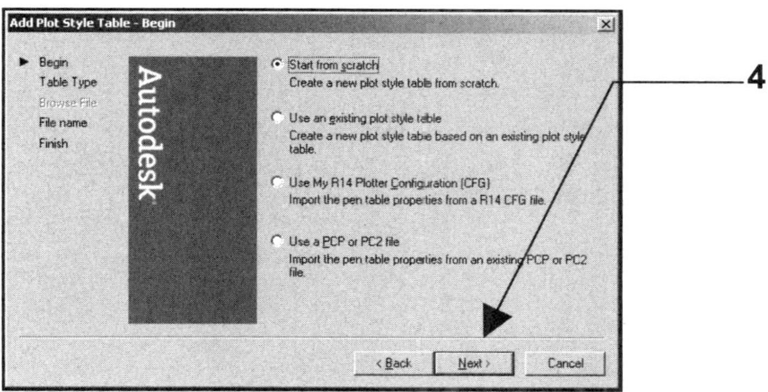

5. Select **"Color-Dependent Plot Style Table"** then the **Next** button.

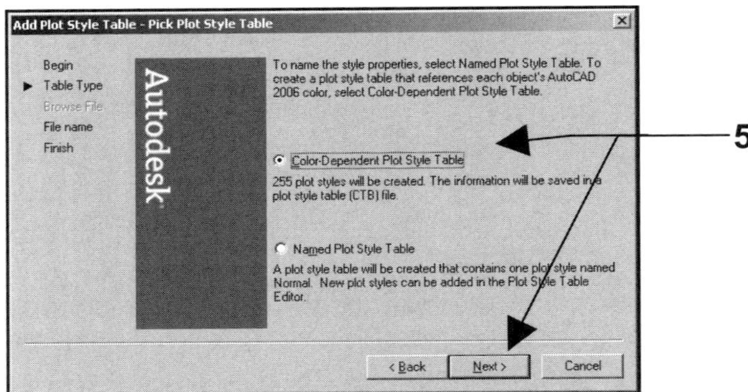

6. Type the new Plot Style Table **name** then select the **Next** button.

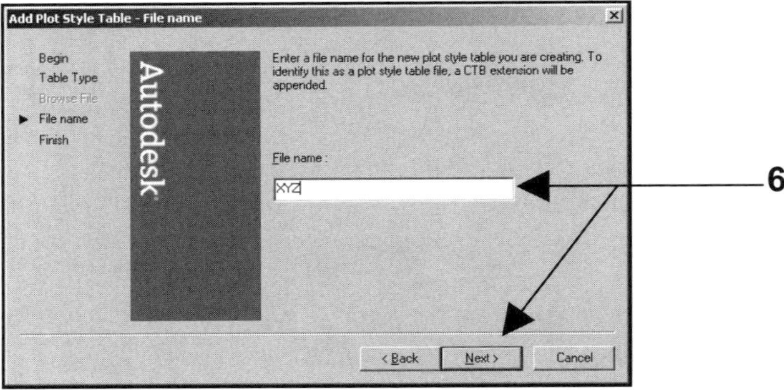

7. Select the **"Plot Style Table Editor"** button.

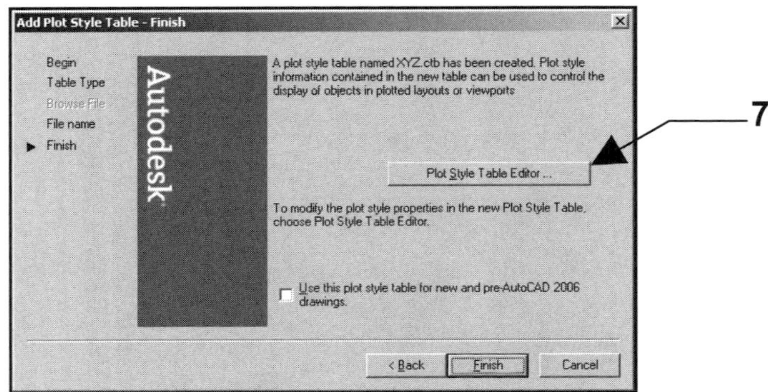

8. Make changes to the **"PROPERTIES"** then select the **Save & Close** button.

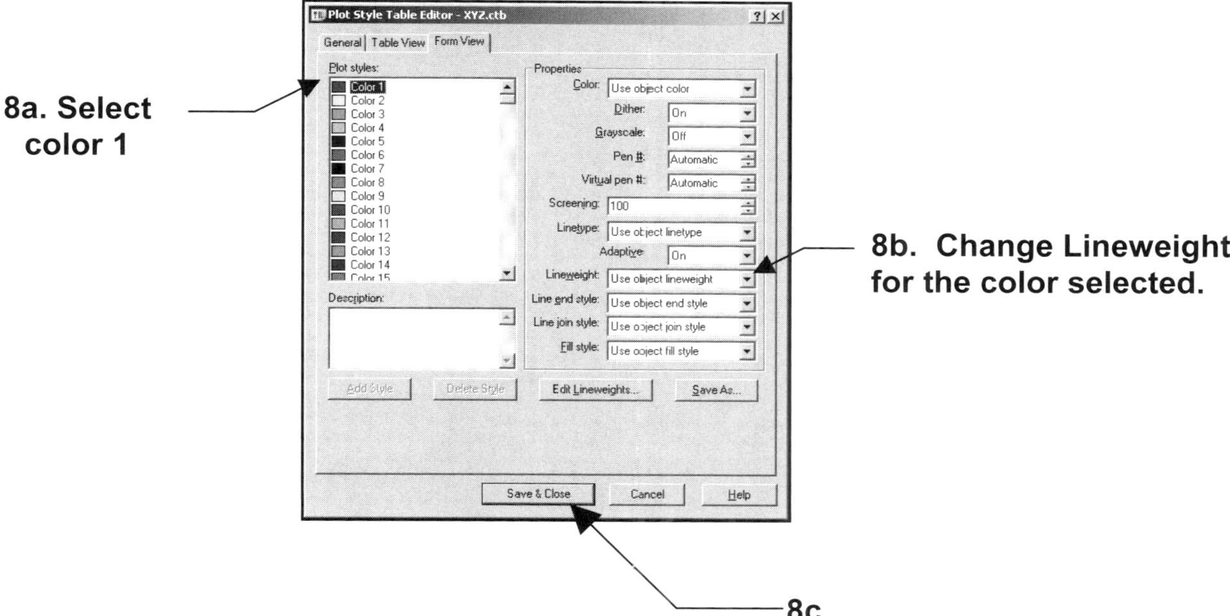

9. Select the **FINISH** button.

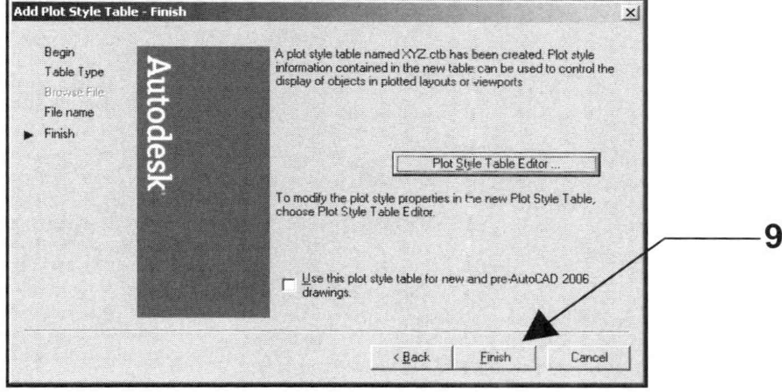

10. Select **File / Plot Style Manager**.

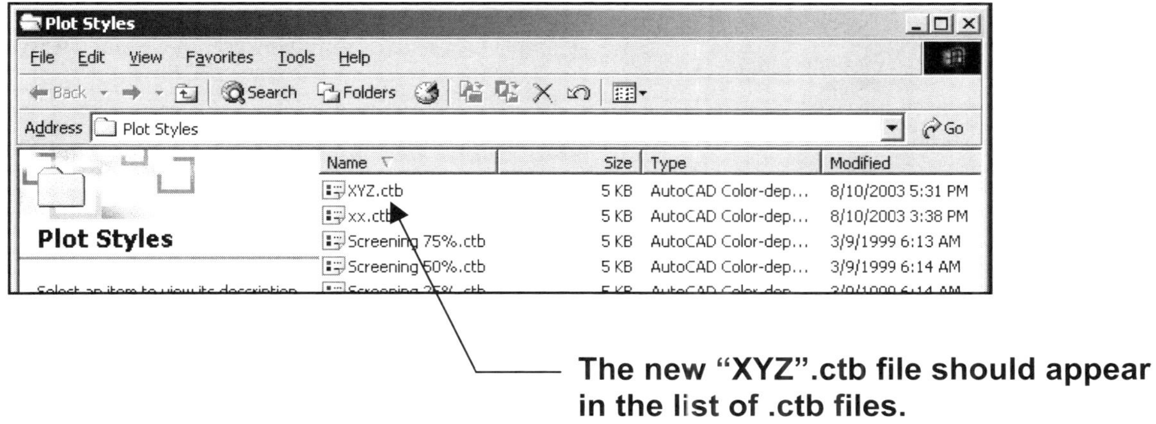

NOTES:

APPENDIX D
METRIC CONVERSION FACTORS

Multiply Length	By	To Obtain
centimeter	0.03280840	foot
centimeter	0.3937008	inch
foot	0.3048a	meter (m)
foot	30.48a	centimeter (cm)
foot	304.8a	millimeter (mm)
inch	0.0254a	meter (m)
inch	2.54a	centimeter (cm)
inch	25.4a	millimeter (mm)
kilometer	0.6213712	mile [U.S. statute]
meter	39.37008	inch
meter	0.5468066	fathom
meter	3.280840	foot
meter	0.1988388	rod
meter	1.093613	yard
meter	0.0006213712	mile [U. S. statute]
microinch	0.0254a	micrometer [micron] (mm)
micrometer [micron]	39.37008	microinch
mile [U.S. statute]	1609.344a	meter (m)
mile [U. S. statute]	1.609344a	kilometer (km)
millimeter	0.003280840	foot
millimeter	0.03937008	inch

Multiply Length	By	To Obtain
rod	5.0292a	meter (m)
yard	0.9144a	meter (m)
Area		
acre	4046.856	meter2 (m2)
acre	0.4046856	hectare
centimeter2	0.1550003	inch2
centimeter2	0.001076391	foot2
foot2	0.09290304a	meter2 (m2)
foot2	929.0304a	centimeter2 (cm2)
foot2	92,903.04a	millimeter2 (mm2)
hectare	2.471054	acre
inch2	645.16a	millimeter2 (mm2)
inch2	6.4516a	centimeter2 (cm2)
inch2	0.00064516a	meter2 (m2)
meter2	1550.003	inch2
meter2	10.763910	foot2
meter2	1.195990	yard2
meter2	0.0002471054	acre
mile2	2.5900	kilometer2
millimeter2	0.00001076391	foot2
millimeter2	0.001550003	inch2
yard2	0.8361274	meter2 (m2)
fathom	1.8288	meter (m)

Symbols of SI units, multiples and sub-multiples are given in parentheses in the right-hand column.

APPENDIX E
DRAWING SCALES

Scale	Drawing Scale Factor	Adjusted Scale times Paperspace
1/16=1'	192	1/192xp
3/32=1'	128	1/128xp
1/8=1'	96	1/96xp
3/16=1'	64	1/64xp
1/4=1'	48	1/48xp
3/8=1'	32	1/32xp
1/2=1'	24	1/24xp
3/4=1'	16	1/16xp
1=1'	12	1/12xp
1-1/2=1'	8	1/8xp
3=1'	4	1/4xp
1=10'	120	1/120xp
1=20'	240	1/240xp
1=25'	300	1/300xp
1=30'	360	1/360xp
1=40'	480	1/480xp
1=50'	600	1/600xp
1=60'	720	1/720xp
1=80'	960	1/960xp
1=100'	1200	1/1200xp
1=200'	2400	1/2400xp
1=10	10	1/10xp
1=20	20	1/20xp
1=16	16	1/16xp
1=30	30	1/30xp
1=40	40	1/40xp
1=50	50	1/50xp
1=100	100	1/100xp
2=1	0.50	2xp
4=1	0.25	4xp
8=1	0.125	8xp
10=1	0.10	10xp
100=1	0.01	100xp

Notes:

Notes:

Notes:

INDEX

AutoCAD 2007 trial software

The enclosed CD contains a fully functioning 30-day trial version of AutoCAD 2007 software. This software has been provided to give you a preview of how easily AutoCAD 2007 installs and is used in conjunction with this Exercise Workbook. After 30 days, from the day you install the software, you will not be able to open the program. At that time you will have the opportunity to activate the software by purchasing the actual software license.

IMPORTANT: The 30-day trial version of AutoCAD can be loaded only once on a single computer. This applies to all 30-day trials regardless of where you obtained them, including from other Exercise Workbooks by Cheryl Shrock.

For example, the 30-day trial from the Advanced Workbook will not install on a computer on which you previously installed the 30-day trial from the Beginning Workbook.

If you are attending a school, check with your instructor for educational discounts.
Or visit: www.autodesk.com for additional purchasing information.

AutoCAD® 2007 software enables highly efficient creation of a single drawing as well as timely coordination of drawing sets. Everyday tools like the table object and tool palettes boost productivity, and the new Sheet Set Manager feature permits content control across entire sets of related drawings. Sheet sets can be composed and collated in a single view, then shared with the entire project team using plots, eTransmit, or DWF™ (Design Web Format™) files.

AutoCAD 2007 system requirements

Refer to page Intro-14

How to install AutoCAD:

Note: AutoCAD 2007 can coexist with earlier versions of AutoCAD and AutoCAD LT.

Two disks are required to install AutoCAD. Insert AutoCAD 2007 Disk 1 of 2 to start the installation. When prompted, insert AutoCAD 2007 Disk 2 of 2 to complete the installation.

1. Insert the AutoCAD 2007 Disk 1 of 2 into your computer's CD-ROM drive.
2. In the AutoCAD Media Browser, on the install tab, click Stand-Alone Installation.
3. Under Install AutoCAD 2007, click Install.
4. On the Autodesk Setup page, click OK to install required support components.
5. In the AutoCAD Installation wizard, follow the directions on each page. When prompted, insert AutoCAD 2007 Disk 2 of 2 to complete the installation.
6. When the installation is complete you will be prompted to: "Authorize AutoCAD" or "run AutoCAD without authorizing."

   Select "**run AutoCAD without Authorizing**"

Note: This trial software will stop working 30 days from the date of installation.